MACHINE DRAWING

MACHINE DRAWING

N D JUNNARKAR

Senior Faculty Member
Mechanical Engineering Department
Chairperson, Quality Assurance Committee
Tolani Maritime Institute
Pune, India

PEARSON

ISBN 978-81-317-0678-7

First Impression, 2007

Published by Pearson India Education Services Pvt.Ltd,CIN:U72200TN2005PTC057128.
Formerly known as TutorVista Global Pvt Ltd, licensees of Pearson Education in South Asia

Head Office: 7th Floor, knowledge Boulevard, A-8(A) Sector-62, Noida (U.P) 201309, India

Registered Office: Module G4, Ground Floor, Elnet Software City, TS -140,Block 2 & 9
Rajiv Gandhi Salai, Taramani, Chennai, Tamil Nadu 600113.,Fax: 080-30461003,
Phone: 080-30461060, www.pearson.co.in email id: companysecretary.india@pearson.com

Digitally Printed in India by Repro India Ltd. in the year of 2015.

To

my wife Madhavi

and

daughter Janhavi

PREFACE

Engineering drawing is the universal language of engineering community and machine drawing is the media through which two technical persons can convey information to get the desired results. This book will not only help the engineering students but will also be useful for practicing engineers.

The book is divided into three parts.

Part A of the book discusses basic principles of technical drawing, dimensioning, representation of standard machine components, limits, fits and tolerances, basic machining processes and surface roughness indications.

Part B of the book ensures a simple approach for the assembly and mainly contains problems on diversified fields with questions on assembly, its sketch, introduction, steps for answer with bill of material and the final answer of the given question. Study of Part II will not only improve visualisation but also provide simple approach on how to assemble components for any assembly.

Part C of the book contains practice problems on assembly of various fields along with hints for the answer.

Any suggestions for the improvement of the book will be highly appreciated.

N D Junnarkar

ACKNOWLEDGEMENTS

I am deeply grateful to all the people who have contributed directly or indirectly to this project in a variety of ways.

I am highly indebted to Professor N K Joshi for his valuable guidance, inspiration and helping nature in all kinds of problems. A special thanks are also due to my organisation, Tolani Maritime Institute (TMI), and to Dr N P Tolani and Mr R N Tolani for their vision and contribution to education which inspired me to do this project. I am thankful to all members of the management of TMI for their support and cooperation.

My gratitude to Mr A K Srivastava, Mr B K Saxena, Mr R K Razdan and Mr S G Dixit from whom knowingly/unknowingly I was inspired for keeping myself busy in this constructive work. I am grateful to Mr Sameer Vanarse for his efforts in DTP work, and Mr Jagdishwaran for his cooperative support.

Last, but not the least, my sincere thanks to all, whose faith in me encouraged me to write this book.

N D JUNNARKAR

CONTENTS

Part B: Assembly Drawings

Part C: Questions on Assembly Drawings

PART A

Theoretical Concepts

CHAPTER 1

Introduction to Engineering Drawing

This chapter contains information about engineering drawing and provides basic knowledge of drawing sheets and drawing instruments

After reading this chapter you will be able to

- identify different drawing instruments
- know about the application of drawing instruments
- select appropriate grade of pencil and drawing sheet size

1.1 DRAWING

A drawing is a graphical representation of a real object. The purpose of a drawing is to define and specify the shape and size of a particular object by means of lines. Other information about the object, which cannot be expressed by lines, is given alongside the drawing in a concise manner. A good drawing gives full information about the object in a precise way. Therefore, a drawing is the shortest of shorthands.

The most important requirement for all technical persons working in an engineering environment is through knowledge of engineering drawing. A drawing worked out by an engineer, illustrating ideas and concepts for engineering purposes, is an engineering drawing. It is the universal graphical language of engineers, a language of use and ever increasing value, and spoken, read, and written in a particular style. As is the case with any language, engineering drawing has also been devised according to certain rules and has its own grammar.

Sketching is a freehand expression in graphical language, and it is the most important tool for engineers engaged in technical work. Technical ideas can be expressed quickly and effectively by sketches, without the use of instruments.

An engineer expresses his ideas on a paper through the medium of drawing. A complete working drawing of a job is followed by giving an accurate shape to the raw materials, according to the drawing. If the engineer commits an error in the initial drawing, it is carried over to the practical work by others, resulting in the loss of time, material, labour, and finally affects the production efficiency of the factory. Therefore, it is extremely important for engineers, designers, supervisors, draughtsmen, mechanics, and others working in an engineering environment to have a thorough knowledge of engineering drawing.

By means of drawings, the shape, size, finish, colour, and construction of any object can be described accurately and clearly in the simplest and shortest way. Best results can be obtained if engineering drawing skills are developed with due consideration to the following:

1. Thorough understanding of original drawing
2. Visualising the job three dimensionally
3. Clearly understanding and appreciating the shape, size, proportion and design
4. Expressing ideas on paper quickly and clearly by sketches
5. Developing speed and accuracy of the drawing using pencil and other drawing instruments

Developing interest in engineering drawing will help in learning it easily. A quick understanding of an original drawing and the ability to make a sketch can be achieved if a person is familiar with the concepts of engineering drawing. The shape, size, and manufacturing methods of a job can be easily understood from the drawing once we develop the necessary skills. Accuracy and speed are developed by practice. Accuracy in the drawing must come first; quality must come before quantity. Pencils and all the drawing instruments should be kept in good condition so that they may give maximum working efficiency. Every care should be taken to make an accurate drawing. At student level to understand this subject through only discussion in the classroom is not easy for which field practice is required.

1.2 DRAWING INSTRUMENTS

A neat and correct drawing is drawn with the help of good drawing instruments, the list of which is given below:

1. Drawing board and stand
2. Tee-square
3. Mini drafter
4. Set square

5. Protractor
6. Instrument box
7. French curves or irregular curves
8. Pencil
9. Eraser and erasing shield
10. Blade, pocket knife, or pencil sharpener
11. Drawing pins, adhesive tape, or clips
12. Drawing paper or drawing sheet, tracing paper, and tracing cloth
13. Duster or dusting brush
14. Sand paper
15. Scales (engineering scales)
16. Sketch book

To record information on paper or any other surface, instruments and equipment are needed. Engineering drawing is entirely a graphical language, hence instruments are essential. Even for a freehand drawing, pencil, eraser, and sometimes coordinate papers are used. To secure satisfactory results in drawing, the equipment should be of high grade to give accuracy and working efficiency. The drawing must be clear, neat, and legible in order to serve its purpose well. Drawing is a means of communication to others, and therefore it must have a good appearance. Hence, it is important for draughtsmen and engineers to have good speed, accuracy, legibility, and neatness.

1.2.1 Drawing Board

Drawing boards are usually rectangular in shape. They are generally made of well seasoned pine wood. In order to prevent warpage, the drawing board is made up of narrow strips glued edge to edge. Care is taken to ensure that the surface of the board is absolutely flat and free from any cracks. On the bottom side of the board, two battens are fixed by screws fitted in slots to allow for seasonal contraction and expansion. On the left hand side of the board, a straight ebony edge, called the *working edge*, is fitted in the groove made on the edge of the board. This work edge facilitates the movement of T-square.

The standard sizes of the drawing boards, according to the Bureau of Indian Standards: IS-1444–1963, are given in Table A1.1.

TABLE A1.1 *Standard Dimensions of the Drawing Boards*

Sl. No.	Designation	Dimension (mm)	Name
1	B_0	1500 × 1000	Antiquarian
2	B_1	1000 × 700	Double Elephant
3	B_2	700 × 500	Imperial
4	B_3	500 × 350	Half Imperial
5	B_4	350 × 250	Quarter Imperial

1.2.2 T-square

T-squares are made of well seasoned hard wood (Fig. A1.1). The two main components of the T-square, namely the *stock* and the *blade* are fastened rigidly at right angles to each other by brass screens and ebony pins. The working edge is at the front and is bevelled to facilitate drawing of lines. T-squares are used for drawing horizontal lines. While drawing, the stock of the T-square is perfectly fitted against the working edge of the drawing board. The stock of the T-square can be moved along the working edge to draw horizontal parallel lines.

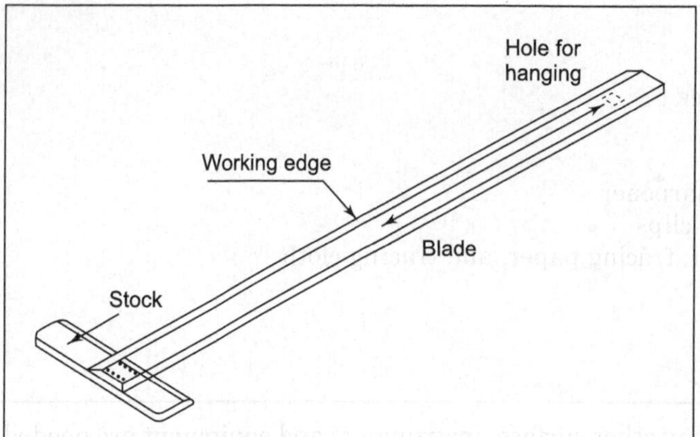

Fig. A1.1 T-square

1.2.3 Mini Drafter

A mini drafter is a very convenient drawing instrument for students since it combines all the functions of a T-square, set squares and a protractor (Fig. A1.2). It consists of two scales at right angles along with a protractor. The entire set up comes as a single unit fitted to one end of the double parallelogram of bars. The farther end of this parallelogram bar is fitted with a clamp. The mini drafter is laid on the drawing board and is clamped normally to the left side of the board. The drawing sheet is then laid on the board in such a manner that the two scales of the mini drafter should align parallel with the horizontal and the vertical printed border lines of the drawing sheet. The sheet is then fixed to the drawing board by means of clips.

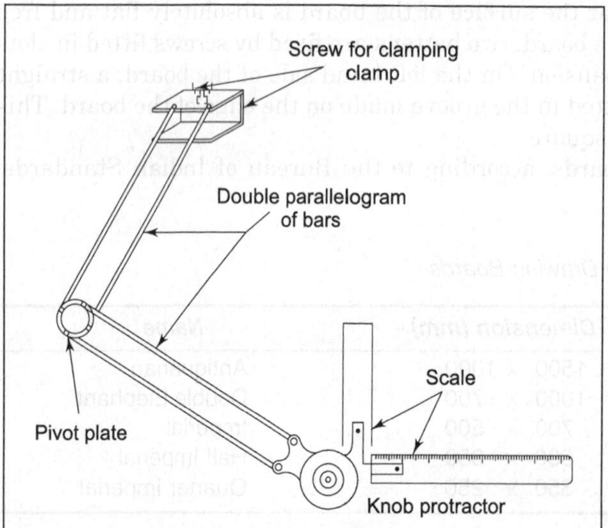

Fig. A1.2 Mini drafter

1.2.4 Set Squares

Two types of set squares are commonly used—a 60° – 30° set square and a 45° set square. They are generally made of transparent celluloid or plastic. Set squares are used to draw vertical, inclined, as well as parallel lines.

1.2.5 Compass

A compass is used to draw circles and arcs of circles. Two types of compasses are used in engineering drawing. One has long legs for drawing big circles. Another type of compass with shorter legs, which can be fixed or adjusted to a particular radius, is used for drawing circles with smaller radii.

1.2.6 Protractor

A protractor is used for setting off angles other than those obtainable using set squares (Fig. A1.3). Normally protractors are made of transparent celluloid.

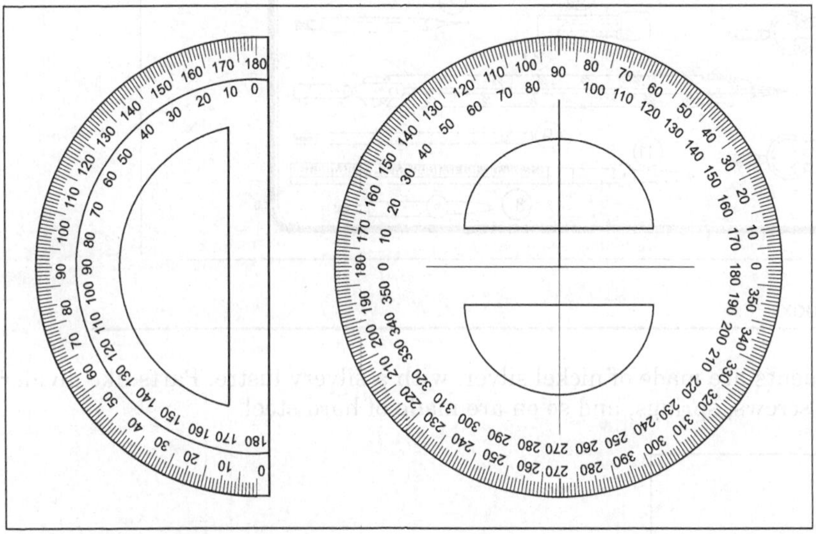

Fig. A1.3 Protractor

1.2.7 Instrument Box

An instrument box consists of the following drawing instruments for making various drawings (Fig. A1.4).

1. Large size compass (150 mm long) with inter-changeable pencil and pen legs
2. Large size divider (150 mm long)
3. Small bow compass (95 mm long)
4. Small ink bow compass (95 mm long)
5. Small bow divider (95 mm long)
6. Lengthening bar
7. Pin point
8. Ink point
9. Ruling pen or liner
10. Holder croquill (for lines)
11. Lead case (for storing lead)

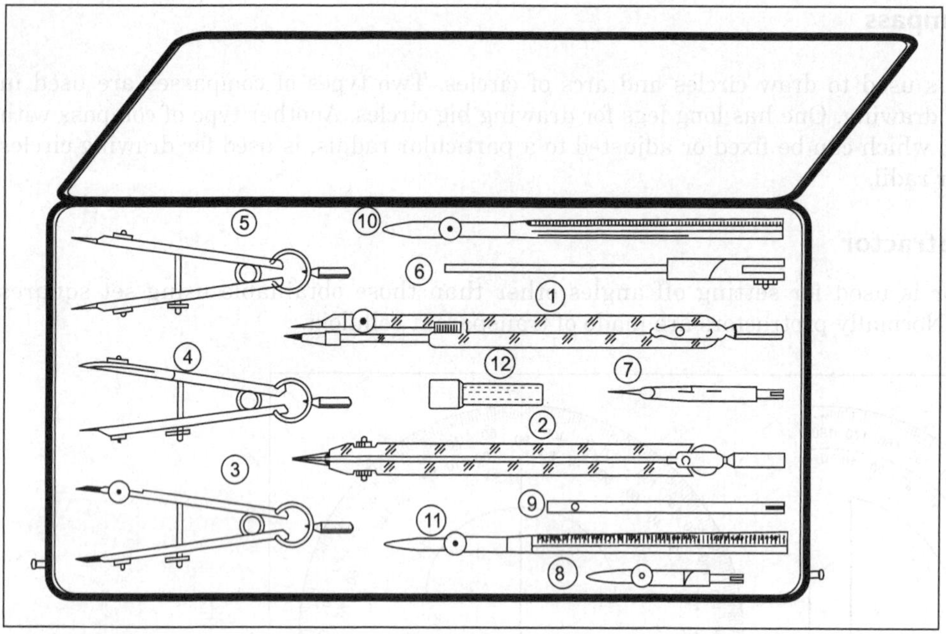

Fig. A1.4 Instrument box

Note: All the instruments are made of nickel silver, with a silvery lustre. Parts like divider points, ruling pen nibs, screws, springs, and so on are made of hard steel.

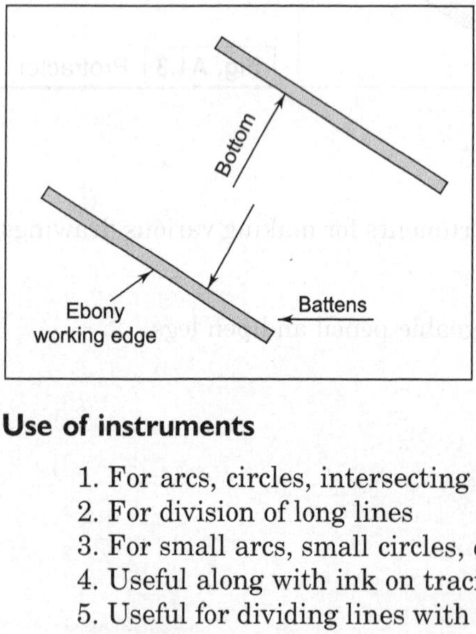

Fig. A 1.5 Drawing board

Use of instruments

1. For arcs, circles, intersecting arcs, and so on
2. For division of long lines
3. For small arcs, small circles, corner radius, and so on
4. Useful along with ink on tracing sheet, for small works like arcs, circles, and so on
5. Useful for dividing lines with ink on tracing sheet
6. Used along with compass for increasing working length
7. Used for locating centres of arcs, circles, and so on
8. Ink point used for ink application on tracing paper
9. Lengthening bar

10+11. Bow pen attachment

12. Lead box.

1.2.8 French Curves

French curves or irregular curves are generally made of wood or plastic. They are used to draw different curves, other than circles of known radius. French curves are available in different shapes and sizes. They are composed of successive segments of geometrical curves, such as ellipse, parabola, hyperbola, involute, logarithmic spiral, ship curves and rail road curves. Proper use of french curves will require skill, especially, when the curves are to be drawn in ink.

1.2.9 Pencils

The neatness, quality and accuracy of a drawing greatly depends upon the type and condition of the pencil used for drawing. Pencil leads are made of graphite with clay added in varying amounts to make 18 grades from 9H to 7B. These grades can be divided into 3 groups:

1. Hard	:	9H to 4H
2. Medium	:	3H to B (3H, 2H, H, HB and B)
3. Soft	:	2B to 7B

Pencil 9H is the hardest and 7B is the softest. Harder pencils have leads of small diameters and softer pencils have larger diameter to give adequate strength. The choice of the grade of a pencil depends upon the type of work, texture of paper, atmospheric humidity, and so on. The following pencils should be used for drawing work in class:

2H Pencil	:	for drawing outlines, centre lines, break lines
H Pencil	:	for dimensioning arrowheads, hatching lines, printing, sketching, extension lines, circles, and arcs

After getting some practice, 3H and 2H pencils can be used for the same work. A good pencil must have the following five attributes:

1. satisfying results in photo printing, technical, artistic, and commercial work
2. very high density and firmness of leads
3. intense blackness, high covering power, and giving clear blue prints
4. compressed leads that resist breakage and wear
5. should be available in many reliable standardised degrees and be capable of producing any desired shade, from the highest grey to the deepest black

Keep your pencil sharp Only a sharp pencil is capable of preparing a neat and accurate drawing. A conical point is used for printing, dimensioning, sketching and drawing circles, arcs, and curves. Straight lines of equal thickness and darkness should be drawn by a chisel point pencil. The drawing of lines by a pencil can be mastered with practice. A perfect draughtsman will be able to tell at once by inspection whether or not a line is correct in width and darkness.

1.2.10 Drawing Sheets

Selection and designation of sizes The original drawing should be made on the smallest sheet maintaining the necessary clarity and resolution. Drawing sheets may be used with their longer sides positioned either horizontally or vertically. The choice of sizes of the original drawing and its reproduction shall be made from the series as shown.

Sheet sizes: The preferred sizes of the trimmed sheets as selected from the main ISO-A series, specified in ISO-16, are given in Table A1.2.

TABLE A1.2 *Drawing Sheets with Standard Dimensions*

Designation	*Dimensions (mm)*
A0	841 × 1189
A1	594 × 841
A2	420 × 594
A3	297 × 420
A4	210 × 297

Special elongated sizes (second choice): When a sheet of greater length is needed, one of the sizes in Table A1.3 should be used.

TABLE A1.3 *Special Elongated Sizes of Drawing Sheets*

Designation	*Dimensions (mm)*
A3 × 3	420 × 891
A3 × 4	420 × 1189
A4 × 3	297 × 630
A4 × 4	297 × 841
A4 × 5	297 × 1051

Exceptional elongated sizes (third choice): When a very large or an extra elongated sheet is essential, one of the sizes in Table A1.4 should be used.

TABLE A1.4 *Drawing Sheets with Exceptional Elongated Sizes*

Designation	*Dimensions (mm)*
A0 × 2	1189 × 1682
A0 × 3	1189 × 2523
A1 × 3	841 × 783
A1 × 4	841 × 2378
A2 × 3	594 × 1261
A2 × 4	594 × 2102
A2 × 5	594 × 2102
A3 × 5	420 × 1489
A3 × 6	420 × 1783
A3 × 7	420 × 2080
A4 × 6	297 × 1261
A4 × 7	297 × 1471
A4 × 8	297 × 1682
A4 × 9	297 × 1892

Layout of a drawing sheet The Bureau of Indian Standards has given standard recommendations regarding the layout of a drawing sheet. A margin of about 20 mm is to be provided on the left side for filing purposes. Two border lines, 5 mm apart, are drawn all around the sheet. The space for the title block is provided on the right hand bottom corner. The title block should have adequate space for recording the technical and administrative details.

Folding of drawing sheets As the technical drawings are in exceptionally extra large sizes, it becomes necessary to know how to fold and preserve these drawings sheets.

Folding marks: Folding marks are made in the drawing sheet. They are helpful in folding of prints correctly and easily. Two methods of folding of prints are in general use. Method I is suitable for prints that are to be filed or bound. It allows prints to be unfolded or refolded without removing them from files.

Fig. A1.6 (a) shows the folding diagram for folding an A1 size sheet by method 1. It is folded in two stages, namely, lengthwise [Fig. A1.6(b)] and crosswise [Fig. A1.6(c)].

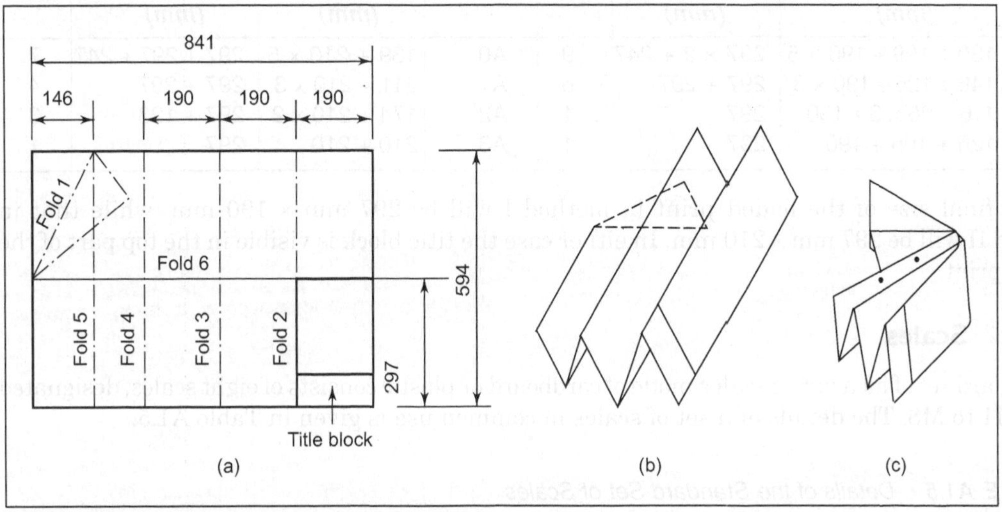

Fig. A1.6 Folding of A1 size drawing sheet by method I

When prints are to be stored and preserved in cabinets they are folded by method II. The folding diagram for folding an A1 size sheet by method II is shown in Fig. A1.7. The two stages of folding are similar to those for method I.

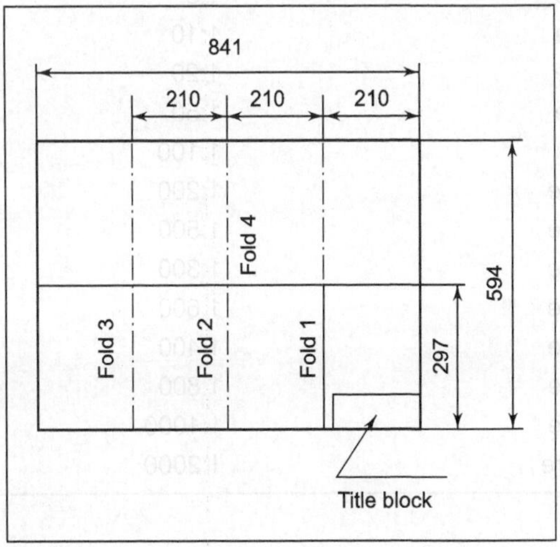

Fig. A1.7 Folding of A1 size drawing sheet by method II

Dimensions for folding of various sizes of drawing sheets by the two methods are given below:

TABLE A1.6 *Folding of Drawing Sheets*

Method I				Method II			
Sheet Size	Horizontal Dimensions from Left (mm)	Vertical Dimensions from Bottom (mm)	No. of Folds	Sheet Size	Horizontal Dimensions from Left (mm)	Vertical Dimensions from Bottom (mm)	No. of Folds
A0	130 + 109 + 190 × 5	297 × 2 + 247	9	A0	139 + 210 × 5	297 + 297 + 247	7
A1	146 + 125 + 190 × 3	297 + 297	6	A1	211 + 210 x 3	297 + 297	4
A2	116 + 96 × 3 + 190	297	1	A2	174 + 210 × 2	297 + 123	3
A3	125 + 105 + 190	297	1	A3	210 + 210	297	1

The final size of the folded print in method I will be 297 mm × 190 mm, while that in method II will be 297 mm × 210 mm. In either case the title block is visible in the top part of the folded print.

1.2.11 Scales

A standard set of measuring scales made of cardboard or plastic consists of eight scales, designated from M1 to M8. The details of a set of scales in common use is given in Table A1.5.

TABLE A1.5 *Details of the Standard Set of Scales*

Designation	Particulars of the Scale	
M1	Full size	1:1
	50 cm to a metre	1:2
M2	40 cm to a metre	1:2.5
	20 cm to a metre	1:5
M3	10 cm to a metre	1:10
	5 cm to a metre	1:20
M4	2 cm to a metre	1:50
	1 cm to a metre	1:100
M5	5 mm to a metre	1:200
	2 mm to a metre	1:500
M6	3.3 mm to a metre	1:300
	6.6 mm to a metre	1.600
M7	0.25 mm to a metre	1:400
	0.125 mm to a metre	1:800
M8	1 mm to a metre	1:1000
	0.5 mm to a metre	1:2000

Review Questions

1. What is engineering drawing?
2. Why is drawing known as the universal language of engineers?
3. What is machine drawing?
4. Enumerate the different types of drawing instruments.
5. How should we use a pencil for drawing work?
6. What are the standard sizes of the drawing sheets?
7. How should we sharpen a pencil?
8. State the uses of french curves.
9. What is the use of drafting machine?
10. State the general precautions to be taken while using drawing instruments.

CHAPTER 2

Lines, Lettering, and Scales

This chapter contains information about lines, letters and scales in engineering drawing

After reading this chapter you will be able to

- use different types of lines
- choose the proper lettering for the text accompanying the drawing
- recognise what precautions should be taken while drawing

2.1 LINES

Engineering drawing uses various types of lines to describe different objects and different purposes. Every line in a drawing has a definite meaning. The thickness and darkness of a line are two different things. To monitor thickness of the line, the draughtsman will have to adjust the pressure applied on the pencil. To vary the darkness of the line, the person will have to select a pencil of the correct grade, that is, H, 2H, B, 2B ... and so on. Information on grades of pencil is given in Chapter 1 (1.2.9).

Only the types and thickness of lines shown in Table A 2.1 shall be used. In cases where lines of different thicknesses are used for special fields (for example, electrical or pipe work diagrams), or the lines specified in the table are used for applications other than those detailed in the last column of the table, the conventions adopted must be indicated according to international standards and explained by notes on the drawing concerned.

TABLE A2.1 *Types of Lines—their Size with Applications*

Line (1)	***Description*** (2)	***General Applications*** (3)	***Thickness*** (4)
A	Continuous thick line	A1 Visible outlines A2 Visible edges	0.7 mm
B	Continuous thin line (Straight or curved)	B1 Imaginary lines of intersection B2 Dimension lines B3 Projection lines B4 Leader lines B5 Hatching B6 Outlines of revolved section in place B7 Short centre lines	0.35 mm
C	Continuous thin line (freehand)	C1 Limits of partial or interrupted views and sections, if the limit is not a chain thin line	0.35 mm
D	Continuous thin line (straight) with zigzags	D1 Line D2 Extra large lines	0.35 mm
E	Dashed thick line	E1 Hidden outlines E2 Hidden edges	0.7 mm
F	Dashed thin line	F1 Hidden outlines F2 Hidden edges	0.35 mm
G	Chain line	G1 Centre lines G2 Lines of symmetry G3 Trajectories	0.35 mm
H	Chain thin line, thick at ends and changes direction	H1 Cutting planes	0.35 mm and .7 mm (at the ends)
J	Chain thick line	J1 Indication of lines or surfaces to which a special requirement applies	0.7 mm

(*Contd.*)

Line (1)	Description (2)	General Applications (3)	Thickness (4)
K	Chain thin double dashed	K1 Outlines of adjacent parts	0.35 mm
		K2 Alternative and extreme positions of moveable parts	0.35 mm
		K3 Centroid lines	0.35 mm
		K4 Initial outlines prior to forming	0.35 mm
		K5 Parts situated in front of the cutting plane	0.35 mm

Typical applications of different types of lines are shown in Fig. A2.1 below.

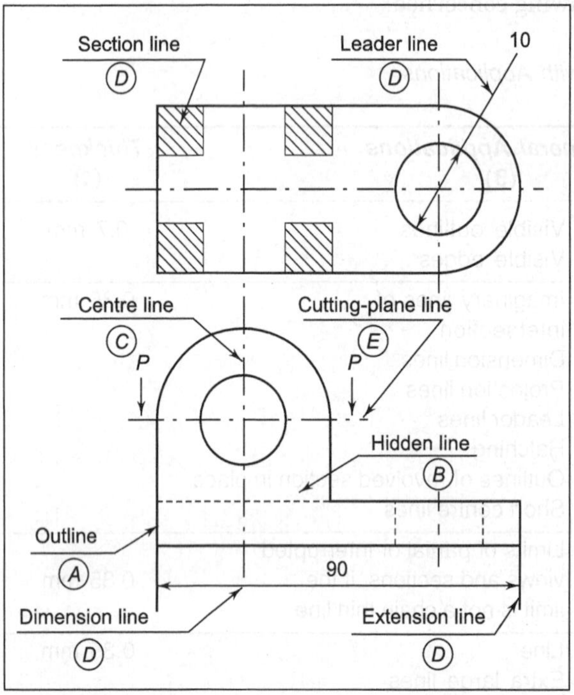

Fig. A2.1 Application of lines

2.1.1 Thickness of Lines

Two thicknesses of lines are used in a drawing. The ratio of the thick to the thin line should not be less than 2 : 1. The thickness of lines should be chosen according to the size and the type of the drawing from the following ranges: 0.18, 0.25, 0.35, 0.5, 0.7, 1, 1.4 and 2 mm. The thickness and the darkness of the lines should be the same in all views to give a good appearence to the drawing.

2.1.2 Spacing of Lines

The minimum space between parallel lines, including hatching, should never be less than twice the thickness of the heaviest line. It is recommended that these spaces should never be less than 0.7 mm.

2.2 LETTERING

Lettering is very important in an engineering drawing to give information about size, surface, finish, and so on. Good lettering improves the quality of the drawing, enhancing its appearance. Freehand lettering should be done on the drawing whenever possible, because it takes less time. To give all the information necessary for a complete drawing, notes must be added to the graphics, describing its shape, dimensions, material, finish, and a descriptive title should also be given, all of which must be lettered freehand in a style that is perfectly legible, uniform, and capable of rapid execution.

2.2.1 Classification of Letter Style

There are various forms of alphabets used in the art of lettering and each is suitable for some particular purpose. The Old Roman style is the oldest of all styles. Though these letters have elementary strokes of even width, Roman letters have elementary strokes accented or consisting of heavy and light lines.

Commercial Gothic is a simplified form of the original Gothic and is mainly used for working drawings. These letters are made in different ratios of width and height, but a ratio of 6 : 5 is preferable.

General proportions Lettering may be divided into three categories, according to the proportion of width and height.

1. Normal letters
2. Compressed or condensed letters
3. Extended letters

Normal letters Normal letters have a height to width ratio of 6 : 5. They are used for general purposes and written with usual spacing.

Compressed or condensed letters The height of these letters is more than their width, making them appear narrow. They are used when the space is limited.

Extended letters The height of these letters is equal or lesser than their width, making them appear wide.

The conventional sizes for some applications are:

1. Main titles in 8 mm
2. Subtitles in 5 mm
3. Notes, dimensions, figures and so on in 3 to 5 mm
4. Drawing number in 10 mm

A good style of single stroke lettering can be ensured by taking into account the following:

1. **Uniformity of thickness** Use of a conical point pencil that is not too hard as it can pierce into the surface of the drawing paper.
2. **Uniformity of shape and size** The width of the stroke of the pencil or pen is the width of the stem of the letter. Light guidelines for both the top and bottom of letters should always be drawn with a sharp pencil.
3. **Uniformity of style** Either the vertical or inclined style should be used, though vertical lettering is prefered. However, both can be used, according to the established practice and the draughts man's convenience. Students should, therefore, learn to execute both forms well. They may give more attention to the style they like and in which they can do better.
4. **Uniformity of spacing** Uniformity in spacing of letters, whether they are vertical or inclined, is a matter of equalising spaces by the trained eye. The background areas between letters, and not the distance between them, should be approximately equal.
5. **Uniformity of shape** If inclined letters are used, all the letters must have the same inclination and should be inclined towards the right hand side, sloping 67.5° to 75° to the horizontal.

2.2.2 Lettering on Drawing Sheets

The main requirements for lettering of engineering drawings are legibility, uniformity, ease, and rapidity in execution. Both, the vertical and sloping styles of letters and numerals are suitable for general use. The recommended sizes of letters and numerals to suit different purposes are given in Table A2.2.

TABLE A2.2 *Common Size of Letters and Numerals for Drawing Work*

Purpose	*Size of Letters and Numerals (mm)*
Main title and drawing number	8 and 10
Subtitles and heading	5
Notes such as legends, schedules, materials, and dimensions	3, 4 and 5

2.3 SCALES

Scale is the ratio of the linear dimension of an element of an object, as represented in the original drawing, to the real liner dimension of this element. The different types of scales are as follows.

Full size scale	: with ratio 1 : 1
Enlargement scale	: with ratio larger than 1 : 1
Reduction scale	: with ratio smaller than 1 : 1

2.3.1 Designation

The complete designation of a scale shall consist of the word 'scale' (or its equivalent in the language used in the drawing) followed by the indication of its ratio, as shown here.

Scale 1 : 1 for full size
Scale X : 1 for enlargement scales
Scale 1 : X for reduction scales

If there is likelihood of misunderstanding, the word 'scale' may be omitted.

2.3.2 Inscription

The designation of the scale used in the drawing shall be inscribed in the title block of the drawing. Where it is necessary to use more than one scale in a drawing, the main scale alone shall be inscribed in the title block, and all other scales are given adjacent to the item reference number of the part concerned or adjacent to the reference letter of a detail view (or section). The recommended scales for use in technical drawings are specified in Table A2.3.

TABLE A2.3 *Types of Common Scales for Technical Drawing*

Category	***Recommended Scales***		
Enlargement scales	50 : 1 5 : 1	20 : 1 2 : 1	0 : 1 1 : 1
Full size scales			1 : 1
Reduction scales	1 : 2 1 : 20 1 : 200 1 : 2000	1 : 5 1 : 50 1 : 500 1 : 5000	1 : 10 1 : 100 1 : 1000 1 : 10000

The scale to be chosen for a drawing will depend upon the complexity of the object to be depicted and the purpose of the representation. In all cases, the selected scale shall be large enough to permit easy and clean interpretation of the information depicted. The scale and the size of the object, in turn, will decide the size of the drawing.

Details that are too small for complete dimensioning in the main representation shall be shown adjacent to the main representation in a separate detail view (or section), which is drawn to a larger scale.

Large scale drawing It is recommended that a full size view be added to the large scale representation of a small object. In this case, the full size view may be simplified by showing the outlines of the object only.

2.4 KEEPING DRAWINGS NEAT

In order to keep a drawing free of blotches and marks, the following need to be kept in mind.

1. Hands should be kept clean at all times during work. Dirty or oily hands should be washed with soap and wiped properly.
2. All the drawing equipment, such as the drawing board, set squares, T-square, scale, and so on should be kept clean using a duster. Dirty set squares should be cleaned with soap and water.
3. Dirt and graphite particles from the pencil will make the drawing dirty. Hence every care should be taken to remove them from the drawing sheet.
4. The sliding equipments on the drawing sheet, such as mini drafter, T-square, and set square must be cleaned properly every time and sliding should be reduced to a minimum.
5. Pencil should be sharpened away from the drawing sheet and other equipments.
6. The sand paper pad, upon which the pencil is to be sharpened, should be kept away from the drawing sheet, otherwise the loose particles of graphite will fly onto the drawing sheet and make it dirty.
7. Direct contact of hands with the drawing sheet should be avoided. In lettering or dimensioning, always place a piece of duster under the hand. Never work with the sleeves or hand resting upon the pencil area.
8. Erasing should be done properly with a soft eraser. An erasing shield should be used where required.
9. Pencil points should always be kept sharp and should be used properly. Extra lines should be rubbed out when the drawing is complete.
10. The drawing sheet should be cleaned by a duster after every erasing.

Review Questions

1. What are the different types of lines used in drawing?
2. State the size of each type of line.
3. What are the different applications of lines in drawings?
4. What is lettering?
5. What do you understand by vertical and inclined lettering?
6. Discuss single stroke lettering.
7. State the procedure for lettering on drawing sheets.
8. What is scale? State its various types.
9. How do you keep a drawing sheet clean?

CHAPTER 3

Projections

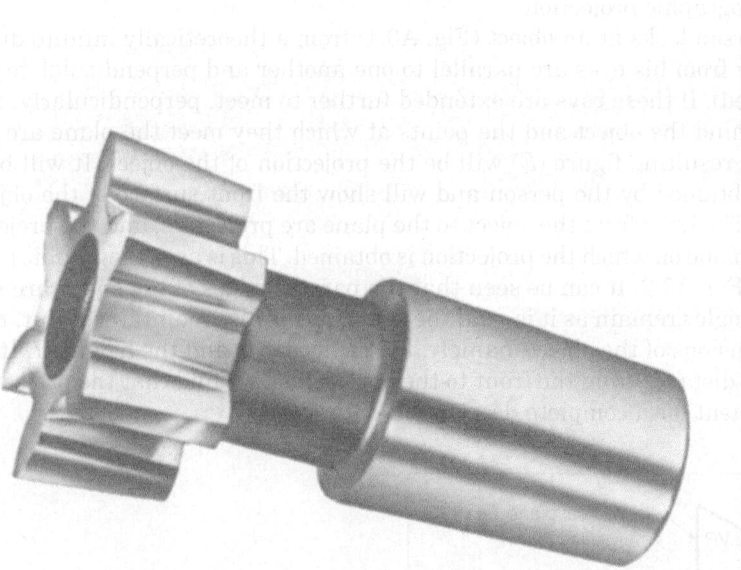

This chapter contains basic information about projections and methods of projection with illustrated sketches

After reading this chapter you will be able to

- differentiate orthographic, isometric, oblique, and perspective drawing
- identify first angle and third angle orthographic projections

3.1 PROJECTIONS

For the purpose of construction, a drawing should show the true shapes of all parts of an object together with their sizes. Orthographic projection is universally adopted for this purpose. By this method, the description of the object regarding its shape and size is conveyed in definite and unmistakable terms so that it can be constructed precisely. The object is shown in two or more views, obtained by looking at it from different angles and arranged in a definite manner.

If straight lines are drawn from various points on the contour of an object to meet a plane, the object is said to be projected on that plane. The figure formed by joining, in correct sequence the points at which the lines meet the plane is called a projection of that object. The lines from the object to the plane are called projectors.

3.1.1 Orthographic Projection

When the projectors are parallel to each other and also perpendicular to the plane, the projection is called an orthographic projection.

Imagine that a person looks at an object (Fig. A3.1) from a theoretically infinite distance so that the lines of view from his eyes are parallel to one another and perpendicular to the front surface (shown shaded). If these rays are extended further to meet, perpendicularly, a vertical plane (*VP*) set up behind the object and the points at which they meet the plane are joined in proper sequence, the resulting figure (*E*) will be the projection of the object. It will be exactly similar to the view obtained by the person and will show the front surface of the object in its true shape and size. The lines from the object to the plane are projectors, and the projectors are perpendicular to the plane on which the projection is obtained. This is an orthographic projection, shown separately in Fig. A3.2. It can be seen that the parallel edges of the object are shown by parallel lines, right angles remain as it is, and the semicircle is also maintained. But, the figure shows only two dimensions of the object, namely, for the length *L* and the height *H*. It does not show the depth—the distance from the front to the back. Thus, we find that the projection from the front is not sufficient for a complete description of the object.

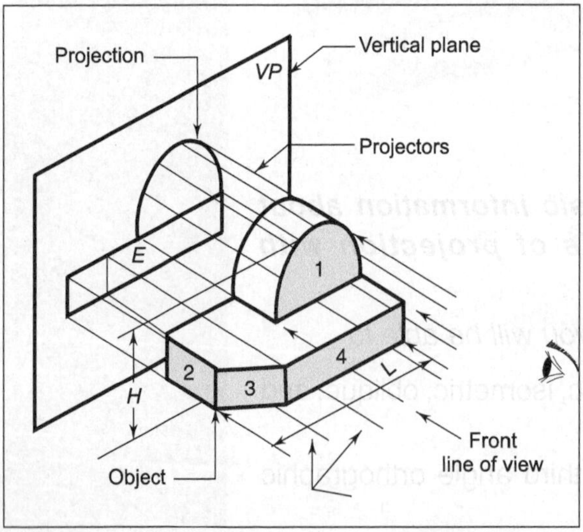

Fig. A3.1 Object in front of the *VP*

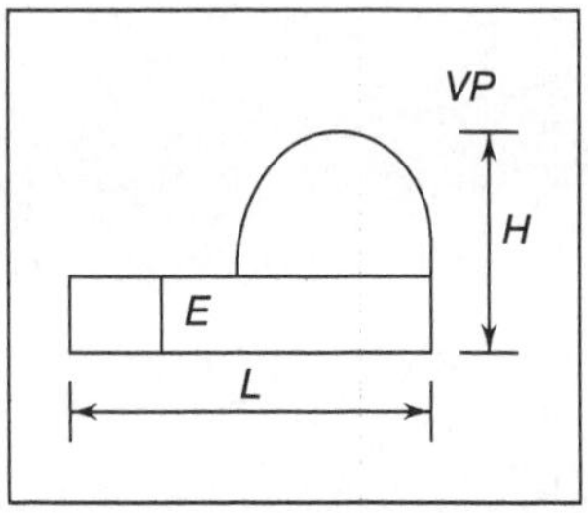

Fig. A3.2 View on the *VP*

Let us further assume that a horizontal plane (*HP*) is winged to the vertical plane so that the object is in front of the vertical plane and above the horizontal plane (Fig. A3.3). The projection on this plane, obtained in a similar manner, is shown in Fig. A3.3. If a person looks at the object from the top surface, as shown shaded in Fig. A3.3, the object will be exactly the same as *P* in Fig. A3.4. This figure shows the length *L* and the depth *D* of the object. It does not show the height *H*.

One of the planes may now be rotated or turned around so that it lies in extension with the other plane. This can be done in two ways, (1) by turning the vertical plane in the direction of the arrow *A* or (2) by turning the horizontal plane in the direction of the arrow *B*. The horizontal plane when turned and brought in line with the vertical plane is shown by dashed lines. The two projections can now be shown on a flat surface in correct relationship with each other, as shown in Fig. A3.4. When studied together the two views supply all the information regarding the shape, and size of the object.

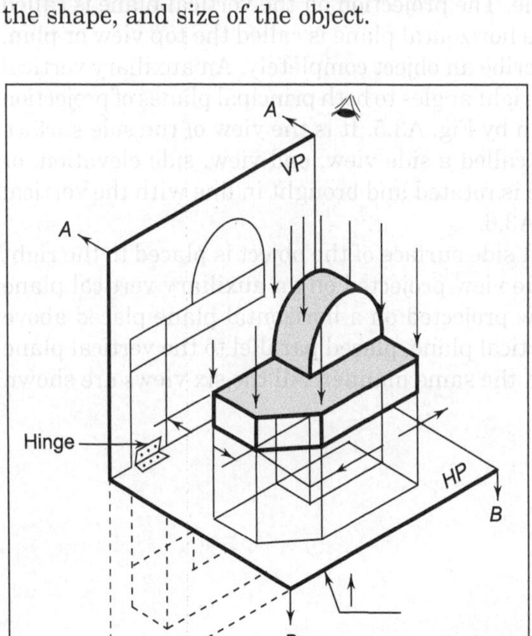

Fig. A3.3 Object above *HP* and in front of *VP*

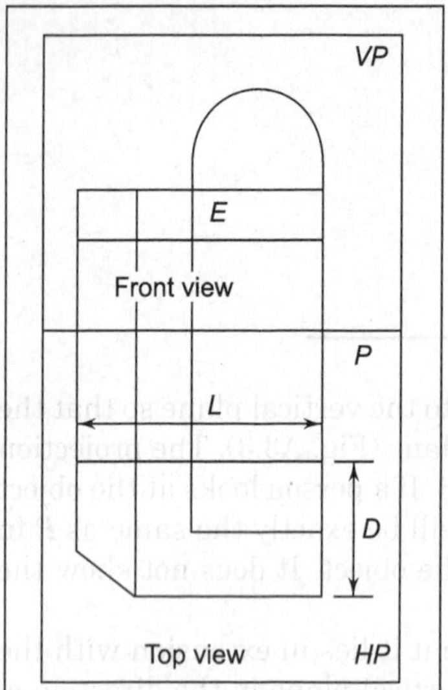

Fig. A3.4 View on *VP* and *HP*

Any object can be represented by means of orthographic projections. The two planes employed for the purpose of orthographic projection are called reference planes or principal planes of projection. They meet each other at right angle. The projection on the vertical plane is called the front view or elevation. The projection on the horizontal plane is called the top view or plan.

Very often two views are not sufficient to describe an object completely. An auxiliary vertical plane (*AVP*) is therefore imagined to be placed at right angles to both principal planes of projection (Fig. A3.5). The projection on this plane is shown by Fig. A3.5. It is the view of the side surface (shown shaded) of the object. This projection is called a side view, end view, side elevation, or end elevation. When the auxiliary vertical plane is rotated and brought in line with the vertical plane, the views seen will be as shown in Fig. A3.6.

It is important to note that the view of the left side surface of the object is placed to the right of the front view. Three additional views, (1) side view projected on an auxiliary vertical plane placed to the left of the object, (2) a bottom view projected on a horizontal plane placed above the object, and (3) a back view projected on a vertical plane, placed parallel to the vertical plane and in front of the object, can also be obtained in the same manner. All the six views are shown projected on the respective planes in Fig. A3.7.

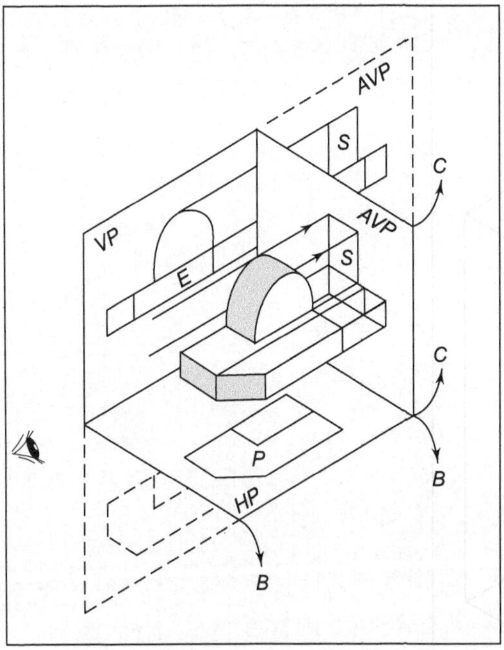

Fig. A3.5 Object infront of *VP*, above *HP*, and in front of *AVP*

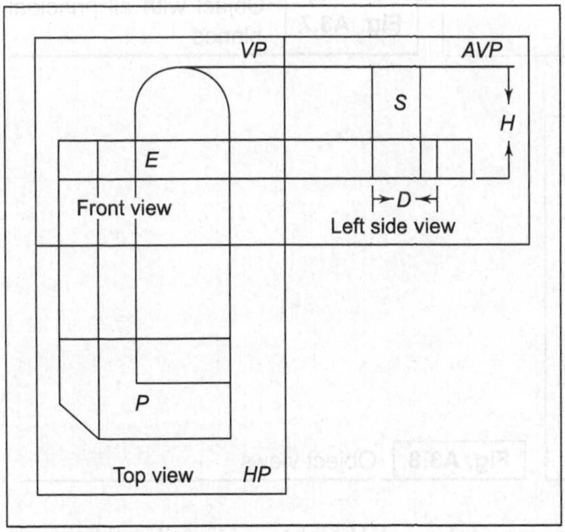

Fig. A3.6 View on *VP*, *HP*, and *AVP*

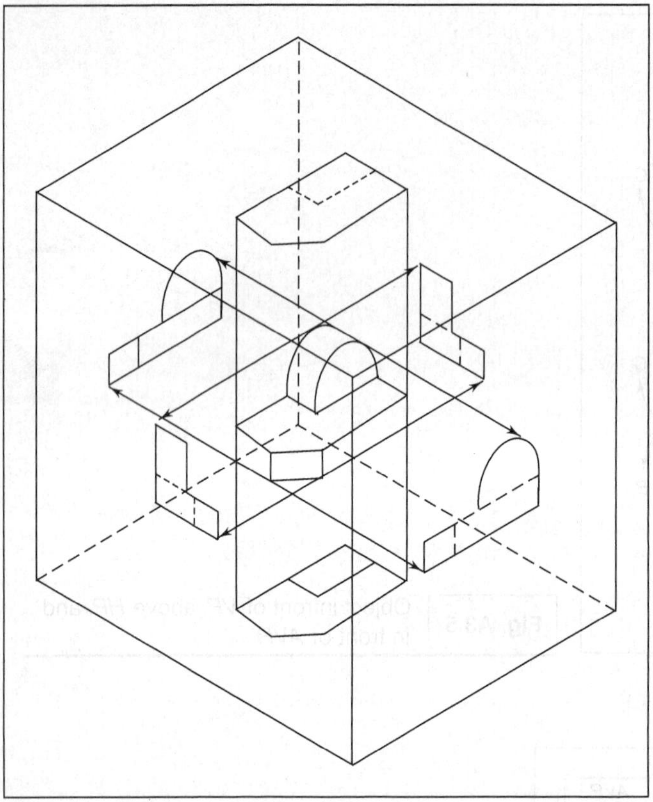

Fig. A3.7 Object with all principal planes

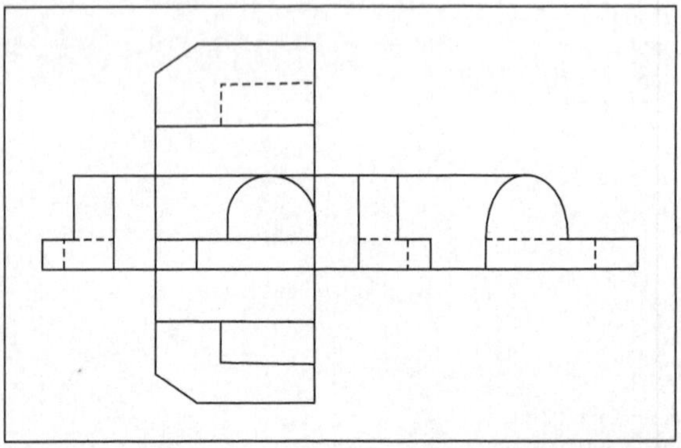

Fig. A3.8 Object views

Fig. A3.8 shows views laid out on a flat surface by rotating all the planes around and bringing them in line with the vertical plane. The front view, top view, and two of the remaining views are required to describe an object completely. Only in exceptional cases, for an object of very complex nature, five or six views may be found necessary. The planes of projection, being imaginary, are usually described when drawing the views. However, care must be taken to place the view in correct relationship, that is, the top view and the bottom view must be vertically in line with the front view. The side views and the back view should be horizontally in line with the front view.

First angle projection If the two principal planes of projection are extended beyond their line of meeting, they form four quadrant or dihedral angles. The object situated in front of the vertical plane and above the horizontal plane, that is, in the first quadrant, is then projected on these planes. This method of projection is known as first-angle projection. The object lies between the observer and the plane of projection. When the views are drawn in their relative positions, the top view comes below the front view. When the observer views the object from the front, the rays of sight intersect the vertical plane. The figure formed by joining the point of intersection in correct sequence is the front view of the object. The top view is obtained in a similar manner by looking from above when the two planes are brought in line with each other. The view will be as shown in Fig. A3.9.

Third angle projection The top view in this case comes above the front view. The side view is obtained by projecting on an auxiliary vertical plane placed perpendicularly to both the horizontal plane and the vertical plane and between the observer and the object (Fig. A3.10).

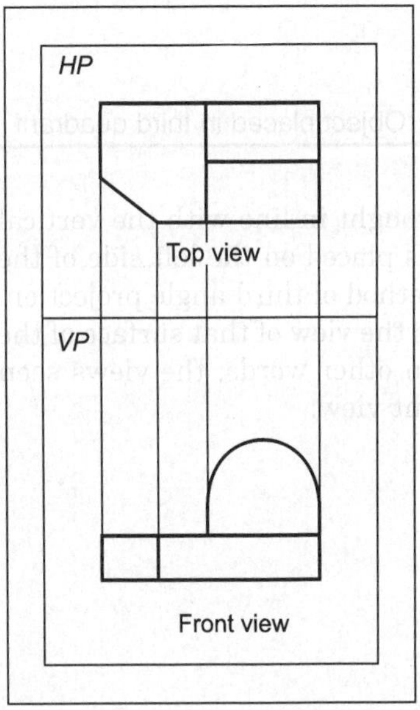

Fig. A3.9 Views with third angle projection

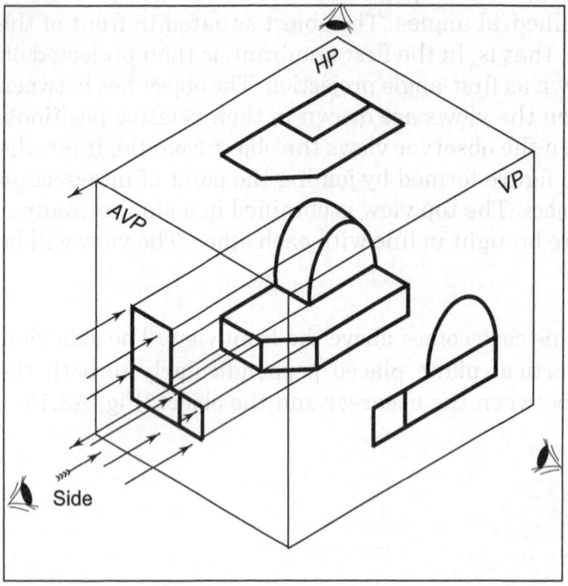

Fig. A3.10 Object placed in third quadrant

The three views obtained after the planes are rotated and brought in line with the vertical plane are shown in Fig. A3.11. The left side view of the object is placed on the left side of the front view. Fig. A3.12 shows all the six views obtained by the method of third angle projection. A careful study of the views will show that each projection shows the view of that surface of the object which is nearest to the plane on which it is projected. In other words, the views seen from any side of the object is placed on the same side of the front view.

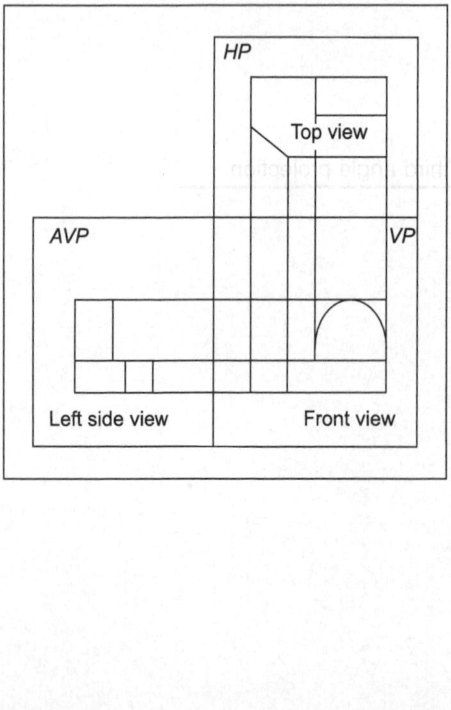

Fig. A3.11 Views with third angle projection

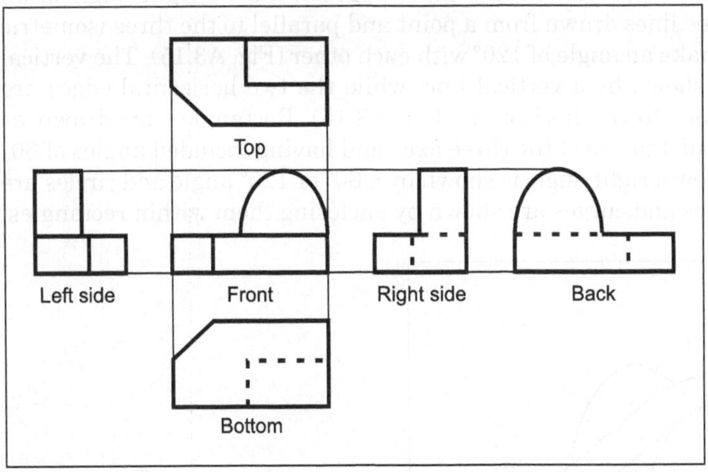

Fig. A3.12 Third angle method of projection

3.1.2 Symbols for Methods of Projection

For every drawing it is absolutely essential to indicate the method of projection adopted. This is done by means of symbolic figures drawn within a little block on the drawing sheet. The symbolic figure for the first angle projection method is shown in Fig. A3.13 and the third angle projection method is shown in Fig. A3.14. These are the views of a frustum of a cone.

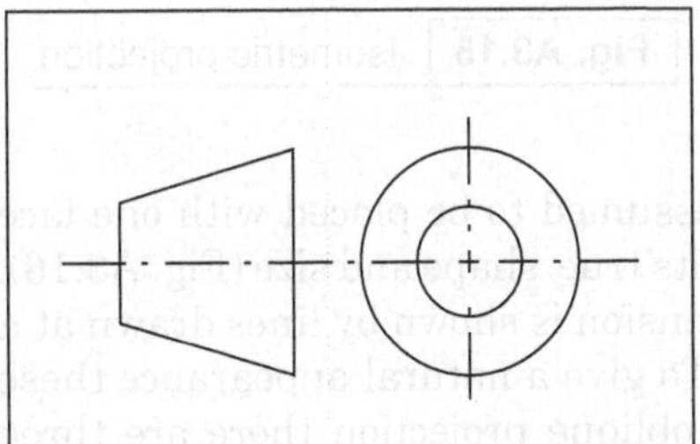

Fig. A3.13 First angle

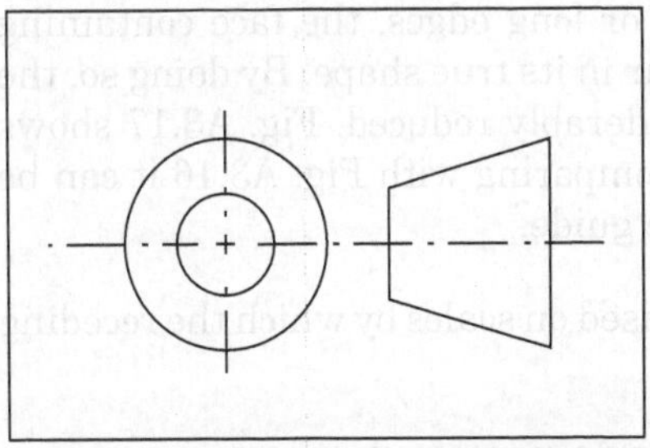

Fig. A3.14 Third angle

3.1.3 Pictorial Projections

In engineering drawing, three methods of pictorial projection are commonly used. They are (1) isometric (2) oblique and (3) perspective projections.

Isometric projection In Isometric projection, the three edges of a solid right angle of an object are shown by means of three lines drawn from a point and parallel to the three isometric axis, which meet at a point and make an angle of 120° with each other (Fig. A3.15). The vertical edges of the solid right angle is shown by a vertical line, while the two horizontal edges are shown by two lines inclined at 30° to the horizontal (Fig. A3.15). Rectangles are drawn as parallelograms, having sides parallel to two of the three axes and having included angles of 60° and 120°. Thus, in an isometric view a right angle is shown by a 60° or 120° angle and circles are shown as ellipses. Irregular figures and curves are shown by enclosing them within rectangles.

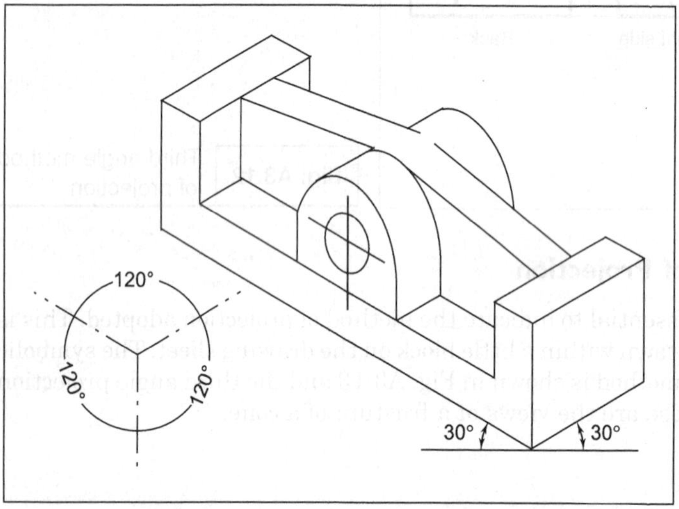

Fig. A3.15 Isometric projection

Oblique projection In oblique projection the object is assumed to be placed with one face parallel to the plane of projection, hence that face appears in its true shape and size (Fig. A3.16). This gives two main dimensions of the object. The third dimension is shown by lines drawn at a convenient angle, generally 30° or 45° with the horizontal. To give a natural appearance these lines are drawn 3/4 or 1/2 the actual length. Thus, in an oblique projection there are three axes—a vertical, a horizontal, and a third included at an angle of 30° or 45° with the horizontal. Rectangular surface and circles parallel to the third axis are shown as parallelograms and ellipses, respectively. When an object has curved surfaces or long edges, the face containing such surfaces or edges is usually so placed that it may appear in its true shape. By doing so, the drawing is simplified and the amount of distortion is considerably reduced. Fig. A3.17 shows the guide with its longer edges parallel to the third axis. Comparing with Fig. A3.16 it can be seen that the former gives a clearer idea of the shape of the guide.

Types of the oblique projection: Oblique projection is based on scales by which the receding lines are drawn.

Cavalier projection: When the receding lines are drawn to full size scale and the projectors are inclined at an angle of 30°, 45°, or 60° to the plane of projection, such an oblique projection is known as a *cavalier projection*. Refer to Fig. A3.18(a).

Cabinet projection: If the receding lines are drawn to half size scale the resultant oblique projection is known as a *cabinet projection*. Refer to Fig. A3.18(b).

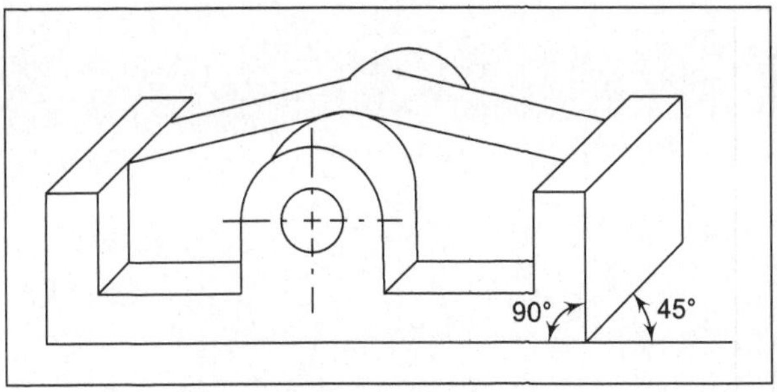

Fig. A3.16 Oblique projection

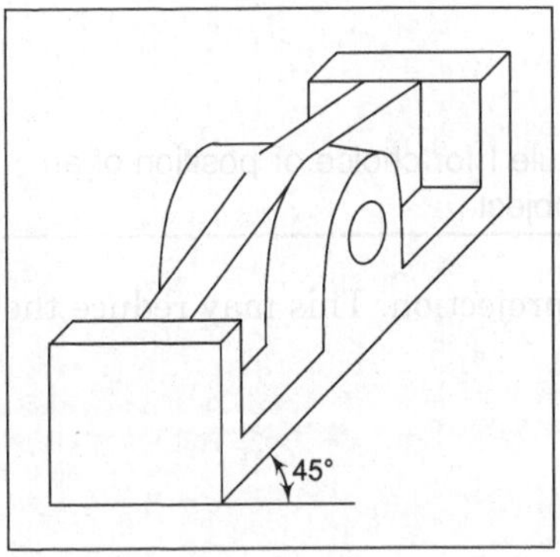

Fig. A3.17 Oblique projection

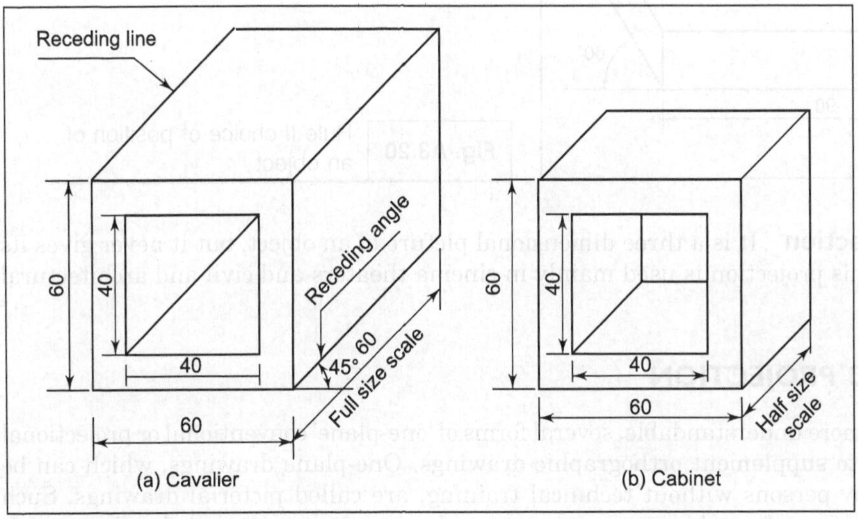

Fig. A3.18 Types of projection

Rules for Choice of Position of an Object:

The chioce of position of an object should be such that minimum distortion of the object can occur in the oblique projection. This can be achieved by observing the following rules:

Rule I The face of an object containing essential contours (i.e., circles, irregular shapes etc.) must be kept parallel to the plane of projection (Fig. A3.19).

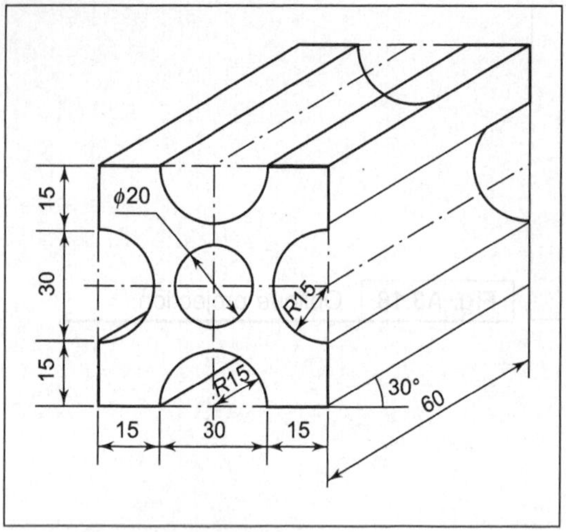

Fig. A3.19 Rule I for choice of position of an object

Rule II Keep the longest dimension parallel to the plane of projection. This may reduce the distortion effect of the object (Fig. A3.20).

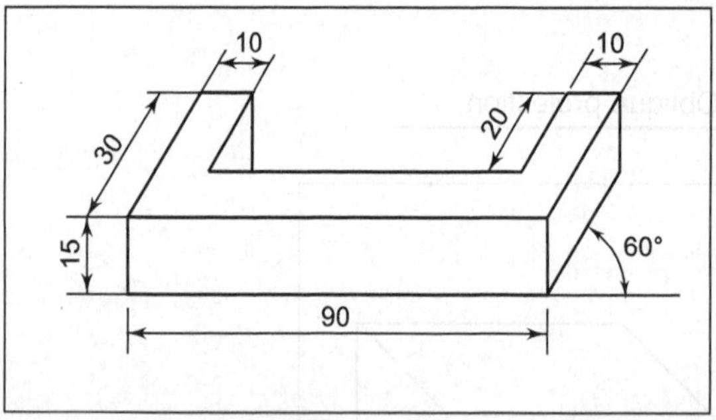

Fig. A3.20 Rule II choice of position of an object

Perspective projection It is a three dimensional picture of an object, but it never gives its true dimensions. This projection is used mainly in cinema theaters and civil and architectural works.

3.2 ISOMETRIC PROJECTION

To make a drawing more understandable, several forms of 'one-plane' conventional or projectional drawings are used to supplement orthographic drawings. One-plane drawings, which can be easily understood by persons without technical training, are called pictorial drawings. Such drawings reveal the face of an object at once, more or less as they appear to the observer. It must be emphasised that as far as engineering is concerned, all one-plane drawing methods are normally used as auxiliaries to the standard method of orthographic projection. An engineering student is required to be proficient at drawing 'pictorial views' of objects and also at converting given orthographic views.

3.2.1 Isometric Drawing

In isometric projection lines do not show the true length of the object edges and are instead foreshortened. In making such projections, either it should be projected from the orthographic views of the object, as already discussed, or a special isometric scale should be used. To avoid this tedious construction, if the foreshortening of the lengths is ignored and the true lengths are laid off along the isometric axes or along the isometric lines, the view obtained is called an isometric drawing.

Isometric scale for isometric drawing

Lines in an isometric projection: The characteristics of lines in an isometric projection are illustrated in Fig. A3.21.

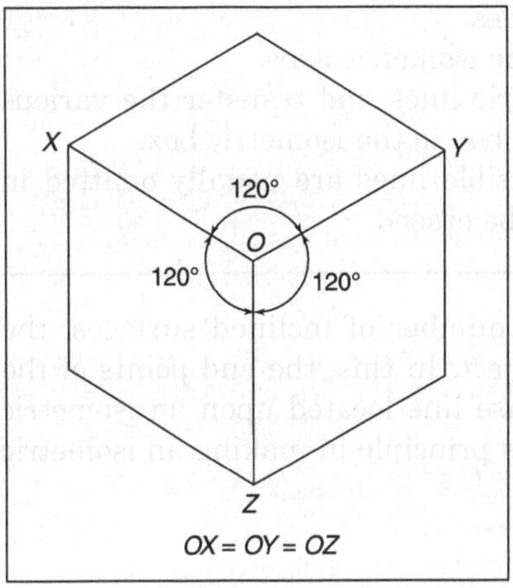

Fig. A3.21 Isometric axes

1. All lines that are parallel on the object are parallel on the isometric projection.
2. Vertical lines on the object remain vertical in the isometric projection.
3. Horizontal lines on the object are drawn at an angle of 30° to the horizontal.

Isometric scale: The projected lengths of the object along and parallel to the isometric axes are approximately 81 per cent of their true lengths. So a proportionate scale called an isometric scale may be used for marking the distances along and parallel to the isometric axes. The construction of the isometric scale is shown in Fig. A3.22.

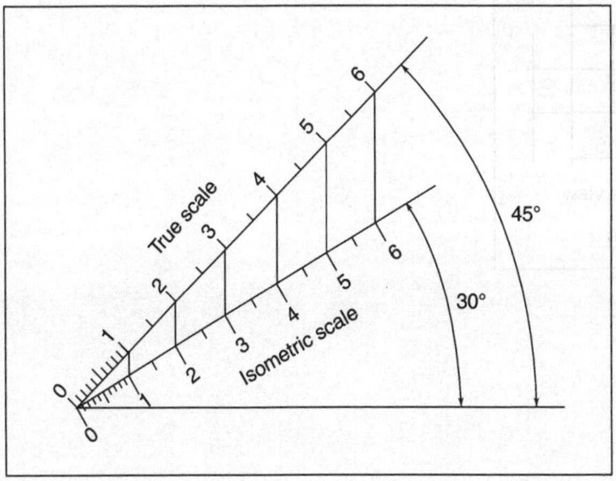

Fig. A3.22 Construction of isometric scale

Method of making an isometric drawing Some important methods of drawing isometric views are discussed below.

Box method: In this method, the object is enclosed in a rectangular box in which it may just fit (Fig. A3.23). The box is drawn around the object in orthographic views. The different steps in making the isometric drawing or projection with the help of this box are as follows.

1. Draw the orthographic views of the object to the desired scale. If an isometric projection is to be drawn, use the isometric scale.
2. Enclose the views in a rectangular box by very light lines.
3. Construct the isometric view of the enclosing box on the isometric axes.
4. Take measurements along isometric axes and isometric lines and transfer the various details of the object from the rectangular orthographic box to the isometric box.
5. Finish the view by brightening the visible lines. Invisible lines are usually omitted in isometric drawings. If drawn very light they need not be erased.

Coordinate construction: When an object contains a number of inclined surfaces, the coordinate construction method is used to represent the object. In this, the end points of the edges are located with respect to an assumed isometric base line located upon an isometric reference plane. Fig. A3.23 illustrates the application of this principle in making an isometric drawing of an object with inclined surfaces.

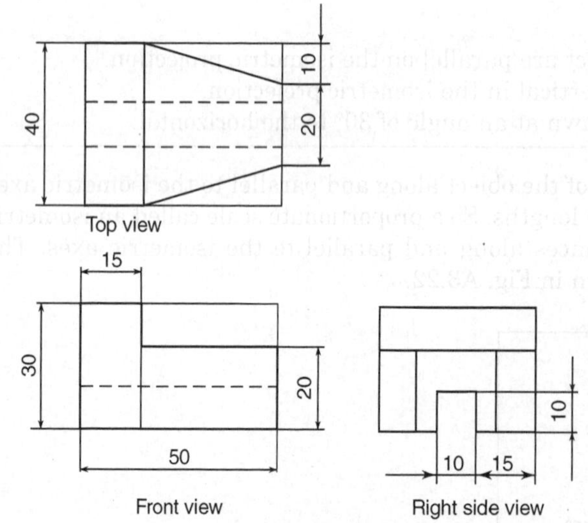

(a) Orthographic Views (Third angle projection)

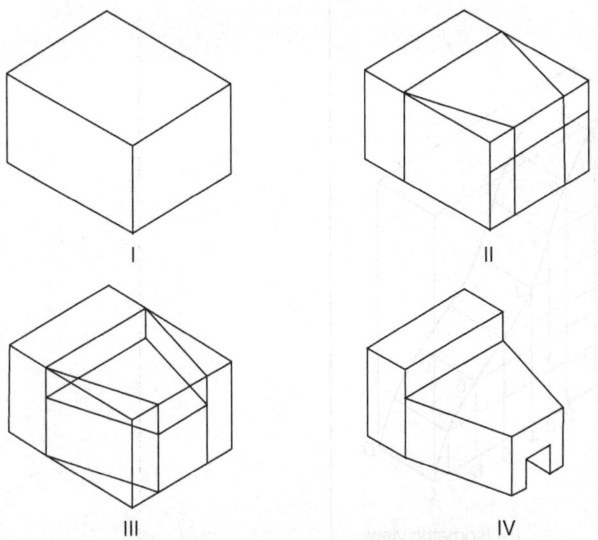

(b) Steps in drawing an isometric view

Fig. A3.23 Box method

Offset construction: Irregular curved surfaces are mostly shown in isometric drawings by the offset method. This is a modification of the coordinate construction method. The curve can be drawn by plotting points on the curve, located by measuring along isometric lines. Fig. A3.24 illustrates the aplication of this method.

(a) Orghographic views (third angle projection)

(b) Isometric view

Fig. A3.24 Coordinate construction method

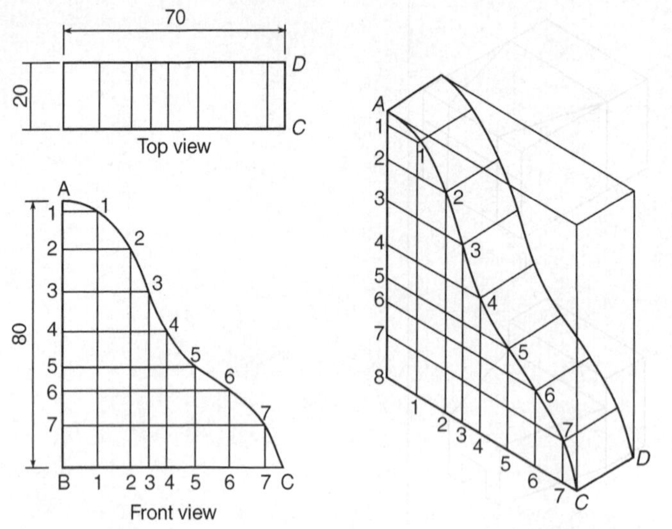

(a) Ohghographic views (third angle projection) (b) Isometric view

Fig. A3.25 Off-set construction method

3.3 PROCEDURE FOR PREPARING A SCALE-DRAWING

A scale drawing must always be prepared from freehand sketches initially prepared from a pictorial view or a real object. In the initial stages of a drawing, always use a soft pencil, namely, HB, and work with a light hand so that lines are thin, faint, and easy to erase if necessary.

1. Determine overall dimensions of the required views. Select a suitable scale so that the views are conveniently placed.
2. Prepare the sheet layout, draw rectangles for the views. Keeping sufficient space between them and from the borders of the sheet is important while preparing the layout.
3. Draw centre lines in all the views. When a cylindrical part of a hole is seen as a rectangle and as a circle, draw two centre lines intersecting each other at right angles to its centre.
4. Draw details simultaneously in all the views in the following order.
   (i) Circles and arcs of circle
   (ii) Sraight lines for the general shape of the object
   (iii) Straight lines, small curves, and so on for minor details
5. After the views have been completed, erase all unnecessary lines completely. Make the outlines so faint that only their impressions exist.
6. Using a 2H or 3H pencil finish the drawing, make the outlines uniform and intensely black (but not too thick).
7. Dimension the views completely. Keep all centre lines.
8. Draw section lines in the view or views that are shown in the section.
9. Fill up the title block and furnish all other necessary particulars.
10. Check the drawing carefully and see that it is complete in all respects.

Review Questions

1. What is projection?
2. What are the different types of projections?
3. What do you mean by orthographic projection?
4. What is first angle projection?
5. What is third angle projection?
6. Why are second and fourth angle projections not in use?
7. Draw symbols for first and third angle methods of projection.
8. What are the different types of pictorial projections?
9. What is isometric projection?
10. What is the procedure for preparing scale?

CHAPTER 4

Dimensioning

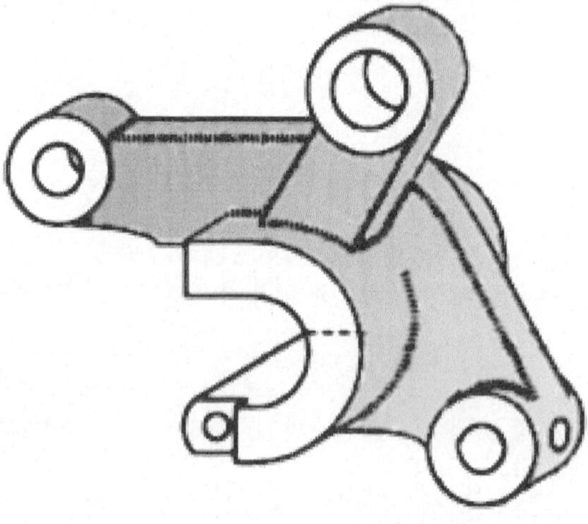

This chapter contains information about dimensioning of different engineering objects

After reading this chapter you will be able to

- dimension the drawing and the views of an object
- implement dimensioning for different drawing works

4.1 GENERAL PRINCIPLES

All dimensions, tolerances, and other similar features are necessary for the correct functioning of the part in its assembly with other parts. Dimensions should be expressed on the drawing together with the information necessary to describe the part completely in its finished form, without losing sight of the manufacturing and inspection requirements. It should not be necessary to deduce a functional dimension from other dimensions or to scale the drawing. No single dimension should be given more than once on the drawing, except where it is unavoidable.

4.2 FUNCTIONAL AND NON-FUNCTIONAL DIMENSIONING

4.2.1 Functional Dimensioning

A functional dimension is a dimension that is essential to the function of a part (Fig. A4.1).

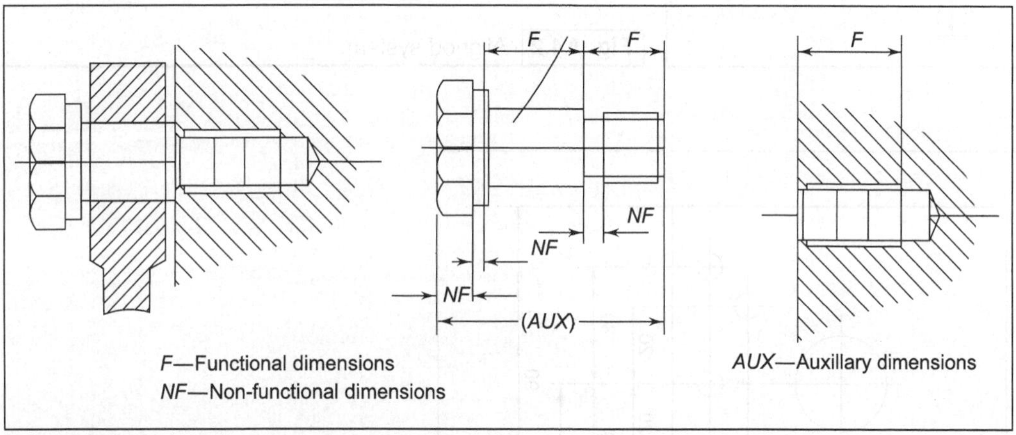

Fig. A4.1 Functional and non-functional dimensions

4.2.2 Datum Dimension

The datum dimension exactly locates a point, line, or plane. It is enclosed in a frame, for example, 30. It may also be used to specify the position of an ordinate, or a point that establishes a true profile.

4.3 PLACING OF DIMENSIONS

1. Dimensions should be placed on a view which shows the relevant features most clearly. The two recommended systems of placing the dimensions are:
   (i) **Aligned system** In this system, all dimensions are placed so that they may be read from the bottom or the right hand edges of the drawing sheet (Fig. A4.2 and Fig. A4.3) and all dimensions should be placed above the dimension lines.
   (ii) **Unidirectional system** In this system, all dimensions are so placed that they may be read from the bottom edge of the drawing sheet (Fig. A4.4 and Fig. A4.5). There are no restrictions in this system controlling the direction of dimension lines. This system is advantageous on large drawings where it is inconvenient to read dimensions from the right side.

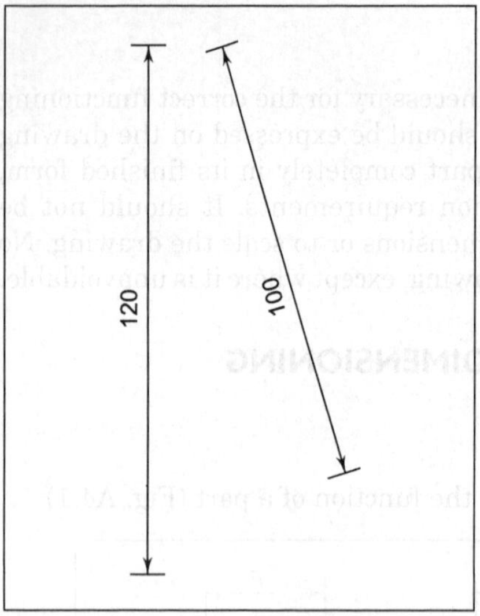

Fig. A4.2 Aligned system

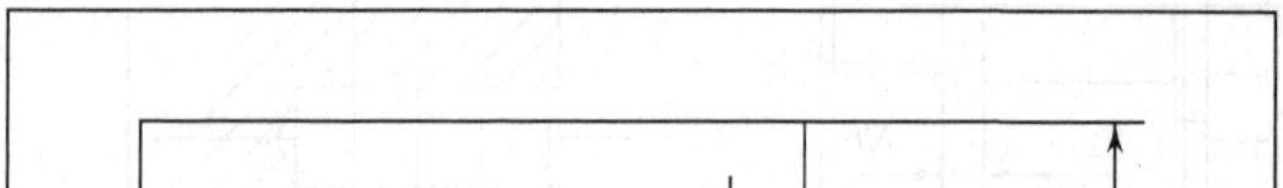

Fig. A4.3 Aligned system

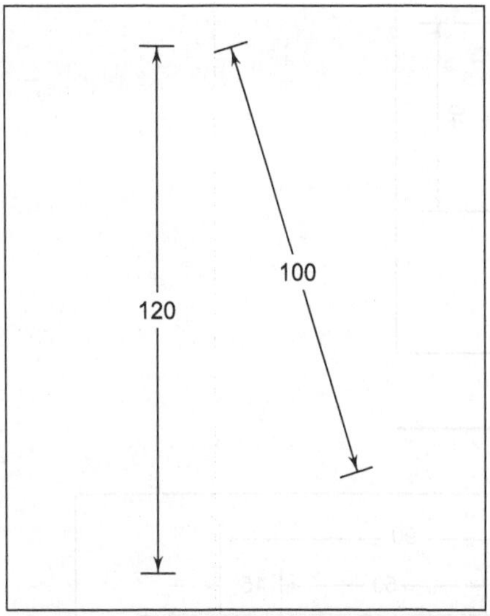

Fig. A4.4 Unidirectional system of dimensioning

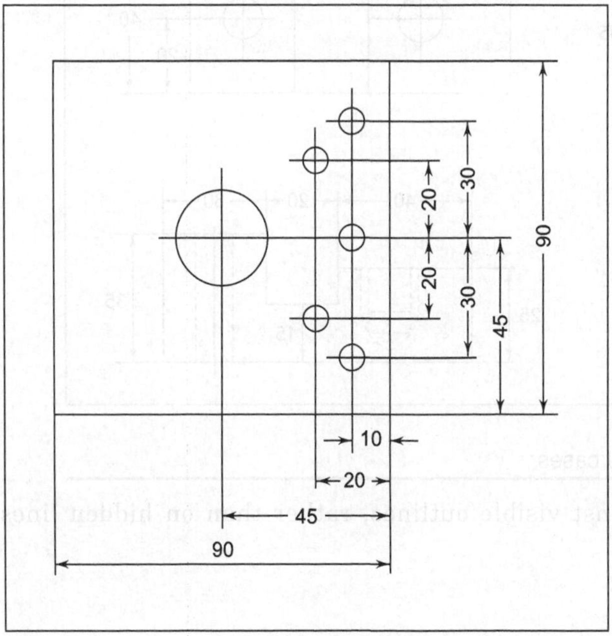

Fig. A4.5 Unidirectional system of dimensioning

2. As far as possible, all the dimensions for a particular operation shall be specified in one view only, such as the diameter and depth of a drilled hole, or the size and depth of a threaded hole, and so on.
3. Normally dimensions should be placed outside the views (Fig. A4.6), but if it is not possible, it may be placed within the view as shown in Fig. A4.7. However, dimensions should not be placed within a view unless the drawing becomes clear by doing so. Dimensions should not be placed too close to each other or to the parts being dimensioned.

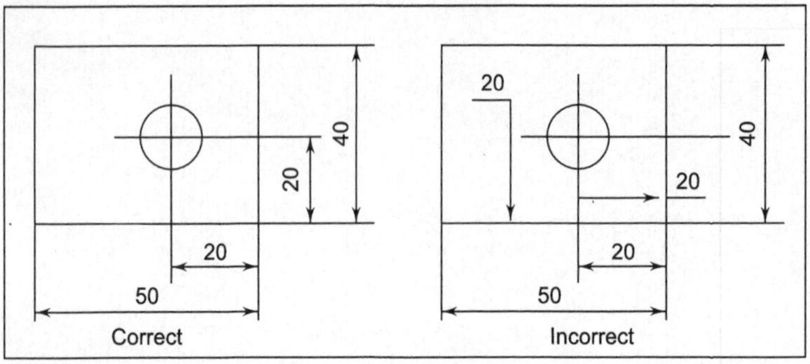

Fig. A4.6 Placing of dimensions

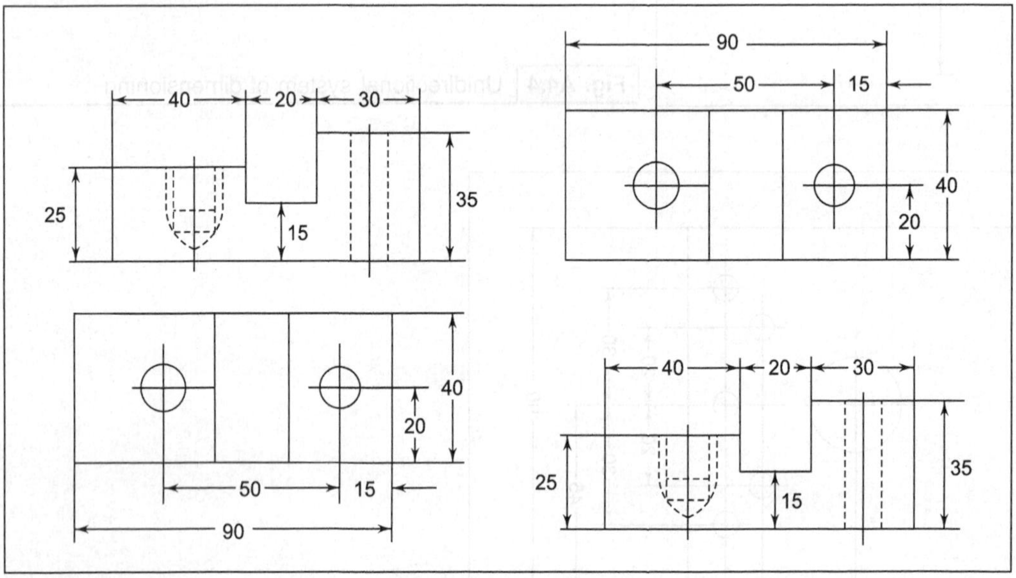

Fig. A4.7 Placing of dimensions in special cases

4. Dimensions are to be given against visible outlines, rather than on hidden lines (Fig. A4.8).

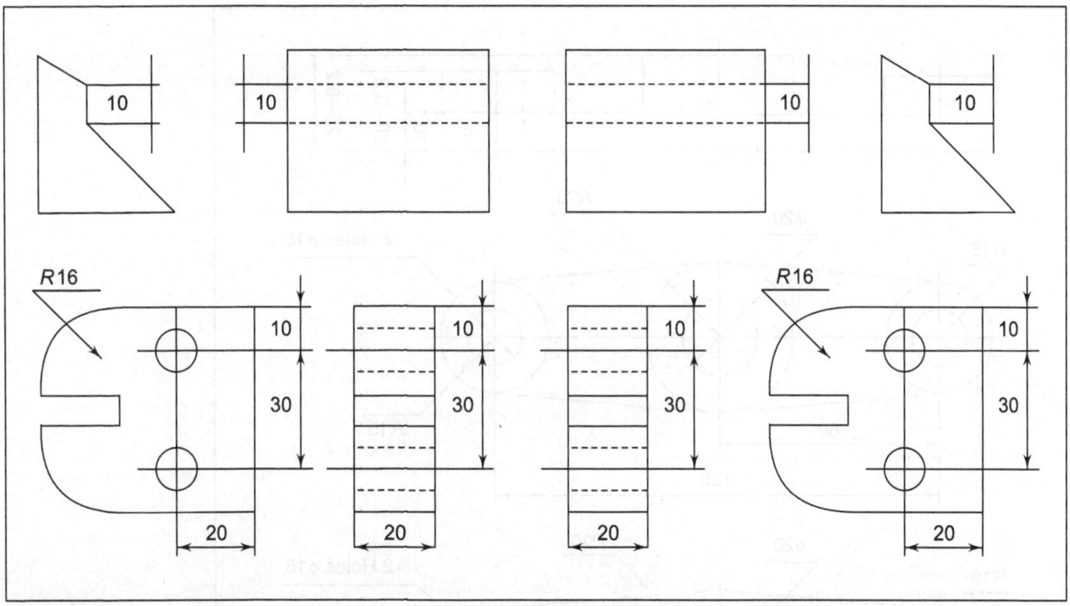

Fig. A4.8 Dimensions from visible outlines of an object

Dimensions are to be given from a base line, the centre line of a hole or a cylindrical part, an important hole, or a finished surface that may be readily established based on design requirements and the relationship to other parts (Fig. A4.9 and A4.10).

5. Dimensioning to a centre line should be avoided, except when the centre line passes through the centre of a hole, or a cylindrical part as shown in Figs. A4.9 and A4.10.

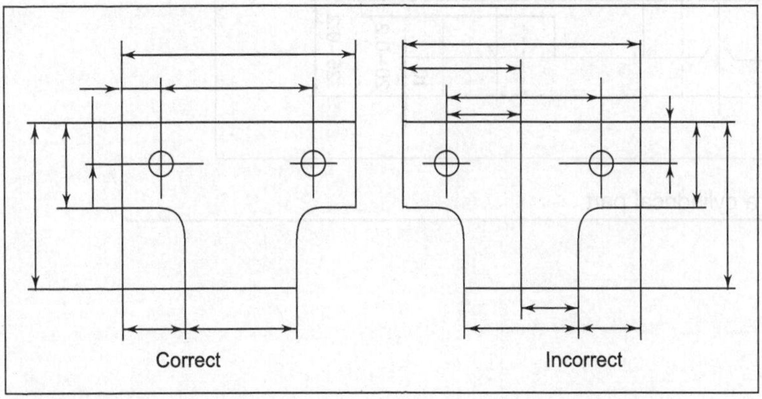

Fig. A4.9 Correct and incorrect method of hole dimensioning

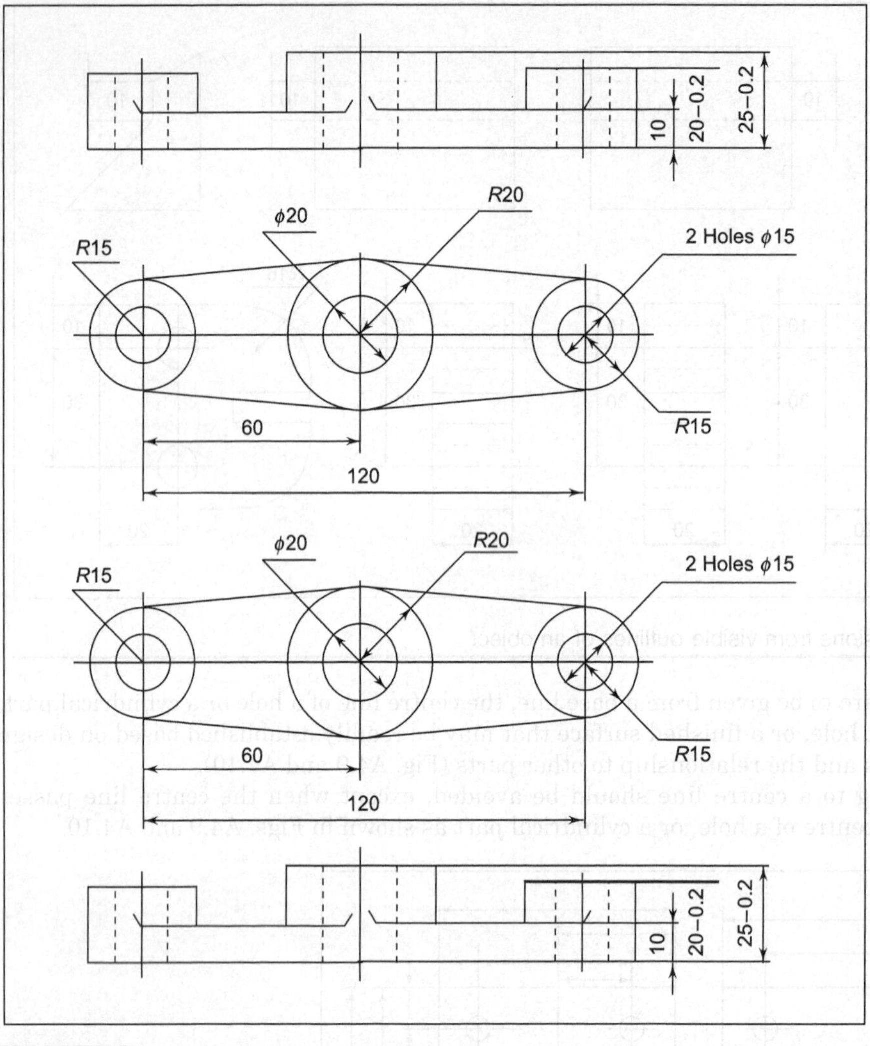

Fig. A4.10 Dimensioning of a cylindrical part

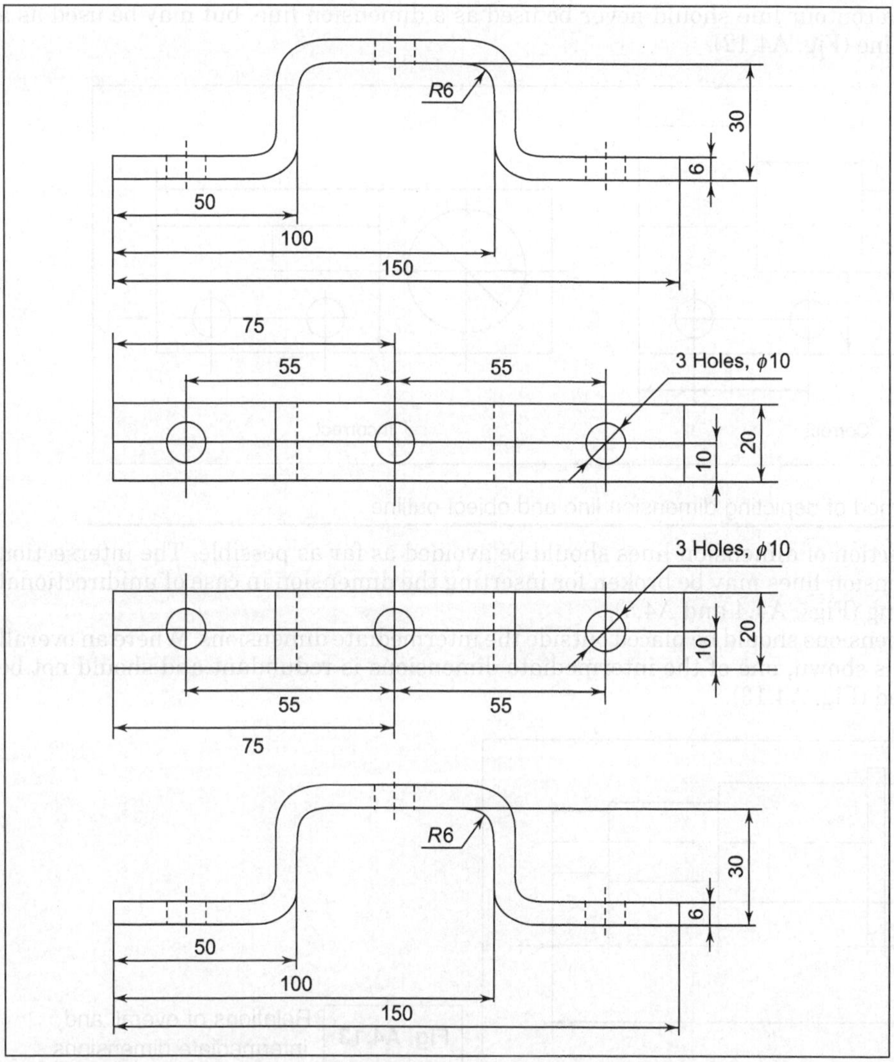

Fig. A4.11 Dimensioning of irregular and machined parts

6. Dimensions for different operations on a part, for example, drilling and bending should be given separately as shown in Fig. A4.11, if permissible by its design.

7. An axis or a contour line should never be used as a dimension line, but may be used as a projection line (Fig. A4.12).

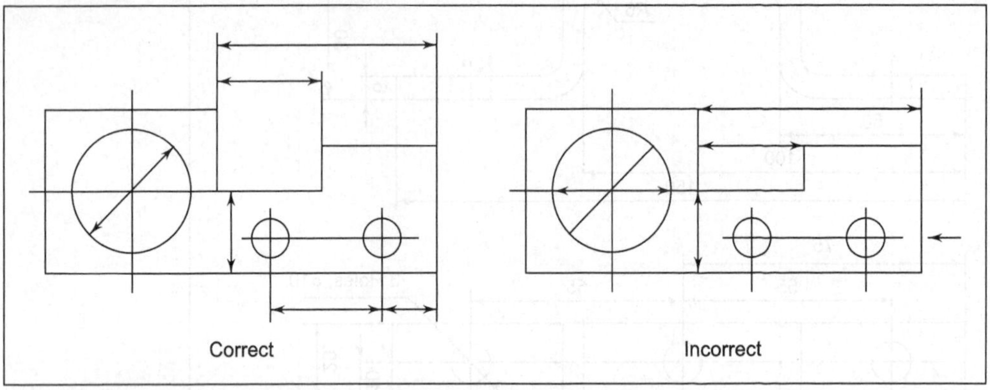

Fig. A4.12 Method of depicting dimension line and object outline

8. The intersection of dimension lines should be avoided as far as possible. The intersection of two dimension lines may be broken for inserting the dimension in case of unidirectional dimensioning (Figs. A4.4 and A4.5).
9. Overall dimensions should be placed outside the intermediate dimensions. Where an overall dimension is shown, one of the intermediate dimensions is redundant and should not be dimensioned (Fig. A4.13).

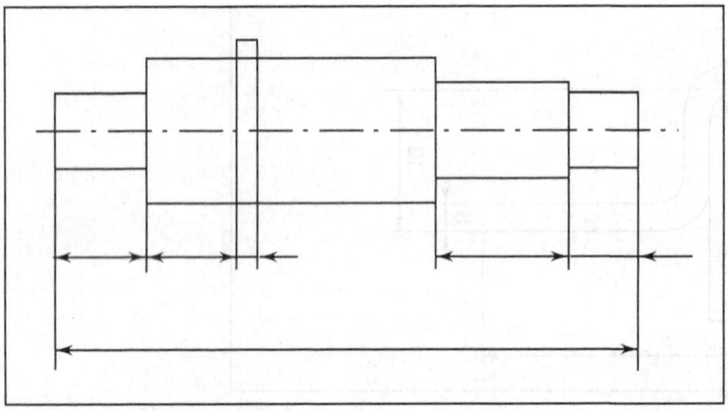

Fig. A4.13 Relations of overall and intermediate dimensions

10. More dimensions than are necessary should not be given to describe the finished product. Exceptions may however be made in the following cases:
    (i) Where it is necessary to give dimesions that apply at intermediate stages of manufacture, for example, for the size of a feature before carburising and finishing to size.
    (ii) Where desirable auxiliary dimensions may be added, enclosed in brackets (Figs. A4.1 and A4.7).

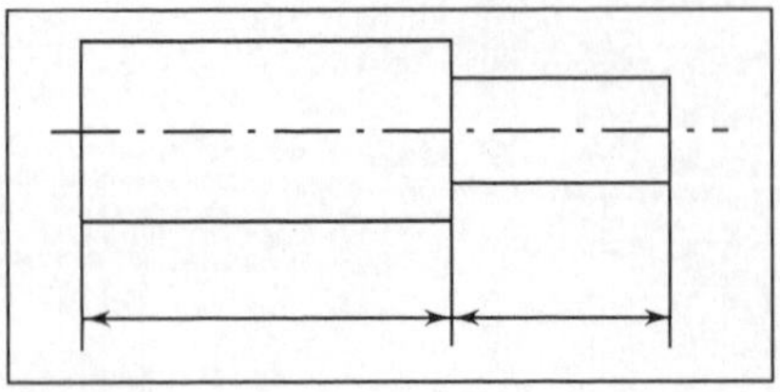

Fig. A4.14 Functional dimensions

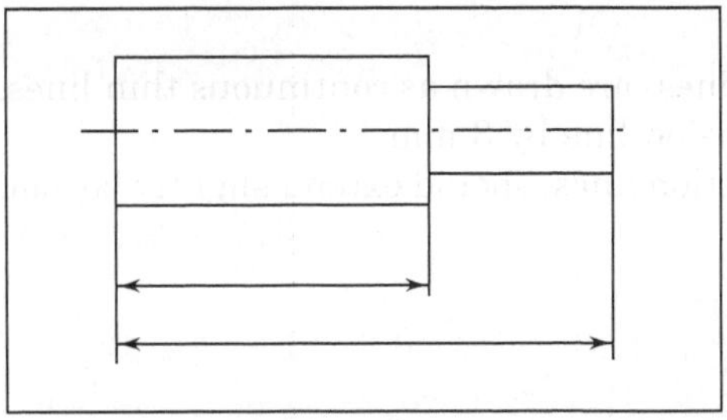

Fig. A4.15 Restructuring of functional dimensions

11. A functional dimension should be expressed directly on the drawing (Figs A4.1 and A4.14). To do otherwise (Fig. A4.15), necessitates a redistribution of tolerances, with the result that tighter tolerances are required to maintain functional requirements. This does not preclude the dimensioning of holes, centre to centre, although the functional dimension may be edge to edge.
12. The non-functional dimension should be placed in the most convenient position to read.
13. Tolerances should be specified for requirements affecting functions or interchangeability, unless workshop practice guarantees the required standard of accuracy. Tolerances should also be used where wide variations are usually permissible.

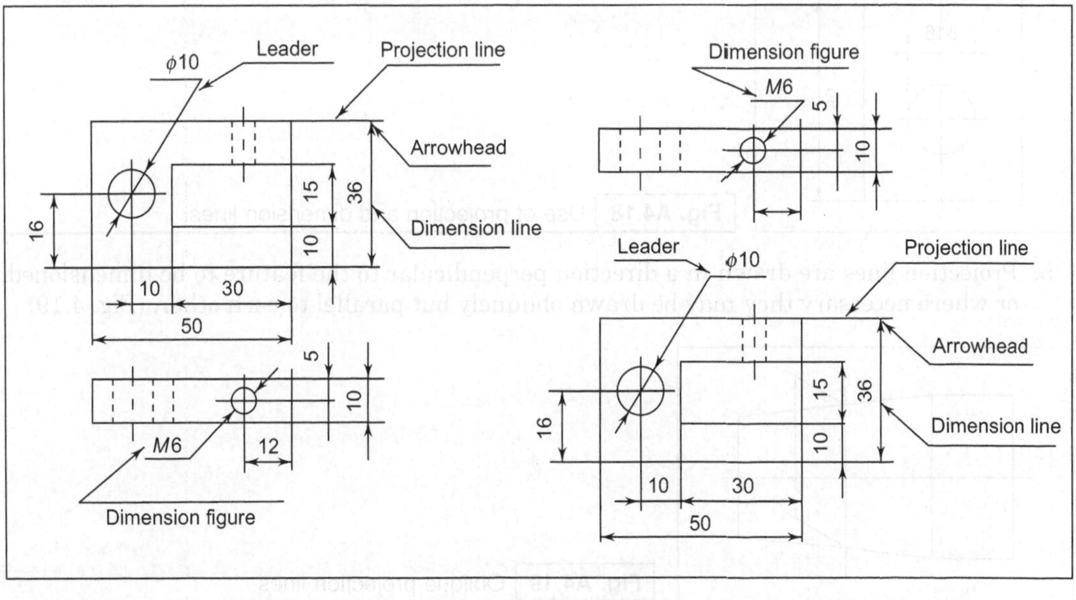

Fig. A4.16 Object showing uses of lines and dimensions

Where it is necessary to limit the overall variation of a series of dimensions to an amount less than the sum of the individual tolerances, a note should be used to call special attention to this requirement.

14. Production process or inspection methods should not be specified unless they are essential to ensure satisfactory functioning or interchangeability. This does not apply to process drawings, nor does it preclude the quoting of drill sizes. (Fig. A4.16)

4.4 METHOD OF EXECUTION

1. Dimension lines and projection lines (extension lines) are drawn as continuous thin lines.
2. Projection lines should extend beyond the dimension line by 3 mm.
3. The construction lines and the intersecting projection lines, should extend slightly beyond their point of intersection. (Fig. A4.17)

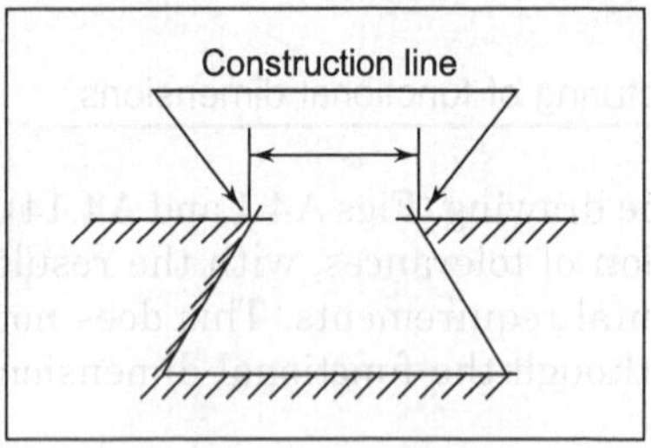

Fig. A4.17 Use of construction and projection lines

4. Projection and dimension lines should not cross other lines, unless this is unavoidable. (Fig. A4.18)

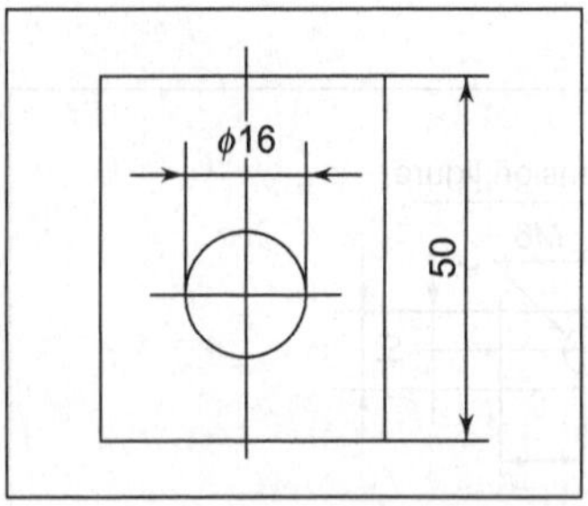

Fig. A4.18 Use of projection and dimension lines

5. Projection lines are drawn in a direction perpendicular to the feature to be dimensioned, or where necessary they may be drawn obliquely but parallel to each other. (Fig. 4.19)

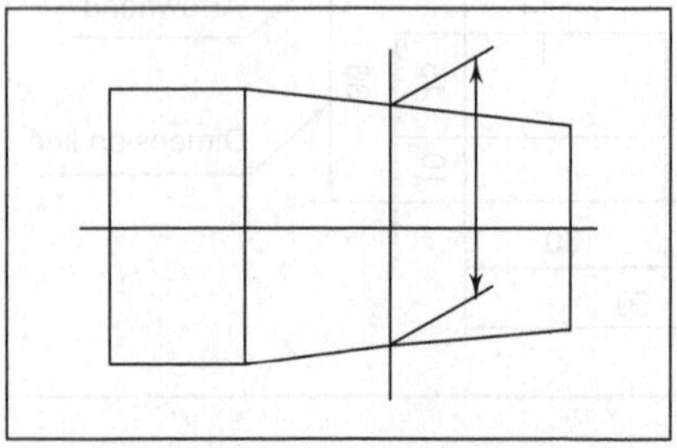

Fig. A4.19 Oblique projection lines

6. In aligned systems of dimensioning, except where unavoidable, dimension lines should not be placed in the 30° zone, as shown in Fig. A4.20.

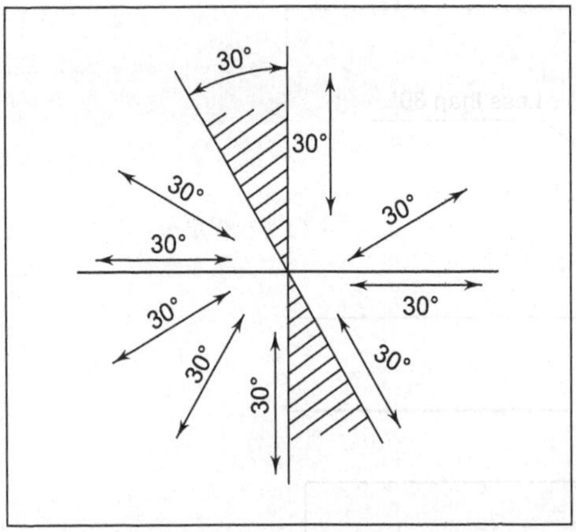

Fig. A4.20 Dimensioning rule

7. Leaders or pointer lines are continuous thin lines drawn from notes and figures to show where they apply. Leaders are terminated either by arrowheads or dots. While arrow heads connected to a leader should always terminate at a line, dots connected to a leader should be within the outline of the object. [Figs. A4.21(a) and 4.21(b)] Leaders should not be curved or drawn free hand. They may also terminate in a short horizontal bar at the bottom level of the lettering of the first or the last line of the note. (Fig. 4.16)

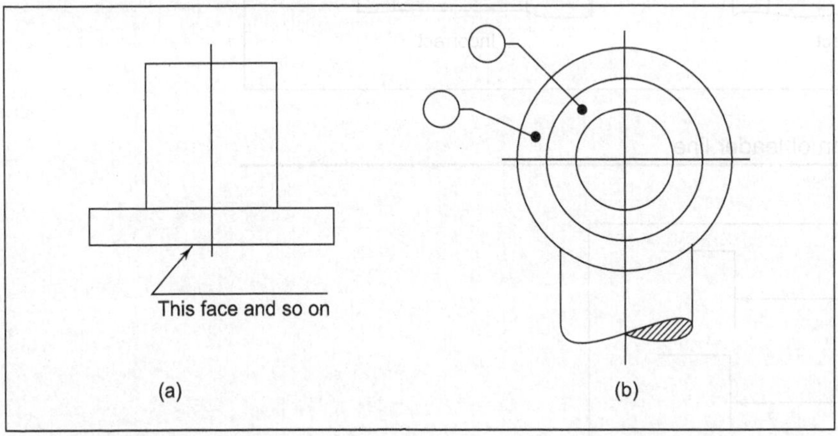

Fig. A4.21 Applications of leader lines

8. Leaders that touch lines should not normally be inclined at an angle less than 30° and they should not be parallel to adjacent dimension or projection line, where confusion might arise. (Fig. A4.22)

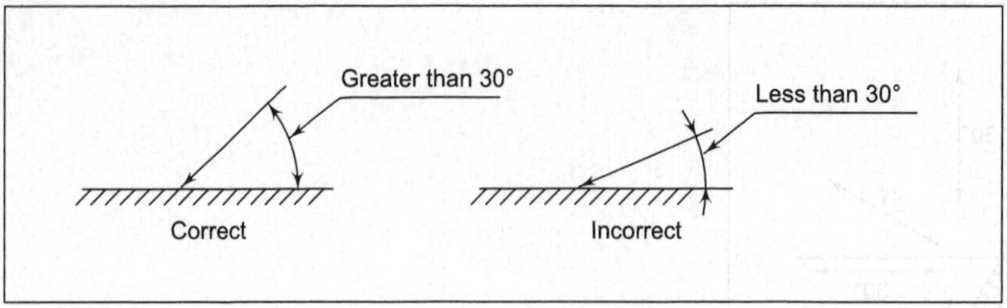

Fig. A4.22 Leader line with greater than 30° angle

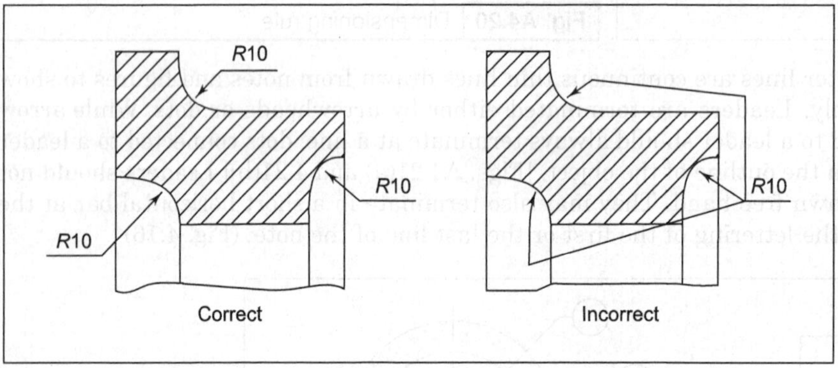

Fig. A4.23 Application of leader line

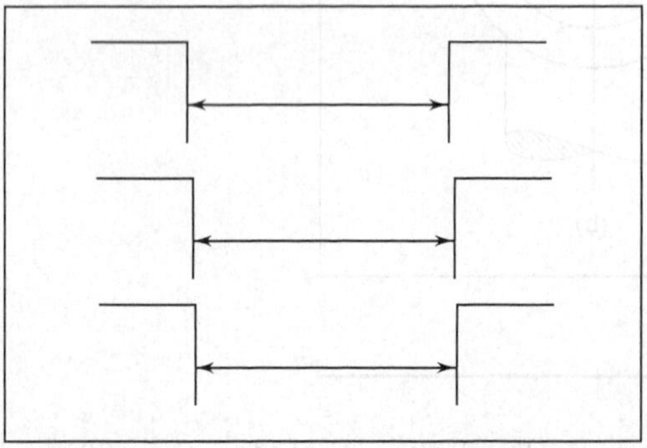

Fig. A4.24 Applications of leader line

9. The use of common leaders should be avoided even if it means repeating the dimensions or notes or using letters symbols.

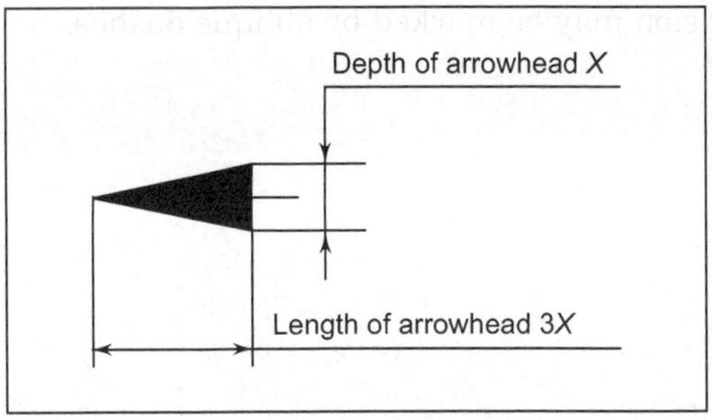

Fig. A4.25 Proportionate dimensions for arrowhead

10. Both ends of a dimension line should be defined by an arrowhead, whose two branches should be at an angle sufficiently open to mark clearly the extremities of the dimension line, or alternatively, the length of the arrowhead may be taken as about three times the depth. The size of arrowheads should be proportionate to the thickness of the lines of the drawing and size of drawings.
11. Arrowheads should be drawn within the limits of the dimensioned feature. When enough space is not available, they should be placed outside. As seen in Fig. A4.26, adjacent arrowheads may be replaced by a clearly marked dot. In Fig. A4.27 a dot replaces a single arrowhead to indicate the common origin of successive dimensions.

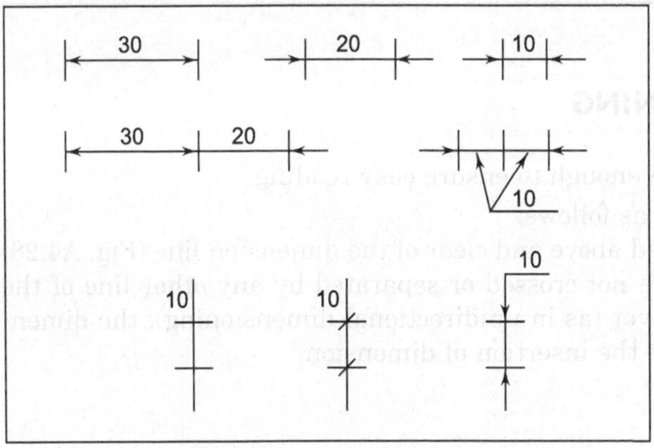

Fig. A4.26 Use of oblique dashes

12. Instead of arrowheads, the ends of the dimension may be marked by oblique dashes.

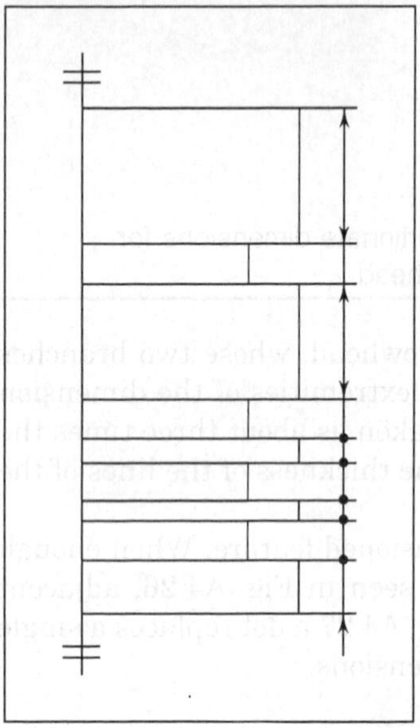

Fig. A4.27 Use of dots in place of arrowheads for small dimensions

4.5 INSCRIPTION OF DIMENSIONING

1. Numerals and letters should be large enough to ensure easy reading.
2. Numerals or letters should be placed as follows:
   (i) Preferably near the middle of and above and clear of the dimension line (Fig. A4.28) and in such a way that they are not crossed or separated by any other line of the drawing. In certain cases, however (as in unidirectional dimensioning), the dimension line may be interrupted for the insertion of dimension.

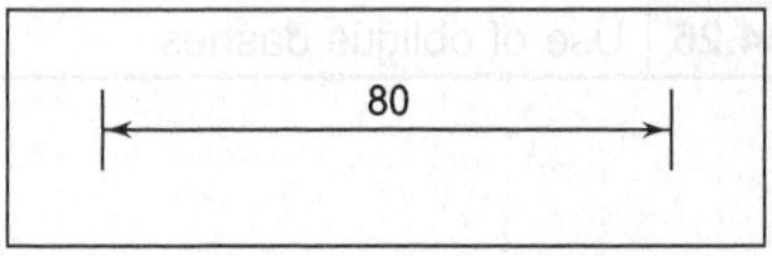

Fig. A4.28 Writing numbers in dimensions

   (ii) In certain cases, nearer to one of the arrowheads, in order to avoid superimposing of numerals or having to follow long dimension lines. Such dimension lines may be shortened (Fig. A4.29).

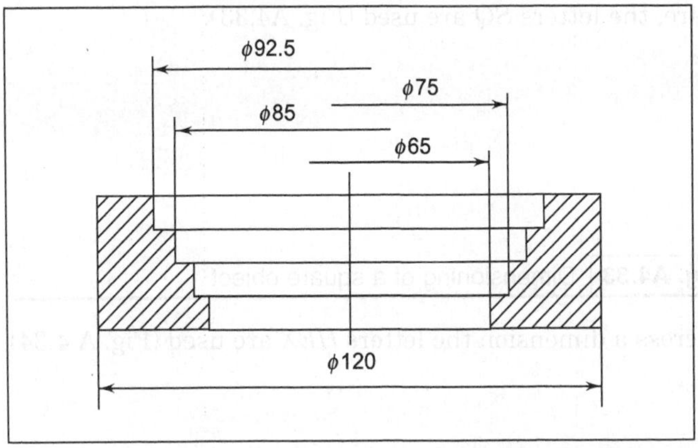

Fig. A4.29 Shortening of dimension line

(iii) Above the horizontal terminating end of a leader (Fig. A4.17).

3. Dimensions of parts that are not drawn to scale should be underlined (Fig. A4.30). However, when it is required to draw the whole drawing not to scale, the abbreviation NTS may be used or the column for scale be scored off.

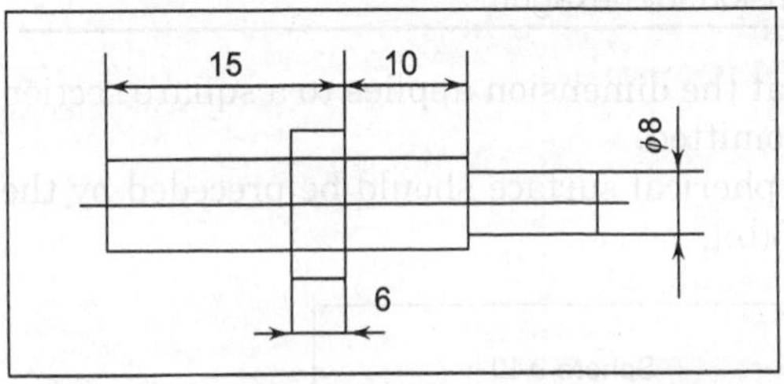

Fig. A4.30 Unscaled figures

4. The addition of letters and symbols is made as follows:
   (i) Before a dimension for a diameter, the symbol ϕ (Fig. A4.31).

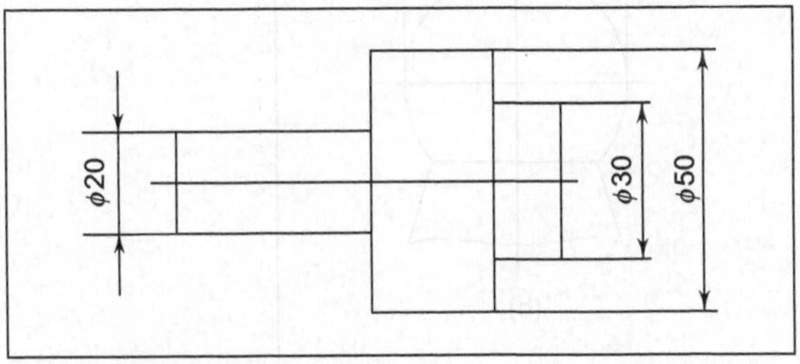

Fig. A4.31 Dimension for diameter

(ii) Before a dimension for a radius, the letter *R* (Fig. A4.32).

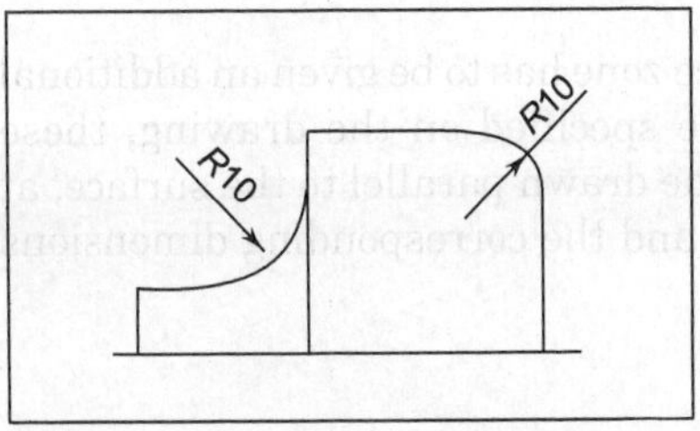

Fig. A4.32 Dimension for radius

(iii) For the dimension of a square, the letters *SQ* are used (Fig. A4.33).

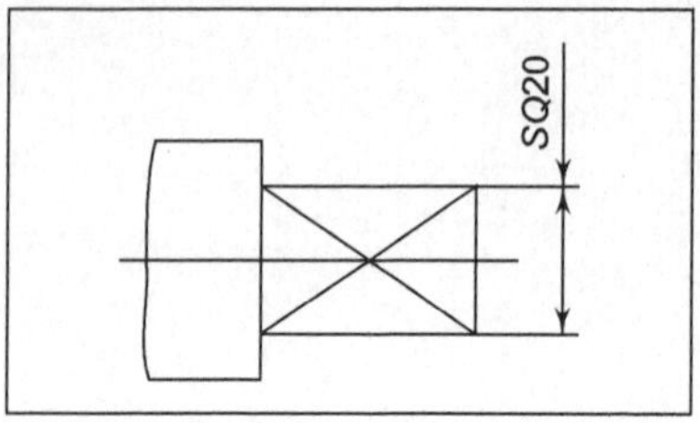

Fig. A4.33 Dimensioning of a square object

(iv) For a hexagon, before and across a dimension the letters *HEX* are used (Fig. A 4.34).

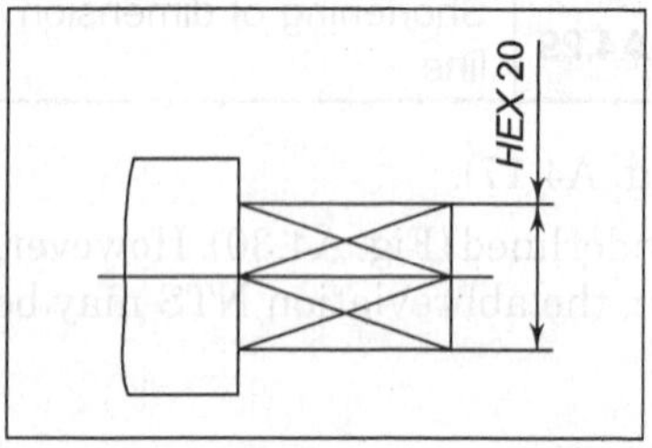

Fig. A4.34 Dimension for hexagon

If it clearly appears from the drawing that the dimension applies to a square section or hexagonal section, the letter may be omitted.
The radius or diameter dimension of a spherical surface should be preceded by the word SPHERE [Figs. A4.35(a) and A4.35(b)].

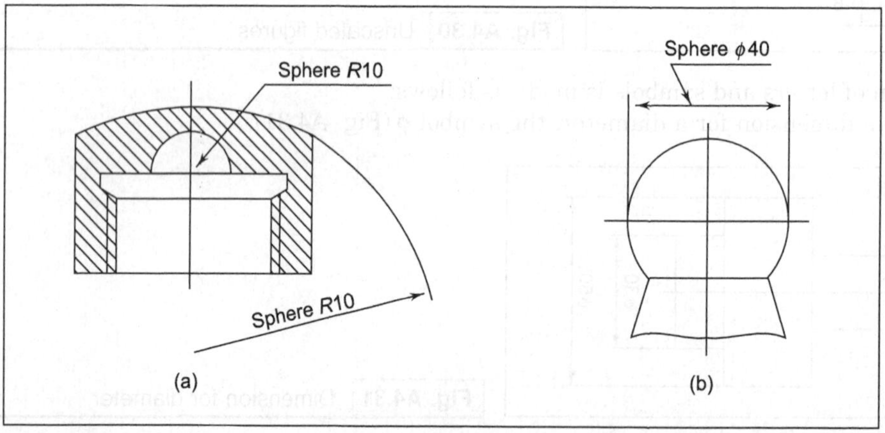

Fig. A4.35 Dimensioning of a sphere

5. When it is desirable to indicate that a surface or a surface zone has to be given an additional treatment, which shall be applied within limits to be specified on the drawing, these limits may be defined by means of a long chain thick line drawn parallel to the surface, at a short distance from it, with locating dimension lines and the corresponding dimensions added (Fig. A4.36).

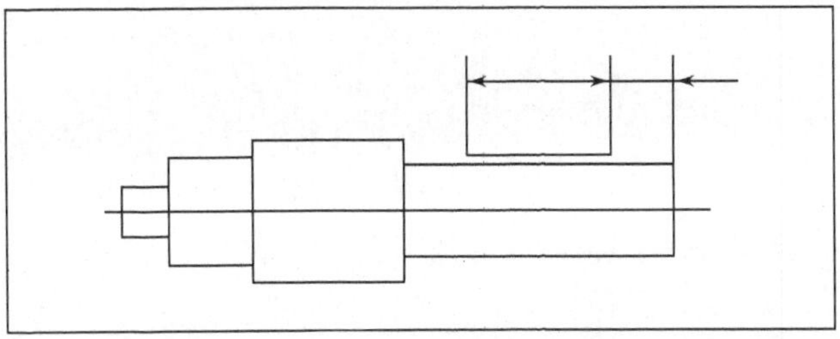

Fig. A4.36 Indicating surface with additional treatment

6. If the location and the extent of the surface to be treated appears clearly from the drawing, it is not necessary to dimension them (Fig. A4.37).

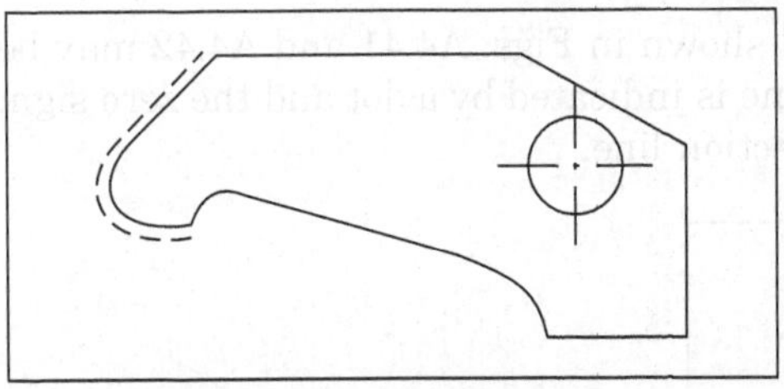

Fig. A4.37 Surface of an object after treatment

4.6 ARRANGEMENT OF DIMENSIONS

1. **Chain dimensioning** Chain dimensioning should only be used where the possible accumulation of tolerances does not affect the functional requirements of the part.

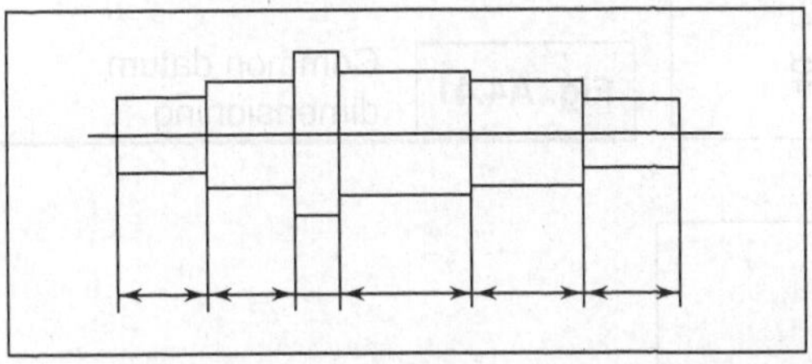

Fig. A4.38 Chain dimensioning

2. **Parallel dimensioning** Where a number of dimensions on the same direction have a common datum feature, the method shown in Figs. A4.39 and A4.40 should normally be used.

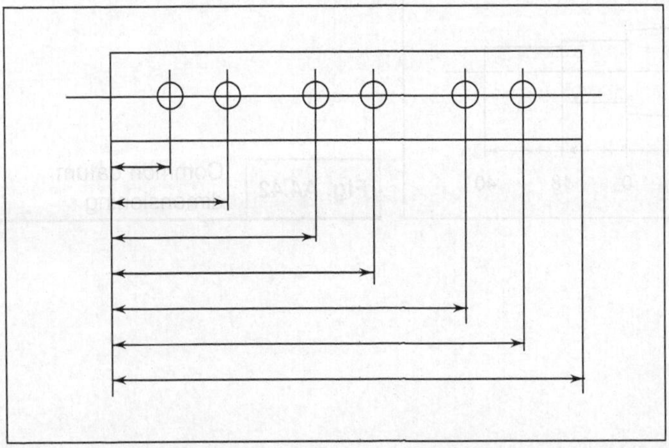

Fig. A4.39 Common datum dimensioning

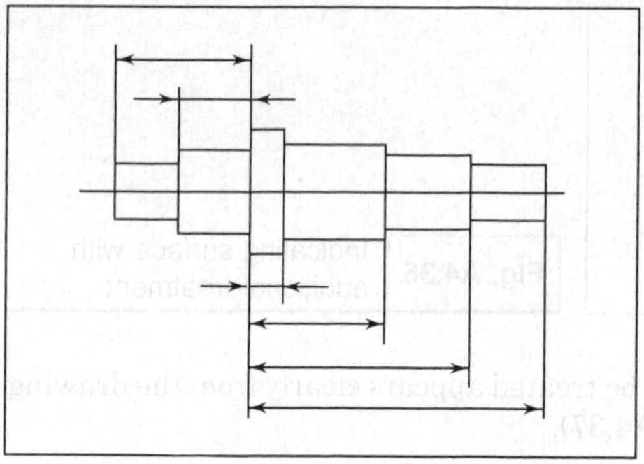

Fig. A4.40 Common datum dimensioning

Where there is no risk of confusion, the method shown in Figs. A4.41 and A4.42 may be used for simplicity. In this method, the datum line is indicated by a dot and the zero sign. The dimensions are placed in line with the projection line.

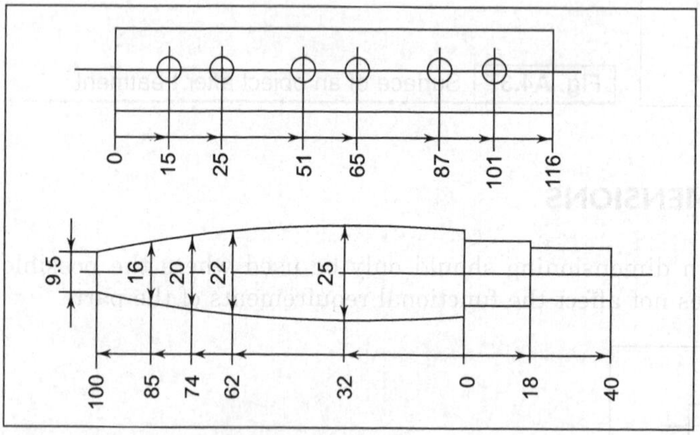

Fig. A4.41 Common datum dimensioning

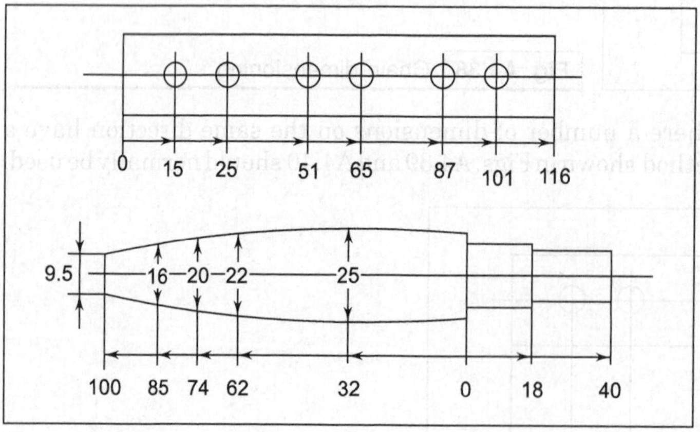

Fig. A4.42 Common datum dimensioning

3. **Combined dimensioning** This results from the simultaneous use of methods indicated below (Fig. A4.43).

1. Chain dimensioning
2. Parallel dimensioning
3. Aligned dimensioning
4. Unidirectional dimensioning

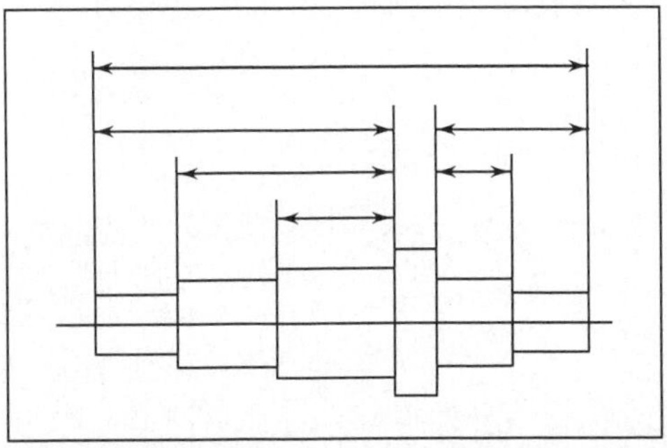

Fig. A4.43 Combined dimensioning

4. **Dimensioning by coordinates** This method may be useful in some cases. Instead of dimensioning as shown in Fig. A4.44 to group dimensions separately from the drawing, refer to the table in Fig. A4.45.

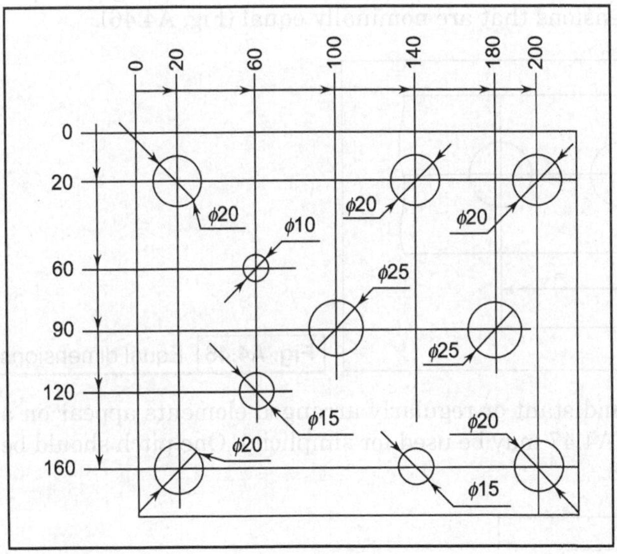

Fig. A4.44 Coordinate dimensioning

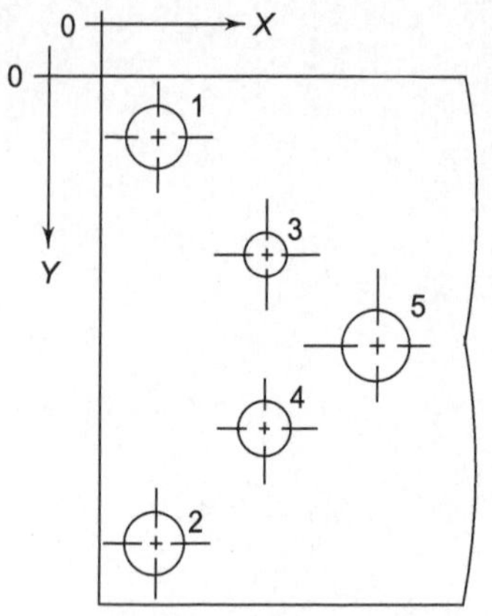

	1	2	3	4	5
X	20	20	60	60	100
Y	20	160	60	120	90
ϕ	20	20	10	15	25

Fig. A4.45 Coordinate dimensioning

5. **Equal dimensions** When a dimension is divided into several parts, the sign is equal to'(=) may be used, to indicate dimensions that are nominally equal (Fig. A4.46).

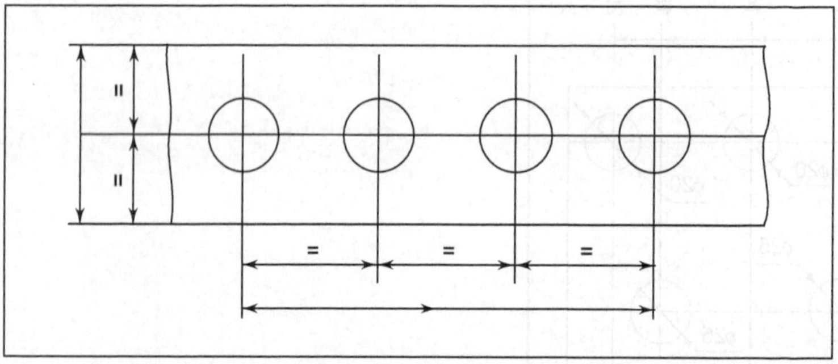

Fig. A4.46 Equal dimensions

6. **Equidistant features** Where equidistant or regularly arranged elements appear on a drawing, the method shown in Fig. A4.47 may be used for simplicity. One pitch should be dimensioned as shown in Fig. A4.48.

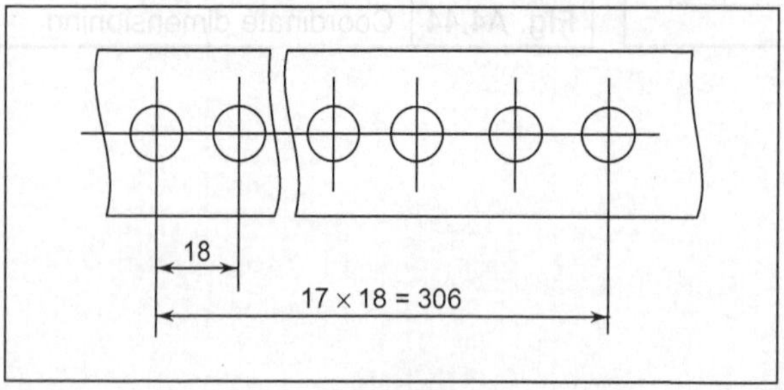

Fig. A4.47 Equidistant dimensioning

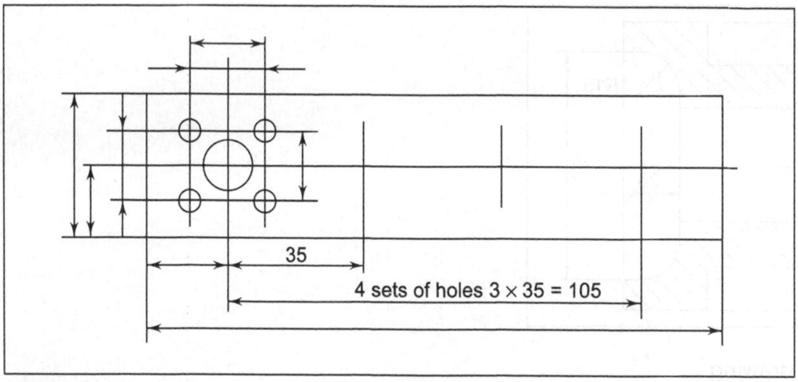

Fig. A4.48 Equidistant dimensioning

7. **Assembled parts** When several parts are drawn in an assembly, the group of dimensions related to each part should be kept separate as much as possible if individual part drawings are not prepared (Fig. A4.49).

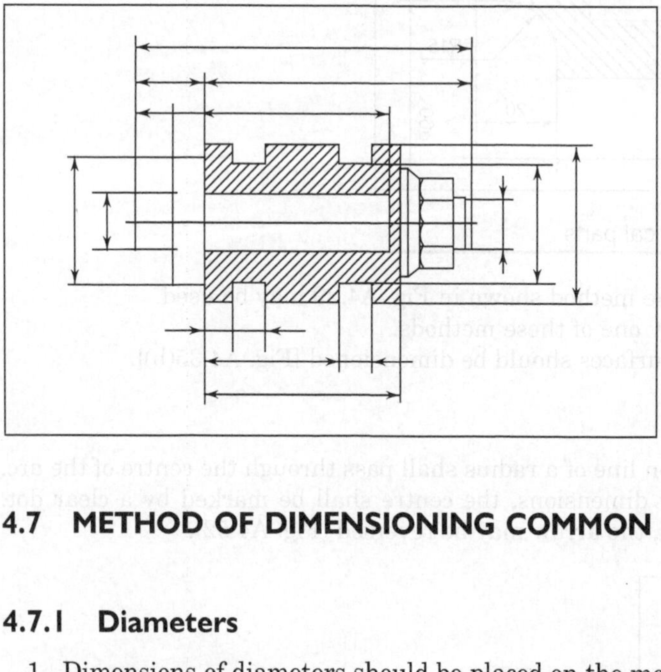

Fig. A4.49 Assembly dimensions

4.7 METHOD OF DIMENSIONING COMMON FEATURES

4.7.1 Diameters

1. Dimensions of diameters should be placed on the most appropriate view to ensure clarity and should be preceded by the symbol (Fig. A4.50).

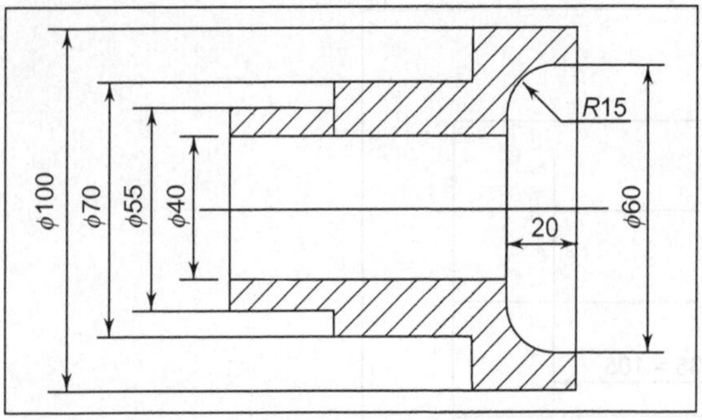

Fig. A4.50 Diameters in drawing

2. Wherever half views of symmetrical parts, such as a cylindrical part, are drawn, the portions of the dimension lines shown should be extended slightly beyond the centre line and the second arrow should be omitted (Fig. A4.51).

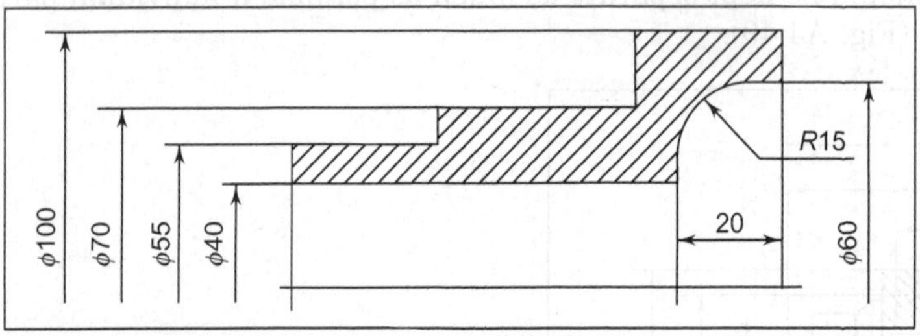

Fig. A4.51 Dimensioning of symmetrical parts

3. Wherever space is restricted, the method shown in Fig. A4.29 may be used.
4. Circle should be dimensioned by one of these methods.
5. The diameter of the spherical surfaces should be dimensioned [Fig. A4.35(b)].

4.7.2 Radii

1. As far as possible, the dimension line of a radius shall pass through the centre of the arc. Where the centre is located by dimensions, the centre shall be marked by a clear dot. While dimensioning small radii, the arrow may be reversed (Fig. A4.52).

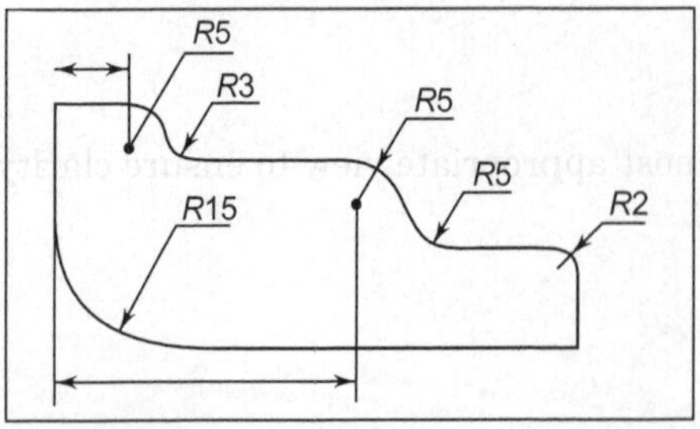

Fig. A4.52 Dimension line of a radius

2. The radii of arcs that need not have their centres located should be dimensioned as shown in Fig. A4.53 and Fig. A4.54.

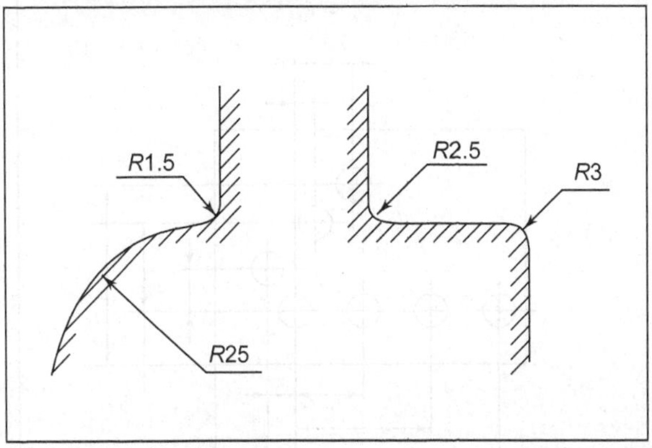

Fig. A4.53 Dimensioning of arcs

3. When the centre of an arc falls outside the limits of the space available, the dimension line of the raidus should be broken or interrupted according to whether or not it is necessary to locate the centre (Fig. A4.54).

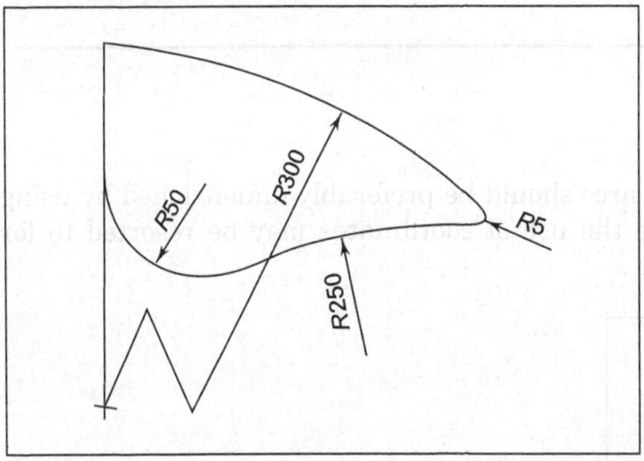

Fig. A4.54 Dimensioning of radius having centre outside the limits

4. The radius of a spherical surface should be dimensioned [Fig. A4.35(a)].

4.7.3 Position of Holes

The position of holes and other features should be defined either by spacing them on pitch circles or giving rectangular coordinates, as shown in Fig. A4.55(a), or by giving centre distances, as shown in Fig. A4.55(b). Where holes symmetrical to edges are to be dimensioned, or when centre distances are given where holes symmetrical to edges are to be dimensioned, the holes are dimensioned, using coordinates, even though they lie on a pitch circle [Fig. A4.55(a)].

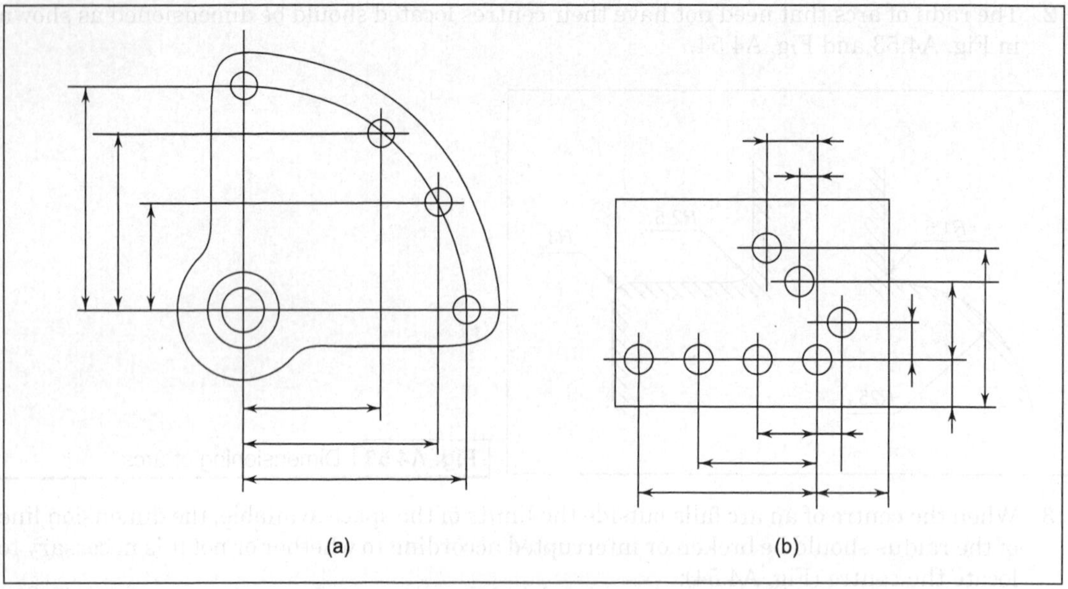

Fig. A4.55 Position and dimensioning of holes

4.7.4 Dimensioning Curved Surfaces

A curved line or surface composed of circular arcs should be preferably dimensioned by using radii. However, if this method is impracticle the use of coordinates may be resorted to for dimensioning curved surfaces (Fig. A4.56).

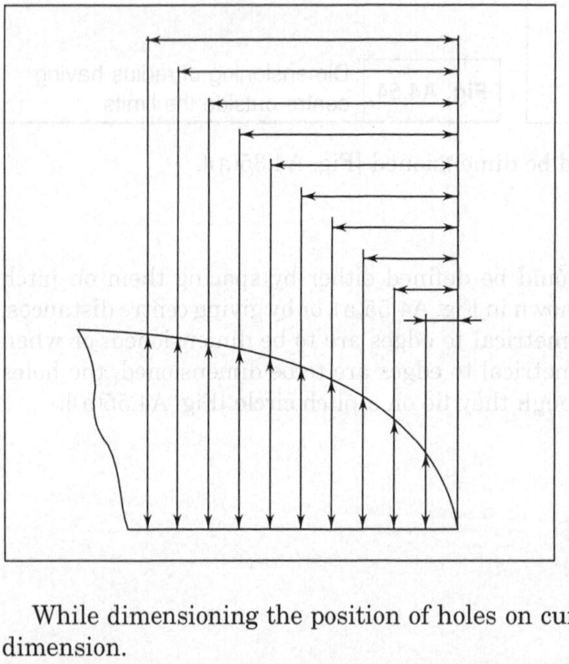

Fig. A4.56 Dimensioning of curved surfaces

While dimensioning the position of holes on curved surfaces, it is preferable to use angular dimension.

4.7.5 Key Way Dimensioning

1. Key ways for parallel and tapered keys in cylindrical shafts and hubs shall be dimensioned according to relevant Indian standards, as shown in Fig. A4.57. In the plan view for slots and key ways, it is sufficient to indicate the length and the width (Fig. A4.58).

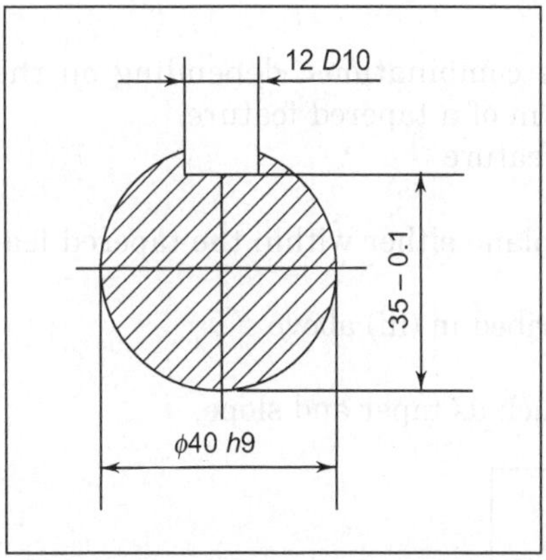

Fig. A4.57 Key way dimensioning

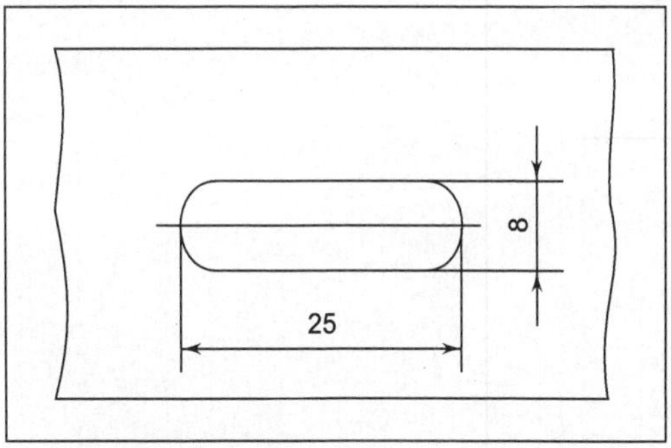

Fig. A4.58 Key way dimensioning

2. Where the bottom of a key way is parallel to the tapered surface, the depth shall be dimensioned as shown in Fig. A4.59. For tapered key ways, the direction of slope, being the direction in which the key is driven, should be indicated by an arrow.

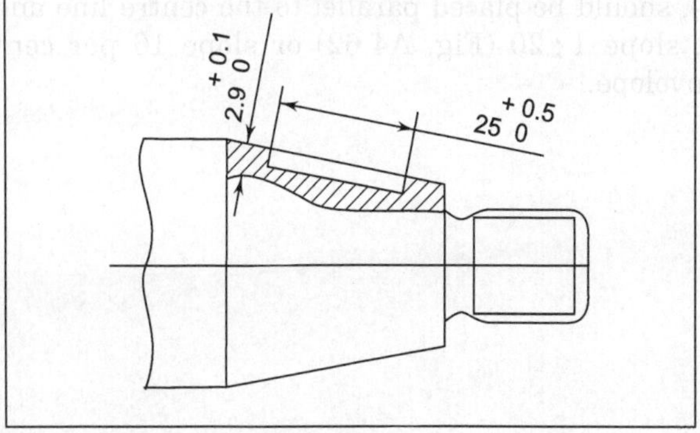

Fig. A4.59 Key way dimensioning

3. The bottom of the key way is dimensioned when it runs parallel to the surface of the taper bore. If the bottom of the key way is parallel to the taper axis, as in tapered shaft ends, the bottom of the key way shall preferably be dimensioned from the surface of a cylinder. In this case, the tolerance of the cylinder diameter (when indicated) should also be considered. In case of taper holes with cylindrical ends, the bottom of the key way shall be dimensioned.

4.7.6 Dimensioning of Tapered Features

1. The following dimensions may be used in suitable combinations, depending on the technological requirements, to define the size and form of a tapered feature.
   (i) Diameters or width at each end of the tapered feature
   (ii) Length of the tapered feature
   (iii) Diameter or width at a selected cross-sectional plane either within the tapered feature or outside
   (iv) Location or position of the cross-section, as described in (iii) above
   (v) Included angle
2. Fig. A4.60 shows the terms used for defining taper, such as taper and slope.

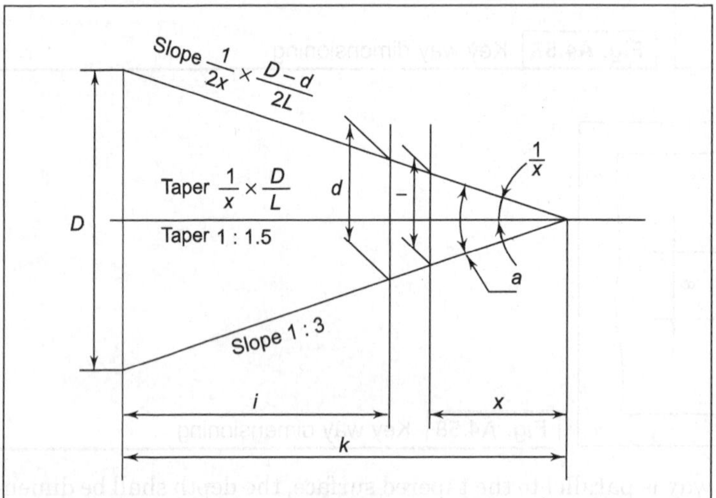

Fig. A4.60 Taper dimensioning

3. Notes, such as taper 1 : 10 (Fig. A4.61), should be placed parallel to the centre line and while indicating slope, for example, slope 1 : 20 (Fig. A4.62) or slope 16 per cent (Fig. A4.63), should be placed to the envolope.

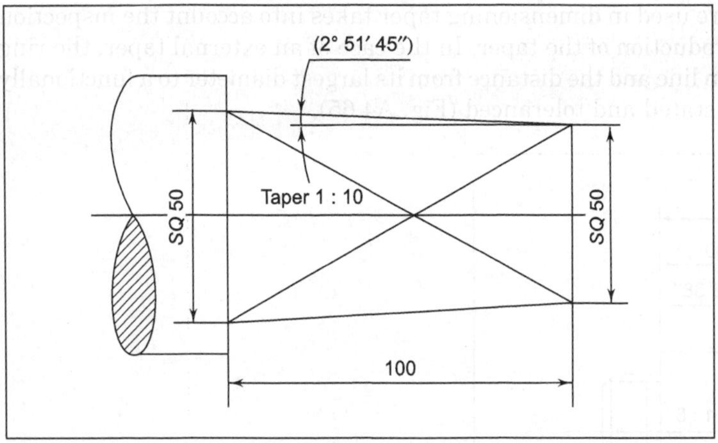

Fig. A4.61 Taper dimensioning

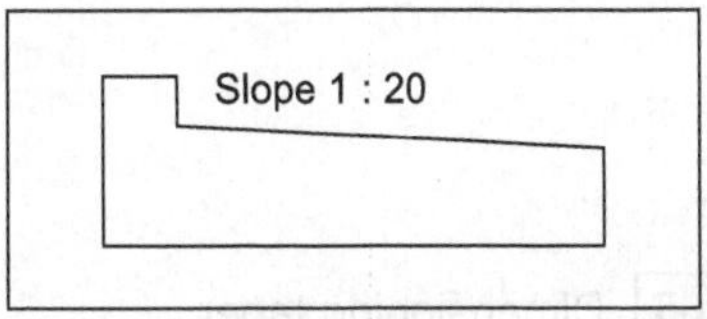

Fig. A4.62 Slope dimensioning

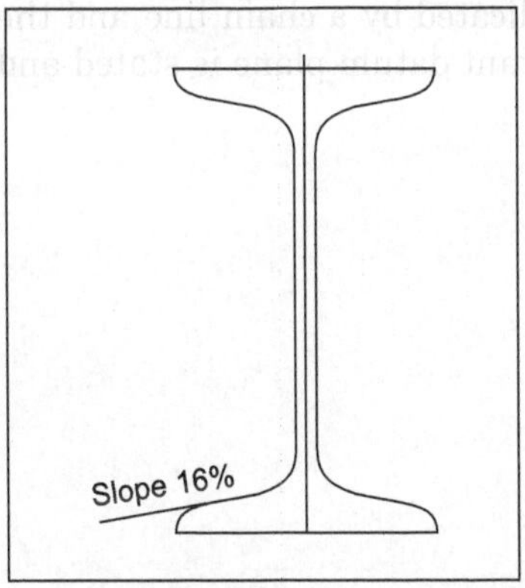

Fig. A4.63 Slope dimensioning

4. In case of tapers, the half angle or the taper angle is indicated in addition in the case of brackets, to facilitate machine settings.
5. If necessary, the direction of taper may be added as shown in Fig. A4.64.

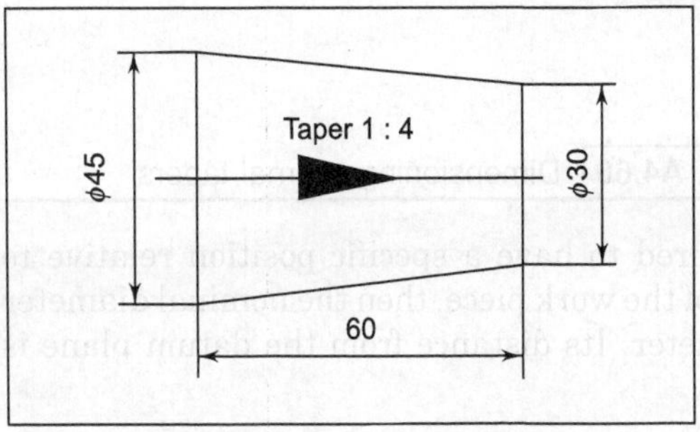

Fig. A4.64 Direction of taper

6. In many cases the procedure used in dimensioning taper takes into account the inspection method employed in the production of the taper. In the case of an external taper, the ring gauge is indicated by a chain line and the distance from its largest diameter to a functionally important datum plane is stated and toleranced (Fig. A4.65).

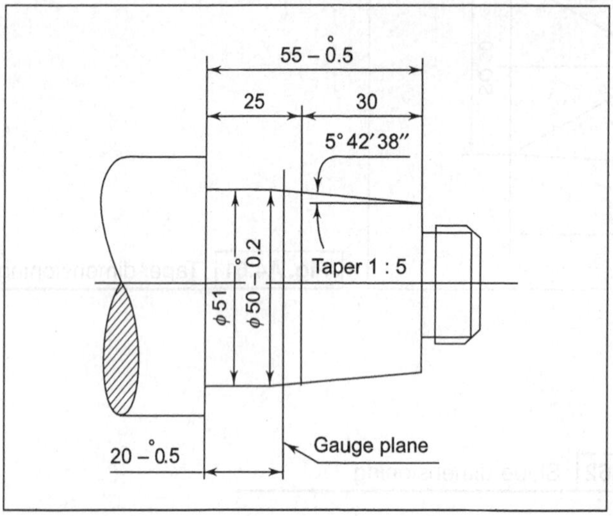

Fig. A4.65 Dimensioning taper

7. In the case of internal tapers, the taper plug gauge is indicated by a chain line and the distance from its largest diameter to a functionally important datum plane is stated and toleranced (Fig. A4.66).

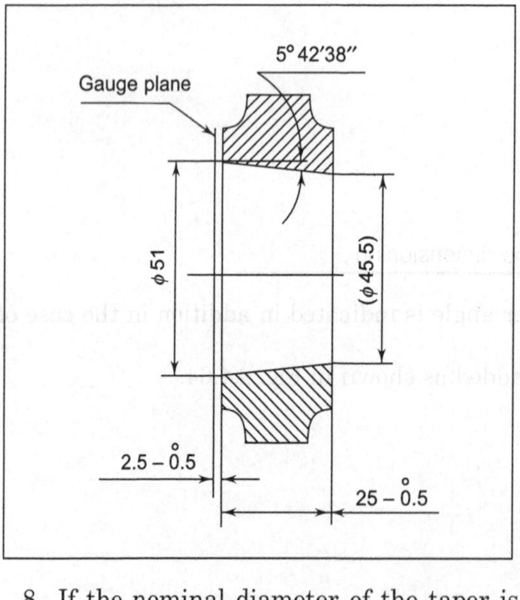

Fig. A4.66 Dimensioning internal tapers

8. If the nominal diameter of the taper is required to have a specific position relative to datum plane, without coinciding with an edge of the work piece, then the nominal diameter of the taper should be given as a gauge diameter. Its distance from the datum plane is toleranced (Fig. A4.67).

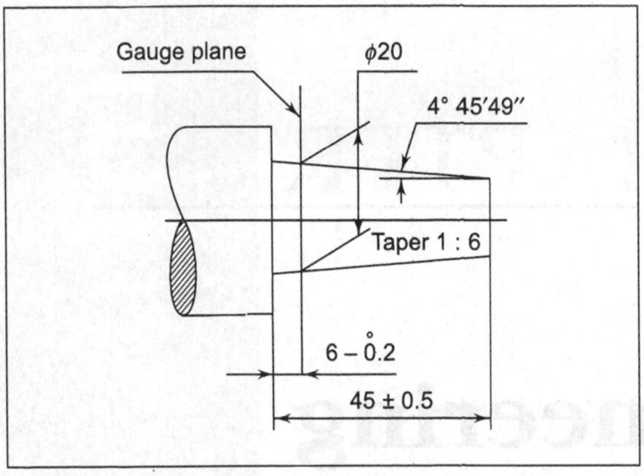

Fig. A4.67 Dimensioning taper

9. For a part having internal and external tapers, the dimensioning is done as shown in Fig. A4.68.

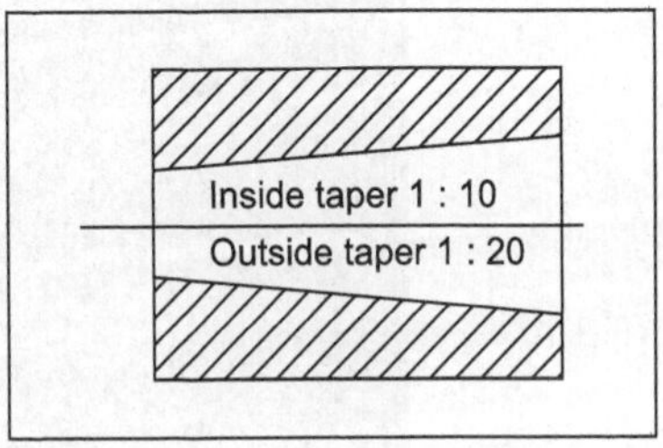

Fig. A4.68 Dimensioning internal and external taper

Review Questions

1. Why dimensioning of an object is important?
2. List the general principles of dimensioning.
3. What is functional dimension?
4. What is non-functional dimension?
5. Give five examples each for correct and incorrect dimensioning.
6. State the types of dimension methods.
7. Explain chain and parallel dimensioning.
8. How do we add letters and symbols in dimensioning?
9. List methods of dimensioning for common features.
10. What are the rules to be adopted for dimensioning?

CHAPTER 5

Engineering Materials

This chapter contains information about different types of engineering materials, their mechanical, physical, and chemical properties along with their engineering applications

After reading the chapter you will be able to

- select appropriate material for manufacturing of assembly components
- prepare a bill of material for assembly questions
- select appropriate material according to the function of the individual component in assembly, considering its mechanical, chemical, and physical properties

5.1 GENERAL DISCUSSION

After the general layout of the machine or its assembly has been made and the mechanisms to be used have been decided upon, it is necessary to select the proper material for each machine member. This involves the consideration of many factors, such as the engineering properties of the available materials, their weight, size, shape; the material cost, fabricating cost, overhead charges; and properties considered peculiar to the member, namely, strength, stiffness, toughness, ductility, fatigue resistance, resilience, shock resistance, corrosion resistance, wear resistance, hardness, frictional qualities, machinability, casting properties, forging properties, temperature properties, and electrical properties.

The selection of a proper material is one of the difficult decisions the designer has to take and it requires experience as the variety of materials available are on an increase. Previously, designers were handicapped by the lack of ferrous and non-ferrous materials suitable for their particular needs, but now the problem is of selecting the most suitable material from the hundreds of materials available. The proper material to be used is one that meets the engineering requirements of the designers, the production and heat-treating requirements of the production department, and the cost requirement of the purchasing agent.

It is important that the materials be used in such a way so as to take full advantage of their natural characteristics. Thus, it is usually desirable to think of change in design when selecting materials. For example, a moulded plastic part may take advantage of the natural colour of the material and of the insulating properties in a design that cannot be employed if sheet metal were used. In the application of moulded plastics or metal castings, it is often desirable to incorporate as many functions as possible in one part with metal stampings; on the other hand, it is sometimes necessary to assemble a part from a number of pieces. Material is considered appropriately used if it is used in a design that takes full advantage of its useful characteristics while avoiding its less favourable properties.

5.2 ENGINEERING MATERIALS

Materials of construction can be classified into two groups.

1. Metallic materials
2. Non-metallic materials

Metallic materials can be further split into ferrous and non-ferrous materials.

Ferrous materials consist mainly of iron and its alloys, such as cast iron, grey cast iron, malleable cast iron, wrought iron, and steels of low and high carbon content.

Non-ferrous materials contain little or no iron. These materials are copper, zinc, tin, lead, and aluminium. The alloys produced by combining these metals are brass, bronze, gun-metal, duralumin, and so on.

Non-metallic materials include plastics, rubber, leather, carbon, wood, glass, and so on.

5.2.1 Properties of Construction Materials

The economical use of construction materials in mechanical design requires considerable knowledge of their mechanical properties. The properties of materials describe this behaviour in mechanical usage.

1. **Homogeneity** A material that exhibits the same properties throughout is said to be homogeneous. Homogeneity is an ideal state, which is not achieved by real materials, particularly metals. However, the variation in properties is so small that calculations for stress and deflection assume that a material is homogeneous throughout.

2. **Isotropy** An isotropic material displays the same elastic properties in all loading directions. The equations of elasticity and strength of materials are based upon this assumption.

3. **Plasticity** When a material is subjected to an external load of such magnitude that deformation continues with no apparent further increase in load, the material is said to have become plastic. Once in this region, the material experiences a permanent set of deformation and will not return to its original size and shape even when the load is removed. Thus, plasticity can be considered as the opposite of elasticity. This definition applies to a theoretically perfect material. Actually material, such as a metal, will continue to deform with only a very small further increase of load.

4. **Proportional limit** Proportional limit defined as that point on a stress-strain curve beyond which stress is no longer proportional to strain. The actual value of the proportional limit is quite difficult to establish because it greatly depends on the quality of measuring instruments.

5. **Proof stress** This stress which will cause a small but permanent strain in the material. This permanent strain is specified as 0.01 per cent in 50 min or called the 0.01 per cent offset.

6. **Modulus of elasticity** It is the ratio of stress to strain within the elastic limit, hence it is really a measure of the stiffness (for tension and compression) or rigidity for the shear of a material. For ductile materials the modulus of elasticity is same both for tension and compression and it is represented by E. In particular, it is called Young's modulus. For brittle materials such as cast iron, this modulus is not the same for tension and compression. In the case of shear or torsion, this modulus is called the shear modulus or modulus of rigidity or transverse modulus and is represented by C.

7. **Hooke's law** This law states that stress is proportional to strain upto the proportional limit of the material.

8. **Elastic limit (proportional unit)** This is the greatest stress that a metal can withstand without permanent elongation, that is, when the load is released, the sample returns to its original length.

9. **Yield point (yield strength)** The stress at which an appreciable elongation occurs without an increase in stress.

10. **Ultimate strength** The ultimate strength is the maximum stress required to break a specimen.

11. **Modulus of elasticity** The modulus of elasticity is the ratio of stress to strain within the elastic limit. It is a measure of stiffness.

12. **Elongation** This is the ratio of the increase in gauge length to the original gauge length, expressed in percentage.

13. **Reduction in area** This denotes is the decrease in the cross-sectional area of the test specimen after fracture to the original area expressed in percentage.

14. **Ultimate strain** This is the unit elongation (elongation per unit of length) at the specimen breaking point. It is a measure of ductility.

15. **Ductility and brittleness** Ductility of a metal is its ability to deform it plastically without fracturing. In general, it means deformation under low stressing instead of sudden impact. Although there are other measures, ductility is most commonly measured by means of elongation and reduction of area in the tensile test.

$$0\ \%\ \text{elongation} = \frac{\text{Final gauge length} - \text{Original gauge length}}{\text{Original gauge length}} \times 100$$

$$0\ \%\ \text{reduction of area} = \frac{\text{Final area} - \text{Final area}}{\text{Original area}} \times 100$$

A material is generally classified as brittle if the percentage elongation is less than 5 in a gauge length of 50 mm.

16. **Stiffness** The ability of a metal to resist bending, stretching, shortening, and twisting is defined by the modulus of elasticity. The modulus applies only within the elastic range, the range within which a metal will return to its original shape if the load is removed.

17. **Indentation hardness** In metal working, hardness generally implies resistance to penetration. It may, however, include resistance to scratching, abrasion, or cutting. Indentation hardness is probably the most widely used mechanical testing procedure. It is a non-destructive test, relatively inexpensive, and can be performed by semi-skilled operators. The Brinell hardness test measures the diameter of the impression of a hard ball under compressive load on the flat surface of the metal sample (unit N/mm^2). The Rockwell hardness test measures the depth of penetration of the diamond point (Rockwell C) of the steel ball (Rockwell B) into the sample.

18. **Fatigue** Fatigue is a description of the behaviour of metals under the action of alternation (cyclic) loads as distinguished from their behaviour under steady loads. Cyclic stressing can occur in bending, tension, compression, torsion, or a combinations of these, and may be found in such diverse applications as axles, connecting rods, springs, aircraft landing gears, and ships hulls.

19. **Impact strength** Impact strength is the ability of a material to withstand shock loading.

20. **Transverse shear strength** The value obtained by dividing the breaking load by the transverse shear area gives the transverse shear strength. It is needed in the design of rivets, bolts, pins and similar parts.

21. **Toughness** Toughness is the amount of energy required upto fracture.

22. **Malleability** It is the property of a material that permits it to be drawn in the form of sheets in cold state. Gold, copper, and aluminium have good malleability.

23. **Damping capacity** The energy dissipated as heat by a unit volume of the material during a completely reversed cycle of stress is its damping capacity. High damping capacity is desirable in order to decrease vibration chatter in machine tools. Cast iron has a high damping capacity.

24. **Creep** At high temperatures such as those encountered in steam boilers, turbines, and piping, the deformation of materials ceases to be elastic and becomes plastic with a continuous increase under a constant load. The equilibrium between stress and load is not established, even after a very long time. The material under tensile stress continues to stretch or creep. Creep is measured in terms of plastic deformation during a certain time. The limiting creep stress for a certain temperature is the maximum stress under which the material will not fail during a prescribed length of time.

5.3 FERROUS MATERIALS

Ferrous materials are composed of iron and carbon together with a number of other elements that are present as impurity elements carried over from the raw materials, or introduced unavoidably during the manufacturing processes, or intentionally added to make an alloy.

5.3.1 Cast Iron

Cast iron is the cheapest of the cast materials and in the plain or alloyed state can be cast almost to any size and form. Cast iron without the addition of alloying elements is weak in tension and shear, strong in compression, and has low resistance to impact. The stress-strain diagram is curved all the way from zero to the rupture stress. There is no well defined proportional limit, the ratio of stress to strain is not constant and the modulus of elasticity decreases as the stress increases. The damping capacity of cast iron is much greater than that of steel.

In grey cast iron carbon is present in the free or graphitic state. It is soft, can be easily machined, and is only moderately brittle. It is not as abrasive wear resistant as chilled iron. It will resist galling wear as in the case of brake shoes, cylinders, and pistons. In general, it may be assumed that the lowest tensile strength is associated with the greatest amount of graphite and higher tensile strength with the smallest amount of graphite. The modulus of elasticity of grey cast iron is somewhat smaller than that of steel. Grey cast iron has high damping capacity.

White cast iron is formed when the casting is rapidly cooled and it has most of the carbon in the combined state, therefore, it is hard and brittle and cannot be machined with ordinary cutting tools, but requires grinding as a shaping process. This material is seldom used for complete casting but is formed on wearing surfaces by the use of chill moulds. White cast iron exhibits white fracture.

Chilled cast iron is formed when composing parts of white cast iron are cooled rapidly to provide hard wear resistant surfaces. The depth of the chill can be controlled by composition and heat treatment. Chilled cast iron is used in such members as dies, car wheels, blow points, and cam followers. In general, chilled cast iron is used for a limited number of engineering applications in which extreme hardness and wear resistance properties are required. It has low impact resistance.

Malleable cast iron is made by heating white cast iron castings for a period of several days in air tight pots filled with an oxide of iron and other packing materials, such as slag, sand, silica, gravel, and pebbles. The action of heat and iron oxide partially removes the carbon and reduces the remainder from a combined state to a globular form of free carbon so that after slow cooling, a strong, soft, and somewhat ductile casting is obtained. Since the carbon change reaches only to a depth of about 10 mm, this process is not suitable for heavy castings. However, castings upto 175 kg have been made. The difference between malleable cast iron and grey cast iron is in a form in which free carbon occurs. In grey cast iron free carbon occurs in the form of flat or plate like particles, whereas in malleable cast iron carbon is in the form of irregularly shaped spherical particles. Malleable cast iron is tough and ductile, resistant to impact, and easily machinable. The higher strength and ductility of malleable cast iron are attributed to the shape of its carbon particles. The applications of malleable cast iron are considerably limited as compared to those of grey cast iron because it is more expensive to produce and better mechanical properties are not required in most instances. The use of malleable cast iron usually involves parts of a complex shape that often need considerable machining to meet specifications such as agricultural implements, automobiles, motor trucks, locomotives, railroad cars, pipe fittings, flanges, valves, chains, and so on.

In general, the damping capacity of cast iron is high. This property makes it suitable for beds, bases, and frames of machinery that are subject to vibration and for crank-shafts and cam-shafts. The effect of mass cooling rates on the structure and properties of costing should be considered.

5.3.2 Alloy Cast Iron

Alloying elements are added to cast iron to improve the mechanical properties or to influence certain characteristics, such as machinability, heat, or corrosion resistance. Alloying elements are seldom added to cast iron unless special properties are required, which cannot be otherwise attained. The more commonly used alloying elements are copper, chromium, molybdenum, nickel, titanium, vanadium, and so on. These alloying elements are added singly or in various combinations.

Nickel is used in amounts ranging from 0.25 per cent to 5.0 per cent in low-alloy cast irons. This promotes machinability in cast iron. In heat and corrosion cast iron—as much as 35 per cent nickel is used. NI-Resist—a typical heat and corrosion resistant cast iron—has a composition of nickel 14 per cent, copper 6 per cent, chromium 4 per cent, silicon 2 per cent, and total carbon 3 per cent. It can withstand a temperature of 800°C and is used in a wide variety of media such as hydraulic, sulphuric, acetic, citric, and formic acids, chlorides, caustics, sea water, and so on.

Chromium in cast iron promotes hardness. Tensile strength and wear resistance are also increased by moderate additions of chromium. Chromium content tends to lower machinability and increase the heat resistance property of cast iron.

Molybdenum is added to cast iron in amounts ranging from 0.25 per cent to 1.25 per cent. The principal function is an appreciable increase in tensile strength, transverse strength, deflection, and hardness. Molybdenum is frequently used in cast iron in combination with nickel or chromium or nickel and chromium.

Vanadium is added to cast iron in amounts ranging from 0.10 per cent to 0.50 per cent. Additions of vanadium increase the tensile strength, transverse strength, and hardness.

Copper is added to cast iron in amounts upto about 1.0 per cent. Its functions are to increase fluidity for improved mould filling ability, impart corrosion resistance, and improve mechanical properties, notably toughness and hardness. Machinability of cast iron is also slightly improved by copper additions.

Silicon in the range upto 2.5 per cent promotes resistance to elevated temperatures. Silicon, in amounts ranging from 11 per cent to 17 per cent, is used in an important class of cast irons. Remarkable resistance to all types of acids, except hydroflouric and not hydrochloric, is an outstanding characteristic of these alloys.

Until quite recently boron received little recognition as an addition to regular grey cast iron. 0.5 per cent boron, 3.5 per cent carbon and 1.0 per cent silicon in cast iron helps to promote roll for rolling mills in order to increase surface hardness and refine the structure.

Quenching Immersion of cast iron in water after it has been heated above the critical range increases its hardness and brittleness.

Baking Castings that have been pickled in acid, to removes and and scale are them baked. Pickling makes casting brittle but baking for a few hours at 150°C removes brittleness.

Drawing or tempering This is done by re-heating the quenched metal to a temperature below the critical temperature. It reduces the brittleness but still leads to an increase of hardness. By such a treatment, a Brinell hardness number ranging from 200 to 400 can be attained, the value of which depends on quenching and tempering temperatures.

5.3.3 Cast Steel

Plain carbon steel is an alloy of iron and carbon varying upto 1.7 per cent. Plain or alloyed cast steel is used where castings of improved properties are required over iron or malleable castings. Cast steel is tougher, stronger, weighs less for the same strength of stiffness, and has a higher endurance limit than cast iron. They are used for machine members of intricate shapes, which require high strength and impact resistance, such as locomotive frames, large internal combustion engine frames, gears, wheels and intricate highly stressed machine parts. Steel castings are more expensive. Corrosion resistance can be improved by the use of alloys or protective coatings. Alloy steel castings have also been developed to meet the demand of industries for greater strength and reliability of cast machine members.

5.3.4 Wrought Steel

The carbon limits for carbon steel are below 0.2 per cent for low-carbon steel, 0.20 per cent to 0.50 per cent in medium-carbon steel, and above 0.5 per cent in high-carbon steel. The mechanical working in the manufacturing process refines the structure and produces a more uniform steel having greater strength, greater toughness, and more durability than is obtained through casting. The properties of steel vary greatly with its carbon content and the form in which the carbon occurs.

Low-carbon steels, containing upto 0.2 per cent carbon, are soft, very ductile, and are easily welded by any process. Low-carbon steels are unresponsive to heat treatment and possess low strength. Medium-carbon steels, machine well and respond to heat treatment.

High-carbon steels, containing over 0.5 per cent carbon, respond readily to heat treatment. In the heated state they may have very high strength combined with hardness, but are not so ductile as the medium-carbon steel. In the higher carbon ranges, the extreme hardness is accompanied by excessive brittleness. The higher the carbon content, the more difficult it is to weld these steels.

5.3.5 Alloy Steels

A plain carbon steel is an alloy of iron and carbon with small amounts of manganese, silicon, sulphur, and phosphorous. The term alloy steel is applied when one or more elements other than carbon are introduced in sufficient quantities to modify its properties substantially. The principal alloying elements used in steel are chromium, nickel, manganese, silicon, molybdenum, vanadium, tungsten, and to a lesser extent copper, cobalt, beryllium, boron, and silver.

Chromium: The addition of chromium results in the formation of various carbides of chromium, which are very hard, yet the resulting steel is more ductile than a steel of the same hardness produced by a simple increase in carbon content. Chromium also refines the grain structure so that these two combined effects result in both increased toughness and hardness. The addition of chromium increases the critical range of temperatures and raises the strength at high temperature. Alloys of chromium resist abrasion and wear.

Nickel: Nickel also increases the critical range of temperature. It is soluble in ferrite and does not form carbides and oxides, and thus increases the strength without decreasing the ductility. Case hardening of nickel steel results in a better core than that obtained in plain carbon steels. Chromium is frequently used with nickel to obtain the toughness and ductility provided by nickel and the wear resistance, and hardness contributed by chromium.

Manganese: Manganese is added to all steels as deoxidising and desulphurising agent, but if the sulphur content is low and the manganese content is high (over one per cent), then it is classified as a manganese alloy. It lowers the critical range of temperatures and increases the time required for transformation so that oil quenching becomes practicable.

Silicon: Silicon is added to all steels as a deoxidising agent. When added to steels, it produces brittleness and leads to high magnetic permeability. The principal use of silicon other than alloying elements, is to improve strength, hardness, corrosion resistance, and electrical conductivity.

Molybdenum: Molybdenum acts very much like chromium but is more powerful in action. It also increases the depth of hardening after heat treatment. Molybdenum is widely used when combined with other alloying elements such as nickel, chromium, or both. Nickel-molybdenum and nickel-chromium-molybdenum steels retain the good features of the nickel-chromium steels and in addition have better machining qualities. Molybdenum increases the critical range of temperatures. Except for carbon, it has the greatest hardening effect and results in the retention of great deal of toughness.

Vanadium: Vanadium is used to toughen and strengthen the steel in order to reduce the grain size and to act as cleaner and degasifer. It has the desirable effect of increasing the life of tools, springs, and other members subjected to high temperatures. As vanadium has very strong tendency to form carbides, it is used only in small amounts.

Tungsten: Tungsten is widely used in tool steels because the tool maintains its hardness even at red heat. Tungsten produces a fine dense structure and adds both toughness and hardness. Its effect is similar to molybdenum, except that greater quantities must be added.

Alloy steel may be forged, swaged, rolled, drawn, and welded, and is especially useful in machining and finishing. It finds numerous applications in members subjected to high stress — either due to external loads, being static or cyclic—to shock loading, or stress concentration, and in parts subjected to wear, high or low temperature, or corrosion. It is also used in applications where special properties, electrical or thermal requirements, are involved and where space and weight must be conserved. Some applications of steel in machinery parts are in crack shafts, connecting rods, piston rods, keys, pins, rivets and bolts, ball and roller bearings, springs, shafting, gears, valves, frames of heavy stationery and transport equipment, tubes, levers, arms, dies, and rolls.

Heat Treatment of Cast Iron Several types of heat treatment are used to alter and enchance some properties of cast iron in order to increase their utility.

Aging: It is applied to relieve the casting without materially affecting the physical properties. It is carried out for 1 to 5 hours in the temperature range of 450°C to 550°C.

Annealing: It is carried out for 1 to 5 hours at 660°C to 870°C, the temperature dependent upon the size of the part. It is intended to reduce hardness and facilitate machining. However, annealing is done at the expense of some strength.

5.3.6 Heat Treatment

Metal is heated in order to give certain desired properties. Some of the properties which may be required and the treatment necessary are as follows.

1. To soften the metal—Annealing
2. To harden the metal to resist wear or to enable it to cut other metals—Hardening
3. To remove some of the extreme brittleness caused by hardening—Tempering
4. To refine the structure after it has been distorted by hammering or working in cold state—Normalising

In addition, there are various other treatments such as torugening the metal to withstand shock, toughening soft steel so that it machines without tearing and treating special steels to increase their strength.

Heat Treatment of Steels

The elastic limit, ultimate strength, and hardness of carbon steel are changed by heating the steel and then allowing it to cool rapidly or slowly. Higher carbon steels and alloy steels are especially responsive to heat treatment. Heat treatment of various steel alloys may be found from Indian Standards and other engineering hand books, and are not be included here. The various processes that may be used are described below.

Annealing: The purpose of annealing is to soften the steel so that it may be more easily machined and to relieve the internal stress that may have been caused by working the metal or by unequal contraction in casting. The process of annealing invloves heating the metal slowly to the required temperature, holding it at that temperature for long enough to enable the internal changes to take place and finally cooling slowly. The annealing temperatures for steels of different carbon contents are given in Table A5.1. Annealing reduces hardness, increases ductility, and usually reduces strength.

TABLE A5.1 *Annealing Temperature is Related to Carbon Content*

Carbon Content (%)	Annealing Temperature (0°C)
Less than 0.12 (Dead mild)	875–925
0.12 to 0.25 (Mild)	840–970
0.3 to 0.5	815–840
0.5 to 0.9	780–810
0.9 to 103 (Tool steels)	760–780

Normalising: When steel is cold worked, that is, hammered, rolled, bent, and so on, in cold condition, its crystal structure is distorted and the metal may become brittle and unreliable. To refine the structure of steel and remove strains that may have been caused by cold working, normalising is resorted to. Normalising is a special kind of annealing in which a material is heated slowly to its annealing temperature below the range (815–840°C), in still air at room temperature.

Quenching: Rapid cooling from above the critical temperature by immersion in cold water and other cooling mediums is called quenching. It is a hardening treatment. The degree of hardness depends on the amount of carbon present and the rate of cooling, which can be varied by using cooling mediums such as ice water, cold water, oil, hot oil, molten lead, and so on. The oil quench is quite slow but prevents quenching cracks caused by the rapid contraction of the object being treated. Quenching in water is used for carbon steels and medium carbon low alloy steels. Some alloy steels, notably those containing tungsten, harden when slowly cooled in air. The quenching treatment raises the strength of the metal and increases its wear resistance, but makes the metal brittle and reduces its ductility.

Tempering or drawing: When a steel specimen has been fully hardened, it becomes very hard, brittle, and has high residual stresses. The steel is unstable and tends to contract on aging. These internal stresses can be relieved by an additional heating process, called tempering or drawing which softens the material and toughens it. After the specimen has been fully hardened by being quenched from above the critical temperature, it is reheated to some temperature below the critical temperature for a certain period of time and then allowed to cool in still air. The temperature to which it is reheated depends upon the composition and the degree of hardness or toughness desired. Tempering reduces the elastic limit and ultimate strength slightly, but they are still higher than they were before drawing.

Case hardening or carburising: This is a process of hardening the outer portion of the metal by prolonged heating, free from contact with air, while packed in carbon in the form of bone char, leather scraps, charcoal, or barium carbonate. The outer material absorbs carbon depending upon the composition of particles, time, and temperature. When the hot metal is quenched, this high carbon steel hardens whereas the low carbon steel of the core remains soft and ductile. Case hardening is applied to soft steels with a carbon content of 0.2 per cent or less, which cannot be hardened by simple heating and quenching.

Cyaniding: For the rapid surface hardening of small parts such as screws, clamps, and washers, this form of surface hardening may be used. This is done by cyanide salt and may be carried out either by placing the article in the fused salt, holding it at the quenching temperature for a few minutes and then quenching out, or by sprinkling the red hot article with powdered salt, heating up again further quenching. Cyaniding produces a thin but a very hard case in a very short time.

Nitriding: Nitriding is surface hardening accompanied by heating certain steel alloys called nitralloys. The treatment consists of heating the part to a temperature of about 560°C inside a chamber through which a stream of ammonia is passed. The part is rough machined before nitriding, and is then heated to the nitriding temperature without ammonia in order to produce distortion, if any. After this the part is finish-machined and nitrided. The nitrided surface has hardness ranging from 730 to 1100 BHN and resists corrosion. Nitriding also raises the endurance limit of the steel.

Flame hardening: It is a process in which a steel part is heated locally above the critical temperature and then quenched. The depth of the flame hardened layer can vary from 1.5 mm to about 3.0 mm, the exact depth depending on the service requirements. After being quenched the piece must be stress relieved by tempering; a temperature of 200°C is usually sufficient. Plain carbon steels with a carbon content of 0.30 to 0.70 per cent are best suited for this method. It is often applied to gear teeth, cams, castings, forgings, and rolled sections, irrespective of their size.

5.3.7 Materials for High Temperature Service

Metal parts for steam engines, internal combustion (IC) engine valves, superheated steam equipment, and similar components are stressed at temperatures ranging from 100°C to 1000°C. The metal used for such parts must be specially selected from materials that retain a large percentage of their strength at high temperatures and those that do not creep. High temperature or heat resisting alloys can be principally of three types:

1. Alloys of the first group contain over 50 per cent iron with additions of 10 to 30 per cent chromium and varying amounts of copper, nickel, tungsten, and silicon.
2. Alloys of the second group contain upto 25 per cent iron, 50 per cent to 60 per cent chromium and nickel, and small amounts of manganese, nickel, tungsten and silicon.
3. Alloys of the third group contain 0.55 per cent to 0.6 per cent iron as an impurity, over 80 per cent nickel, considerable chromium, and small amounts of manganese and silicon. The alloys of this group have excellent resistance to corrosion and temperatures upto 1000°C.

5.4 NON-FERROUS MATERIALS

5.4.1 Aluminium and Aluminium Alloys

Aluminium having a specific gravity of only 2.7 is approximately two-fifths as heavy as iron and one-third as heavy as copper. Pure aluminium resists oxidation, is highly ductile, and has good forming properties, but has poor casting characteristics and machining properties. Pure aluminium has a tensile strength of about 77 MN/m^2, which can be increased considerably by cold working and alloying with other materials. The modulus of elasticity of aluminium and its alloys is 70 GN/m^2, which means that it has about one-third the stiffness of steel.

The most useful alloying elements for aluminium are copper, silicon, manganese, magnesium, and iron. Copper as an alloying element raises the ultimate strength, endurance limit, improves casting characteristics and machinability, but lowers its resistance to corrosion. Aluminium-copper alloys are used in applications such as crank cases and transmission housings and fittings. Aluminium-silicon alloys have better corrosion resistance and mechanical properties but poorer machinability in comparison to aluminium-copper alloys. They are suitable for marine castings, water jackets, housings, and castings, where a minimum of machining is required. The (Cu 0.4 per cent, Mg 5 per cent, Mn 0.5 per cent) aluminium-magnesium-manganese alloy known as "duralumin" is suitable for parts that require severe working and has good corrosion resistance and strength. It is obtainable in sheets, plates, tubes, rods, wires, extruded sections, bolts, and rivets. The plates can be spot welded and are widely used in aircraft construction.

Aluminium alloy castings made in sand moulds, permanent metal moulds, and die casting machines are generally stronger than the poorer grades of cast iron. Better surface finish, closer dimensional tolerances, and better mechanical properties save machining and finishing costs and make the use of permanent moulds and die castings, which is desirable when the quantity justifies the extra cost of the equipment.

Typical uses Carburettor bodies, refrigerator shelf brackets and fittings, cooking utensils, general purpose castings of thin sections, ornamental grills, reflectors used for castings require leak tightness and moderate strength. Vaccum cleaner housing, washing machine agitators, hydraulic brake pistons, diesel cylinder, head covers with pressure requirements, meter parts, bushings, bearing caps, automotive pistons, pulleys and sheaves, automotive and diesel pistons, sole plates for electric hand irons, aircraft generating housing, motor cycles, diesel and aircraft pistons, air cooled heads, aircraft fittings, aircraft gun control parts, aircraft wheels, railroad car seat frames, compressor connecting rods, fuel pump bodies, aircraft gear housings, pipe fittings for marine as well as general use, typewriters, frames, engine parts, water cooled cylinder heads, regular parts of internal combustion engine cylinder heads, aircraft super charge cover, fuel pump bodies, supercharger impellers, gear for liquid cooled aircraft cylinder heads, machine parts, handwheels, aircraft pump parts.

5.4.2 Copper

Copper is easily distinguished from all other metals on account of its red colour. Commercial copper is a tough, ductile, and malleable metal containing less than 5 per cent of impurities such as tin, lead, nickel, bismuth, arsenic, and antimony. Because of its high electrical conductivity, it is used extensively in all parts of electrical apparatus that conduct current. Copper is also a good conductor of heat and is highly resistant to corrosion by liquids. For this reason it is used in locomotive fire boxes, water heating apparatus, water pipes, vessels in breweries, chemical plants, and condenser tubes. For its high heat conduction it is used for soldering iron bits.

Alloys of copper form an important group of materials with a wide variety of properties. Some have high strengths, some are excellent bearing materials, some retain their strength at high temperature and others are valuable for their corrosion resistance. Three groups of copper alloys used in machine design are the brasses, the bronzes, and monel metal. Copper is also used as an alloy in steel in order to increase its resistance to corrosion.

Alloys of copper and zinc (brass) An important use of zinc as an alloy with copper is to give rise to various classes of brass. These alloys are important due to the wide range of production process they lend themselves to, their high resistance to atmospheric effects and corrosion and the wide variety of their mechanical properties. Suitable types of brass lend themselves to casting, hot forging, cold rolling into sheets, drawing into wires and being extruded through dies to give special shaped bars.

By adding small quantities of other elements (aluminium, iron, manganese, and tin) the strength of the bars may be greatly increased from their normal strength of 300–375 MN/m^2, and a range of high tensile brasses is available with ultimate strengths as high as 600 MN/m^2.

The melting point of brass varies according to its composition, but most brasses in the common range liquify between temperatures of 850°C and 960°C. Hard brass may be softened by heating to about 750°C.

To improve the machining quality of brass, 1 per cent or 3 per cent of lead is often added. In its ordinary condition the metal is soft and ductile and tends to drag under the tool. Small amounts of tin are sometimes added to brass in order to increase its hardness and resistance against the corrosive action of sea water.

Copper-tin alloys (bronzes) Copper alloyed with tin forms a set of alloys called bronzes, which are important metals in engineering practices.

Gunmetal contains 88 per cent copper, 10 per cent tin, and 2 per cent zinc. Zinc is added to increase its fluidity. The metal is used chiefly for castings that must be strong and resistant to corrosion by water and atmosphere. Gunmetal is not suitable for cold working, but may be forged at about 600°C, (just below red heat). Another common bronze is made of 5 per cent tin, 5 per cent zinc, 5 per cent lead, and 85 per cent copper.

Bronzes are more costly than brasses and, hence, are used only when the cheaper alloys do not prove to be satisfactory. They are used in the highest quality thermostatic bellows and other parts that require resistance to severe stretching together with good tensile and elastic strength.

Copper and tin alone do not readily alloy and small quantites of phosphorous are generally added, giving rise to common trade name phosphorous bronze. Bronzes are particularly useful because of corrosion resistance, good fatigue properties, and useful spring characteristics, the latter being due to high elastic limit and creep resistance at room temperatures. As tin increases upto 8 per cent, the strength and ductility increase rapidly. More than 8 per cent tin increases the strength, but then the ductility reduces.

Copper-nickel alloys (monel metal) Alloys containing less than 50 per cent nickel are called cupro nickles or nickel silvers. These alloys also contain a small percentage of zinc. These alloys are silver in appearance and tarnish little under atmospheric condition. They are malleable and can be worked without annealing.

Monel metals are the most important nickel copper alloys, having high strength and toughness, and excellent resistance to salt water, brine, caustic soda, hydrochloric acid, sulphuric acid, and other corrosive agents.

Monel metal contains 67 per cent nickel and 30 per cent copper with a small amount of iron, manganese, silicon, and carbon. It is a general purpose material used under corrosive temperatures upto 260°C, and for still higher temperatures if sulphur is not present. It holds its strength at elevated temperatures better than bronzes and some steels and is used for turbine blades, high temperature valves and for springs subjected to temperatures upto 200°C. Unlike plain steel, it increases in strength, hardness, and fatigue strength as the temperature decreases to as low as –150°C, the elongation and impact strength remaining practically constant.

Copper-aluminium alloy (aluminium bronze) Aluminium bronze is copper with 2 per cent to 11 per cent aluminium. Its tensile strength increases as the aluminium content becomes greater. 10 per cent aluminium bronze is used in cast and hot working conditions for parts requiring high tensile strength and resistance to corrosion, wear, and alternating stress. Iron upto 0.5 per cent increases the strength and hardness without reducing the ductility appreciably. An alloy containing 11 per cent aluminium, 5 per cent nickel, 5 per cent iron, 79 per cent copper has extreme hardness and wear resistance at high temperatures, a quality essential in aircraft engine valve seats. In general, aluminium bronzes are used widely in aircrafts for bushings, gears, bearings, valve guides, shock absorbers, pistons, and other similar parts.

Bearing metal: It has been found by experience that for an efficient bearing combination, the following conditions are necessary.

1. The shaft and the bearing should be different in their nature with the bearing being softer than the shaft.
2. The most efficient bearing metal is the one consisting of small pieces of a comparatively hard metal embedded in the softer body of another metal.

5.5 NON-METALLIC MATERIALS

The commonly used non-metallic materials are wood, glass, rubber, leather, carbon, and plastics.

Wood: Wood has some uses in machine members. It is used where light weight parts subjected to moderate shock loading are required, for example, in circuit breaker operating rods, where non-metallic bearing material is desirable.

Glass: It is usually thought of as a weak brittle material. Glass parts can be moulded by heat and pressure or by heating and blowing. They can be cut and finished by grinding and can be machined by carbide tools, but the surface will be too rough for ordinary use. Glass is used for all parts of centrifugal pumps for acids, piping of mechanical processes, pipe fittings, exchangers, and lining for tanks and fittings.

The tensile strength of glass ranges from 42.0 to 84.0 MN/m^2 with small surface scratches reducing the strength by 50 per cent. The compressive strength of glass is over 700 MN/m^2. The modulus of elasticity varies from 45 GN/m^2 to 70 GN/m^2. Allowable design stresses range from 3.5 to 7.0 MN/m^2.

Rubber: Rubber and similar synthetic materials, such as neoprene, have a variety of applications in machinery. Rubber should be protected from high temperature, oil, and sunlight. It is an excellent material for seats and diaphragms, water lubricated bearings, parts subjected to vibrations, tubes, and hose. In industry, hard rubber is used for electric insulation, switch handles, bearings and so on.

Leather: Leather is very flexible and withstands considerable wear under suitable conditions. Its modulus of elasticity varies according to the load. It is used in belt drives and as a washer.

Carbon: Since long, carbon has been used in electrical insulations and has not been generally used in mechanical insulations. Use of carbon gives low friction losses, low wear rates when operating against metals, and is used in chemical handling equipments. Pump rotors, vanes, and gears of carbon have been used with good results to replace similar parts of bronze and laminated plastics. Carbon parts have replaced clutch plates and rings of metal and cork.

Plastics: Plastics are used in different fields, for example, buttons, fountain pens, telephones, dials, knobs, and so on. The word 'plastic' was used originally to indicate a material that could be formed or moulded by pressure at moderately elevated temperatures. Plastics are available in thin sheets, coatings, petroleum laminates, and filaments for weaving. Raw materials are mainly derived from agriculture and petroleum. Plastic moulding is a low cost-high quality production method that is now replacing many parts formerly made by wood, glass, or metal.

Review Questions

1. Why is the study of materials important?
2. Classify engineering materials.
3. State the properties of construction materials.
4. Define: (a) elasticity (b) plasticity (c) impact strength.
5. State the types of ferrous materials and their uses.
6. Why is heat treatment necessary?
7. State the materials used for high temperature service.
8. State the types of non-ferrous materials and their uses.
9. Which are the commonly used non-metallic materials?
10. What is the method of providing information about materials on a drawing sheet?

Chapter 6

Manufacturing Processes

This chapter contains basic information about the shop processes and the machines used for carrying out these processes

After reading the chapter you will be able to

- judge the importance of machine shop processes in assembly drawings
- plan the processes for different components
- select proper machining operation for manufacturing of individual components in any assembly

6.1 MANUFACTURING PROCESSES

For manufacturing processes and machines, or equipments used for these processes, knowledge of shop floor drawings is essential. Shop drawings are made specifically for use by the workman on the shop floor. Shop drawings should contain information in brief, in the text form, about the relevant processes and tools.

Knowledge of machines, processes, tools, and measuring instruments will certainly help students in making shop floor drawings.

6.1.1 Machine Shop

The various parts manufactured in a machine shop, forging shop, and press shop are shown in Fig. A6.1, from which students can get an idea about the different shop processes.

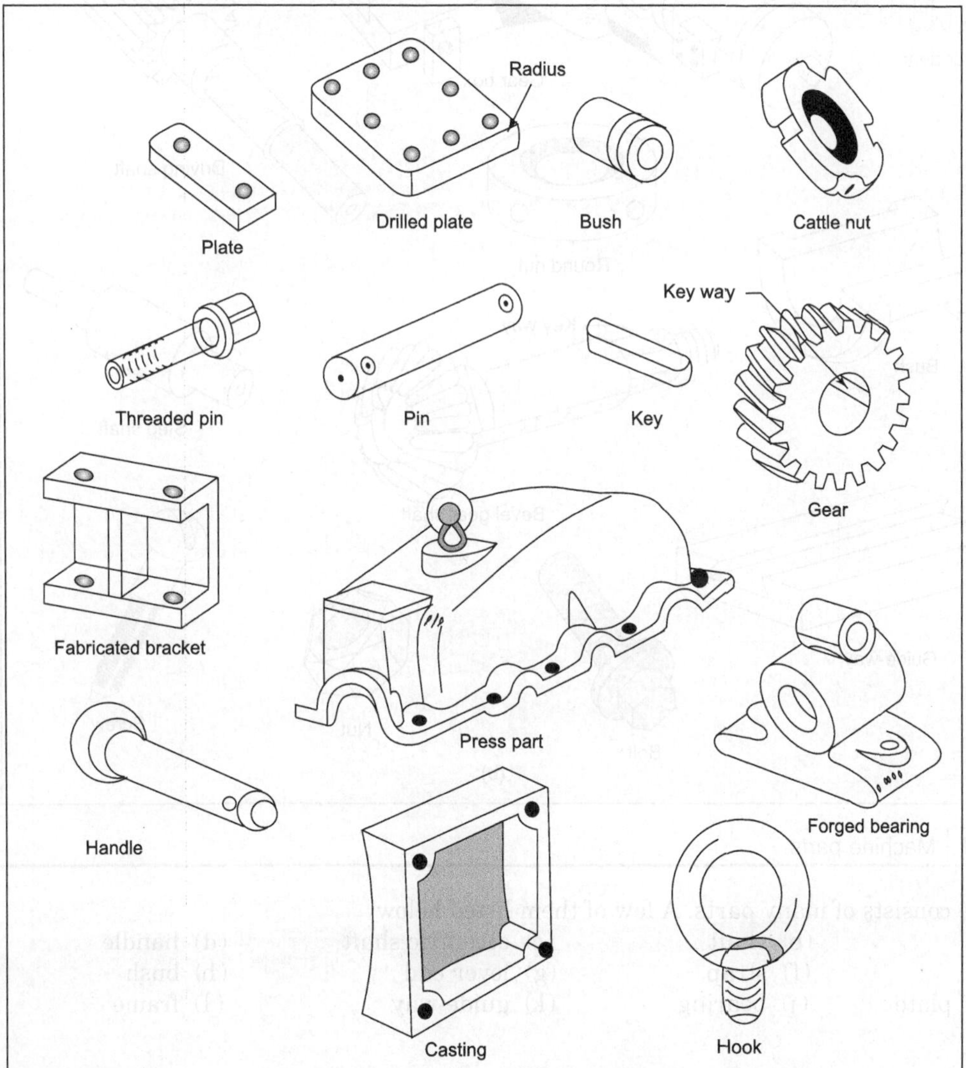

(a)

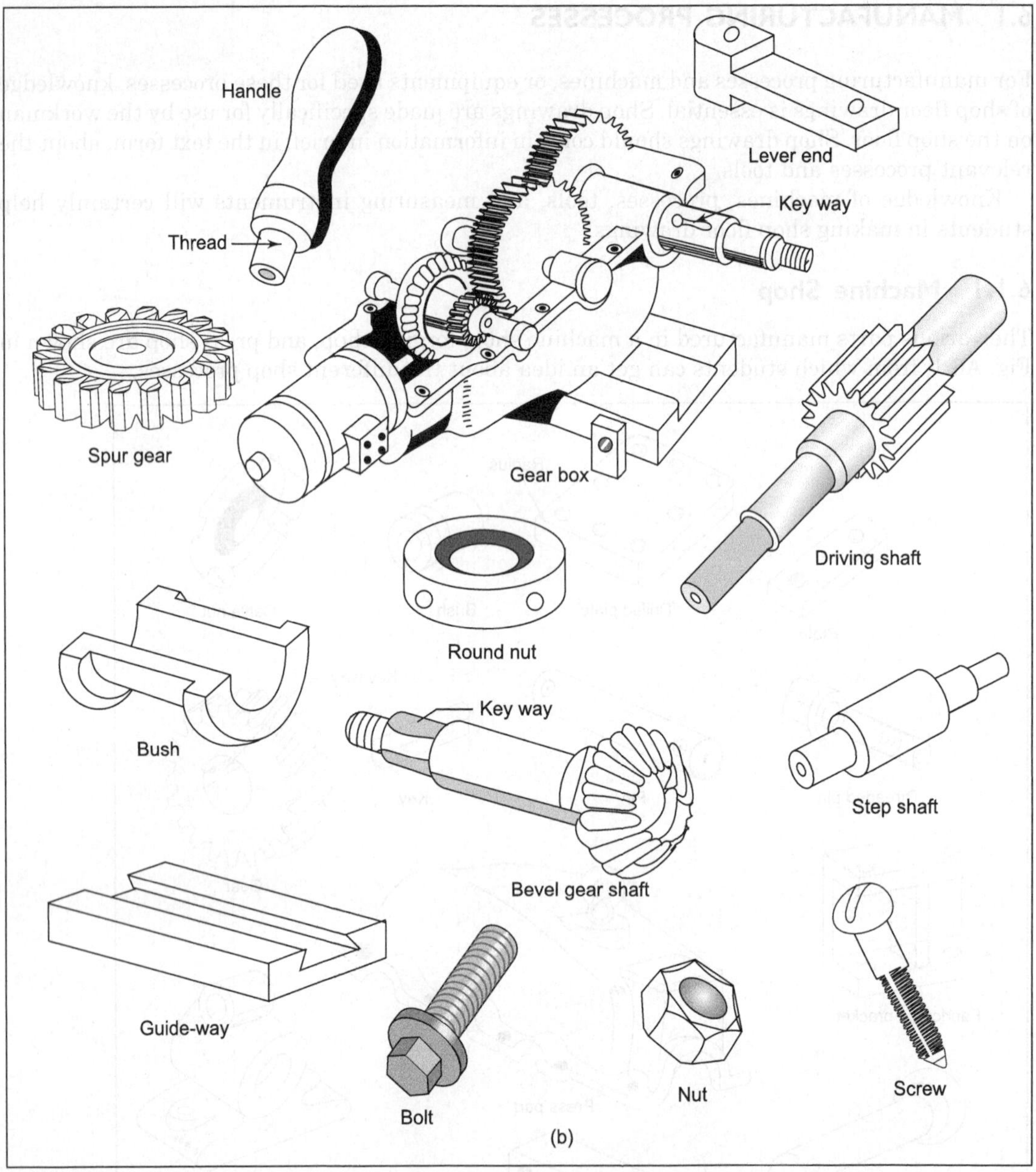

Fig. A6.1 Machine parts

A machine consists of many parts. A few of them listed below.

(a) bolt (b) shaft (c) eccentric shaft (d) handle
(e) butt (f) strap (g) lever end (h) bush
(i) cover plate (j) bearing (k) guide-way (l) frame
(m) gear

All processes are basically divided into two groups.

1. Machining processes
2. Non-machining processes

Fig. A6.2 depicts the basic operation of all processes.

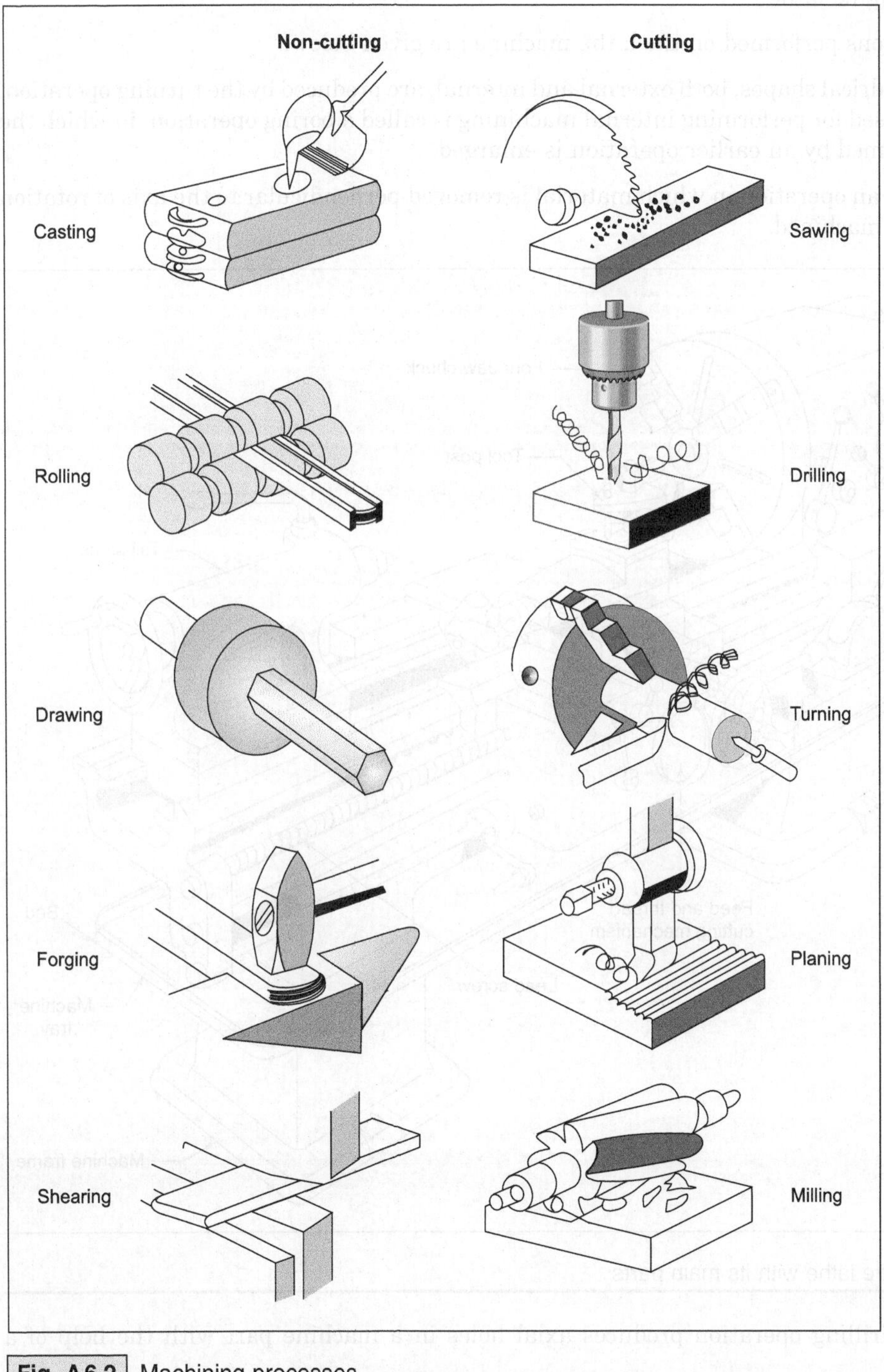

Fig. A6.2 Machining processes

Lathe Cylindrical shapes can be achieved by various turning processes on a lathe machine. The machine is most commonly known as 'centre lathe'. Fig. A6.3 shows a centre lathe and indicates its main parts. While machining extra long jobs, the centre of the job is made to coin-cide with the centre of the machine, hence the name 'centre lathe'. The centre lathe is also known as an 'engine lathe'.

Some operations performed on the lathe machine are given below.

Turning Cylindrical shapes, both external and internal, are produced by the turning operation. The operation used for performing internal machining is called a boring operation, in which the hole already formed by an earlier operation is enlarged.

Facing This is an operation in which material is removed perpendicular to the axis of rotation of the job being machined.

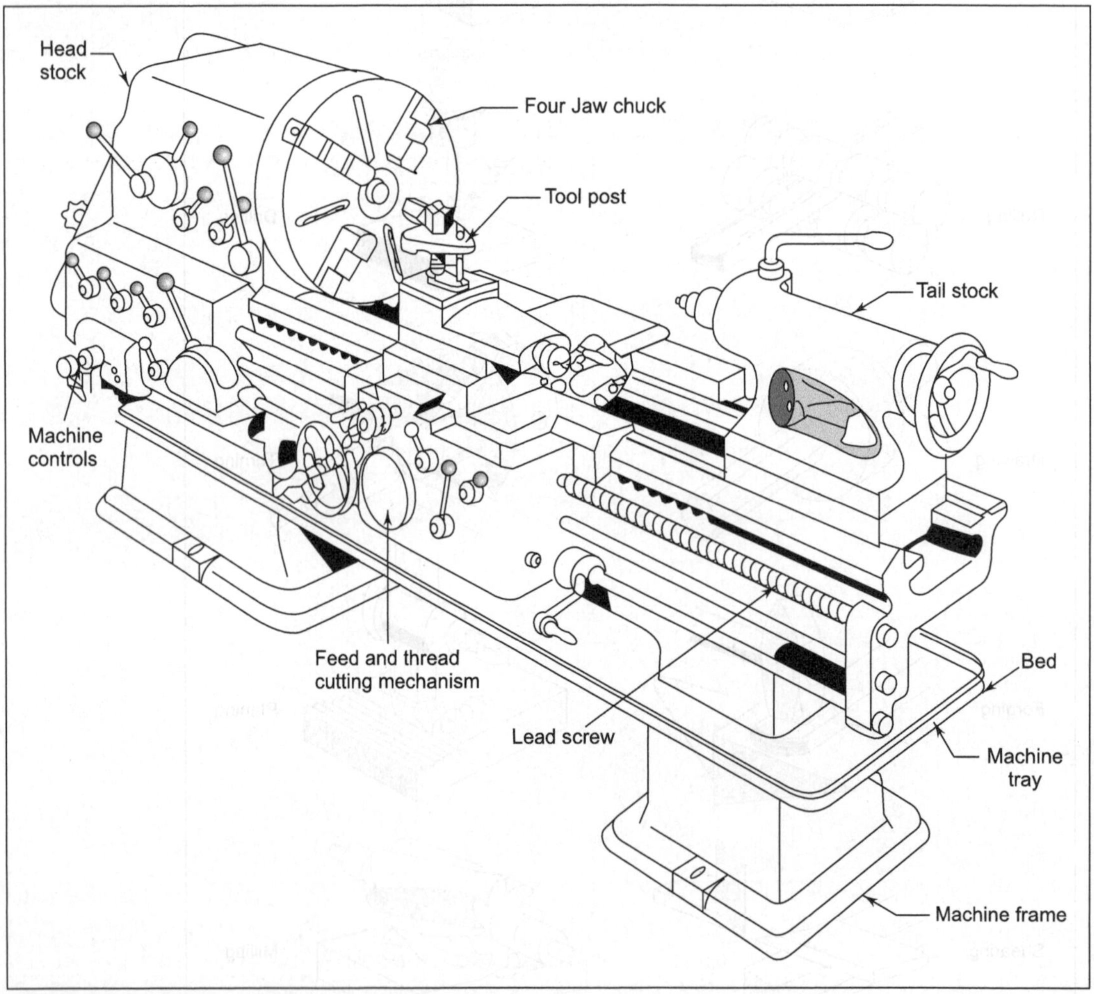

Fig. A6.3 Centre lathe with its main parts

Drilling The drilling operation produces axial holes in a machine part with the help of a twist drill.

Threading This operation produces threads, either on the external or internal surfaces, with the help of threading tools.

Taper turning This operation produces tapers, both on the internal and external surfaces, by using a special attachment.

Knurling The knurling operation produces rough surfaces required for grip, by using a knurling tool.

Drilling machine This machine produces round holes by cutting the material. Fig. A6.4 shows the most common pillar drilling machine, and indicates its main parts.

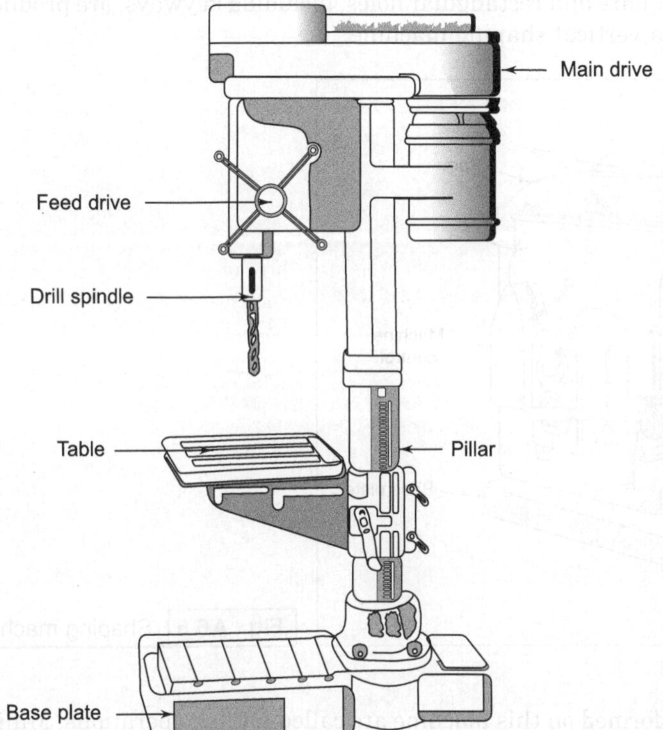

Fig. A6.4 Pillar drilling machine

Some of the operations performed on a drilling machine are given below.

Drilling The operation of producing holes by cutting the material at any inclination by mounting the job properly on the machine table is called drilling.

Reaming The reaming operation produces accurate and smooth holes after the drilling operation by using the 'reamer' as a tool.

Counter boring This operation is used in assembly to join two parts by counter head screws. Counter boring means enlarging the diameter of a hole for the accommodation of the head of the screw.

Counter sinking This operation produces a flat surface, after assembly, with the help of counter sunk screws.

Thread cutting This operation can be done manually as well as with a machine. External threading on pipes and cylindrical jobs is done by using a tool 'die', while holding job in a vice.

Tapping This operation cuts internal threads using a tool known as 'tap'.

Shaping machine An operation performed on this machine is called a shaping operation. Shaping is the machining process to produce generally flat surfaces using a single point cutting tool. Fig. A6.5 shows a shaping machine and indicates its main parts. A similar machine, in which the tool moves vertically up and down, is the slotting machine or vertical shaping machine. To produce flat surfaces in small size jobs, a shaping machine is used. To produce flat surfaces in large sized jobs, a planing machine is used. To produce flat surfaces, planing machine is used. In a shaping machine, the workpiece is held fixed and the tool reciprocates over the surface of the job to be machined. Square and rectangular holes, including keyways, are produced with the help of a slotting tool on a vertical shaping machine.

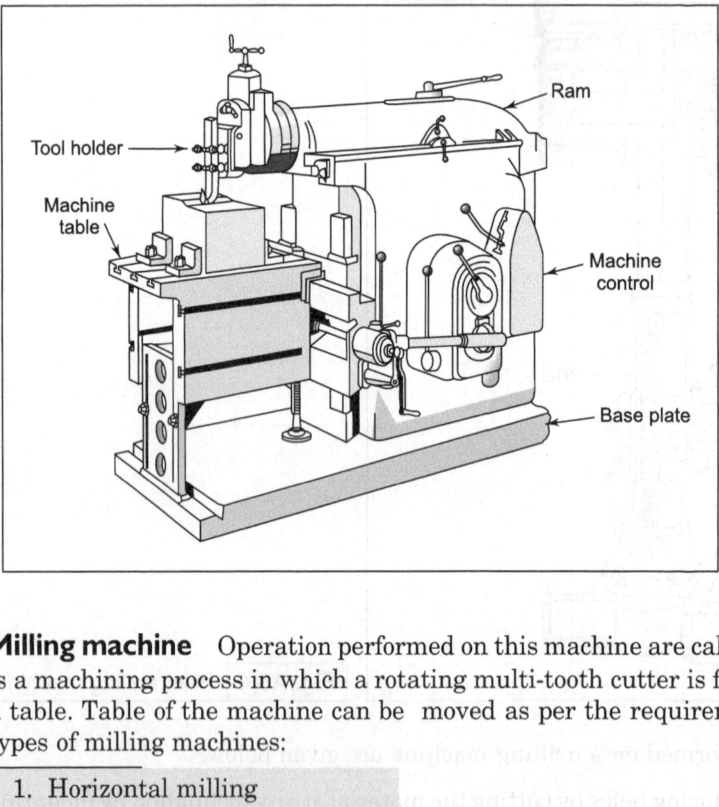

Fig. A6.5 Shaping machine

Milling machine Operation performed on this machine are called milling operations. Milling is a machining process in which a rotating multi-tooth cutter is fed against a work mounted on a table. Table of the machine can be moved as per the requirement of the job. There are two types of milling machines:

1. Horizontal milling
2. Vertical milling

Surface milling slotting operation, keyway milling, and end milling operations can be performed on these machines as per the requirement of the job (Fig. A6.6).

Broaching machine In case of mass scale production of, shapes such as keyways, square, round, hexagonal, or irregular holes, a tool 'Broach' is used and the machine used is called a Broaching machine. These are of two types:

1. Horizontal
2. Vertical

Grinding machine In this machine, operations are performed for getting a smooth surface on the machined job by removing a very small amount of material from the original job. There are two basic types of grinding operations, cylindrical and surface grinding. In cylindrical grinding, both the workpiece and the grinding wheel rotate at a very high speed, whereas in surface grinding the grinding wheel rotates while the workpiece reciprocates.

Lapping Lapping is a super finishing process whereby a perfect type of fit is obtained in between the two mating parts. In this process a very fine abrasive is mixed with oil to make a paste which is applied to the mating surfaces. The surfaces are then rubbed against each other to get the desired fit and surface finish.

Honing This is also a surface finishing operation where extremely fine abrasives are applied to the finishing tool. The finishing tool is normally in the form and shape of the surface to be finished.

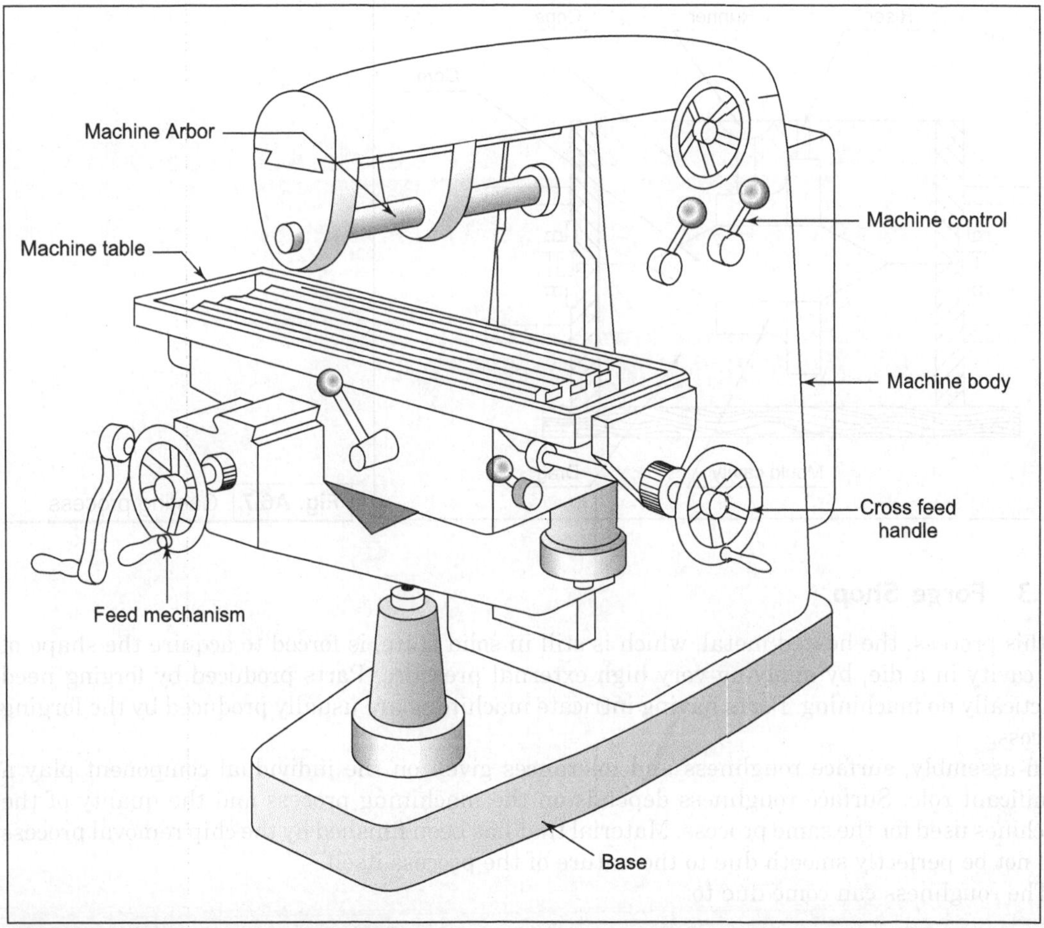

Fig. A6.6 Horizontal milling machine

6.1.2 Foundry Shop

Casting Castings are produced by pouring molten metal into the cavities of a mould or die. The mould or die is produced with a cavity in the shape of the desired part. Sand mould casting is one such process of casting. In this process, a sand mould is obtained by ramming the specially prepared sand around the pattern. After the removal of the pattern, the cavity which is a replica of the pattern, is left in the mould. A pattern is an exact copy of the component to be produced. To produce a job with the holes, sand cores are placed in the mould cavity. When the molten metal solidifes in the mould, it leaves the desired openings in the casting around the cores. This will eliminate drilling and boring operations while machining. Parts made of cast iron and non-ferrous alloys are produced by the sand moulding process. For mass scale production of any part by the sand casting process, sand moulds are replaced by permanent metal moulds, also known as dies. Parts that are made of aluminium, zinc and copper alloys are generally produced by this process. Die casting is also one of the type casting processes. In this process molten metal is forced into the mould or die under very high pressure which, is maintained until the metal solidifies. Die casting parts are superior to sand casting parts.

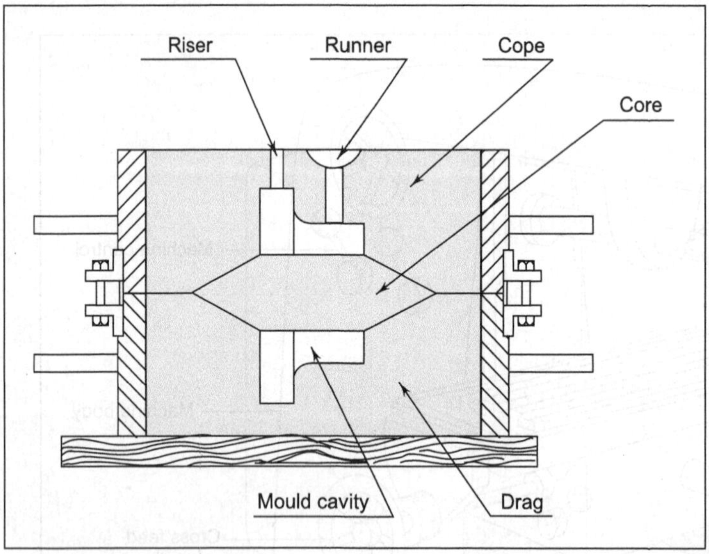

Fig. A6.7 Casting process

6.1.3 Forge Shop

In this process, the heated metal, which is still in solid state, is forced to acquire the shape of the cavity in a die, by applying very high external pressure. Parts produced by forging need practically no machining. Parts having intricate machining are usually produced by the forging process.

In assembly, surface roughness and tolerances given on the individual component play a significant role. Surface roughness depends on the machining process and the quality of the machines used for the same process. Material that has been finished by the chip removal process will not be perfectly smooth due to the nature of the process itself.

The roughness can come due to:

- Irregularities arising out of inaccuracies in the machine tool itself
- Irregularities due to vibrations which that cause chatter marks on the component
- Irregularities due to feed marks of the tool
- Irregularities due to the rupture of the material in the process of chip removal

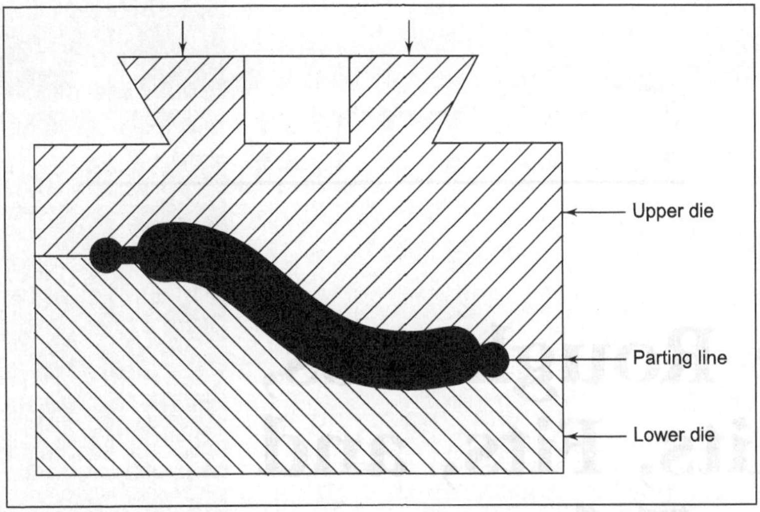

Fig. A6.8 Forging process

Review Questions

1. Why is the knowledge of manufacturing processes important?
2. What is a machine shop?
3. What is a foundry shop?
4. What is a forge shop?
5. What are the cutting and non-cutting processes?
6. Discuss with applications — lathe, shaping, milling, and grinding machines.
7. What is the casting process?
8. What is the forging process?
9. How will you identify a machining process from a given drawing?
10. Do you feel that the knowledge of production engineering is important for machine drawing?

CHAPTER 7

Surface Roughness, Limits, Fits, and Tolerances

This chapter contains information about the selection of fits between two components of an assembly, and surface roughness, which can be achieved through different machining processes

After reading the chapter you will be able to

- Understand the importance of limits, fits, and tolerances for assembly work
- select the proper fit for assembling two machined components

7.1 SURFACE ROUGHNESS

Surface roughness limits the performance and the properties of a component in the following manner:

- Roughness increases friction and the wear between unlubricated surfaces that are in contact with each other.
- In lubricated systems the danger of seizure is reduced due to smooth surfaces.
- Roughness will tend to reduce the effectiveness of interference fits.
- Fatigue strength of the component increases with an increase in smoothness.
- Roughness reduces fluid flow through small sections.

To understand the concept of roughness, consider Fig. A7.1(a) which shows a surface profile at a 1 : 1 scale. However, when the same surface is magnified 20 times in the vertical direction, keeping the same scale horizontally, one finds that the surface is full of hills and valleys, as shown in Fig. A7.1(b).

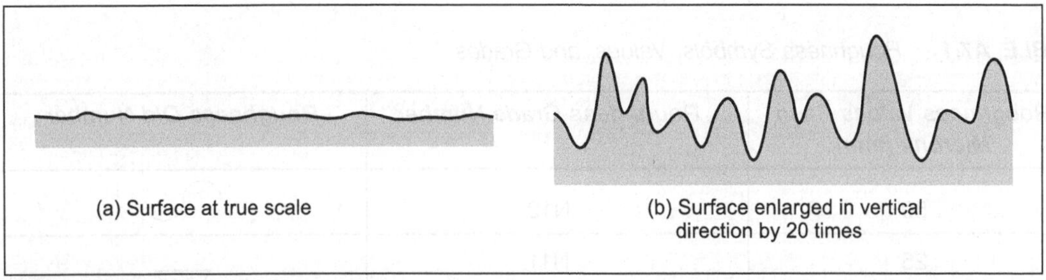

Fig. A7.1 Surface roughness

7.1.1 Machining Symbols

Surface roughness is indicated in the form of symbols on an engineering drawing. The basic symbol consists of two legs of unequal lengths making an angle of about 60° between the legs. The symbols are as follows.

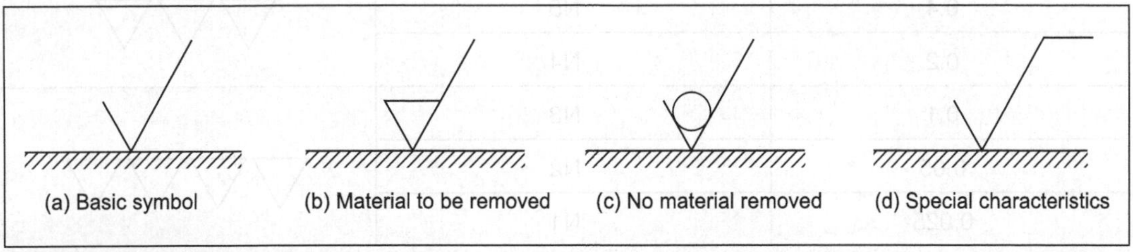

Fig. A7.2 Machining symbols

The surface roughness symbol can be placed on the extension line of the component as shown in Fig. A7.3.

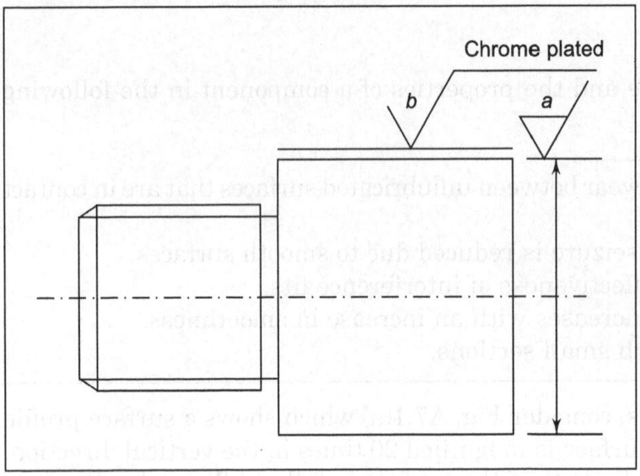

Fig. A7.3 Surface roughness with special characteristics

The roughness values *Ra* can be expressed by the corresponding grade numbers, as given in the following table.

TABLE A7.1 *Roughness Symbols, Values, and Grades*

Roughness Values Ra in Microns (mm)	Roughness Grade Number	Roughness Old Number
50	N12	∼
25	N11	▽
12.5	N10	
6.3	N9	▽▽
3.2	N8	
1.6	N7	
0.8	N6	▽▽▽
0.4	N5	
0.2	N4	
0.1	N3	▽▽▽▽
0.05	N2	
0.025	N1	

Values of *Ra* are obtainable from the manufacturing process. It is important for the designer as well as the machine shop supervisor to know how machining processes generate the desired surface finish, as they need to decide the manufacturing process for each component.

Table A7.2 shows the possible surface finishes derived through different manufacturing processes

TABLE A7.2 *Relation between Surface Finish and Manufacturing Process*

	Method of Manufacturing	**Values of Surface Finish Ra(μm)**											
		50	**25**	**12.5**	**6.3**	**3.2**	**1.6**	**0.8**	**0.4**	**0.2**	**0.1**	**0.05**	**0.02**
Casting	Sand casting									▨			
	Permanent mould casting						▦	▨	▨				
	Die casting						▨	▨					
Shaping	Forging							▦	▨	▨	▨		
	Rolling					▦	▨	▨					
	Extrusion					▦	▨	▨	▨				
Machining—General	Turning				▦	▦	▦	▨	▨				
	Boring				▦	▦	▦	▨	▨				
	Cylindrical grinding	▦	▦	▦	▦	▨	▨	▨					
	Centreless grinding		▦	▦	▦	▨	▨	▨					
	Surface grinding	▦	▦	▦	▦	▨	▨	▨					
	Internal grinding	▦	▦	▦	▦	▨	▨	▨					
	Planning							▨	▨	▨			
	Broaching				▦	▦	▨	▨					
	Drilling								▦	▨	▨		
	Reaming					▦	▨	▨	▨				
	Face milling						▦	▨	▨	▨			
	Shell milling							▦	▨	▨	▨		
	Lapping	▦	▨	▨	▨								
	Honing	▦	▦	▨	▨								
	Super finishing	▦	▦	▨	▨								
Machining-manual	Filing						▦	▨	▨	▨			
	Hand polishing with emery paper			▦	▦	▨	▨						
	Cutting with hacksaw									▨	▨	▨	
	Cutting with flame											▨	
Surface processing	Grinding with abrasive belt			▨	▨	▨	▨	▨	▨				
	Buffing with cloth	▦	▨										
	Buffing with fibre wheel			▦	▨	▨							
	Shot blasting							▦	▨	▨	▨	▨	
	Tumbling				▨	▨	▨	▨	▨	▨			

Note: ▦ With Difficulty ▨ Normally Achieved

7.2 LIMITS, FITS, AND TOLERANCES

In a mass scale production system, as number of components and number of assemblies are manufactured on mass scale, many a time it is difficult, and even not viable, to manufacture all parts under one roof. In this case, many manufacturers offload jobs to outsiders, sometimes even those situated in other parts of the country. Producing the exact size given by the designer in a drawing is a difficult task, especially in case of machining operations. To overcome this problem, the designer gives a range of limits to the basic size, within which he expects the finished size of the job offloaded by his factory. The assembly designer will finalise the limits depending upon the kind of fit required for the assembly of those components. Dimensional variations occur due to the following reasons:

- Variations in material properties during machining
- Inaccuracies and errors in manufacturing machines
- Human errors and errors at the time of setting of machines.

Assembly procedures, manufacturing constraints, and other such requirements define the permissible variations from the basic size. There are two types of assemblies.

1. Selective assembly
2. Non-selective assembly

Selective assembly is slow, costly, and difficult to maintain, due to the requirement of specific spares. Non-selective assembly is rapid, low cost, and easy to maintain and provides interchangeability between components. When components are mass-produced, unless they are interchangeable, the purpose of mass production is not fulfilled. By interchangeability, we mean that the identical components, manufactured by different personnel under different environments, can be assembled and replaced without any further rectification during the assembly stage, without affecting the functioning of the component when assembled.

Necessity of the limit system If components are to be interchanged, they need to be manufactured to the same size, which is not possible in mass production. Hence, it becomes necessary to permit the operator, a small deviation from the exact size, which is not possible in the case of all components. At the same time, the deviated size should not affect the quality of the assembly. This sort of dimensioning is known as limit dimensioning. Various standard systems of limits and fits are followed by different countries based on the International Organisation for Standardisation (ISO) specifications. The system of limits and fits followed in our country has been stipulated by the Bureau of Indian Standard (BIS). The other systems for limits and fits are the British Standard System (BSS) and the German Industrial Standard (GIS).

7.2.1 The Indian Standard System of Limits and Fits

A machine component has a large number of dimensional details. It is impossible to manufacture an absolute dimension as specified by the design department. Hence, it becomes important to study the different aspects of dimensions and their details, like size, basic size, actual size, limits of size, and so on.

Size It is a number expressed in a particular unit in the measurement of length.

Basic size It is the size based on which the dimensional deviations are given (Fig. A7.4).

Actual size It is the size of the component by actual measurement after it is manufactured. It should lie between the two limits of size if the component is to be accepted.

Limits of size These are the extreme permissible sizes within which the operator is expected to make the component, that is, maximum and minimum limits (Fig. A7.5).

Maximum limit of size It is the greater of the two limit sizes (refer Fig. A7.5(a) and Table A7.3).

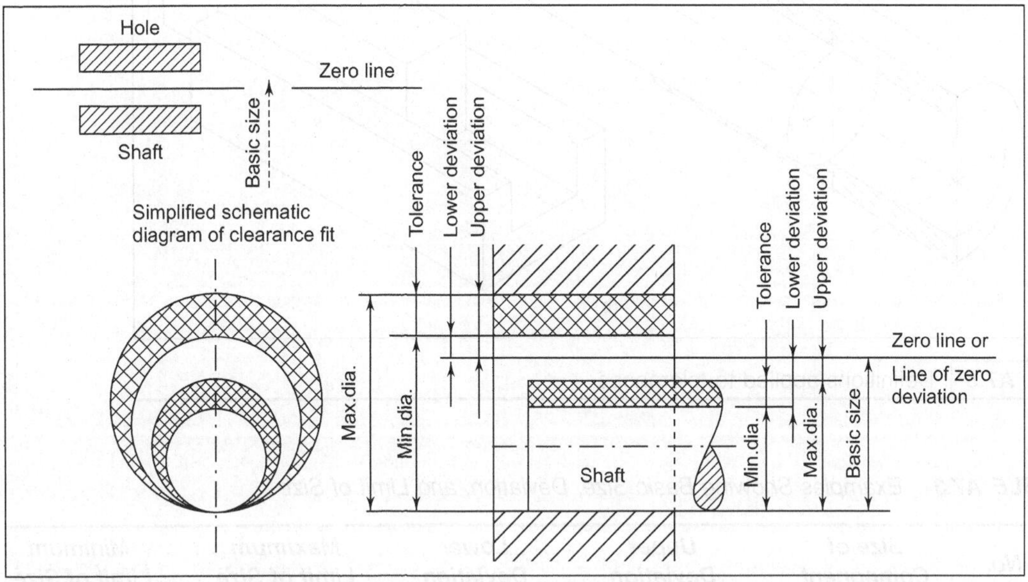

Fig. A7.4 Definitions applied to tolerances

Minimum limit of size It is the smaller of the two limits of size [refer Fig. A7.5(a) and Table A7.3].

Hole In the BIS system of limits and fits all the internal features of a component, including those which are not cylindrical, are designated as holes [Fig. A7.5(b)].

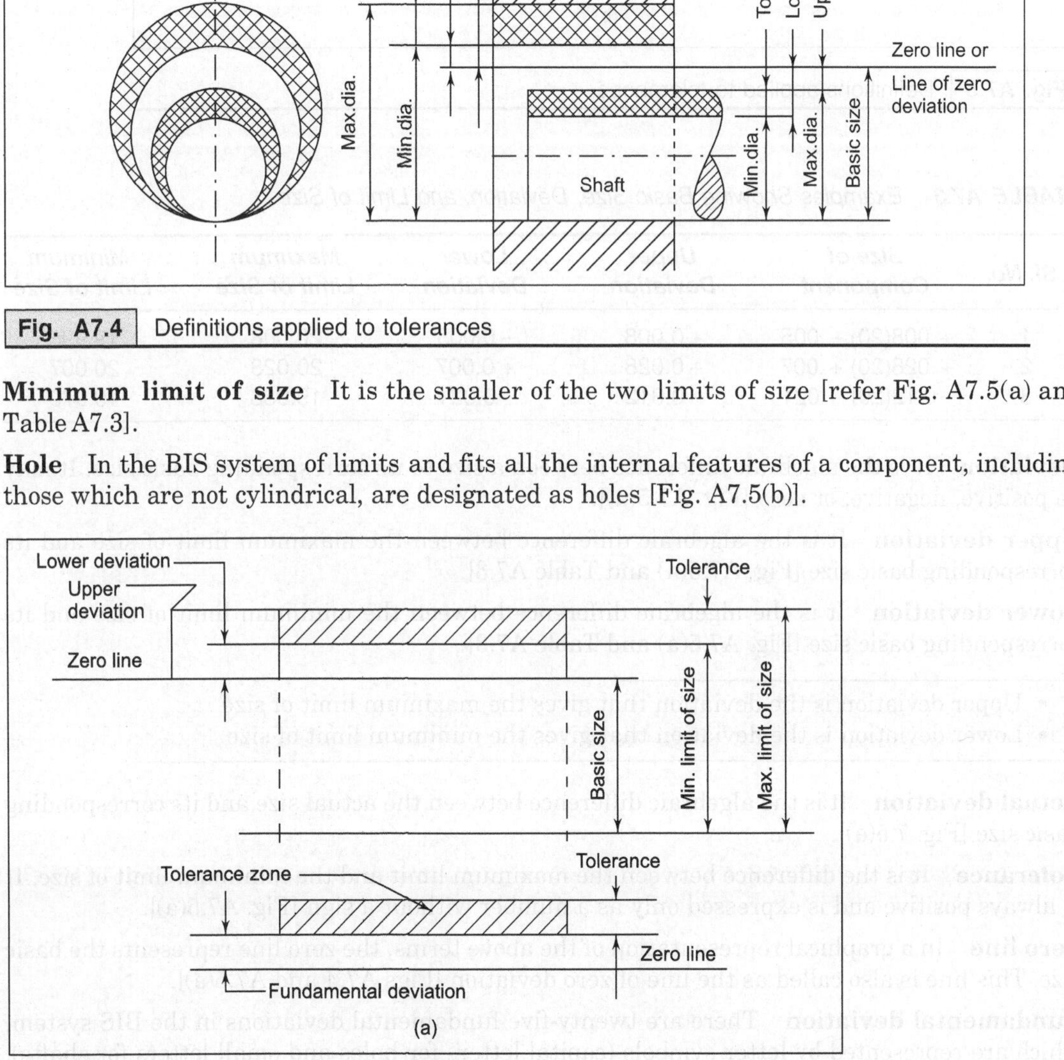

Fig. A7.5 Definitions applied to tolerances

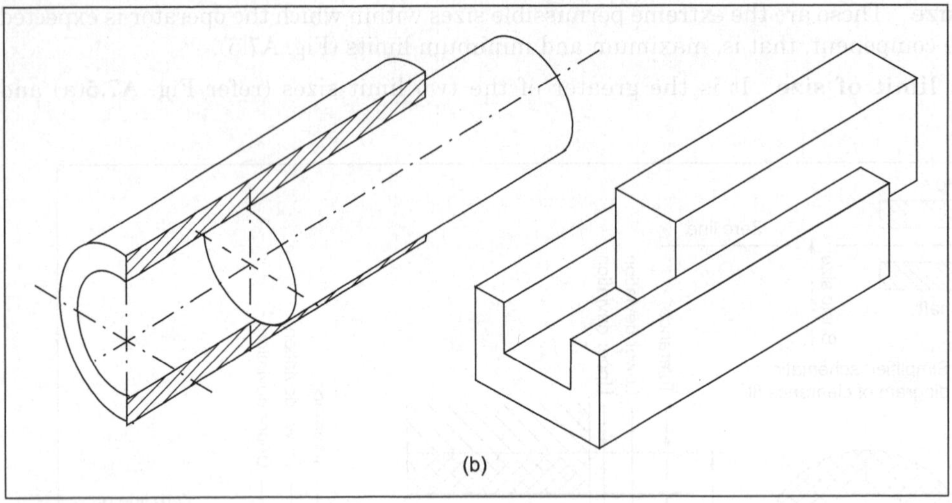

Fig. A7.5 Definitions applied to tolerances

TABLE A7.3 Examples Showing Basic Size, Deviation, and Limit of Size

Sl. No.	Size of Component	Upper Deviation	Lower Deviation	Maximum Limit of Size	Minimum Limit of Size
1.	+.008(20) −.005	+0.008	−0.005	20.008	19.995
2.	+.028(20) +.007	+0.028	+0.007	20.028	20.007
3.	−0.12(20) −.021	−0.012	−0.021	19.988	19.979

Deviation It is the algebraic difference between a size to its corresponding basic size. It may be positive, negative, or zero [Fig. A7.5(a)].

Upper deviation It is the algebraic difference between the maximum limit of size and its corresponding basic size [Fig. A7.5(a) and Table A7.3].

Lower deviation It is the algebraic difference between the minimum limit of size and its corresponding basic size [Fig. A7.5(a) and Table A7.3].

- Upper deviation is the deviation that gives the maximum limit of size.
- Lower deviation is the deviation that gives the minimum limit of size.

Actual deviation It is the algebraic difference between the actual size and its corresponding basic size [Fig. 7.5(a)].

Tolerance It is the difference between the maximum limit and the minimum limit of size. It is always positive and is expressed only as a number without a sign [Fig. A7.5(a)].

Zero line In a graphical representation of the above terms, the zero line represents the basic size. This line is also called as the line of zero deviation [Figs A7.4 and A7.5(a)].

Fundamental deviation There are twenty-five fundamental deviations in the BIS system, which are represented by letter symbols (capital letters for holes and small letters for shafts), that is, for holes, ABCD … Z excluding I, L, O, Q, and W (Fig. A7.6).

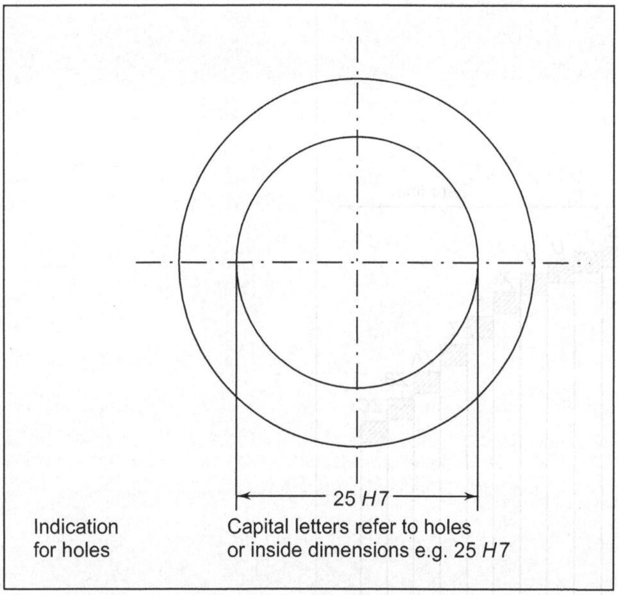

Fig. A7.6 Example of fundamental deviation of holes

In addition to the above, four sets of letters JS, ZA, ZB, and ZC are included. For shafts, the same 25 letter symbols, but in small letters, are used (Fig. A7.7).

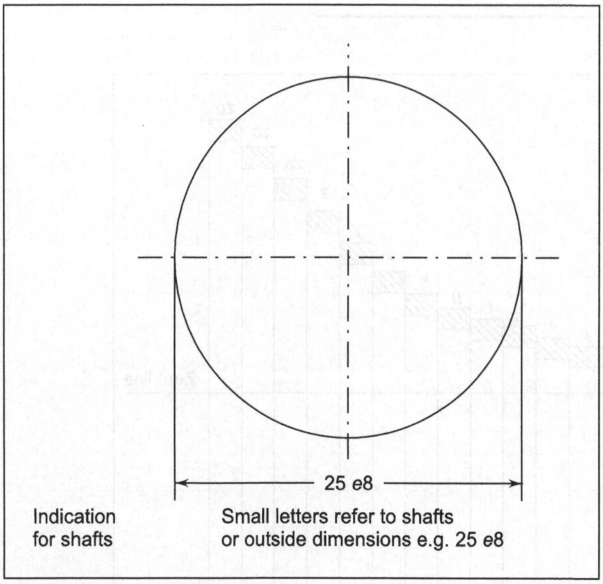

Fig. A7.7 Example showing fundamental deviation for shafts

The position of the tolerance zone with respect to the zero line is shown in Figs A7.8 and A7.9.

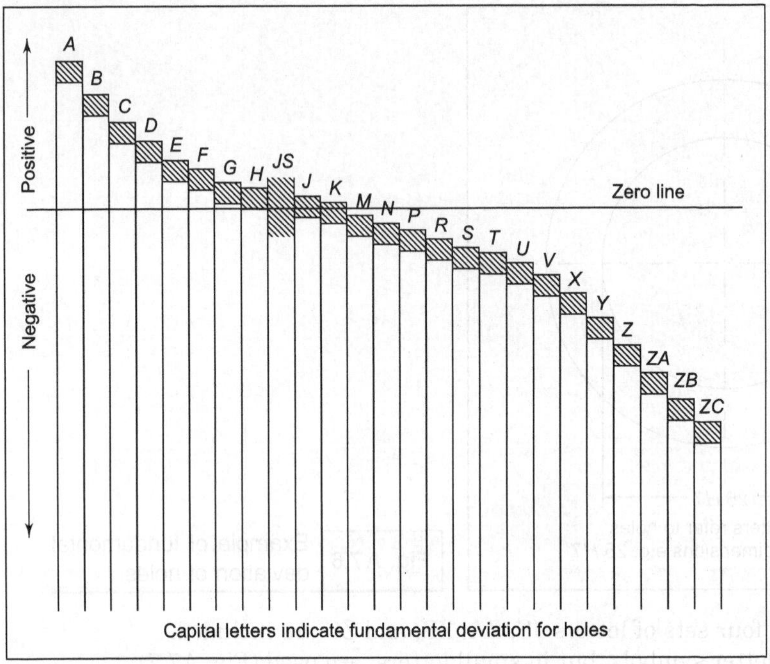

Fig. A7.8 Tolerance zone

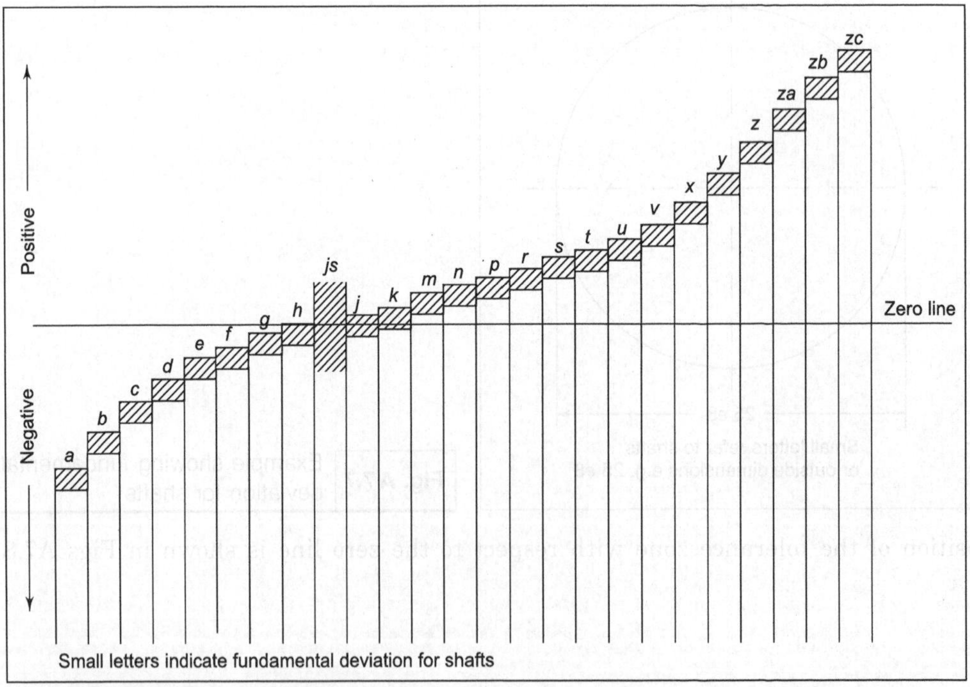

Fig. A7.9 Tolerance zone

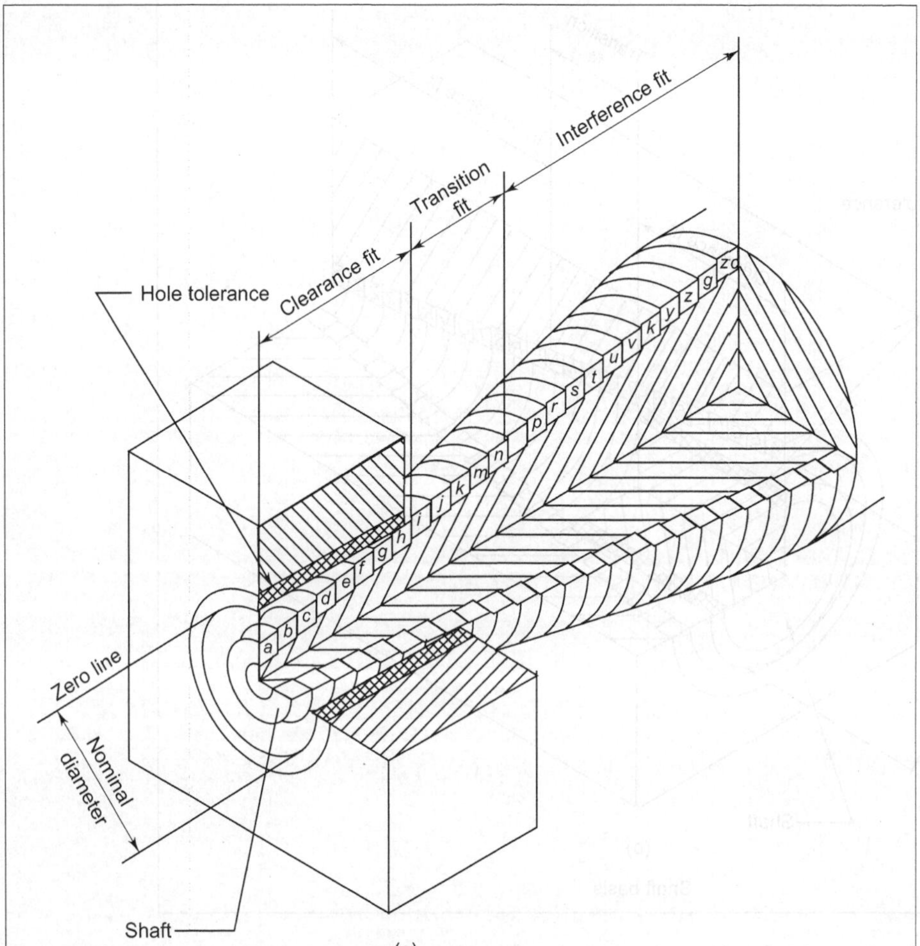
Interference fit
Transition fit
Clearance fit
Hole tolerance
a
b
c
d
e
f
g
h
i
j
k
m
n
p
r
s
t
u
v
k
y
z
g
zc
Zero line
Nominal diameter
Shaft

(a)

Hole basis

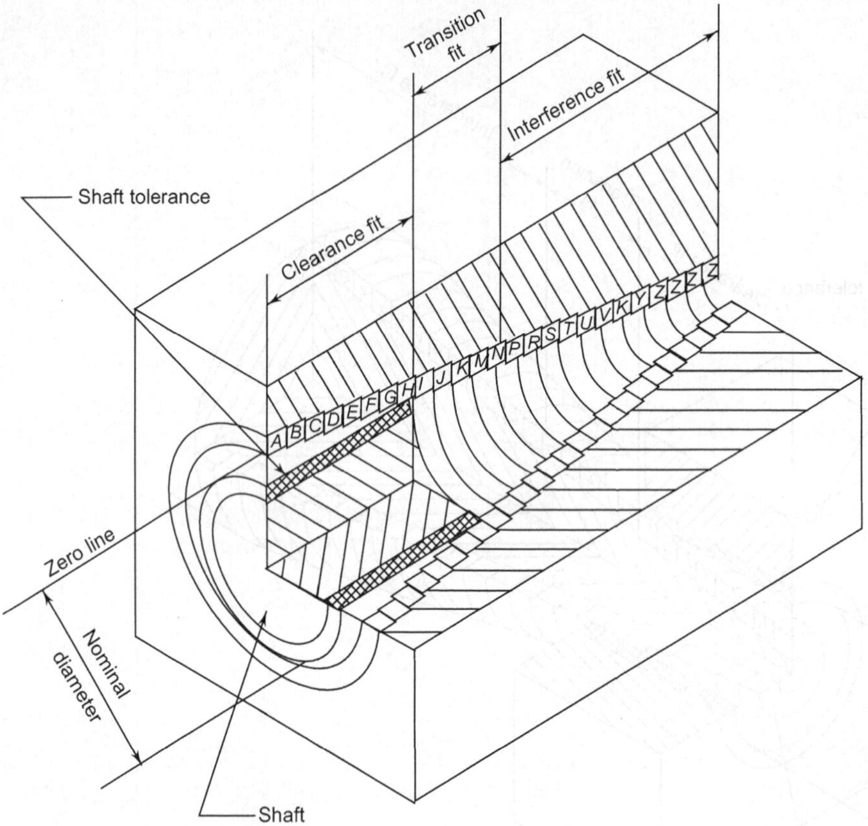

(b)

Shaft basis

Fig. A7.10 Fundamental deviations for achieving the different classes of fits

Fundamental tolerance This is also known as grade of tolerance. In the Indian Standard System, there are 18 grades of tolerances represented by number symbols, both for hole and shaft, such as IT01, IT0, IT1, upto IT16 (Fig. A7.11). A high number gives a large tolerance zone.

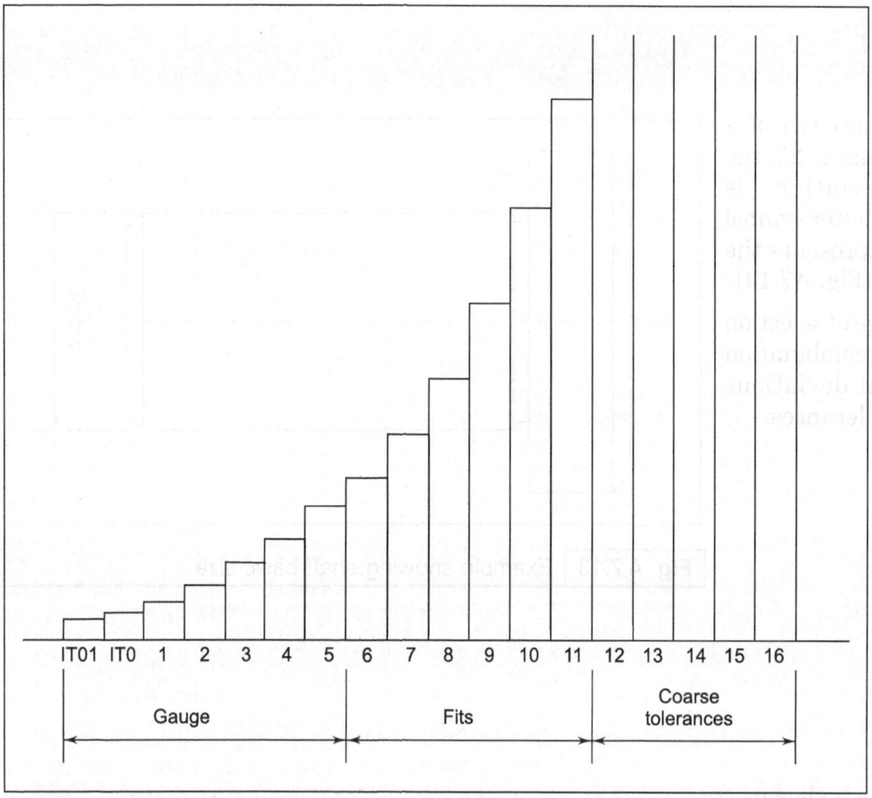

Fig. A7.11 Tolerance zone

The grade of tolerance refers to the accuracy of manufacture. In a standard chart, the upper and lower deviations for each combination of fundamental tolerance are indicated for sizes ranging up to 500 mm (refer to IS 919).

Toleranced size This includes the basic size, the fundamental deviation, and the grade of tolerance.

Example A7.1

If 25 *H*7 is the tolerance size of a hole whose basic size is 25, the fundamental deviation is represented by the letter symbol *H* and the grade of tolerance is represented by the number symbol 7 (Fig. A7.12).

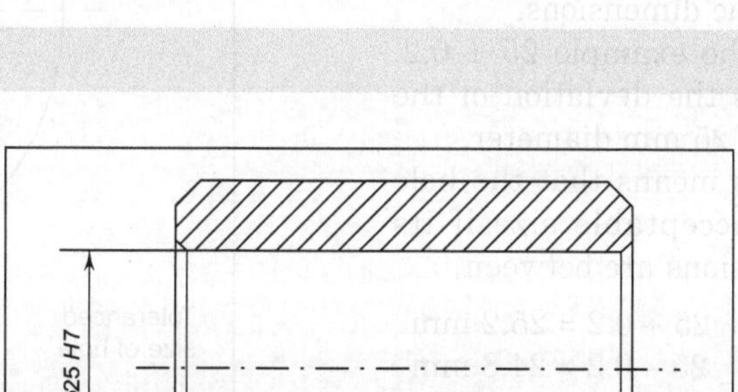

Fig. A7.12 Example showing hole basic size

Example A7.2

If 25 *e*8 is the tolerance size of a shaft whose basic size is 25, the fundamental deviation is represented by the letter symbol *e* and number 8 represents the grade of tolerance (Fig. A7.13).

A very wide range of selection can be made by the combination of 25 fundamental deviations and 18 grades of tolerances.

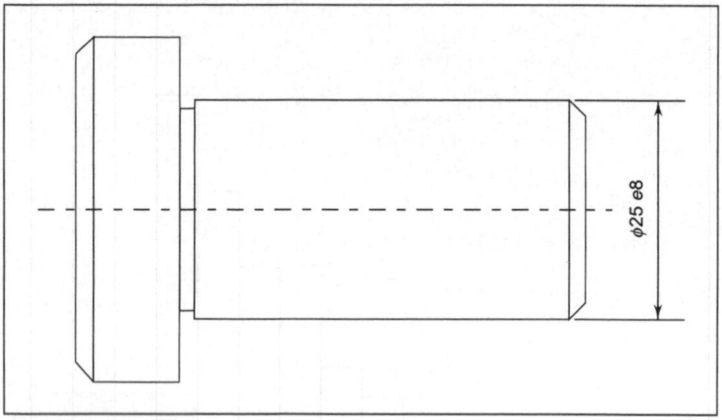

Fig. A7.13 Example showing shaft basic size

Example A7.3

In Fig. A7.14, a hole is shown as 25 ± 0.2, which means that 25 mm is the basic dimension and ±0.2 is the deviation.

As pointed, out earlier, the permissible variation from the basic dimension is called 'deviation'. The deviation is mostly given on the drawing with the dimensions.

In the example 25 ± 0.2, ±0.2 is the deviation of the hole of 25 mm diameter.

This means that the hole is of acceptable size if its dimensions are between

25 + 0.2 = 25.2 mm

or 25 – 0.2 = 24.8 mm

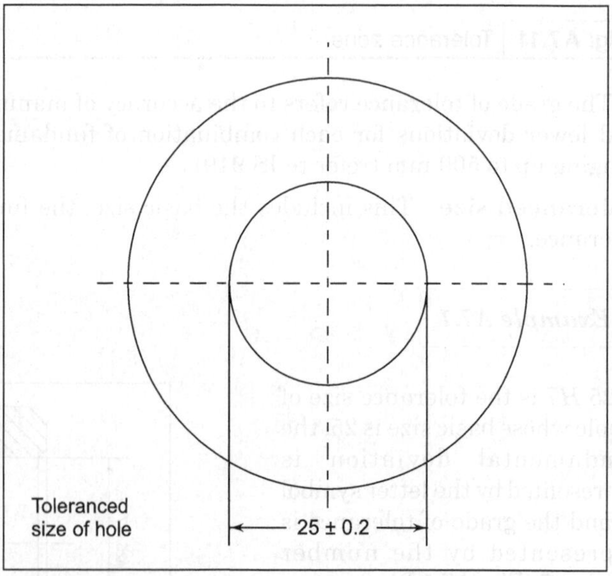

Fig. A7.14 Example based on hole size

25.2 mm is known as the maximum limit (Fig. A7.15) and 24.8 mm is known as the minimum limit (Fig. A7.16).

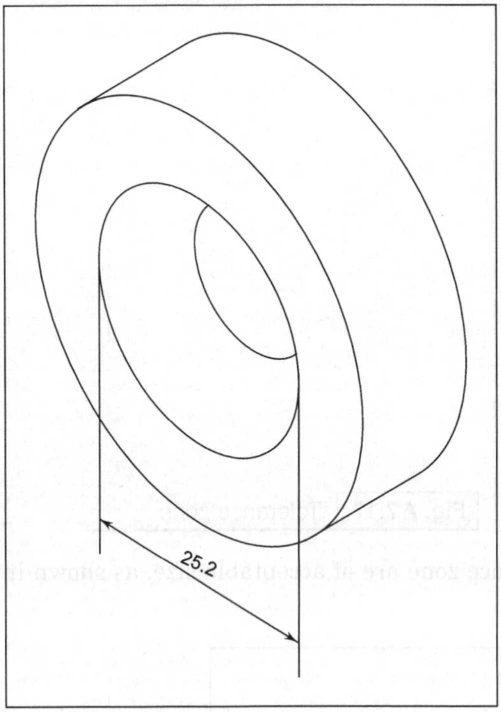

Fig. A7.15 Maximum limit of hole

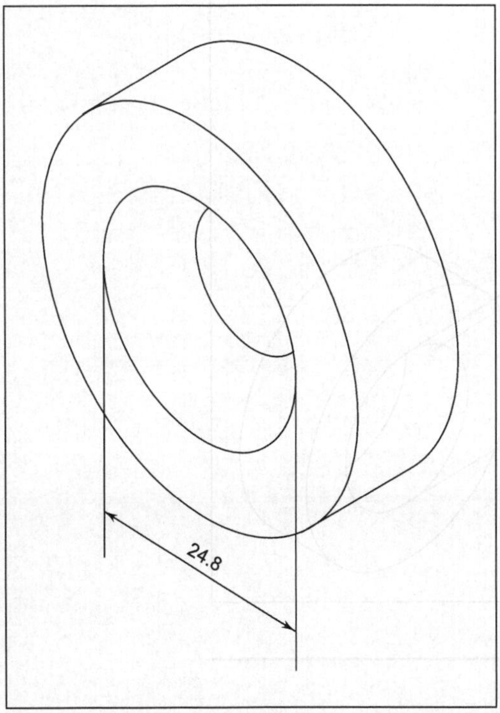

Fig. A7.16 Maximum limit of hole

The difference between the maximum and minimum limits is the 'tolerance'. Here, tolerance is 0.4 mm (Fig. A7.17).

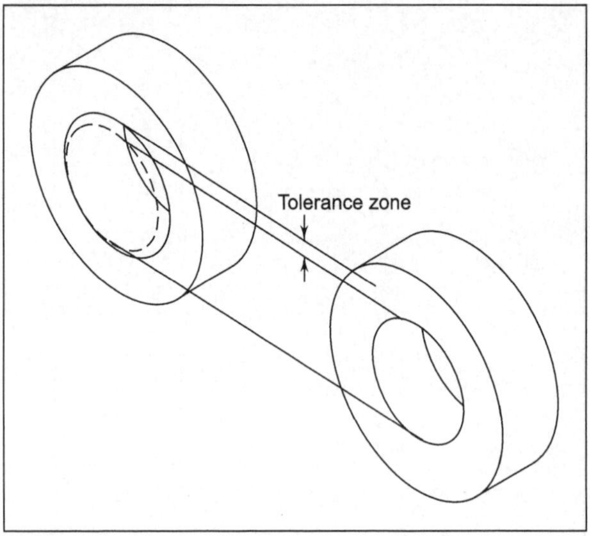

Fig. A7.17 Tolerance zone

All the dimensions of the hole within the tolerance zone are of acceptable size, as shown in Fig. A7.18.

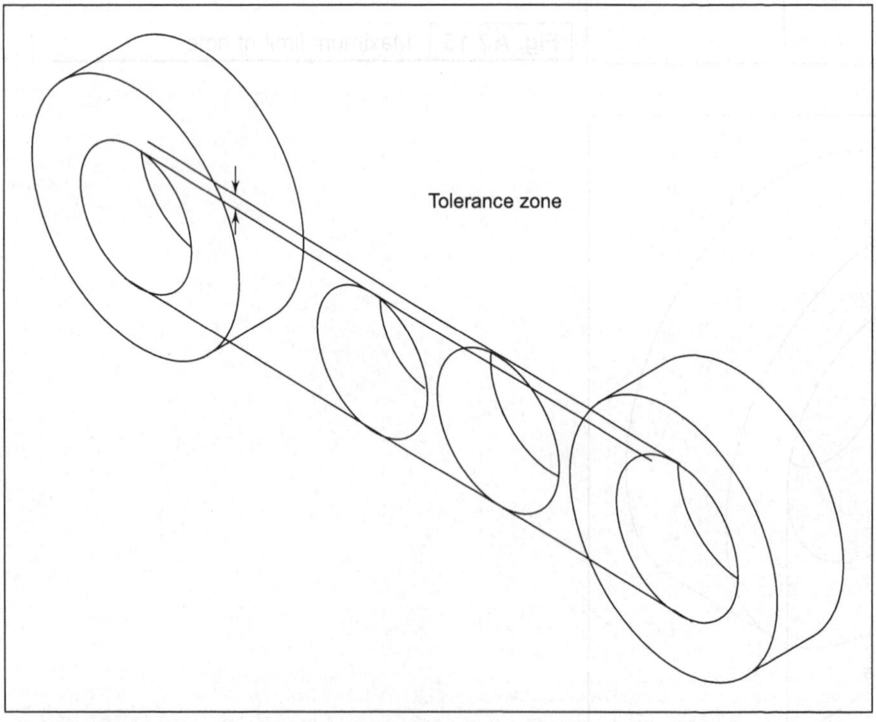

Fig. A7.18 Example of acceptable size

As per IS 696, while dimensioning the components as a drawing convention, deviations are expressed as tolerances.

7.2.2 Fits and their Classification, as Per the Indian Standard

There are a wide variety of combinations of tolerances and deviations, it is not practical to accommodate so many combinations, as they may not all be required. Hence, brief information has been given on fits and types of fits (Clearance, Interference, and Transition fit).

Fit It is the relationship that exists between two mating parts, a hole and a shaft, with respect to their dimensional differences before assembly.

Expression of a fit A fit is expressed by writing the basic size of the fit first, (the basic size that is common to both the hole and the shaft) followed by the symbol for the hole, and the symbol for the shaft. For example,

30 *H*7 – *g*6 or 30 *H*7 – *g*6 or 30 *H*7/*g*6

Clearance In a fit, the clearance is the difference between the size of the hole and the size of the shaft, which is always positive.

Clearance fit This fit always provides clearance. Here the tolerance zone of the hole will be above the tolerance zone of the shaft (Fig. A7.19).

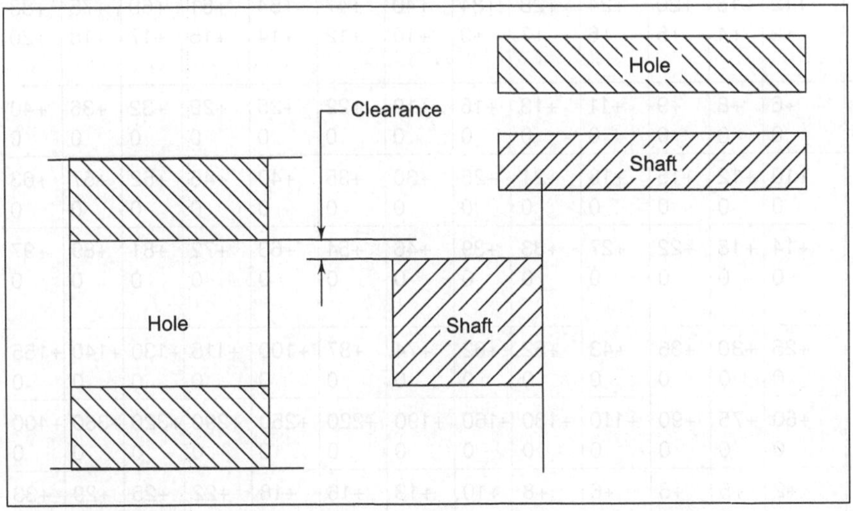

Fig. A 7.19 Clearance fit

TABLE A7.4 *Diameter Limits of Fits and Clearances*

System Basic Shaft		Symbol and Grade	Diameter Limits in mm												
			Over 1	3	6	10	18	30	50	80	120	180	250	315	400
			Upto 3	6	10	18	30	50	80	120	180	250	315	400	500
Running Fit Holes	With large minimum clearance	D10	+60 +20	+78 +30	+98 +40	+120 +50	+149 +65	+180 +80	+220 +100	+260 +120	+305 +145	+355 +170	+400 +190	+440 +210	+480 +230
Running Fit Holes	With medium minimum clearance	E7	+24 +14	+32 +20	+40 +25	+50 +32	+61 +40	+75 +50	+90 +60	+107 +72	+125 +85	+146 +100	+162 +110	+182 +125	+198 +135
Running Fit Holes	With medium minimum clearance	E8	+28 +14	+38 +20	+47 +25	+59 +32	+73 +40	+89 +50	+106 +60	+126 +72	+148 +85	+172 +100	+191 +110	+214 +125	+232 +135
Running Fit Holes	With small minimum clearance	F7	+16 +6	+22 +10	+28 +13	+34 +16	+41 +20	+50 +25	+60 +30	+71 +36	+83 +43	+96 +50	+108 +56	+119 +62	+131 +68
Running Fit Holes	With small minimum clearance	F8	+20 +6	+28 +10	+35 +13	+43 +16	+53 +20	+64 +25	+76 +30	+90 +36	+106 +43	+122 +50	+137 +56	+151 +62	+165 +68
Running Fit Holes	With smallest minimum clearance	G7	+12 +2	+16 +4	+20 +5	+24 +6	+28 +7	+34 +9	+40 +10	+47 +12	+54 +14	+61 +15	+69 +17	+75 +18	+83 +20
Transition-slide Fit	Basic hole size	H6	+6 0	+8 0	+9 0	+11 0	+13 0	+16 0	+19 0	+22 0	+25 0	+29 0	+32 0	+36 0	+40 0
Transition-slide Fit	Basic hole size	H7	+10 0	+12 0	+15 0	+18 0	+21 0	+25 0	+30 0	+35 0	+40 0	+46 0	+52 0	+57 0	+63 0
Transition-slide Fit	Basic hole size	H8	+14 0	+18 0	+22 0	+27 0	+33 0	+39 0	+46 0	+54 0	+63 0	+72 0	+81 0	+89 0	+97 0
Transition-slide Fit	Basic hole size	H9	+25 0	+30 0	+36 0	+43 0	+52 0	+62 0	+74 0	+87 0	+100 0	+115 0	+130 0	+140 0	+155 0
Transition-slide Fit	Basic hole size	H11	+60 0	+75 0	+90 0	+110 0	+130 0	+160 0	+190 0	+220 0	+250 0	+290 0	+320 0	+360 0	+400 0
Transition	Push Fit	J6	+2 −4	+5 −3	+5 −4	+6 −5	+8 −5	+10 −6	+13 −6	+16 −6	+18 −7	+22 −7	+25 −7	+29 −7	+33 −7
Transition	Push Fit	J7	+4 −6	+6 −6	+8 −7	+10 −8	+12 −9	+14 −11	+18 −12	+22 −13	+26 −14	+30 −16	+36 −16	+39 −18	+43 −20

TABLE A7.5 Diameter Limits for Interference and Fits

System Basic Shaft		Symbol and Grade	Diameter Limits in mm: Over 1 Upto 3	3–6	6–10	10–18	18–30	30–50	50–80	80–120	120–180	180–250	250–315	315–400	400–500
Transition	Warning Fit	K6	0 −6	+2 −6	+2 −7	+2 −9	+2 −11	+3 −13	+4 −15	+4 −18	+4 −21	+5 −24	+5 −27	+7 −29	+8 −32
		K7	0 −10	+3 −9	+5 −10	+6 −12	+6 −15	+7 −18	+9 −21	+10 −25	+12 −28	+13 −33	+16 −36	+17 −40	+18 −45
	Secure against turning	M6	−2 −8	−1 −9	−3 −12	−4 −15	−4 −17	−4 −20	−5 −24	−6 −28	−8 −33	−8 −37	−9 −41	−10 −46	−10 −50
Transition		M7	2 −12	0 −12	0 −15	0 −18	0 −21	0 −25	0 −30	0 −35	0 −40	0 −46	0 −52	0 −57	0 −63
	Tight Fit	N6	−4 −10	−5 −13	−7 −16	−9 −20	−11 −24	−12 −28	−14 −33	−16 −38	−20 −45	−22 −51	−25 −57	−26 −62	−27 −67
		N7	−4 −14	−4 −16	−4 −19	−5 −23	−7 −28	−8 −33	−9 −39	−10 −45	−12 −52	−14 −60	−14 −66	−16 −73	−17 −80
Interference	Fit Holes	P7	−6 −16	−8 −20	−9 −24	−11 −29	−14 −35	−17 −42	−21 −51	−24 −59	−28 −68	−33 −79	−36 −88	−41 −98	−45 −108
		P9	−6 −31	−12 −42	−15 −51	−18 −61	−22 −74	−26 −88	−32 −108	−37 −124	−43 −143	−50 −165	−56 −186	−62 −202	−68 −223

System Basic Hole		Symbol and Grade	Diameter Limits in mm: Over 1 Upto 3	3–6	6–10	10–18	18–30	30–50	50–80	80–120	120–180	180–250	250–315	315–400	400–500
Running Fit Shafts	With large minimum clearance	d11	−20 −80	−30 −105	−40 −130	−50 −160	−65 −195	−80 −240	−100 −290	−120 −340	−145 −395	−170 −460	−190 −510	−210 −570	−230 −630
	With medium minimum clearance	e7	−14 −24	−20 −32	−25 −40	−32 −50	−40 −61	−50 −75	−60 −90	−72 −107	−85 −125	−100 −146	−110 −162	−125 −182	−135 −198
		e8	−14 −28	−20 −38	−25 −47	−32 −59	−40 −73	−50 −89	−60 −106	−72 −126	−85 −148	−100 −172	−110 −191	−125 −214	−135 −232
	With small minimum clearance	f7	−6 −16	−10 −22	−13 −28	−16 −34	−20 −41	−25 −50	−30 −60	−36 −71	−43 −83	−50 −96	−56 −108	−62 −119	−68 −131
		f8	−6 −20	−10 −28	−13 −35	−16 −43	−20 −53	−25 −64	−30 −76	−36 −90	−43 −106	−50 −122	−56 −137	−62 −151	−68 −165
	With smallest minimum clearance	g5	−2 −6	−4 −9	−5 −11	−6 −14	−7 −16	−9 −20	−10 −23	−12 −27	−14 −32	−15 −35	−17 −40	−18 −42	−20 −47
		g6	−2 −8	−4 −12	−5 −14	−6 −17	−7 −20	−9 −25	−10 −29	−12 −34	−14 −39	−15 −44	−17 −49	−18 −54	−20 −60

TABLE A 7.6 *Standard Charts for Tolerances*

System Basic Hole	Symbol and Grade	Diameter Limits in mm: Over 1 Upto 3	3–6	6–10	10–18	18–30	30–50	50–80	80–120	120–180	180–250	250–315	315–400	400–500
Transition-slide Fit — Basic shaft size	h5	0	0	0	0	0	0	0	0	0	0	0	0	0
		−4	−5	−6	−8	−9	−11	−13	−15	−18	−20	−23	−25	−27
	h6	0	0	0	0	0	0	0	0	0	0	0	0	0
		−6	−8	−9	−11	−13	−16	−19	−22	−25	−29	−32	−36	−40
	h8	0	0	0	0	0	0	0	0	0	0	0	0	0
		−14	−18	−22	−27	−33	−39	−46	−54	−63	−72	−81	−89	−97
	h9	0	0	0	0	0	0	0	0	0	0	0	0	0
		−25	−30	−36	−43	−52	−62	−74	−87	−100	−115	−130	−140	−155
	h11	0	0	0	0	0	0	0	0	0	0	0	0	0
		−60	−75	−90	−110	−130	−160	−190	−220	−250	−290	−320	−360	−400
Transition Push Fit	j5	+2	+3	+4	+5	+5	+6	+6	+6	+7	+7	+7	+7	+7
		−2	−2	−2	−3	−4	−5	−7	−9	−11	−13	−16	−18	−20
	j6	+4	+6	+7	+8	+9	+11	+12	+13	+14	+16	+16	+18	+20
		−2	−2	−2	−3	−4	−5	−7	−9	−11	−13	−16	−18	−20
Transition Wring Fit	k5	+4	+6	+7	+9	+11	+13	+15	+18	+21	+24	+27	+29	+32
		0	+1	+1	+1	+2	+2	+2	+3	+3	+4	+4	+4	+5
	k6	+6	+9	+10	+12	+15	+18	+21	+25	+28	+33	+36	+40	+45
		0	+1	+1	+1	+2	+2	+2	+3	+3	+4	+4	+4	+5
Secure against turning	m5	+6	+9	+12	+15	+17	+20	+24	+28	+33	+37	+43	+46	+50
		+2	+4	+6	+7	+8	+9	+11	+13	+15	+17	+20	+21	+23
	m6	+8	+12	+15	+18	+21	+25	+30	+35	+40	+46	+52	+57	+63
		+2	+4	+6	+7	+8	+9	+11	+13	+15	+17	+20	+21	+23
Transition Tight Fit	n5	+8	+13	+16	+20	+24	+28	+33	+38	+45	+51	+57	+62	+67
		+4	+8	+10	+12	+15	+17	+20	+23	+27	+31	+34	+37	+40
	n6	+10	+16	+19	+23	+28	+33	+39	+45	+52	+60	+66	+73	+80
		+4	+8	+10	+12	+15	+17	+20	+23	+27	+31	+34	+37	+40
Interference Fit	p6	+12	+20	+24	+29	+35	+42	+51	+59	+68	+79	+88	+98	+108
		+6	+12	+15	+18	+22	+26	+32	+37	+43	+50	+56	+62	+68
	r6	+16	+23	+28	+34	+41	+50	–	–	–	–	–	–	–
		+10	+15	+19	+23	+28	+34	–	–	–	–	–	–	–
	s6	+20	+27	+32	+39	+48	+59	–	–	–	–	–	–	–
		+14	+19	+23	+28	+35	+43	–	–	–	–	–	–	–

* For complete values of tolerances refer relevant standards, which are available with the Indian Standards Institution.

(All Tolerances in microns and Shaft Diameters in mm)

Example A7.4

With 20*H*7/*g*6 being the given fit, we can find out that the deviation for a hole 20*H* in the standard table of diameter limits for basic shaft systems is + 21

These numbers indicate the deviations in microns, where 1 micrometer = 0.001 mm

The limits of the hole are 20 + 0.021 mm and 20 + 0 = 20.000 mm (Fig. A7.20).

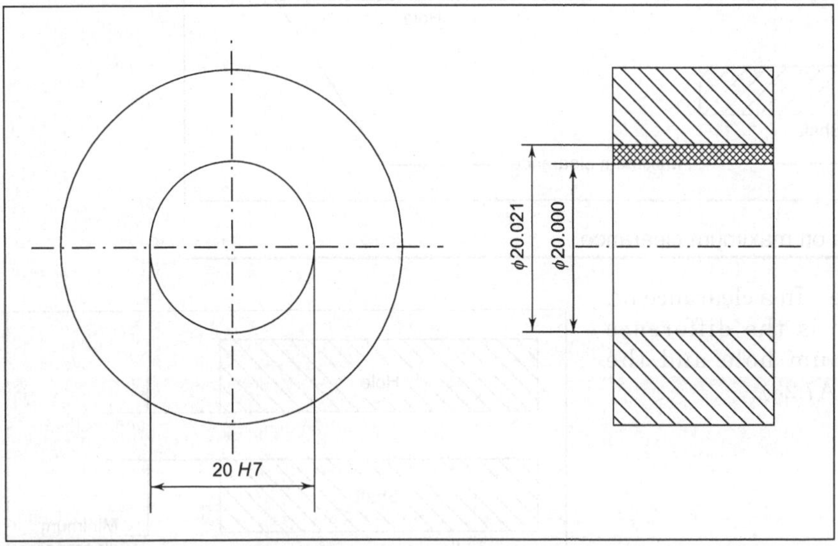

Fig. A7.20 Example showing deviation on hole basic size

For a shaft 20*g*6, we find in the standard table of diameter limits the deviation for basic hole system is –7 and – 20.

So the limits of the shaft are

20 – 0.007 = 19.993 mm

and 20 – 0.020 = 19.980 mm (Fig. A7.21).

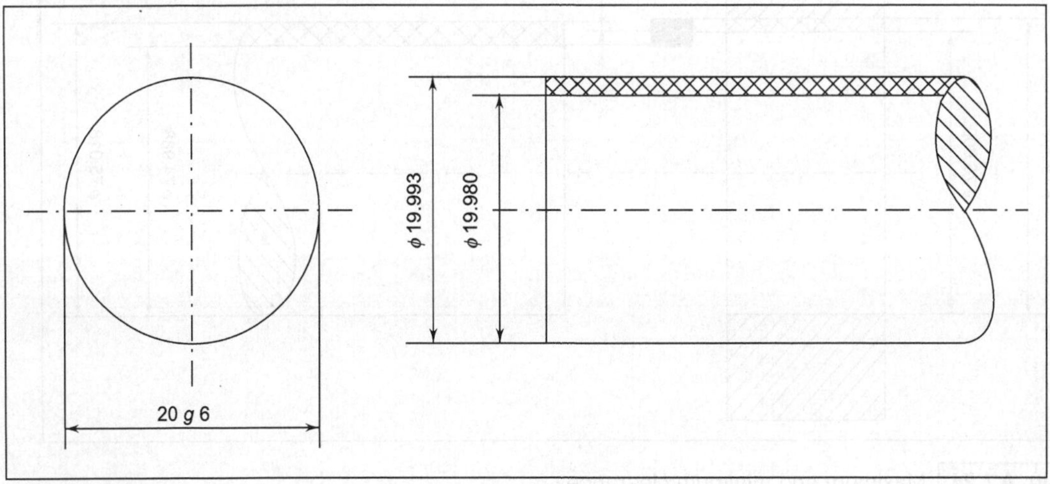

Fig. A7.21 Example showing deviation on shaft basic size

Maximum clearance In a clearance fit or transition fit, the maximum clearance is the difference between the maximum hole and minimum shaft (Fig. A7.22).

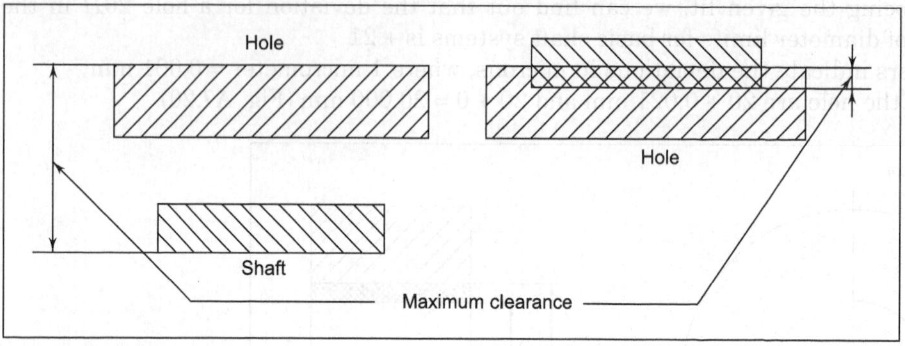

Fig. A7.22 Example on maximum clearance

Minimum clearance In a clearance fit, minimum clearance is the difference between the minimum hole and the maximum shaft (Fig. A7.23).

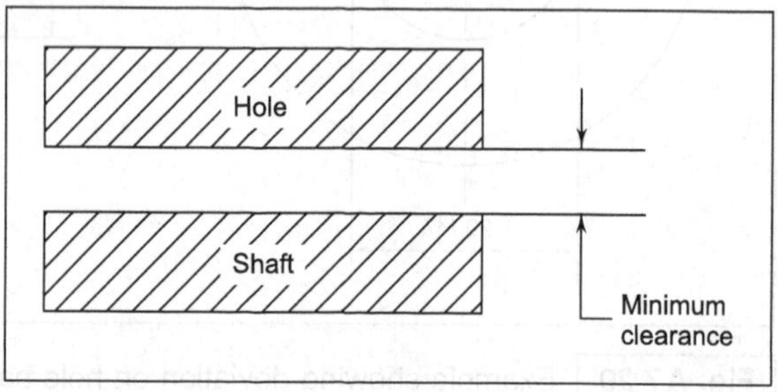

Fig. A7.23 Example of minimum clearance

The minimum clearance is 20.000 – 19.993 = 0.007 mm (Fig. 7.24).
The maximum clearance is 20.021 – 19.980 = 0.041 mm (Fig. 7.25).

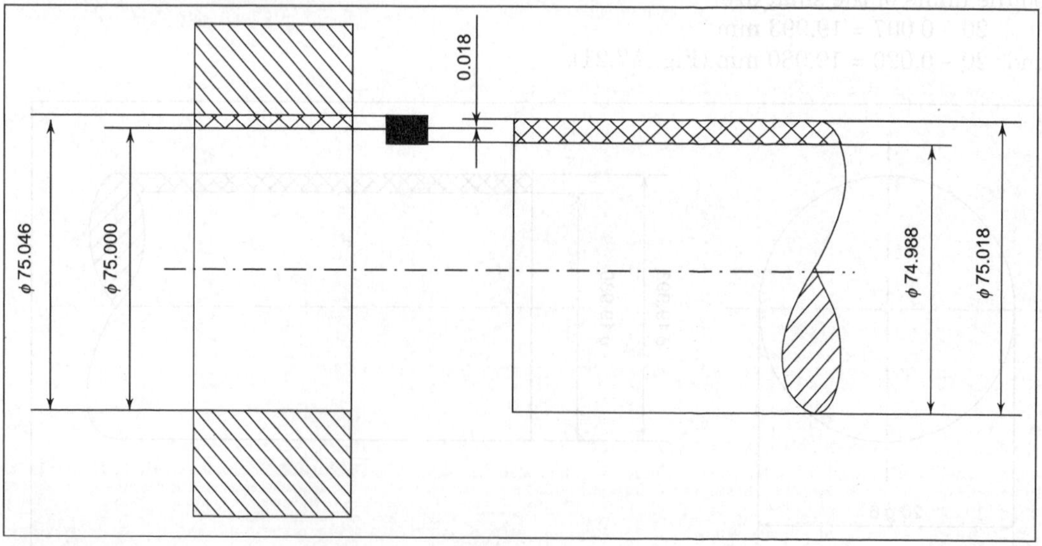

Fig. A7.24 Maximum and minimum clearance

There is always a clearance between the hole and the shaft. This is the clearance fit.

Interference It is the difference between the size of the hole and the shaft before assembly, and is negative. In this case, the shaft is always larger than the hole size.

Interference fit It is a fit that always provides interference. Here, the tolerance zone of the hole will be below the tolerance zone of the shaft (Fig. A7.25).

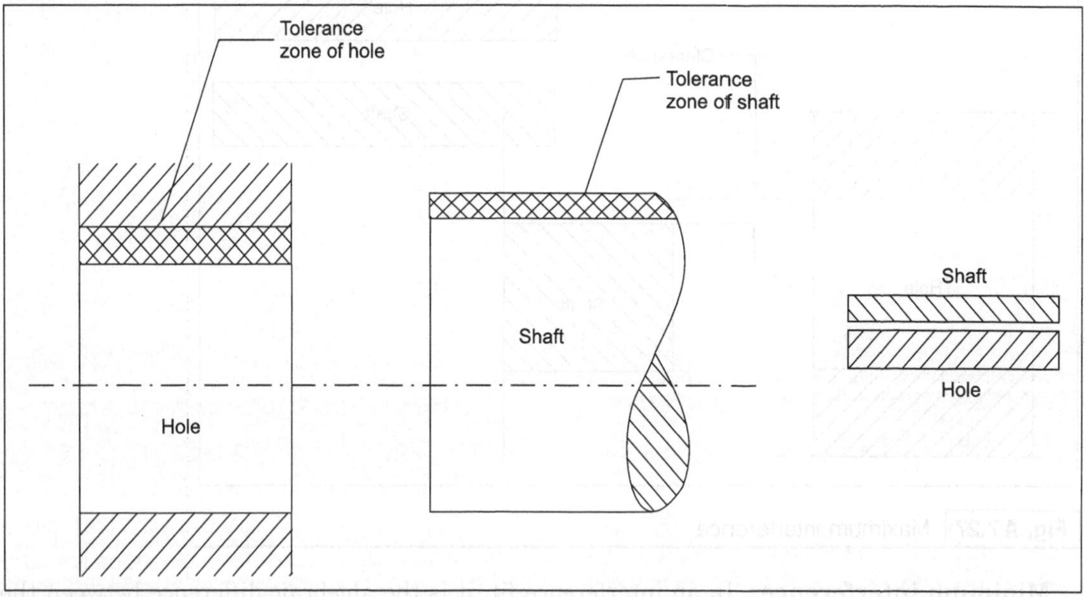

Fig. A7.25 Interference fit

Example A7.5

Fit 25 *H7*/*p6* (Fig. A7.26).

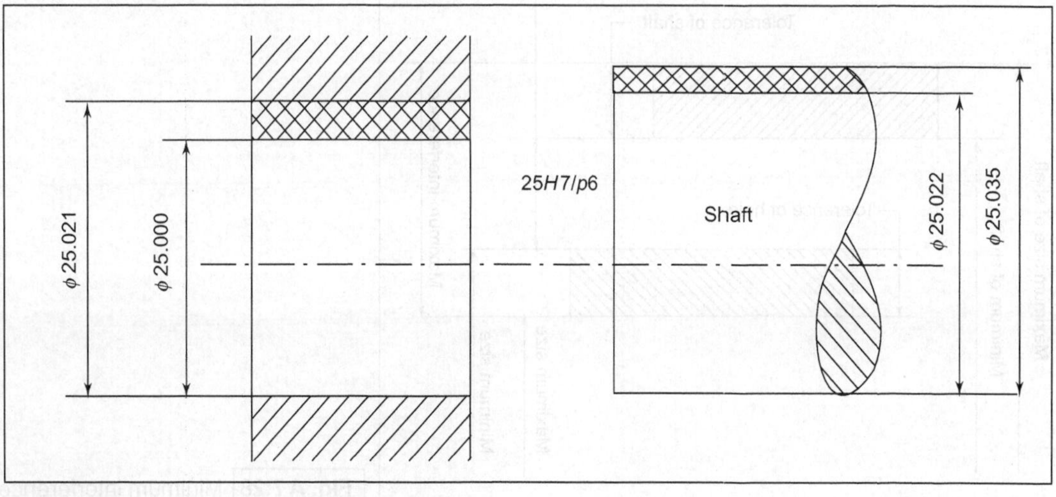

Fig. A7.26 Example on interference fit

The limits of the hole are 25.000 and 25.021 mm and the limits of the shaft are 25.022 and 25.035 mm. The shaft is always bigger than the hole. This is an interference fit.

Maximum interference In an interference fit or transition fit, it is the algebraic difference between the minimum hole and the maximum shaft (Fig. A7.27).

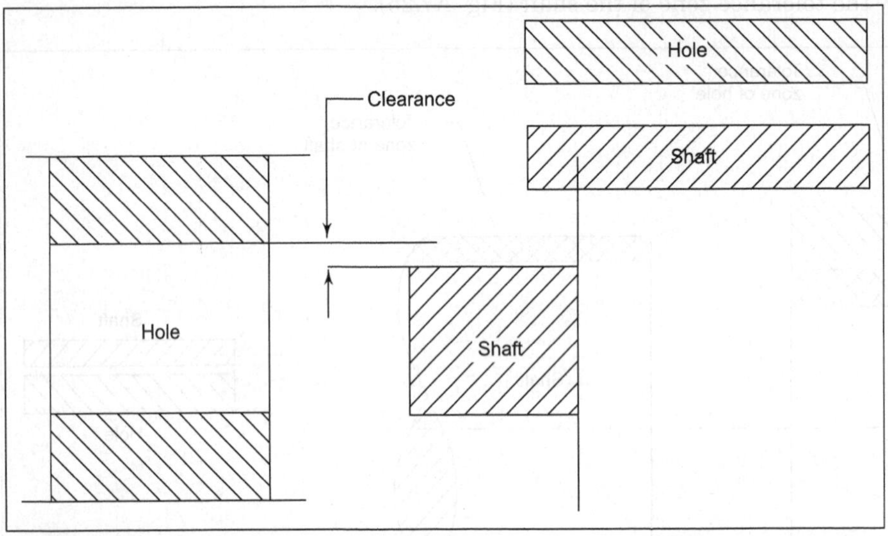

Fig. A7.27 Maximum interference

Minimum interference In an interference fit, it is the algebraic difference between the maximum hole and the minimum shaft (Fig. A7.28).

In Example 7.5,

The maximum interference is equal to 25.035 – 25.000 = 0.035

The minimum interference is equal to 25.022 – 25.021 = 0.001

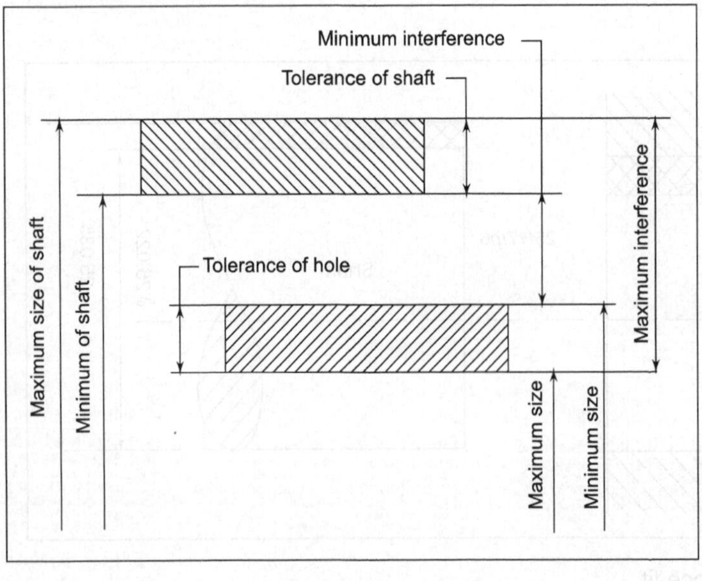

Fig. A7.28 Minimum interference

Transition fit It is a fit that may sometimes provide clearance and sometimes interference. When this class of fit is represented graphically, the tolerance zones of the hole and shaft will overlap each other (Fig. A7.29).

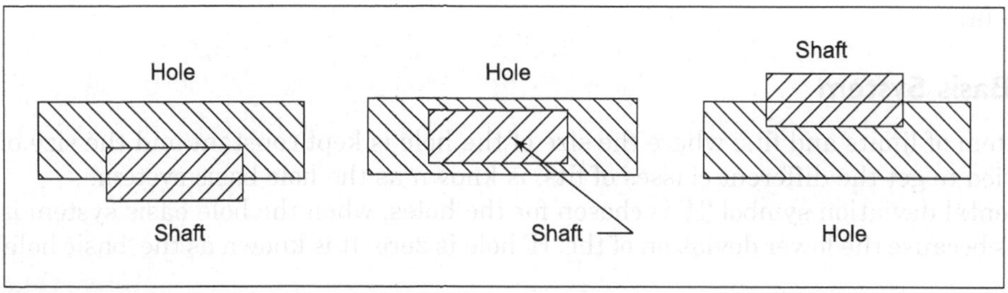

Fig. A7.29 Transition fit

Example A7.6

Fit 75 *H*8/*j*7 (Fig. 7.30).

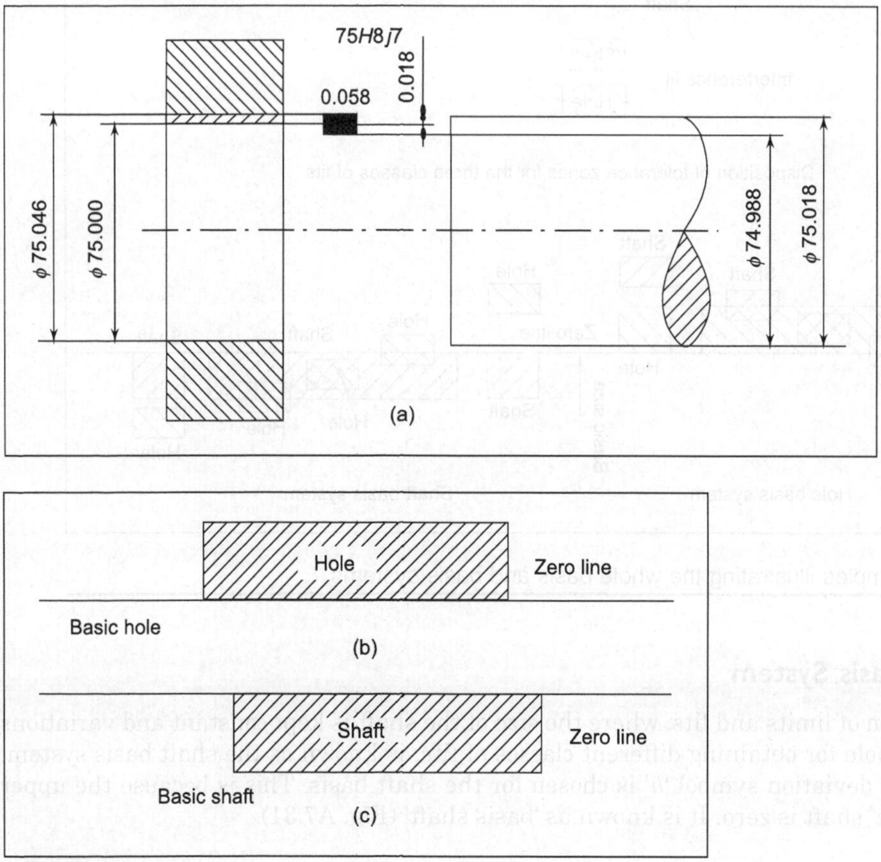

Fig. A7.30 Example of transition fit

The limits of the hole are 75.000 and 75.046 mm and those of the shaft are 75.018 and 74.988 mm.

If the hole is 75.000 and the shaft 75.018 mm, the shaft is 0.018 mm bigger than the hole.

This results in interference. This is a transition fit, because it can result in a clearance fit or an interference fit.

7.2.3 Hole Basis System

A standard system of limits and fits, where the size of the hole is kept constant and the size of the shaft is varied to get the different classes of fits, is known as the hole basis system.

The fundamental deviation symbol 'H' is chosen for the holes, when the hole basis system is followed. This is because the lower deviation of the 'H' hole is zero. It is known as the 'basic hole' (Fig. A7.31).

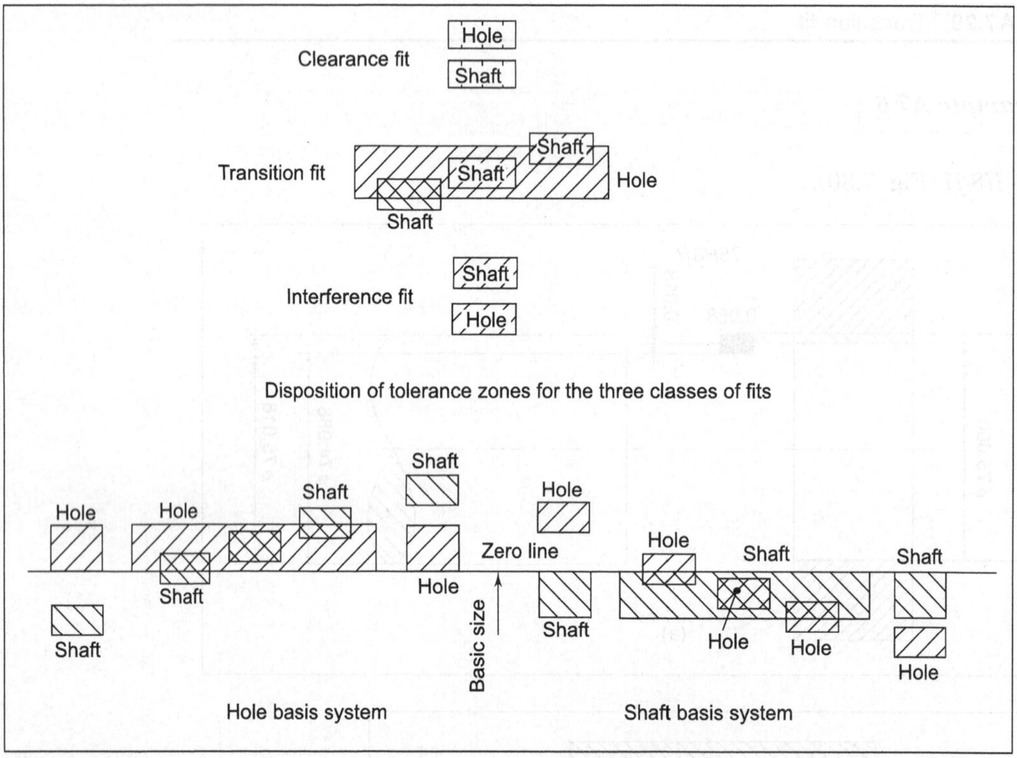

Fig. A7.31 Examples illustrating the whole basis and basis systems

7.2.4 Shaft Basis System

A standard system of limits and fits, where the size of the shaft is kept constant and variations are given to the hole for obtaining different classes of fits, is known as the shaft basis system. The fundamental deviation symbol '*h*' is chosen for the shaft basis. This is because the upper deviation of the '*h*' shaft is zero. It is known as 'basis shaft' (Fig. A7.31).

The hole basis system is mostly followed, because, depending upon the class of fit, it is always easier to alter the size of the shaft as it is external, but it is difficult to make minor alterations to a hole. Moreover, the hole can be produced by using standard tooling.

The three classes of fits, both under the hole basis and shaft basis, are illustrated in Fig. A7.31.

Review Questions

1. How is the interchangeability of parts important?
2. Define unilateral and bilateral tolerances.
3. What are hole based system and shaft based systems?
4. What is fit? State the types of fits.
5. Define (a) basic size, (b) limits, (c) allowance, (d) tolerance, (e) deviation.
6. What is surface roughness? How is the value of surface roughness determined?
7. How are the values of surface roughness indicated on a drawing?
8. Draw machining symbols and state their uses.
9. Why are tolerances required for the machining processes?

CHAPTER 8

Conventional Representation of Machine Components

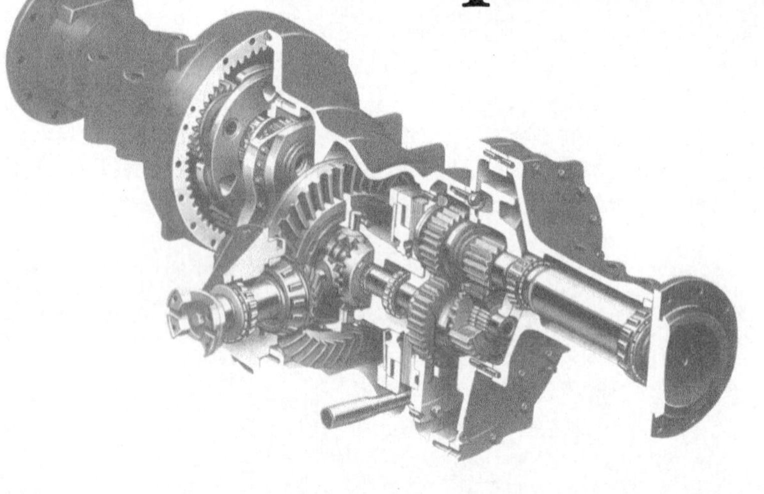

This chapter contains symbolic representation of different commonly used mechanical components for assembly drawing work

After reading the chapter you will be able to

- select proper symbols for mechanical components
- interpret technical drawings that have symbols
- save effort and time in assembly, where different machine parts are repeated

8.1 GENERAL DISCUSSION

A machine drawing is a graphical representation of assemblies and subassemblies of machine components. Here, either two or more than two components are involved in assembly. In many cases, common machine components like grears, screws, nuts, bolts, and so on are repeated in drawings. Therefore, in order to minimise efforts and ensure easiness in making drawings, common conventional representations have been developed as a part of graphical language. These are nothing but a conventional representation of machine parts.

A machine drawing consists of many standard features and components such as threads, tapped holes, knurlings, splines, springs, gears, welded joints, structural sections, welded joints, and so on. Since these features or components are common to almost all machine drawings, they can be represented by common conventions. This reduces detailing in drawing of each such feature or component.

8.2 CONVENTIONAL REPRESENTATION OF MACHINE PARTS

8.2.1 Conventional Representation of Screw Threads

Whatever may be the kind of thread profile like *V*, square, acme and so on, conventional representation for external and internal threading operations remains the same, as shown in Table A8.1.

For external threads Conventionally the outer circle should be full and inner one representing minor diameter is 3/4th.

For internal threads Conventionally the inner circle should be full and outer one representing the major diameter of joining bolt/pin is 3/4th.
The various screw/threads and their conventions are given in Table A8.1

TABLE A8.1 *Screw Threads and Their Conventions*

Actual Projection	***Convention***
Visible screw threads 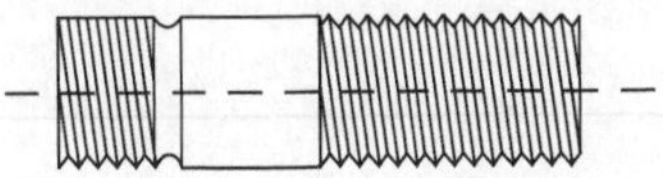	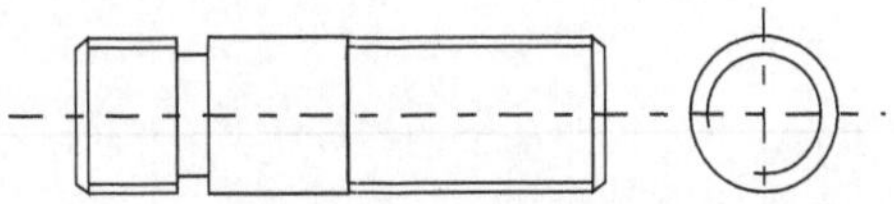
Section in external screw threads 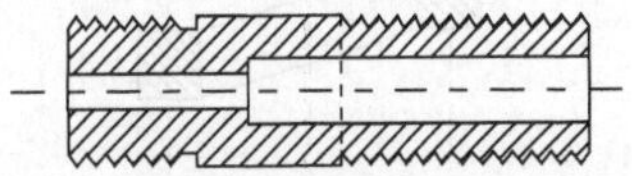	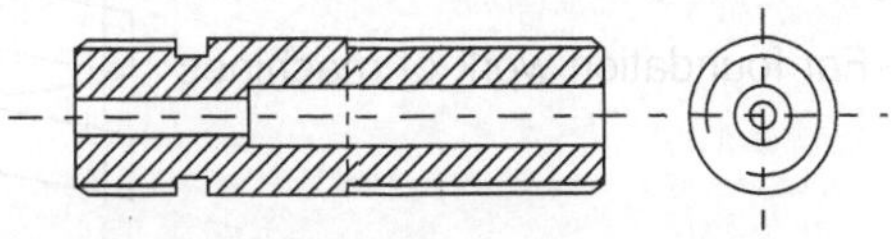
Section in internal screw threads 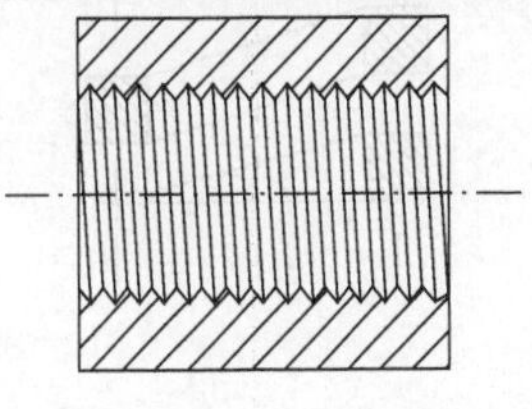	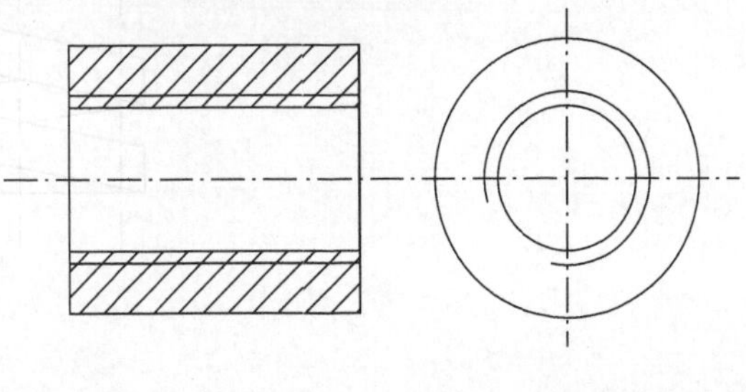

Actual Projection	Convention
Tapped hole 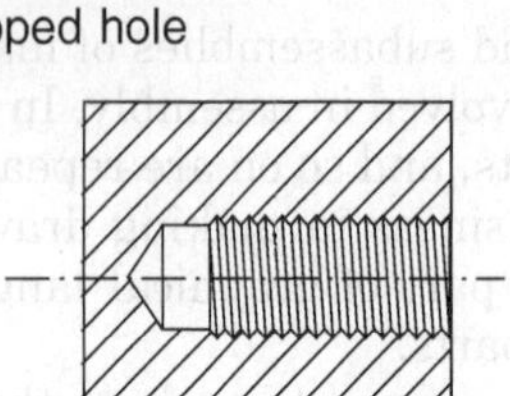	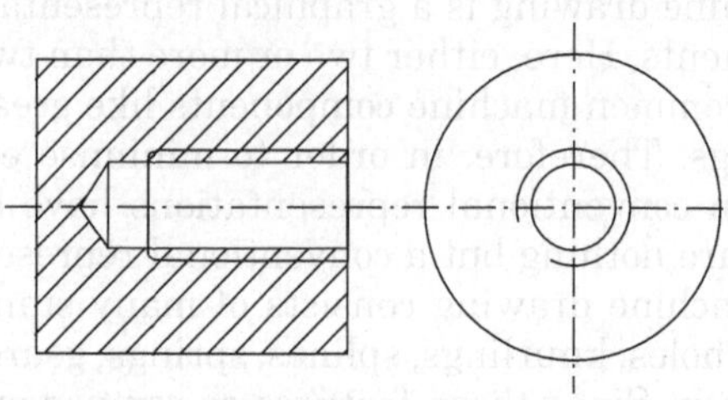

8.2.2 Conventional Representation of Springs

Conventional representation for commonly used springs is given below. Along with this, on drawings some more details can be used, like material used for spring, number of coils, diameter of coil wire, and so on.

The representation of different types of springs is shown in Table A8.2.

TABLE A8.2 *Springs and Their Conventions*

Actual Projection		Convention
Cylindrical helical compression spring in circular section **Use**–For foundation work of machines, automobile suspension	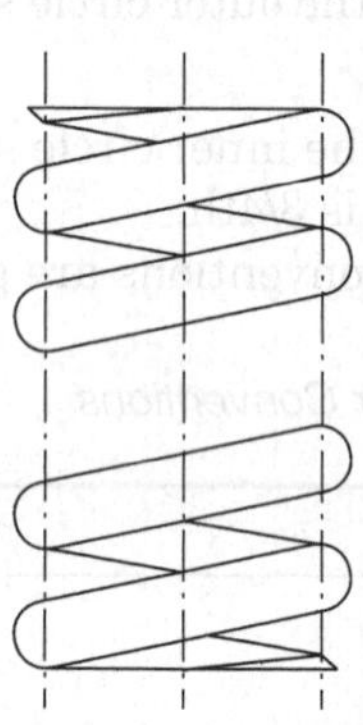	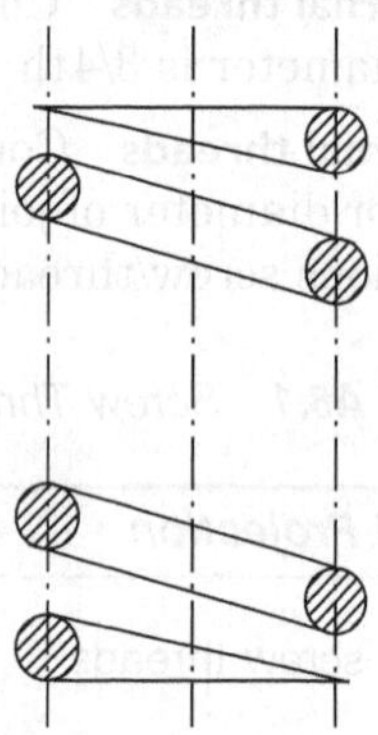
Cylindrical helical compression spring in rectangular section **Use**–For foundation work of machines	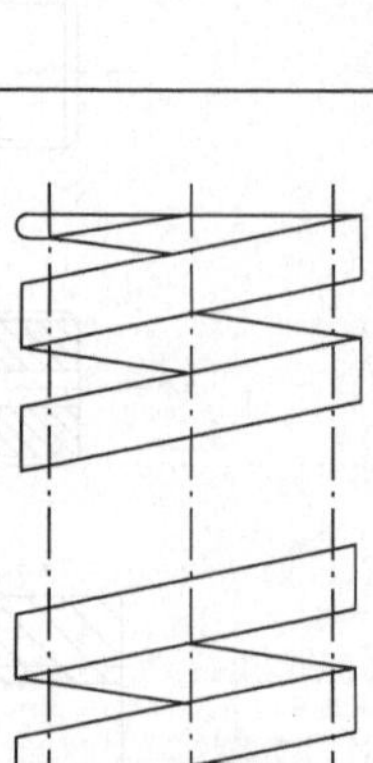	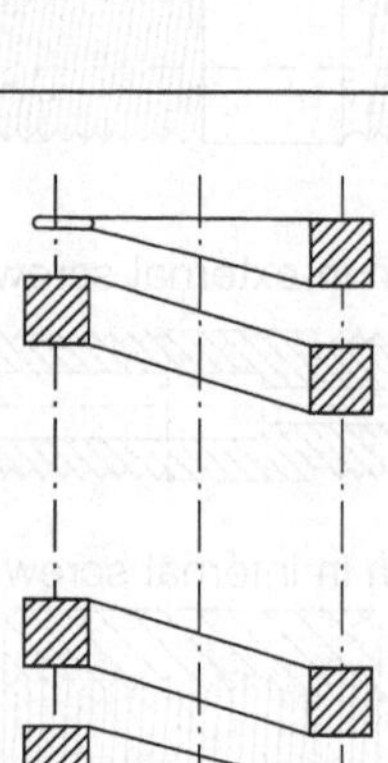

Actual Projection		***Convention***
Coonical helical compression spring in circular section **Use**–Special application in machine tool manufacturing	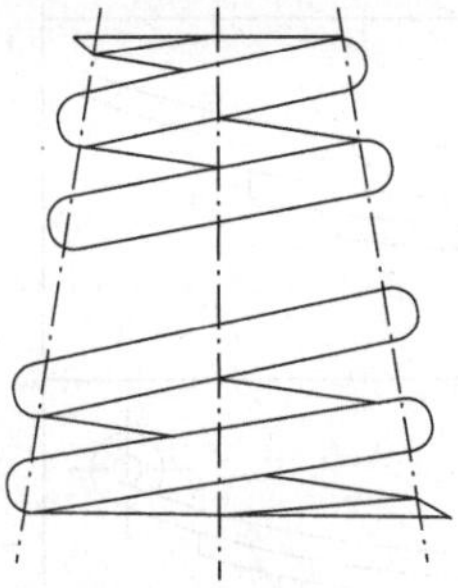	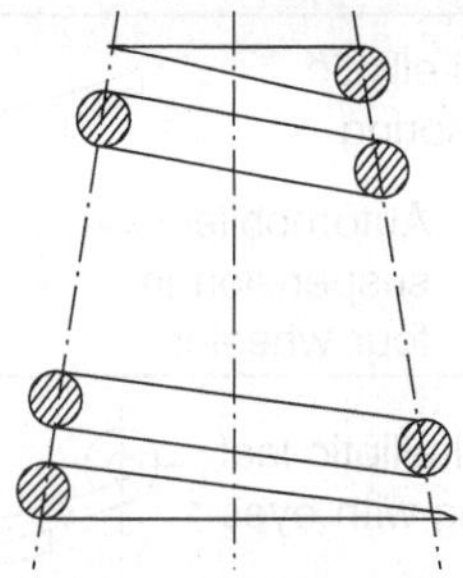
Volute springs **Use**–In machine tool manufacturing	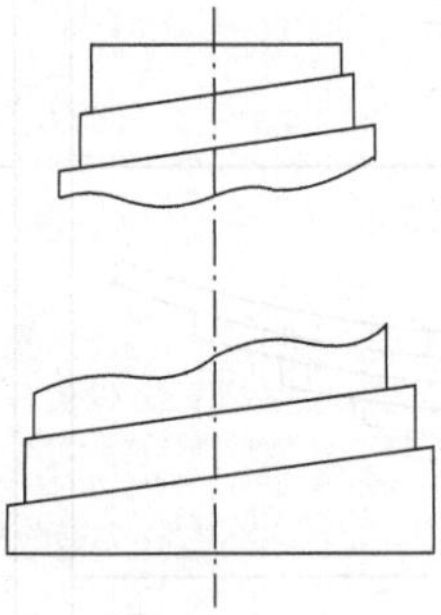	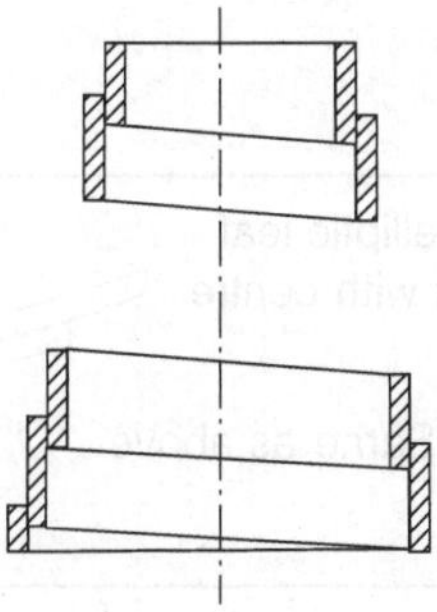
Cylindrical helical tension spring in circular section **Use**–In spring balance, bull worker, and so on	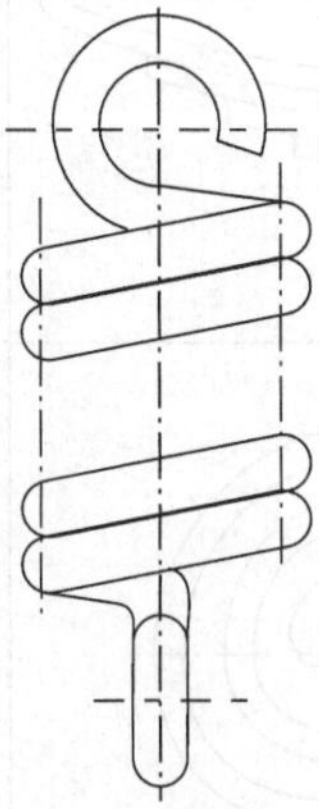	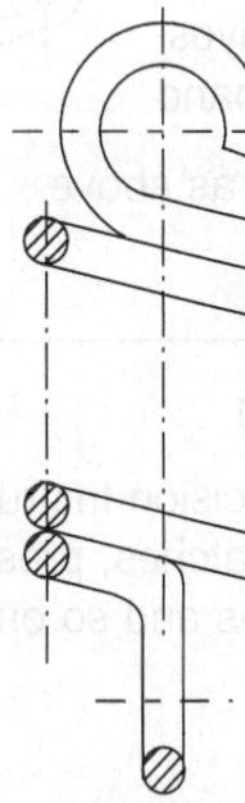
Cylindrical helical torsion spring in circular section (Right hand wound) **Use**–Toys 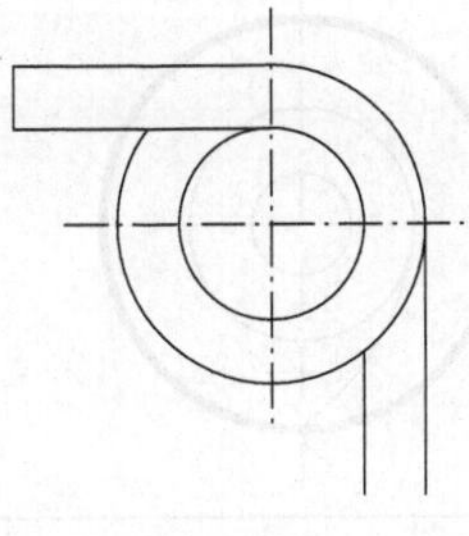	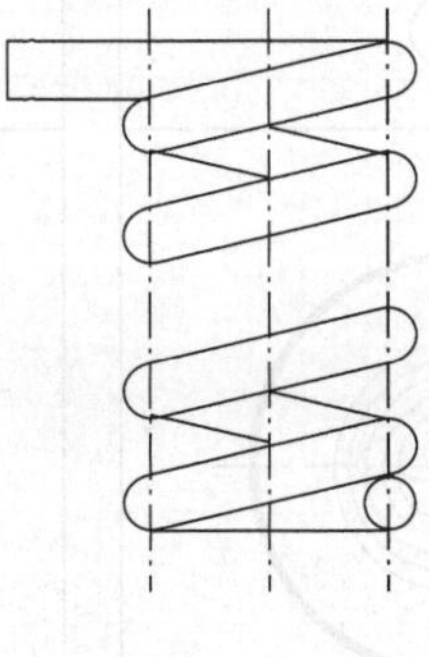	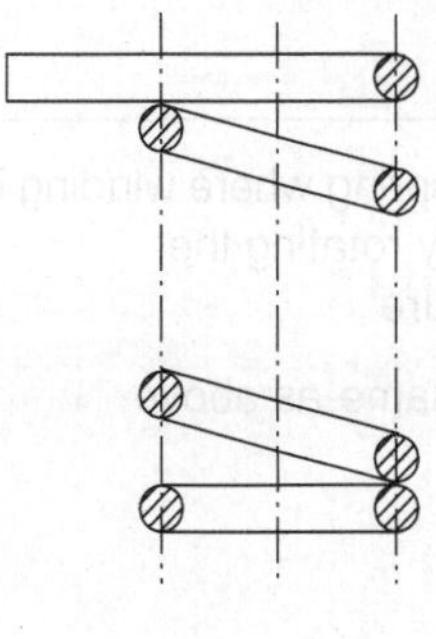

Actual Projection		Convention
Semi-elliptic leaf spring **Use**– Automobile suspension in four wheeler	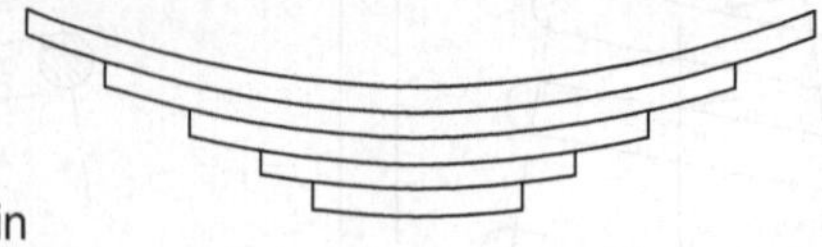	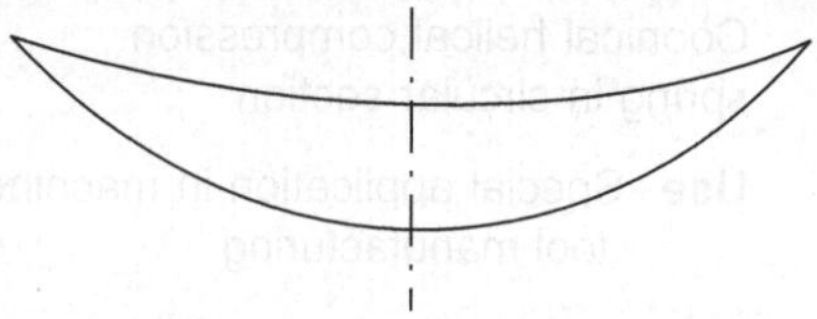
Semi-elliptic leaf spring with eyes **Use**–Same as above	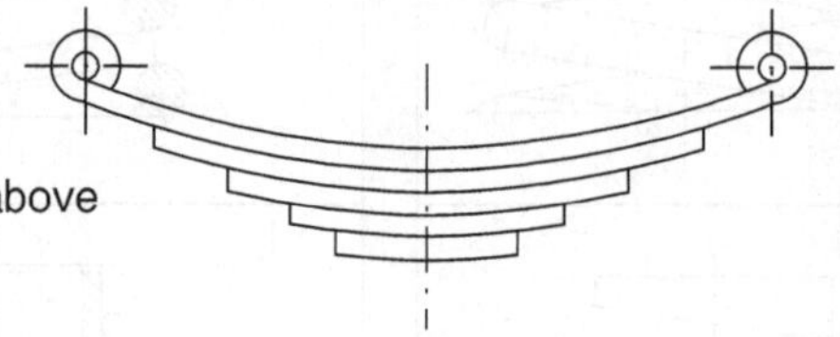	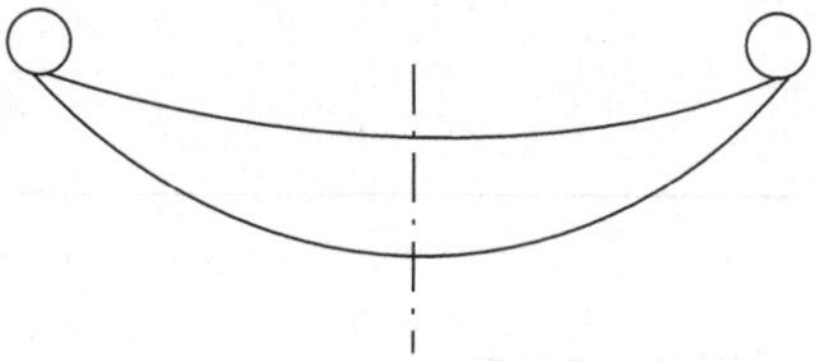
Semi-elliptic leaf spring with centre band **Use**–Same as above	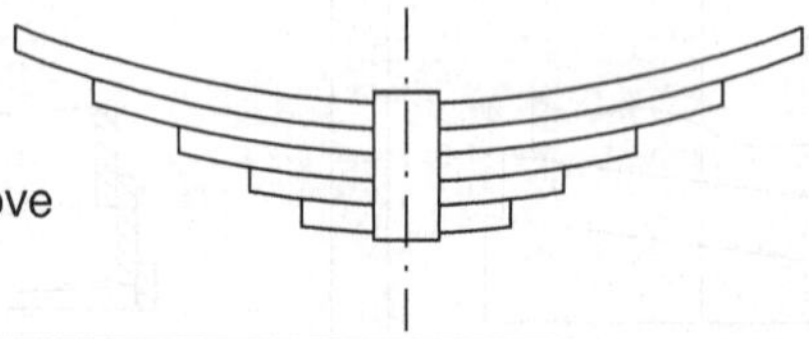	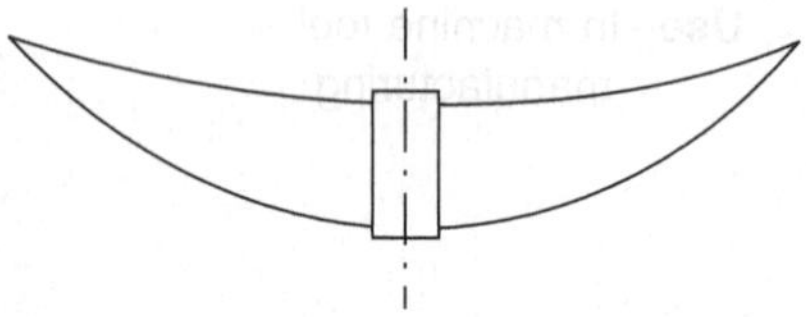
Semi-elliptic leaf spring with eyes and centre band **Use**–Same as above	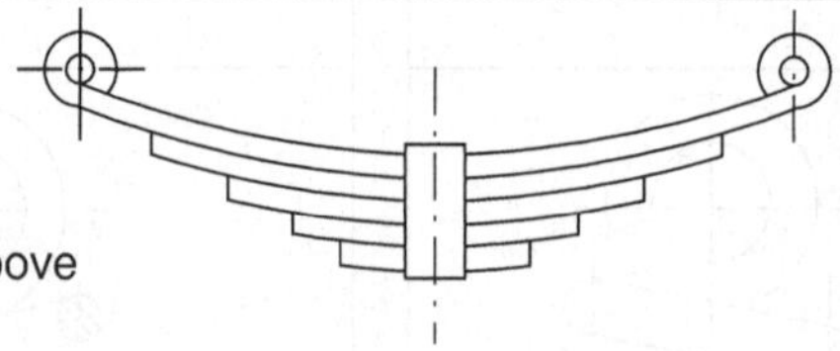	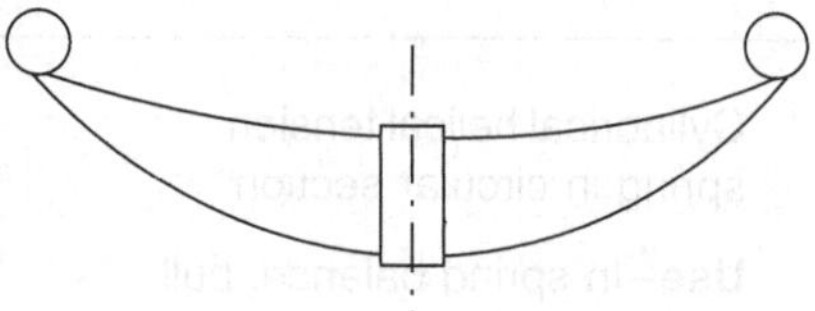
Spiral spring **Use**– In precision Instruments like watches, pressure gauges and so on	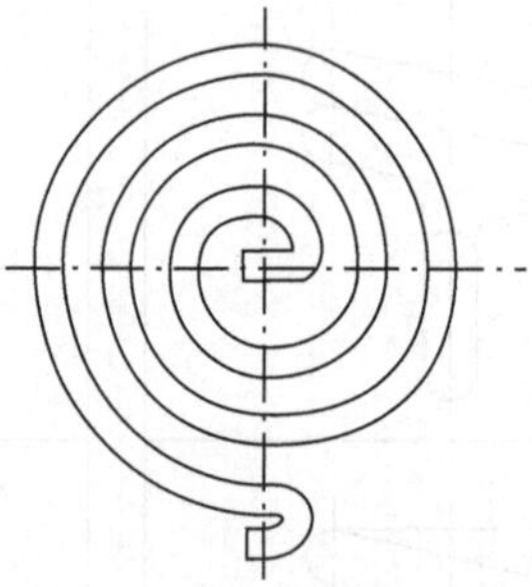	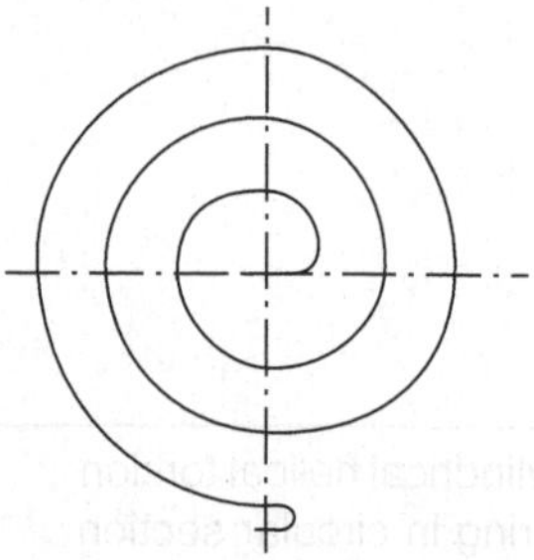
Spiral spring where winding is done by rotating the enclosure **Use**–Same as above	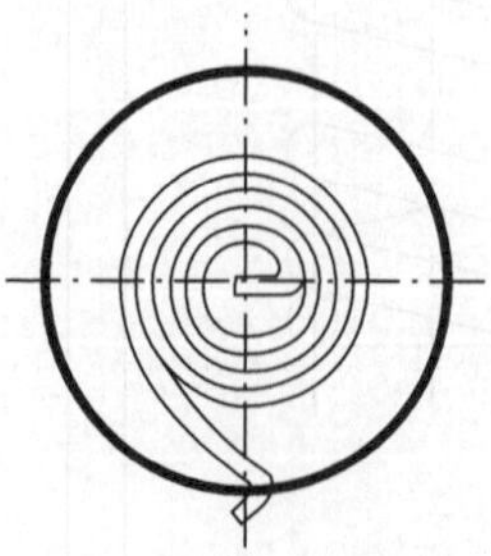	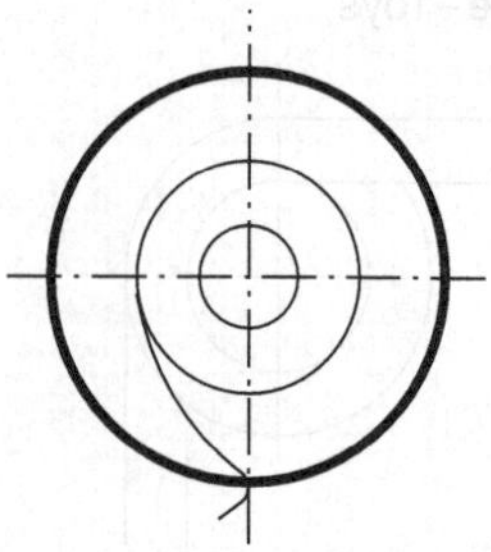

8.2.3 Conventional Representation of Gears

For gears, which are symmetrical components about their axis, there is no need to draw a complete sketch. Instead an equal sign is marked on the axis, implying that the part shown is replicated on the other side of the axis.

Different types of gears are used in engineering drawing. They can be either spur gears, helical gears, bevel gears, or worm gears. The representation of spur gears, bevel gears, and worm wheel gears are shown in Fig. A8.1.

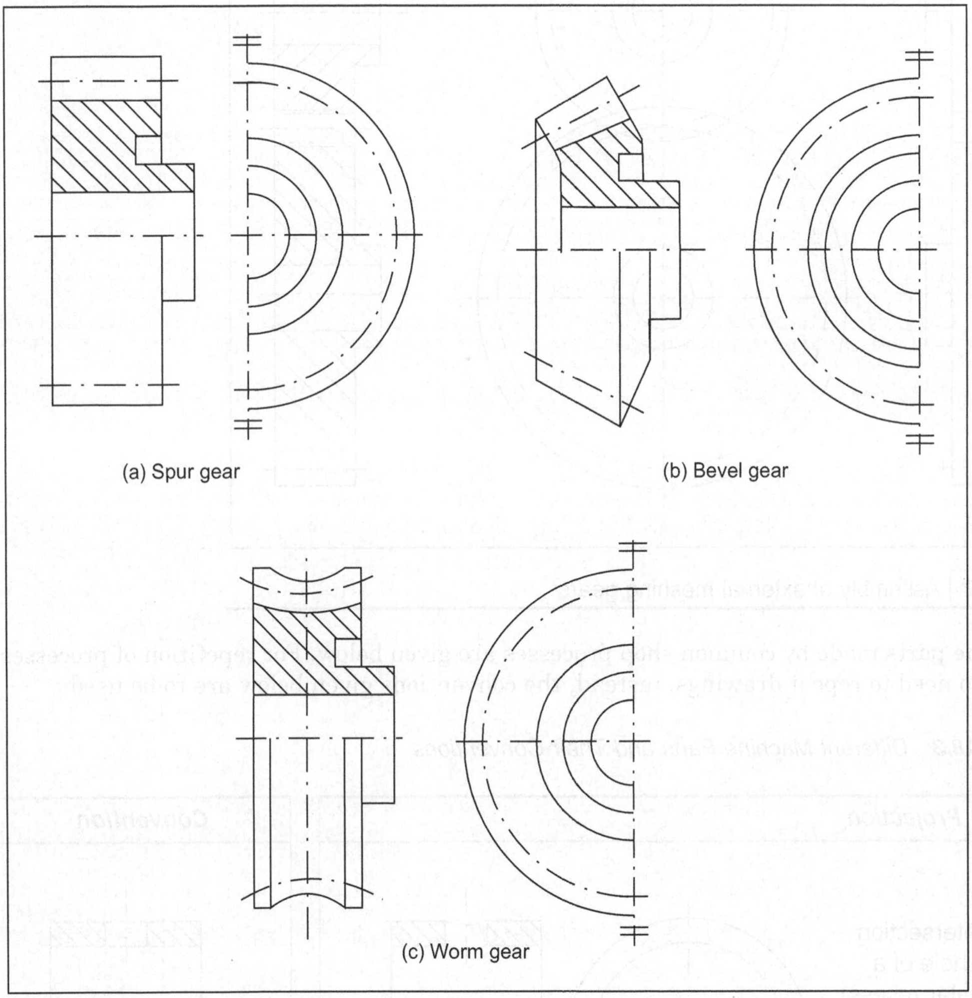

Fig. A8.1 Types of gears

In cases where assemblies of gears in meshed position are to be indicated, they can be represented as shown in Fig. A8.2.

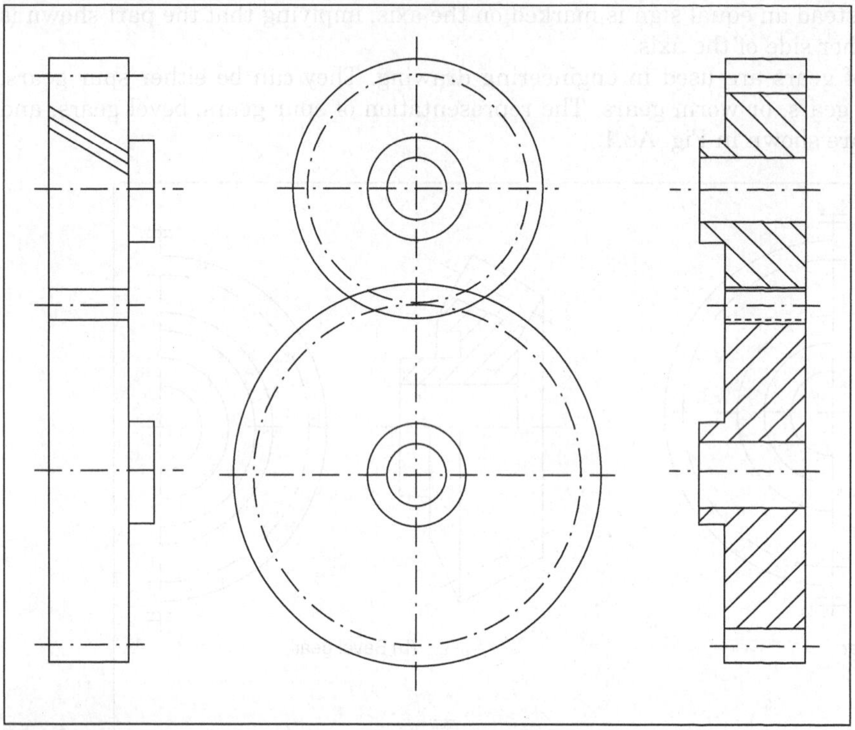

Fig. A8.2 Assembly of external meshing gears

Machine parts made by common shop processes are given below. For repetition of processes there is no need to repeat drawings, instead, the conventions given below are to be used.

TABLE A8.3 *Different Machine Parts and Their Conventions*

Actual Projection		*Convention*
Small intersection (Drilled hole of a rectangular recess)	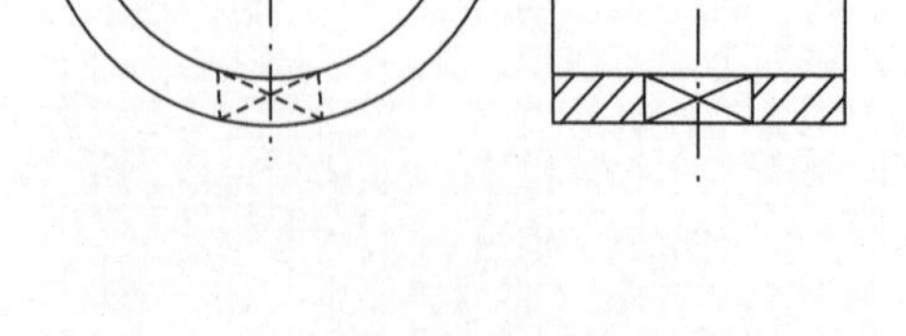	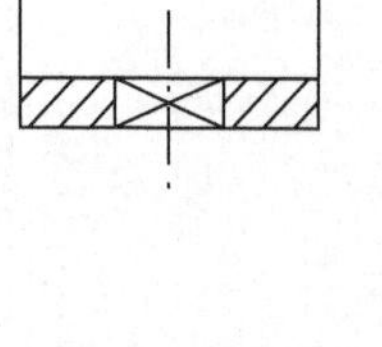

Actual Projection		***Convention***
Extended cylindrical pin and rectangular piece	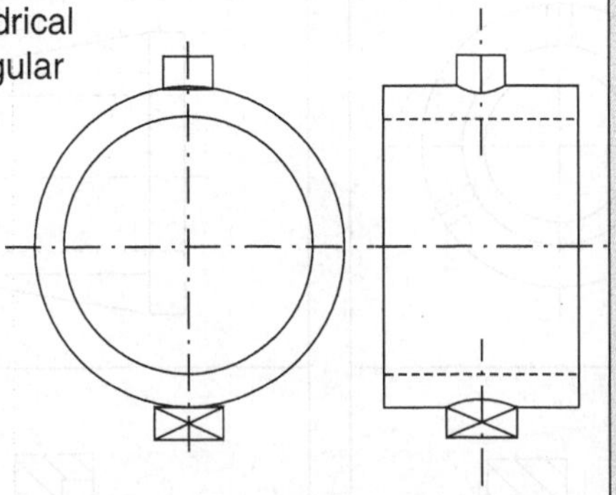	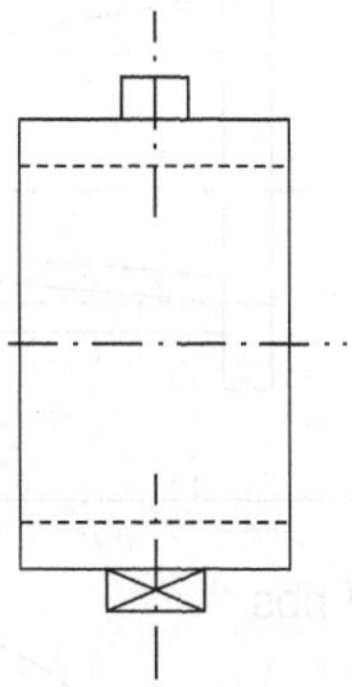
Holes on linear	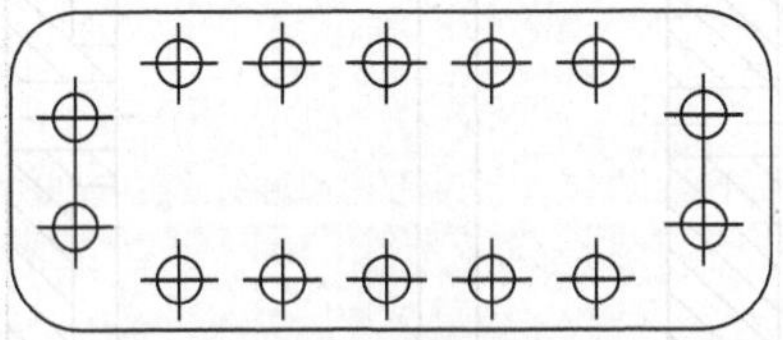	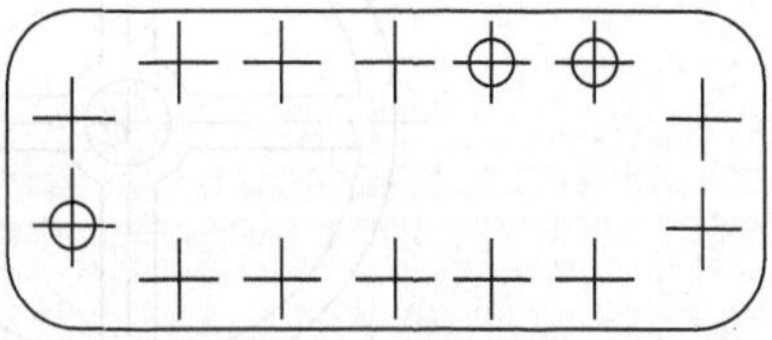
Holes on circular pitch	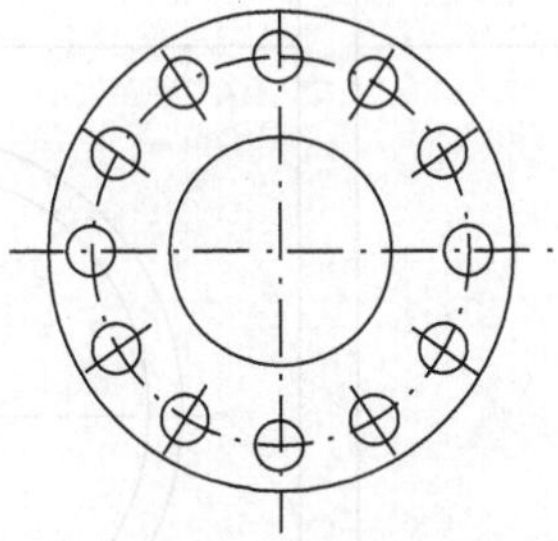	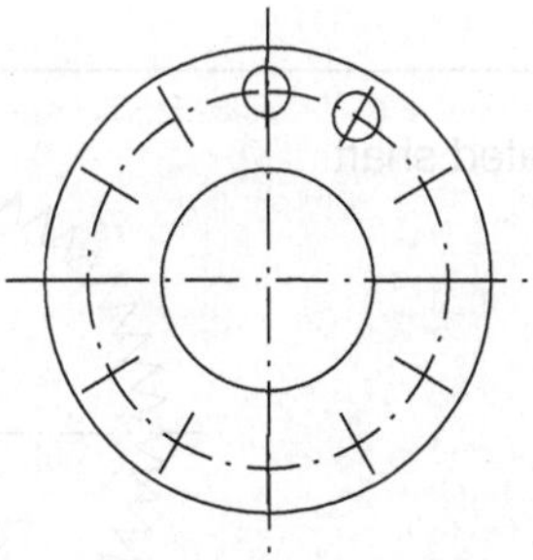
Repeated parts	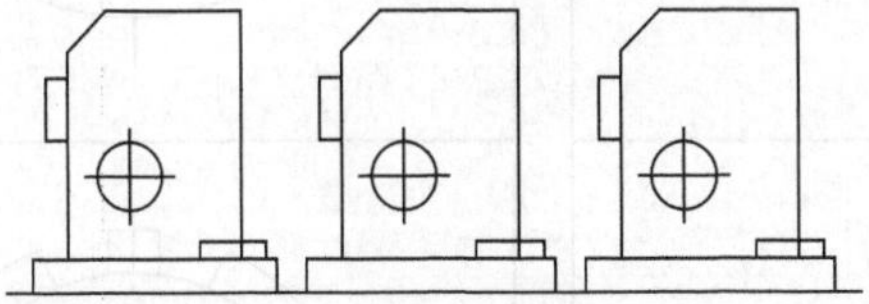	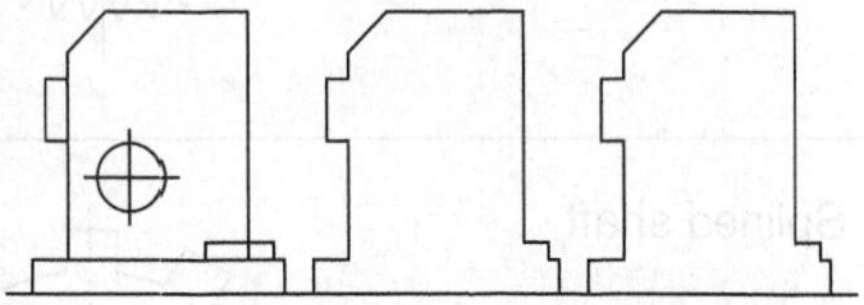
Slotted head	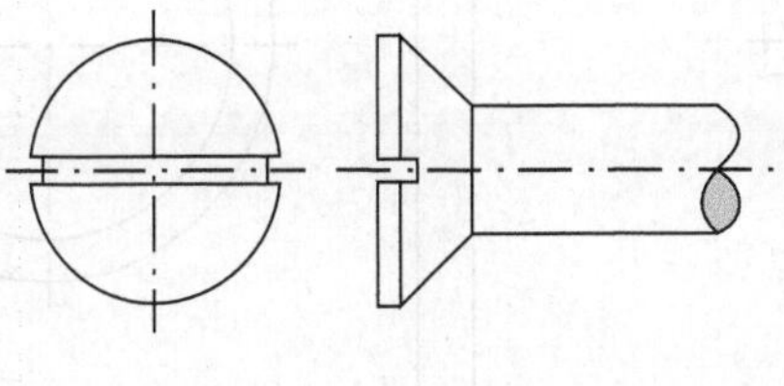	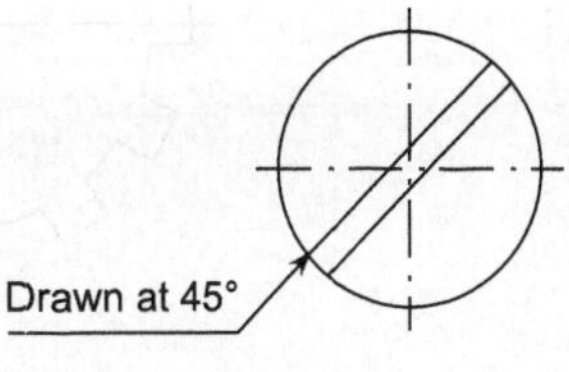

Actual Projection		Convention
Radial ribs	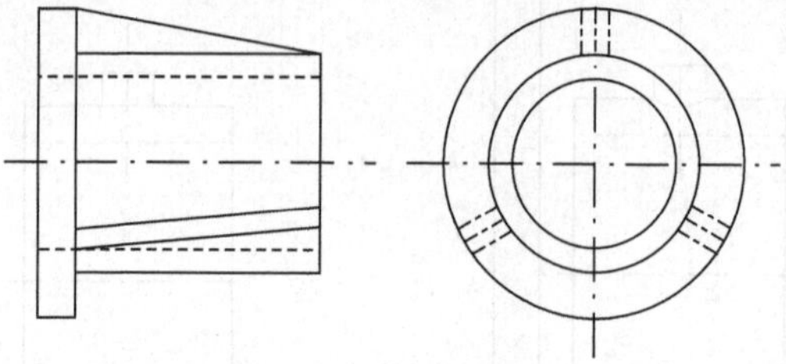	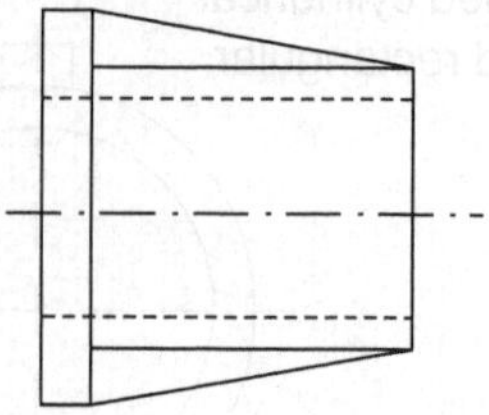
Section of ribs	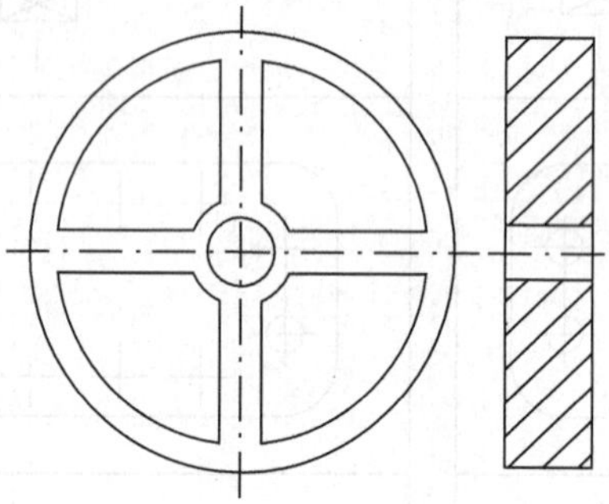	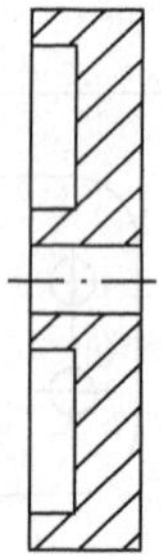
Serrated shaft	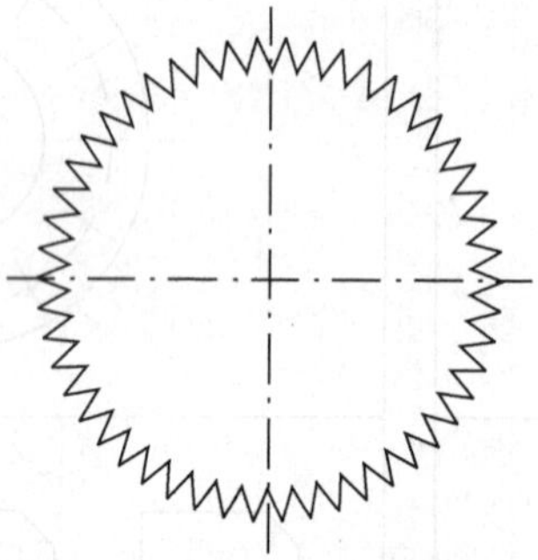	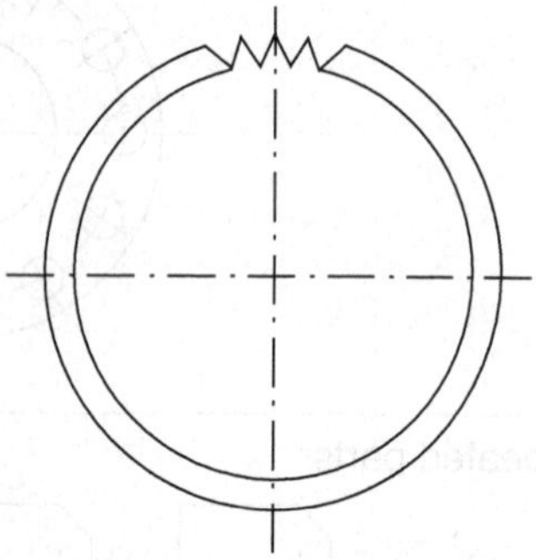
Splined shaft	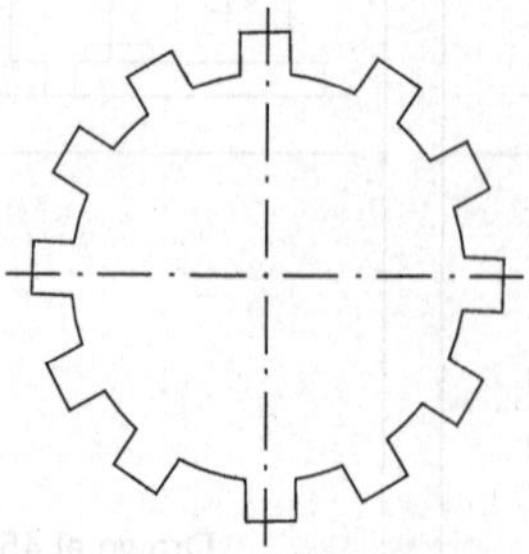	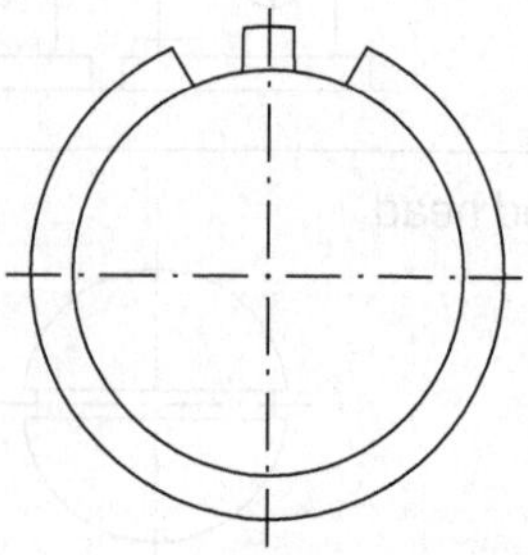

Actual Projection	***Convention***
Ratchet and pinion 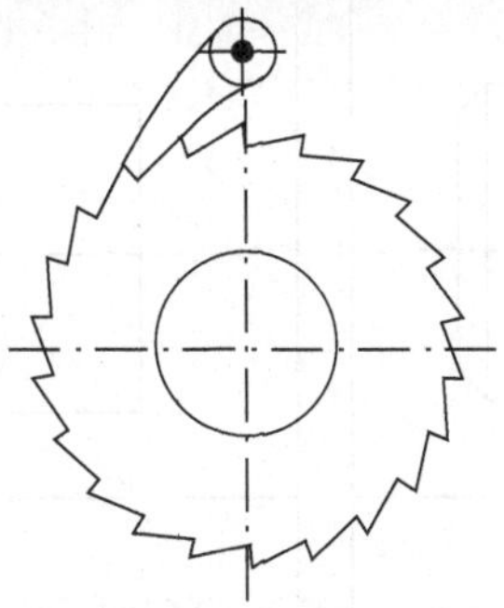	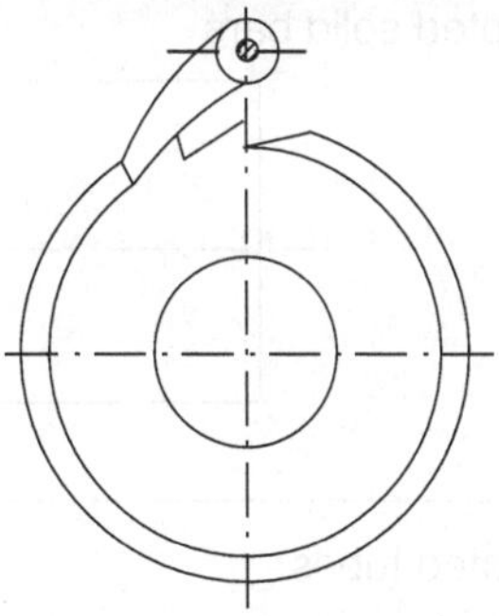
Bearings 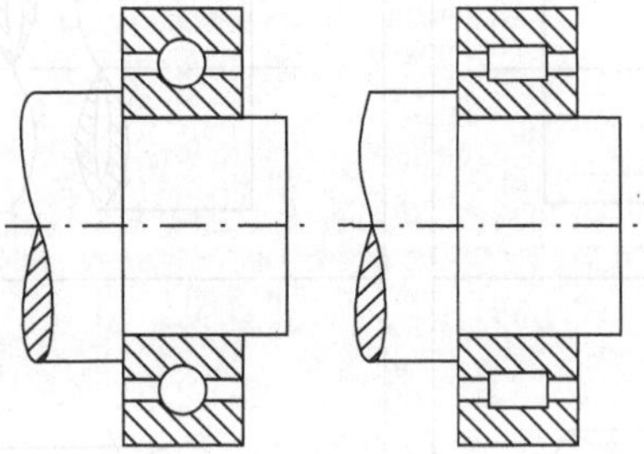	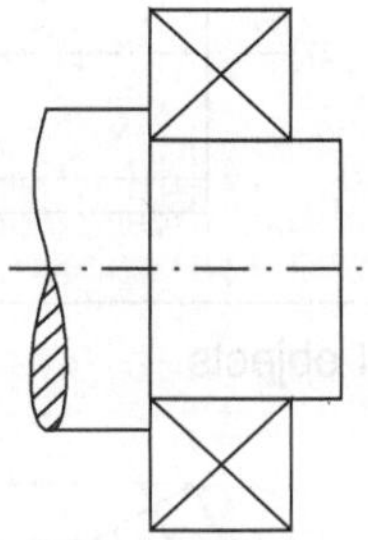
Chain wheels 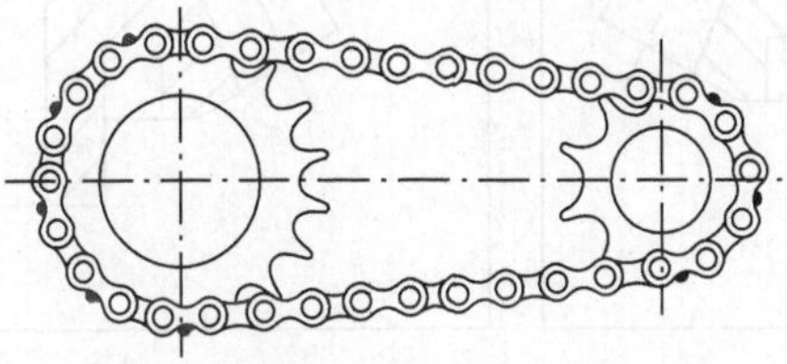	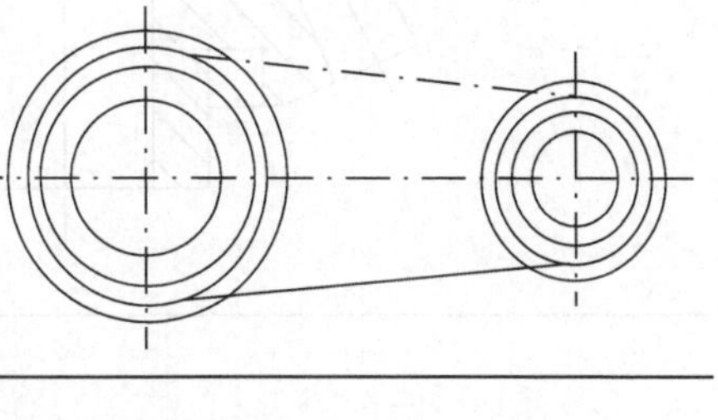
Straight knurling 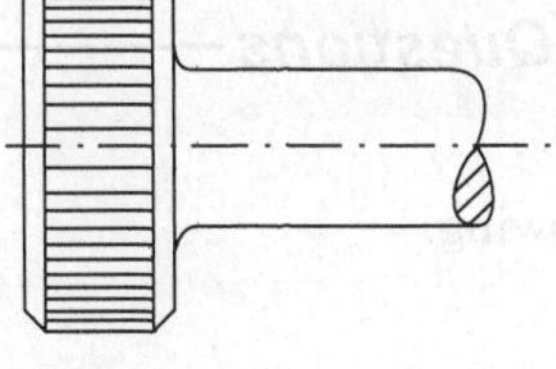	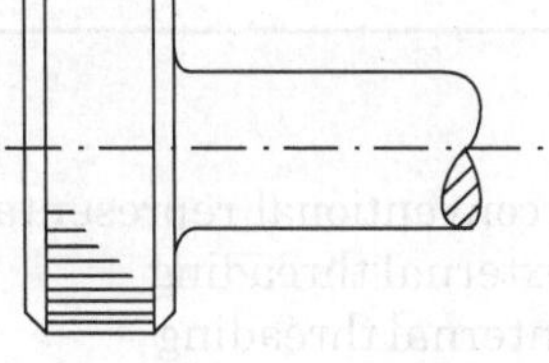
Diamond knurling 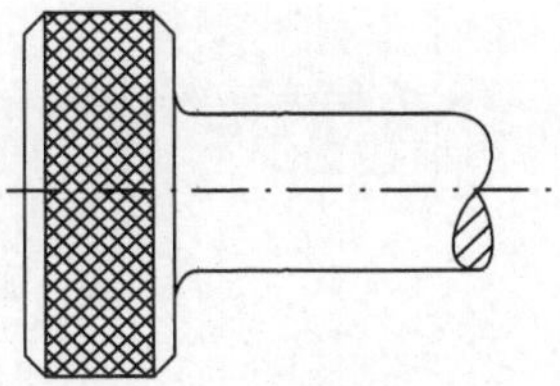	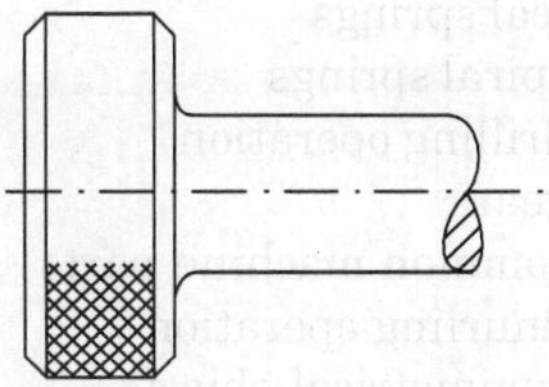

Actual Projection	Convention
Interrupted solid bars 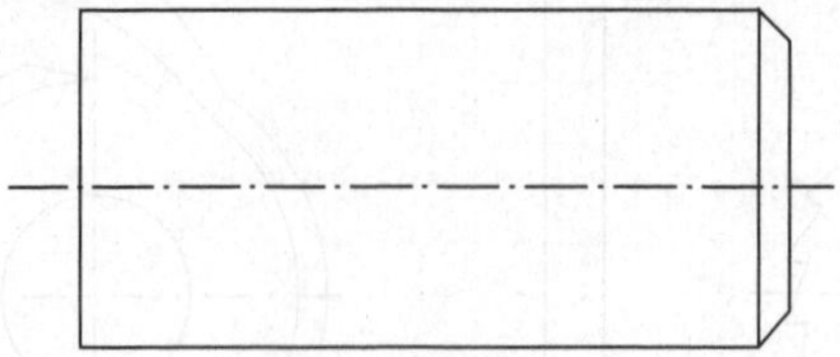	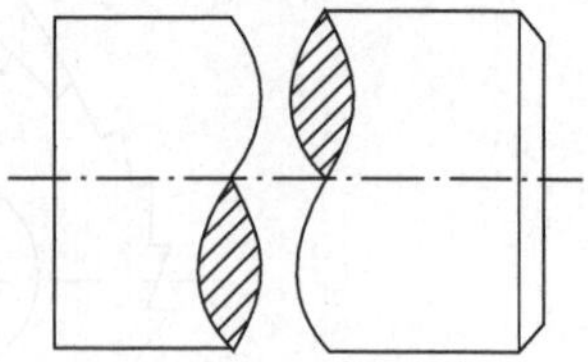
Interrupted tubes 	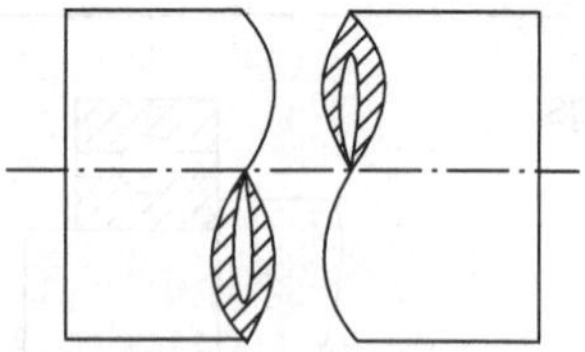
Symmetrical objects 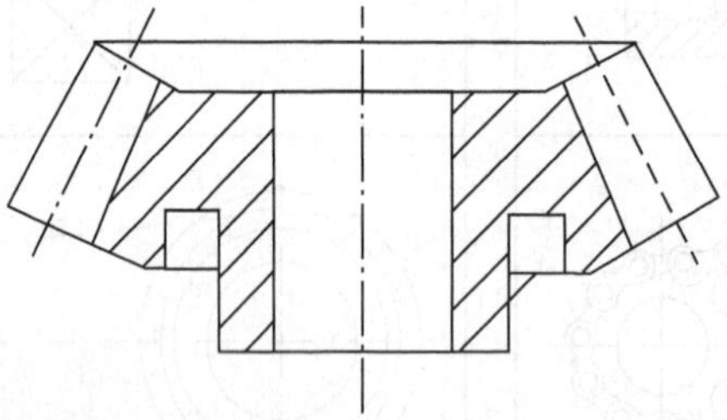	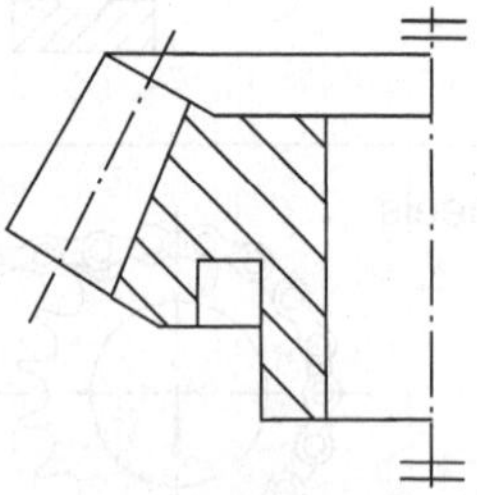

Review Questions

Study conventional representations for the following.

(a) external threading
(b) internal threading
(c) helical springs
(d) leaf springs
(e) spiral springs
(f) drilling operation
(g) gears
(h) common machine parts
(i) knurling operation
(j) symmetrical objects

CHAPTER 9

Screw Fasteners

This chapter contains information on screw fasteners

After reading the chapter you will be able to

- understand screw features
- identify thread forms
- identify right and left hand threading
- specify nuts and bolts with their applications

9.1 SCREW FASTENERS

A screw thread is used for temporary locking of two components. By cutting a helical groove on a cylindrical surface, screw thread can be generated. The external threaded rod is the screw and internal threaded hole is the nut. The screw thread can be cut on a lathe machine or by means of a die set and a nut can be made by tap set. A small hole is first drilled in the nut and then it is threaded by a tool called tap. This process is known as tapping process. The various features and definitions of screw member are as below.

Crest and major diameter: The crest is the outermost part of a thread. The diameter associted with the crest is called the major diameter.

Root and minor diameter: The root is the innermost part of a thread. The diameter associated with the root is called as the minor diameter, core diameter, or root diameter.

Pitch: It is the distance measured parallel to the axis, between a point on one thread form and a corresponding point on the adjacent thread form. For example, the distance from root to root or crest to crest. The pitch can mathematically be expressed by the following equation

$$Pitch = \frac{1}{Number\ of\ threads/Unit\ length}$$

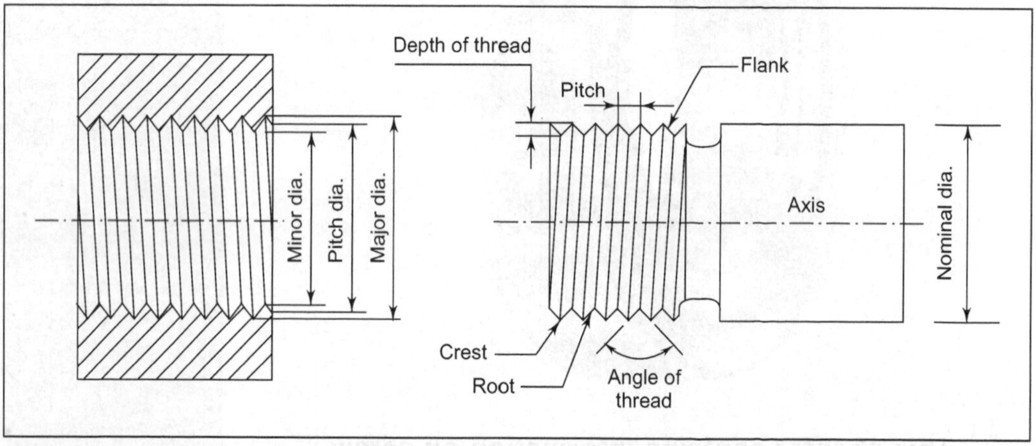

Fig. A9.1 Nomenclature of screw threads

Pitch diameter: This is the diameter of an imaginary cylinder, the surface of which would pass through the thread at such points, so as to make equal the width of the thread and the width of the space between the threads. The pitch diameter is also called the effective diameter.

Lead: It is the distance between two corresponding points on the same helix. It can also be defined as the advancement of the screw linearly for one rotation of the nut. In cases where the threads are single start, that is, there is only one helix, the pitch and lead would be the same.

9.1.1 Thread Forms

Screw threads are normally used as temporary fastening components for joining parts in assemblies. They are basically used for adjustment in assemblies and for transmission of power from one part to another. Different forms of screw threads are used, such as

1. Triangular or *V* form.
2. Square form

Other types are modified from these two main types.

Forms of *V*-threads

1. British Standard Whitworth Threads (*BSW*)
2. British Association Threads (*BA*)
3. Sellers Threads
4. Unified Standard Threads
5. ISO Metric Thread
6. ISO Metric Trapezoidal Threads

Forms of square threads

7. Square Threads
8. Acme Threads
9. Knuckle Threads
10. Buttress Threads

British standard whitworth threads (*BSW*) These threads have a coarse pitch and are found on screwed fastenings for several applications. British Standard Pipe (*BSP*) threads with fine pitches are used for steel pipes and tubes carrying fluids. These threads have large effective core diameters and are specified by the bore of the pipe rather than the outside diameter. They are designated as *G*1/2, *G*1 or *BSP* 1/2″, *BSP* 1″, and so on. The dimensional details of *BSP* threads are given in Table A9.1. These threads give an effective leak-proof joint due to their fine pitches and form.

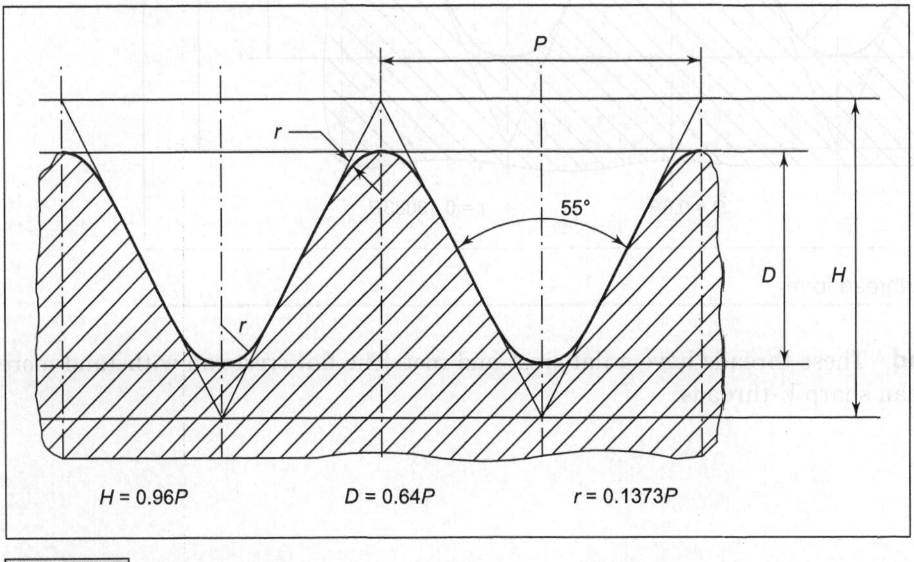

Fig. A9.2 BSW thread form

TABLE A9.1 Standard BP Threads/Pipe Threads

BSP Size (inch)	Pitch	TPI	Diameters Major (mm)	Effective (mm)	Minor (mm)	Tap Drill (mm)
1/8	0.907	28	9.728	9.147	8.566	8.8
1/4	1.337	19	13.157	12.301	11.445	11.75
3/8	1.337	19	16.662	15.806	14.940	15.25
1/2	1.814	14	20.955	19.793	18.631	19
3/4	1.814	14	26.441	25.279	24.117	24.25
1	2.309	11	33.249	31.770	30.291	30.75
$1\frac{1}{4}$	2.309	11	41.910	40.431	38.952	39
$1\frac{1}{2}$	2.309	11	47.803	46.324	44.845	45
2	2.309	11	59.614	58.135	56.656	57

British association thread (*BA*) This is a form of *BSW* thread with fine pitches and is used for precision work.

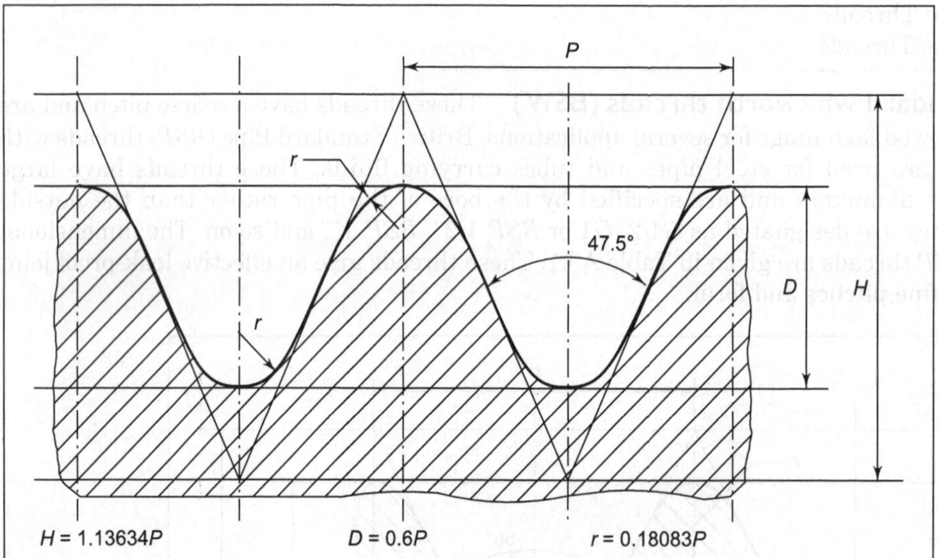

Fig. A9.3 BA thread form

Sellers thread These threads have a flat crest and root. The flat crest can withstand more rough usage than sharp *V*-threads.

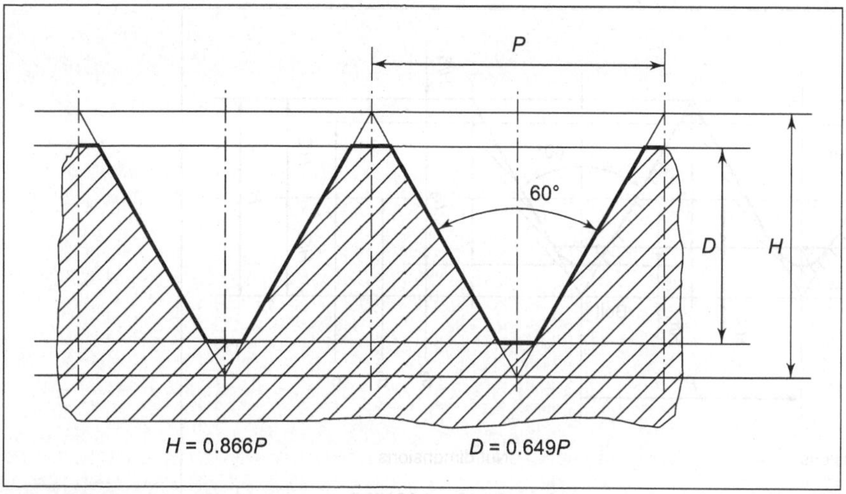

Fig. A9.4 Sellers thread form

Unified standard thread These threads have been standardised by the ISO and are available in inch and metric series. This thread has rounded crest and roots.

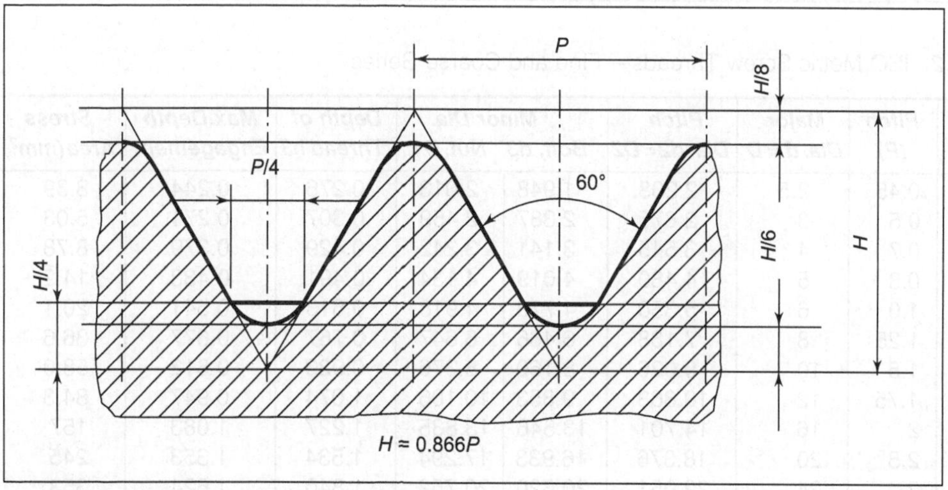

Fig. A9.5 Unified standard thread form

ISO metric thread The ISO metric threads are completely standardised and are used commonly in any engineering industry. The dimensional details of the threads for the bolt and the nut are illustrated in Fig. A9.6. The standard dimesions of different sizes are listed in Table A9.2.

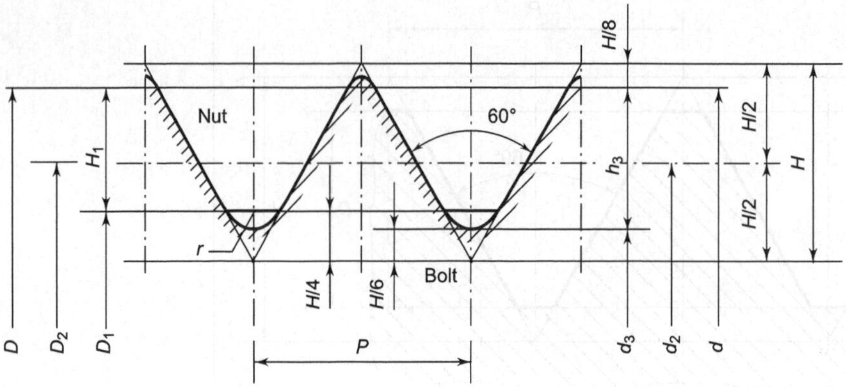

Bolt dimensions

$H = 0.86603P$
$d_2 = d - 0.64952P$
$d_3 = d - 1.22687P$
$h_3 = 0.61343P$

Nut dimensions

$D_1 = D - 1.08253P$
$D_2 = D - 0.64952P$
$H_1 = 0.54127P$
$R = 0.14434P$

Fig. A9.6 ISO metric thread form

TABLE A9.2 ISO Metric Screw Threads—Fine and Coarse Series

	Series	Pitch (P)	Major Dia. d = D	Pitch Dia. d2 = D2	Minor Dia. Bolt, d3	Minor Dia. Nut, D1	Depth of Thread h3	Max.Depth Engagement	Stress Area (mm^2)
Coarse Series	M2	0.45	2.5	2.208	1.948	2.013	0.276	0.244	3.39
	M3	0.5	3	2.675	2.387	2.459	0.307	0.271	5.03
	M4	0.7	4	3.545	3.141	3.242	0.429	0.379	8.78
	M5	0.8	5	4.480	4.019	4.134	0.491	0.433	14.2
	M6	1.0	6	5.350	4.773	4.918	0.613	0.541	20.1
	M8	1.25	8	7.188	6.466	6.647	0.767	0.677	36.6
	M10	1.5	10	9.026	8.160	8.376	0.920	0.812	58.0
	M12	1.75	12	10.863	9.853	10.106	1.074	0.947	84.3
	M16	2	16	14.701	13.546	13.835	1.227	1.083	157
	M20	2.5	20	18.376	16.933	17.294	1.534	1.353	245
	M24	3	24	22.051	20.320	20.752	1.840	1.624	353
	M30	3.5	30	27.727	25.706	26.211	2.147	1.894	561
	M33	3.5	33	30.727	28.706	29.211	2.147	1.894	694
	M136	4	36	33.402	31.093	31.670	2.454	2.165	817
Fine Series	M 8x1	1	8	7.350	6.773	6.918	0.613	0.541	39.2
	M10x1.25	1.25	10	9.188	8.466	8.647	0.767	0.677	61.2
	M12x1.25	1.25	12	11.188	10.466	10.647	0.767	0.677	92.1
	M16x1.5	1.5	16	15.026	14.160	14.376	0.920	0.812	167
	M20x1.5	1.5	20	19.026	18.160	18.376	0.920	0.812	272
	M24x2	2	24	22.701	21.564	21.835	1.227	1.083	384
	M30x2	2	30	28.701	27.546	27.835	1.227	1.083	621
	M36x3	3	36	34.051	32.320	35.752	1.840	1.624	865

ISO metric trapezoidal thread These threads are also standardised by the ISO and are similar to acme threads.

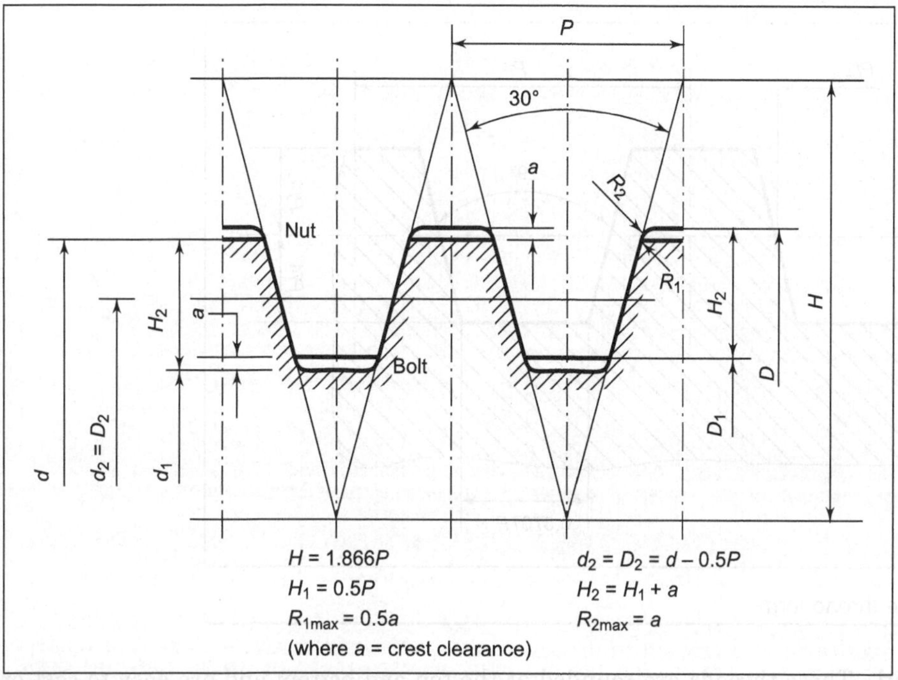

Fig. A9.7 ISO Metric trapezoidal thread form

Square thread As the name suggests, these threads have their flanks at right angles to the axis. They are generally used for power transmission. (e.g. lead screws of lathe). The pitch of the square threads is often assumed to be twice that of a *BSW* thread of a same diameter.

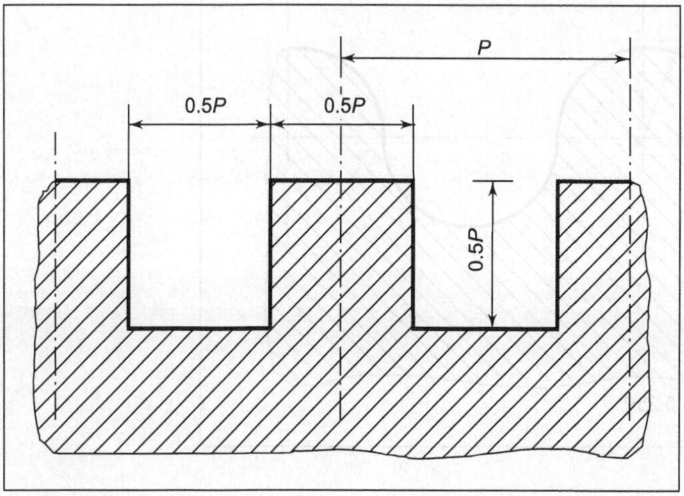

Fig. A9.8 Square thread form

Acme thread This is a variation of a square thread. It is much stronger than the square thread and is easy to manufacture. These threads are used in brass valves, lead screws of lathes, and bench vices.

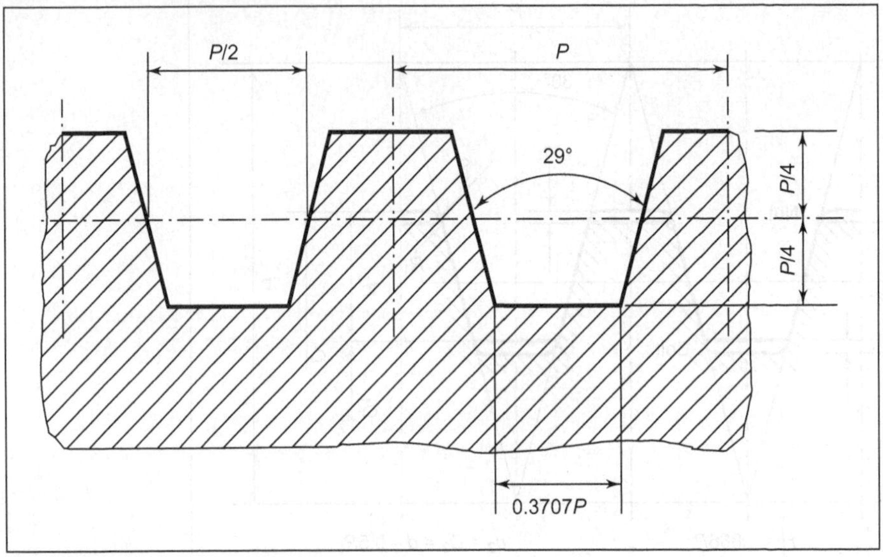

Fig. A9.9 Acme thread form

Knuckle thread These threads are rounded at the top and bottom and are easy to cast or roll. However, they cannot be manufactured economically on machines. These threads are used in railway carriage couplings and necks of bottles.

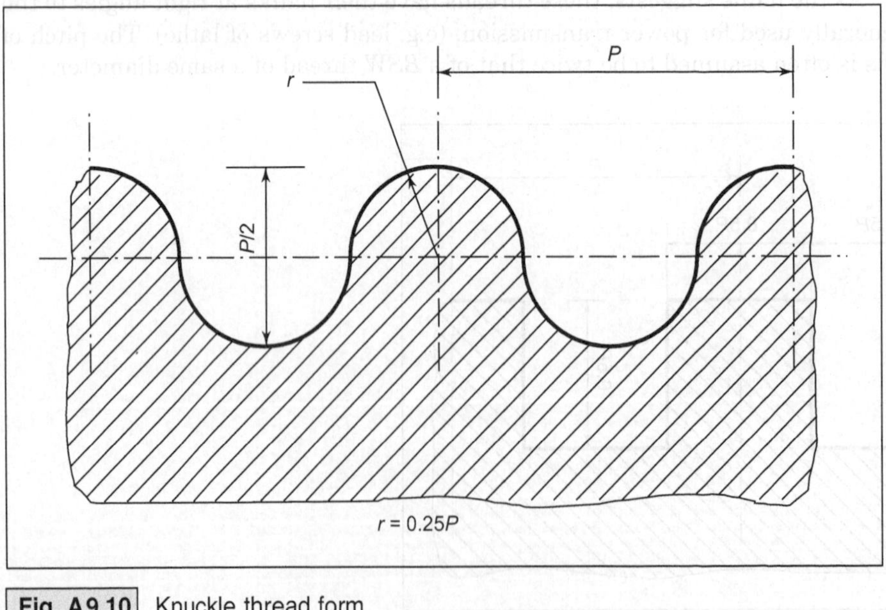

Fig. A9.10 Knuckle thread form

Buttress thread These threads are used while transmitting power in one direction. This force is transmitted almost parallel to the axis. It combines the low frictional resistance of square threads and the strength of *V*-threads.

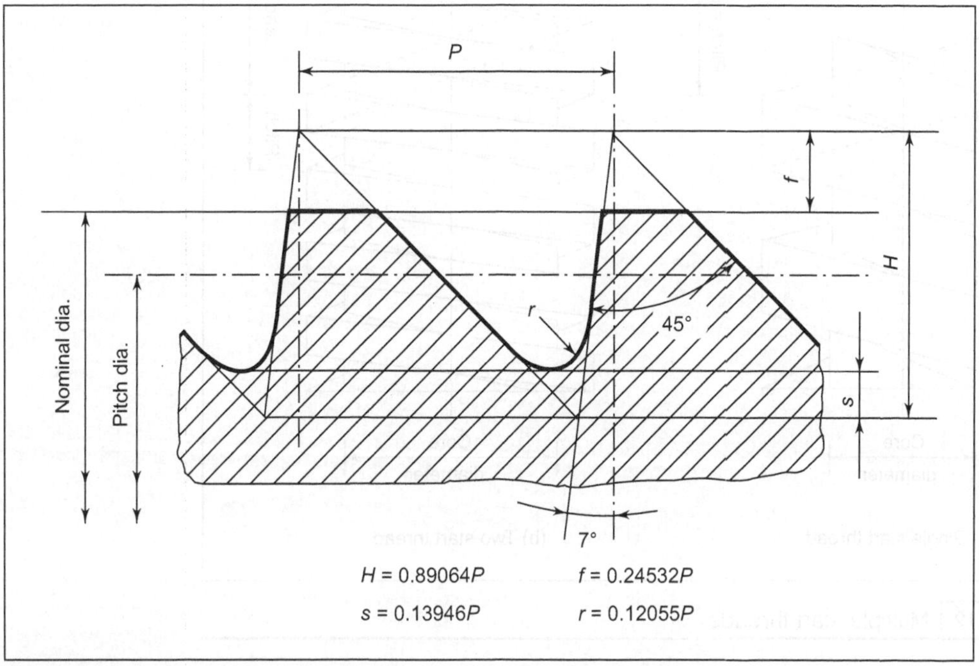

Fig. A9.11 Buttress thread form

9.1.2 Multiple Start Threads

In a single start thread, the pitch and the lead are the same. The depth of the thread is dependent on the pitch. If, for an application, one requires a large axial movement for a given rotation of the nut, the lead has to be large. A large lead would mean that the core diameter would cut on the same screw.

In multiple start threads where two or three helix occur, the same pitch is offset parallel to each other, resulting in a large lead. The pitch being small, the screw member is strong and at the same time, the larger lead ensures a large axial movement for a given rotation of the nut. In general, the lead and pitch are related by the equation

$$Lead = Pitch \times Number\ of\ starts$$

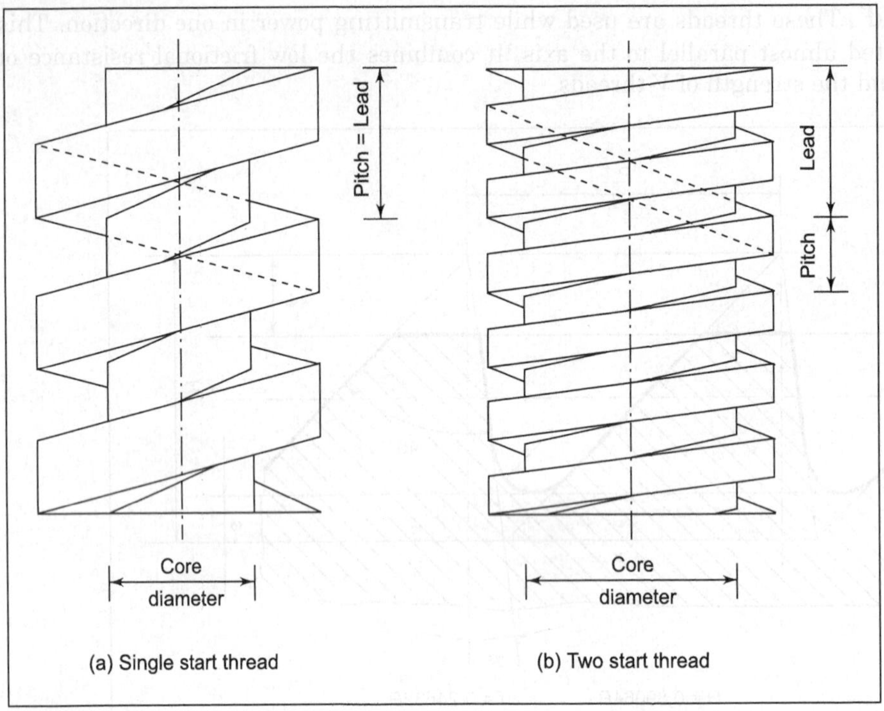

Fig. A9.12 Multiple start threads

9.1.3 Left-hand and Right-hand Thread

A nut, on a screw bolt, that needs to be turned in clockwise direction is a right-hand thread. However, if the nut screws off the bolt, then the thread is a left-hand thread, as shown in Fig. A9.13. If the right-hand is curled in the direction of rotation of the nut, the thumb points to the axial direction of displacement of the nut. This would give a right-hand, thread. If the thread is left-hand, then using the left-hand, the direction of axial motion of the nut can be determined.

It is noted that the thread is always assumed to be right-hand, unless it is indicated otherwise on a drawing. The indication is shown in Fig. A9.13. When the axis of the screw is vertical, the lines slope downwards from right to left in case of right-hand thread. The slope is in the reverse direction for a left-hand thread.

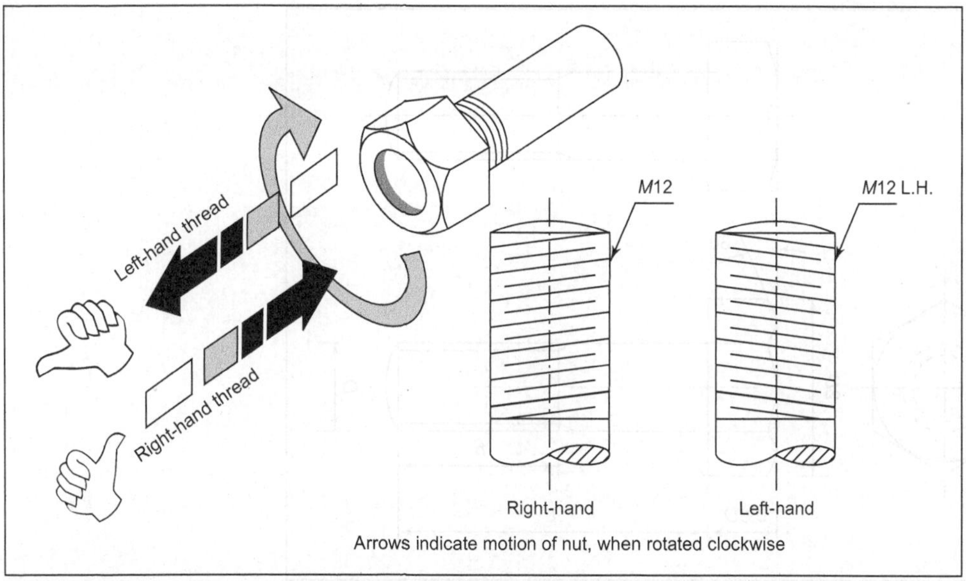

Fig. A9.13 Left-hand thread and right-hand thread

9.2 BOLTS

A cylindrical piece/job with a head on one side and threaded length on the other side is called a bolt. The function and the purpose of a bolt decides the shape of head of the bolt. The bolt is used alongwith the nut to tighten two parts. The bolt head is chamfered to avoid sharp corners.

The different types of bolts generally used in the engineering field are:

1. Hexagonal-headed bolt
2. Square-headed bolt
3. Cup-headed bolt
4. Cylindrical-headed bolt
5. T-headed bolt
6. Eye bolt
7. Lifting eye bolt

Hexagonal-headed bolt This is one of the most commonly used bolts. The size of the bolt is *M*8 and its length is 24 mm. It is made of steel with mechanical properties confirming to symbol 4*D*. This bolt is denoted as *M*8 × 24 — IS:2389-*P*-4*D*.

Here 4*D* indicates the minimum tensile strength of 37 kgf/mm^2 and 25 per cent elongation. *P* shows the precision grade, the other grades being *T*—Turned and *B*—Black.

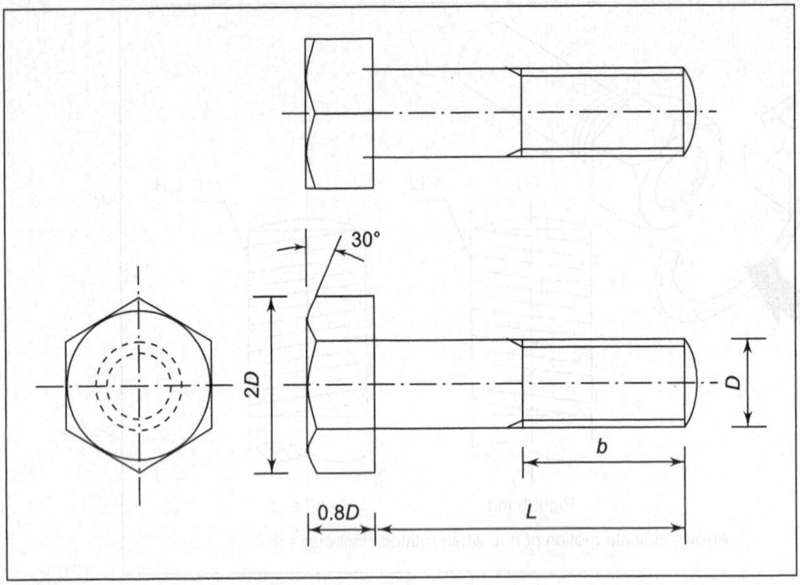

Fig. A9.14 Hexagonal-headed bolt

Square-headed bolt This bolt is used where a head is to be accomodated in a recess. The recess is also in the form of a square shape. This prevents the bolt from turning when the nut is screwed on or off it.

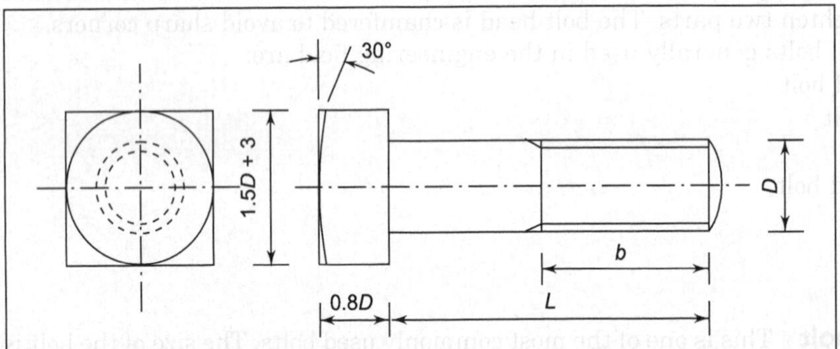

Fig. A9.15 Square-headed bolt

Cup-headed bolt It is a square neck bolt which fits in a square recess. This fit prevents the bolt from rotating.

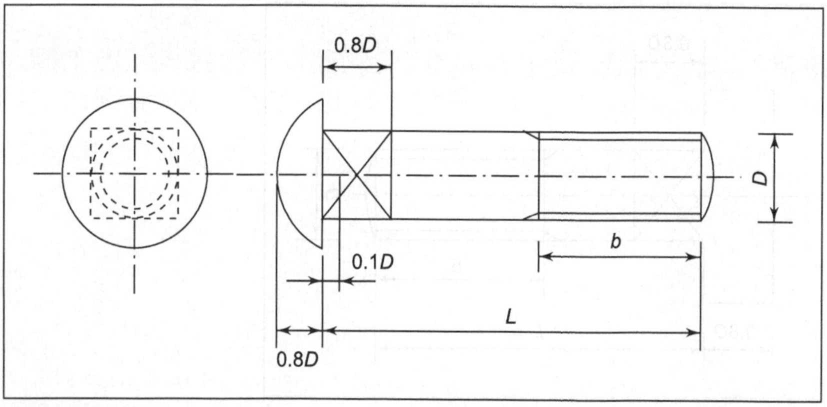

Fig. A9.16 Cup-headed bolt

Cylindrical-headed bolt This bolt is used when limited space is available and it also avoids the use of spanner. The rotation of the bolt is prevented by means of a pin inserted into the shank, just below the head. These bolts are commonly used in big ends of connecting rods and eccentrics.

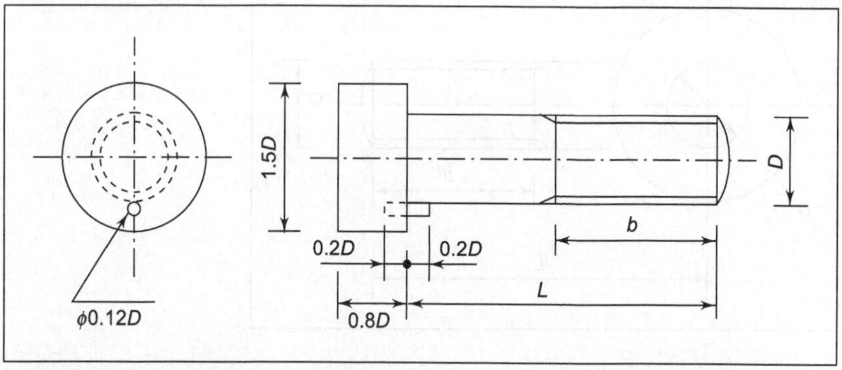

Fig. A9.17 Cylindrical-headed bolt

***T*-Headed bolt** This bolt is mainly used in machine tool tables that are provided with *T*-slots. This makes it possible to mount jigs and fixtures anywhere on the slide by using *T*-headed bolts. The neck of the bolt is made square, fitting into the slot provided on the slide. The square shape prevents the bolt from rotating.

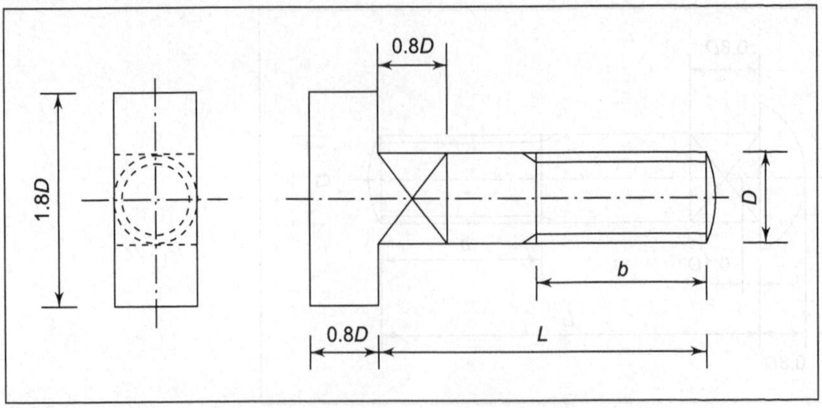

Fig. A9.18 T-headed bolt

Eye bolt This bolt has a circular ring at one of the ends. It is rectangular in cross-section at its head, aiding in holding it to prevent rotation.

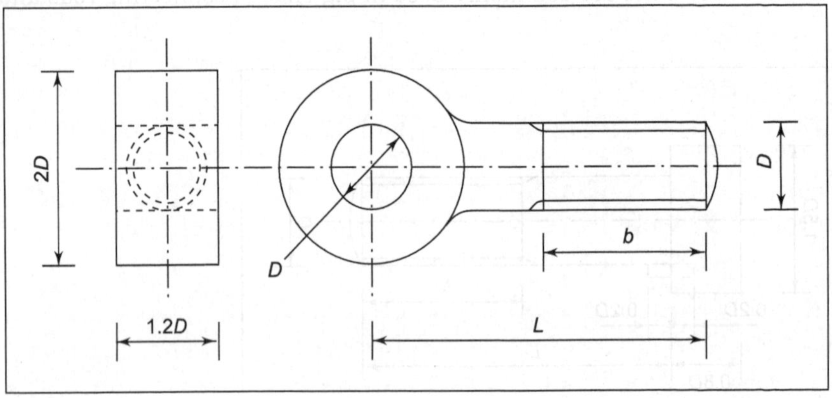

Fig. A9.19 Eye bolt

Lifting eye bolt This type of bolt is mainly used as an aid to lift heavy machinery. Hooks can be placed at the ends, inside the ring.

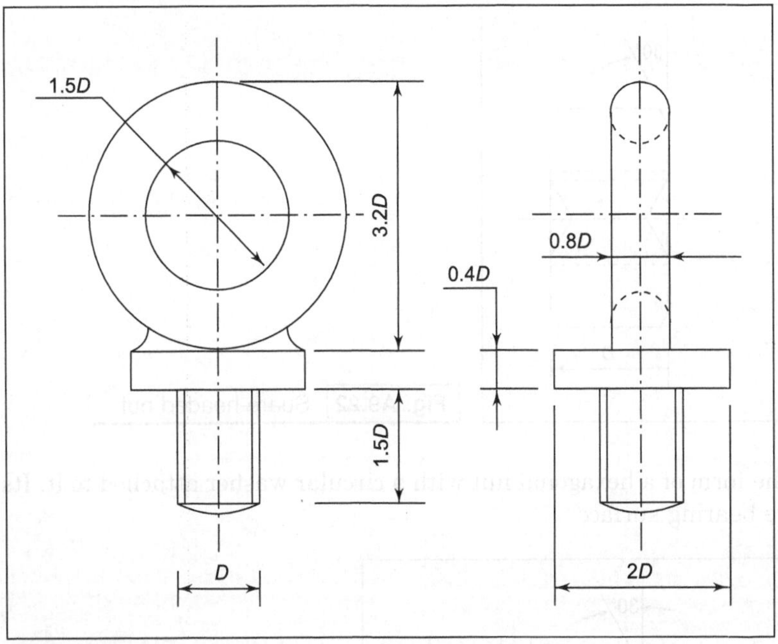

Fig. A9.20 Lifting eye bolt

9.3 NUTS

The nut is a machine component used with a bolt, stud, or pin to join two or more parts together temporarily. It is known by the outer shape of its body. It consists of internal threads of different forms, according to its application.

The different types of nuts generally used in engineering application are:

1. Hexagonal-headed nut
2. Square-headed nut
3. Flanged nut
4. Cup/Domed nut
5. Capstan nut
6. Wing nut
7. Ring nut

Hexagonal-headed nut This form of nut finds application in almost any form of industry. Adequate dimensions of a hexagonal nut are shown in Fig. A9.21.

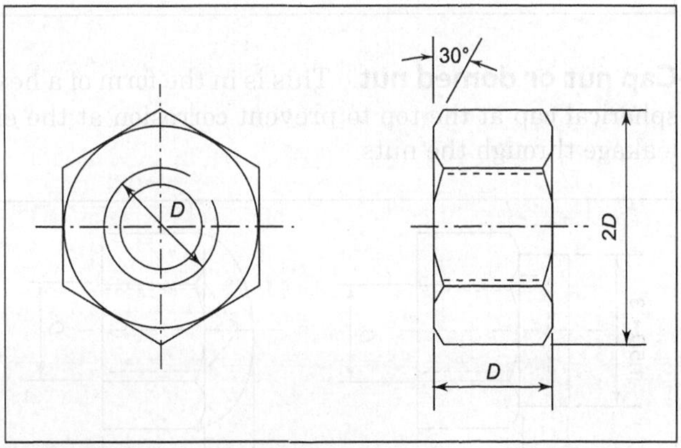

Fig. A9.21 Hexagonal-headed nut

Square-headed nut The square-headed nut is used where it can be placed in a square recess, thereby preventing its rotation.

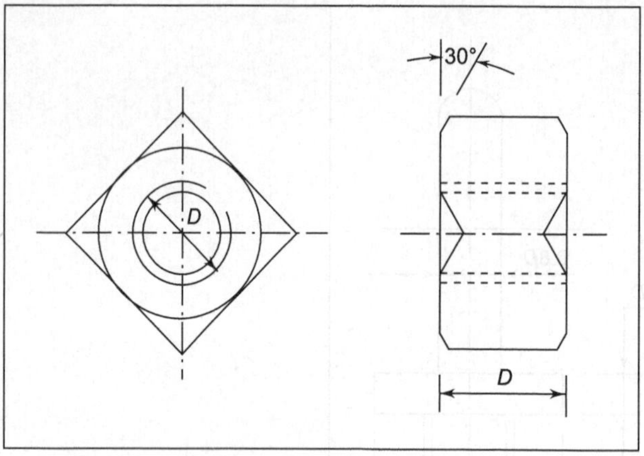

Fig. A9.22 Suare-headed nut

Flanged nut This is in the form of a hexagonal nut with a circular washer attached to it. Its flat surface provides a large bearing surface.

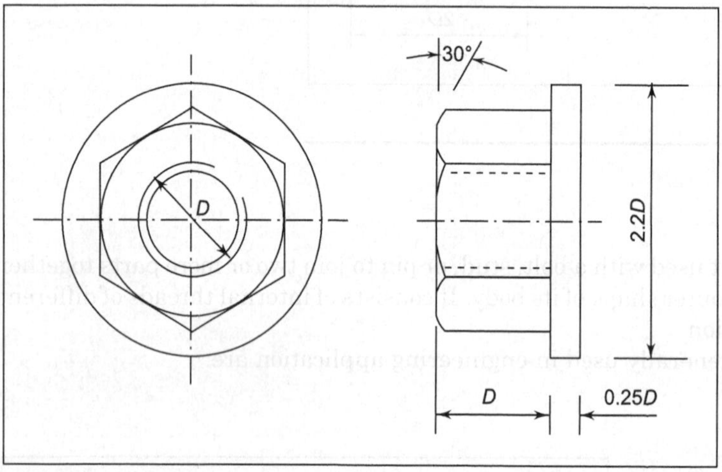

Fig. A9.23 Flanged nut

Cap nut or domed nut This is in the form of a hexagonal nut provided with a cylindrical or spherical cap at the top to prevent corrosion at the end of the bolt. This form of nut prevents leakage through the nuts.

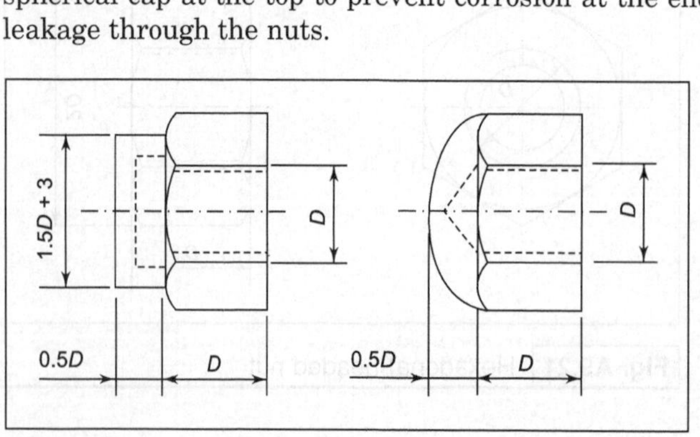

Fig. A9.24 Cap/Domed nut

Capstan nut This is in the form of a cylinder with holes drilled in the curved surface. These holes can be used to turn the nut by placing a tommy bar in it.

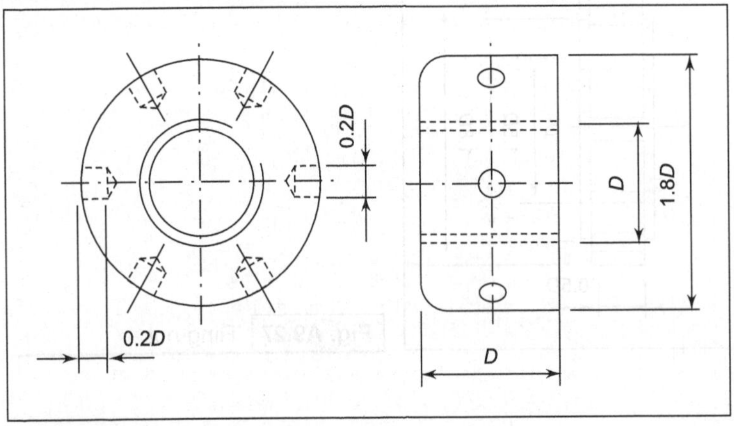

Fig. A9.25 Capstan nut

Wing nut This nut has two projections with which the nut can be easily rotated by hand.

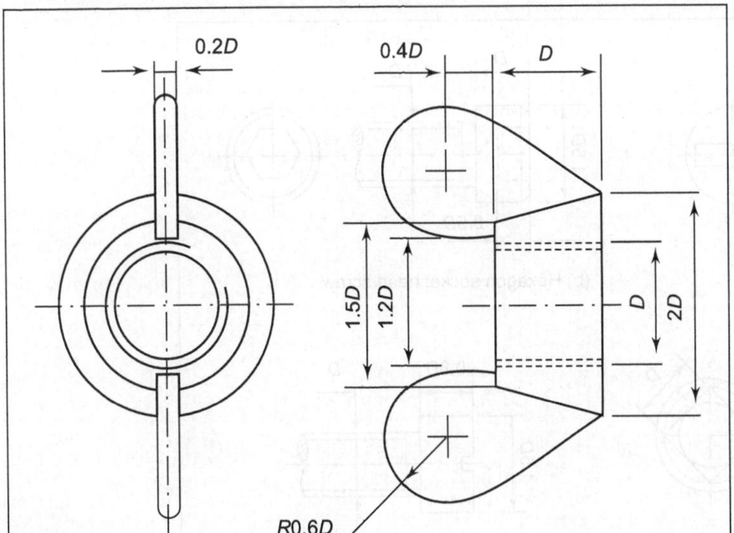

Fig. A9.26 Wing nut

Ring nut This nut is in the form of a ring provided with slots in the curved surface for a special spanner. These nuts are often used in mounting bearings.

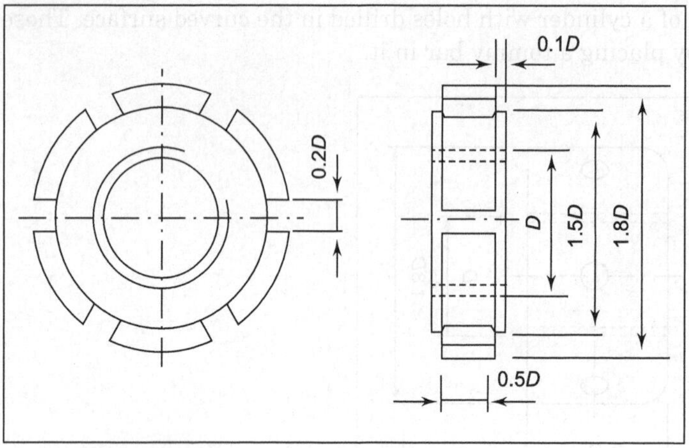

Fig. A9.27 Ring nut

9.4 SCREWS

Screws and bolts are similar, except in threading length. Screws are threaded till the end. They are used to prevent relative motion between two parts. The heads of screws have screw driver slots or they can be turned by wrenches or spanners (Refer Fig. A9.28).

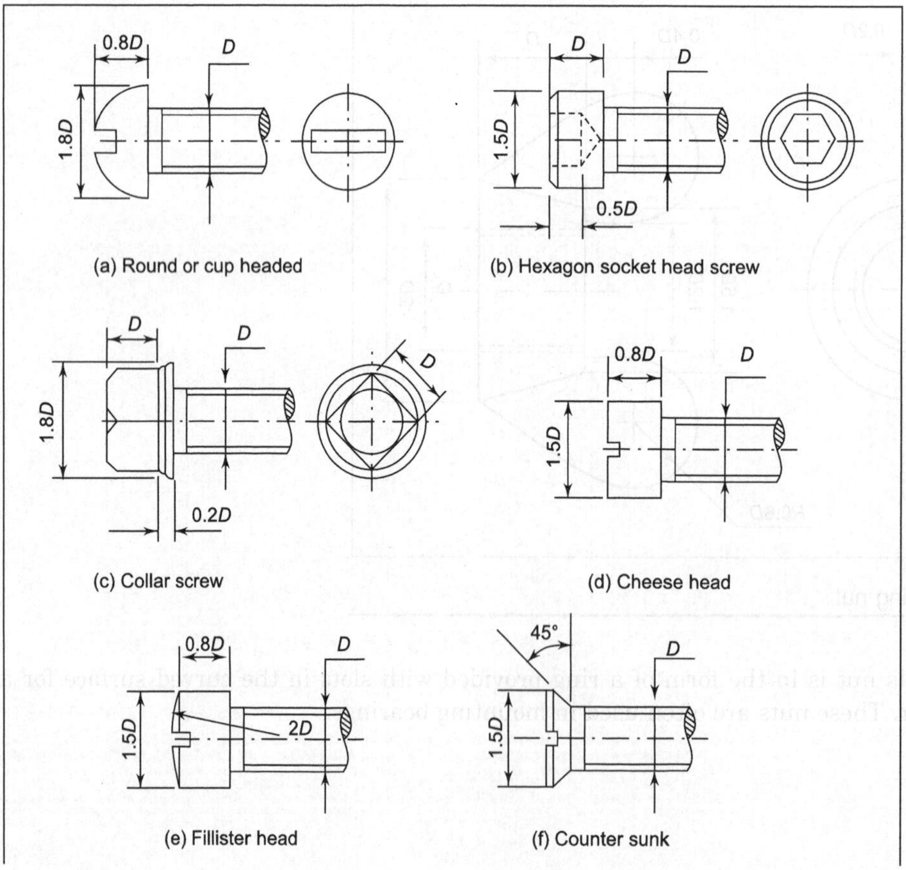

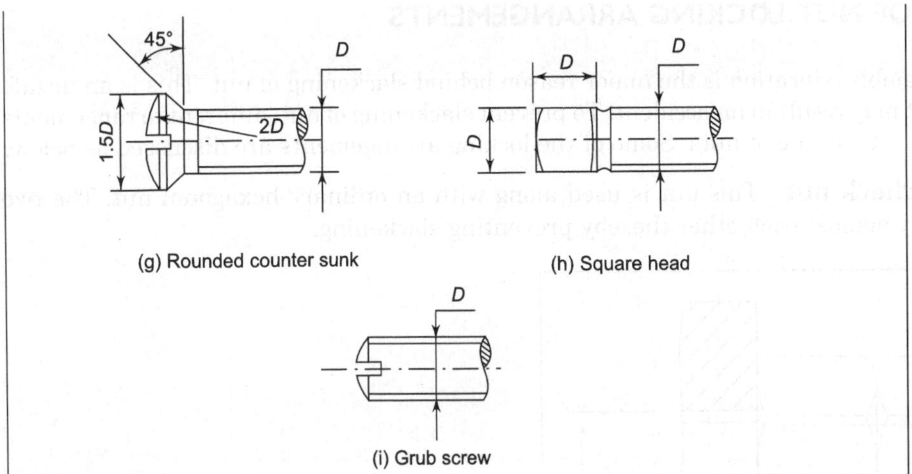

(g) Rounded counter sunk

(h) Square head

(i) Grub screw

Fig. A9.28 Screw heads

A special form of set-screw is a grub screw [Fig. A9.28 (a)] which has no head. Another important form of a screw is the hexagon socket head cap screw [Fig. A9.28 (b)]. These are used with a counter boring in the top plate, which makes the bolt head flush with the surface. They can be tightened using allan keys.

Allan keys are right angled to one side and extra long in order to get leverage for tightening and loosing of allen bolts. The cross-section of the key is set according to the cross-section of the head of the bolt.

The set-screw can have different ends, as shown in Fig. A9.29.

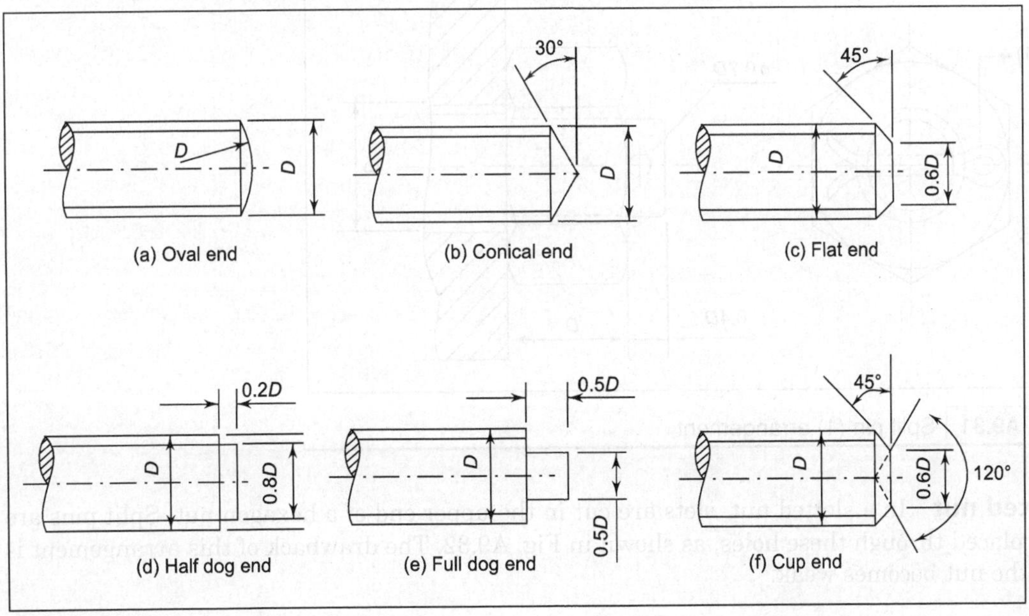

(a) Oval end

(b) Conical end

(c) Flat end

(d) Half dog end

(e) Full dog end

(f) Cup end

Fig. A9.29 Screw ends

9.5 TYPES OF NUT LOCKING ARRANGEMENTS

In working assembly, vibration is the major reason behind slackening of nut. This is an unsafe situation, and it may result in an accident. To prevent slackening of nut, different arrangements are used to achieve locking of nuts. Some of the locking arrangements are discussed as below.

Lock nut or check nut This nut is used along with an ordinary hexagonal nut. The two nuts lock tightly against each other thereby preventing slackening.

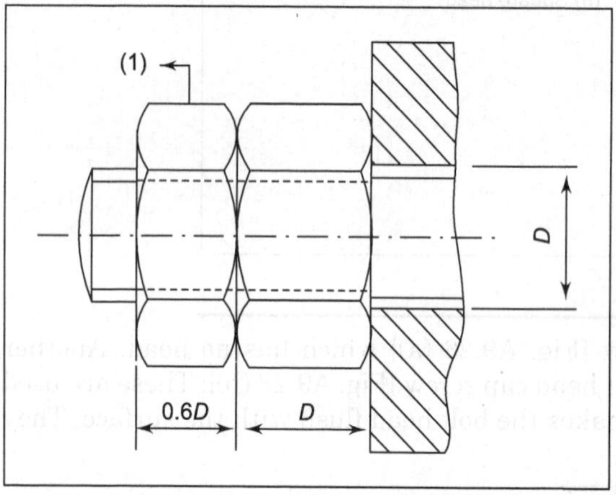

Fig. A9.30 Lock/Check nut(1) arrangement

Split-pin This method utilises inserting of a split-pin through a hole provided in the bolt end. This safely secures the nut in its position resulting in locking of the nut.

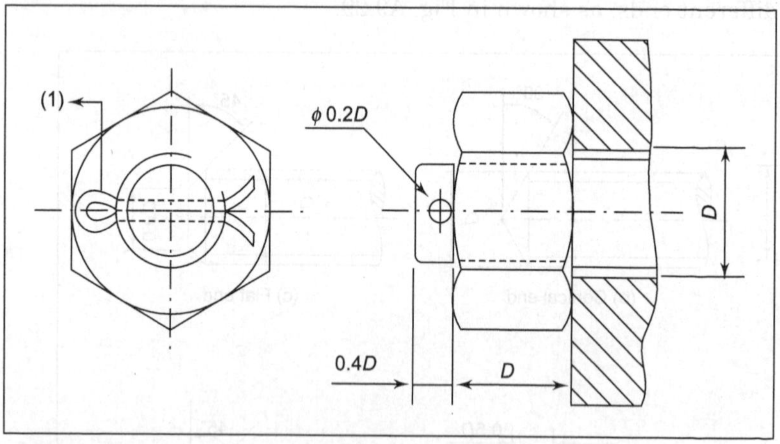

Fig. A9.31 Split pin (1) arrangement

Slotted nut In a slotted nut, slots are cut in the upper end of a hexagon nut. Split pins are then placed through these holes, as shown in Fig. A9.32. The drawback of this arrangement is that the nut becomes weak.

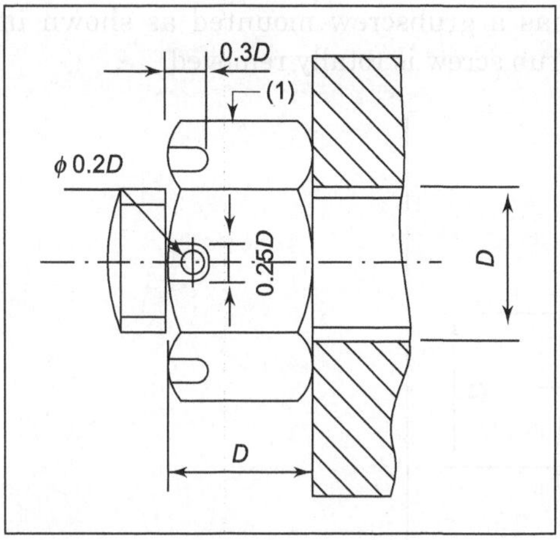

Fig. A9.32 Slotted nut (1) arrangement

Castle nut Castle nuts are an improvement over slotted nuts as they have a cylindrical collar in which slots are provided for inserting split-pins.

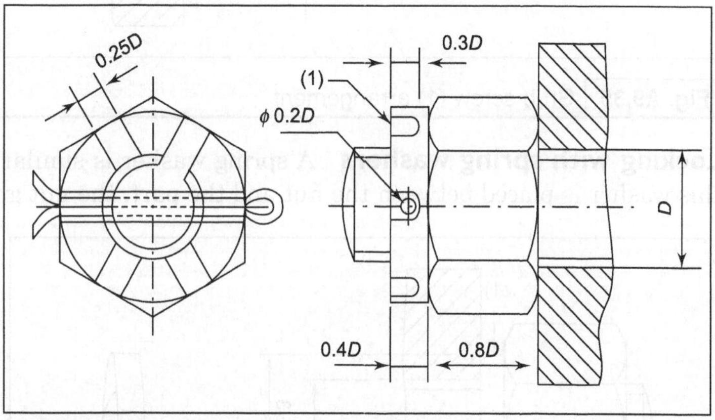

Fig. A9.33 Castle nut (1) arrangement

Sawn nut or wiles nut This nut is an ordinary hexagonal nut with a slot cut half way through it. A cap screw is passed through a clear hole in the upper part and is tightened in the lower tapped part of the nut. This causes tightening of the threads on the bolt, resulting in locking due to friction.

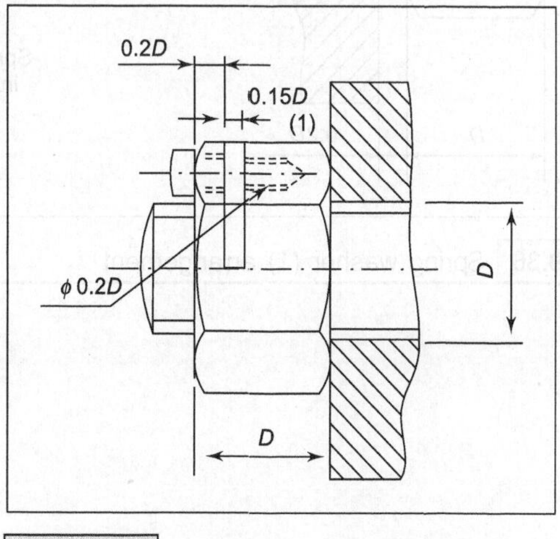

Fig. A9.34 Sawn nut (1) arrangement

Locking with grub screw This arrangement has a grubscrew mounted as shown in Fig. A9.35. This prevents rotation of the nut till the grub screw is totally removed.

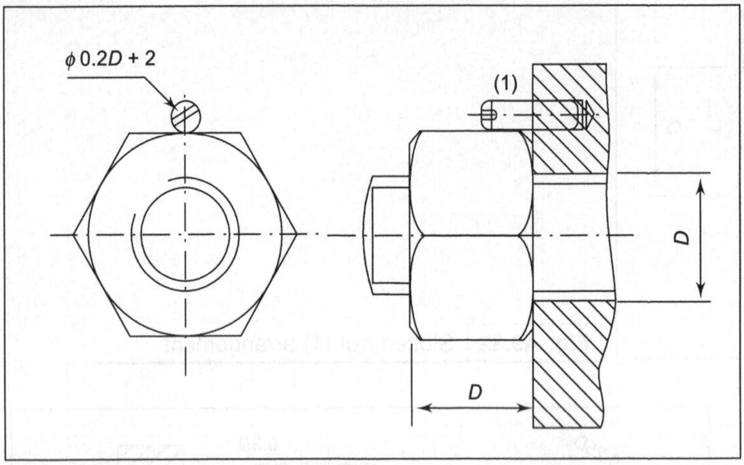

Fig. A9.35 Grub screw (1) arrangement

Locking with spring washers A spring washer is similar to a single coil of spring. When this washer is placed between the nut and the part, the nut grips the thread on the bolt.

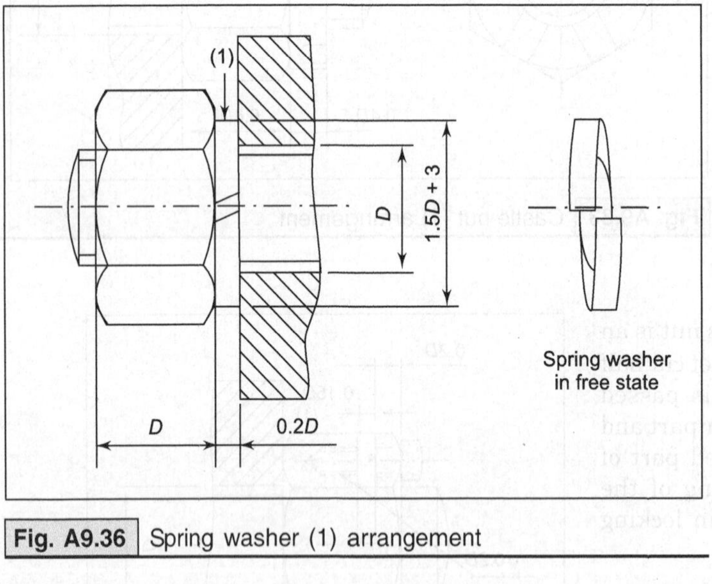

Fig. A9.36 Spring washer (1) arrangement

Review Questions

1. What is screw fastener?
2. Define important parameters of screw threads.
3. State and draw different types of thread forms.
4. How are pipe threads different from normal threads?
5. What is multiple start thread?
6. What are the pitch and lead in the case of single start and double start threads?
7. How is it possible to identify right-hand and left-hand threading?
8. State and draw the types of bolts, nuts, and screws.
9. Differentiate between (a) stud bolt and hexagonal bolt, (b) bolt and screw.
10. Draw two types of nut locking arrangements.

CHAPTER 10

Pipe Joints

This chapter contains information on pipe joints

After reading the chapter you will be able to

- classify types of joints
- draw symbols for piping joints
- identify commonly used pipe fittings

10.1 GENERAL DISCUSSION

Hollow cylinders used to carry water, oil, steam, gas, and so on from one location to another are called pipes. The size of a pipe is stated by its nominal bore. The thickness of the pipe is not considered in it. For example, 50 *NB* pipe means the diameter of its bore is 50 mm.

In an industry,pipes are the main source of supply of fluid from one place to another. Pipes carry fluids such as water, solvents, steam, gases, air, oil, and so on to various places. They are available in standard lengths and are made up of different materials such as cast iron, wrought iron, steel, brass, copper and poly vinyl chloride (PVC). They are available in standard lengths and it is necessary to join various pipes to achieve the desired network. Generally, pipes are designated by their bore diameter, which is called Nominal Bore (*NB*). Depending on the material of pipe and the fluid it carries, different types of joints are required to connect various pipes. Some of these are discussed below.

TABLE A10.1 Varieties of Pipes and Their General Use

Sl. No.	*Pipe Name*	*Size in mm*	*Application*
1	Cast iron pipe	50 to 1200	For water, gas, and steam
2	Wrought iron pipe	6 to 150	For domestic water and gas at low pressure
3	Steel pipes	10 to 1800	For water, steam, sewage or air at high pressure
4	Lead pipes	5 to 50	For domestic purpuses or pipelines with several bends
5	Copper and brass pipes	3 to 50	For hot water in engines

10.2 PIPE JOINTS

The temperature and pressure of the fluid decides the type of joint to be used.

Cast iron pipe joints Cast iron pipes are used if the pipe diameters are not less than 50 mm. The pipes are cast with integral flanges at the ends, as shown in Fig. A10.1. The faces of the flanges are machined at right angles to the axis of the pipe, so as to achieve perfect alignment when they are joined. To make a joint leak-proof, a packing ring or gasket of soft material, such as rubber or leather, is placed between the two flanges. The joint is then secured by tightening the nuts and bolts. For high pressure applications, ribs may be provided on the flange for strength, as shown in Fig. A10.2. Sometimes flanges are provided with a facing of 3 to 5 mm thickness, between which the packing ring is placed, as shown in Fig. A10.3.

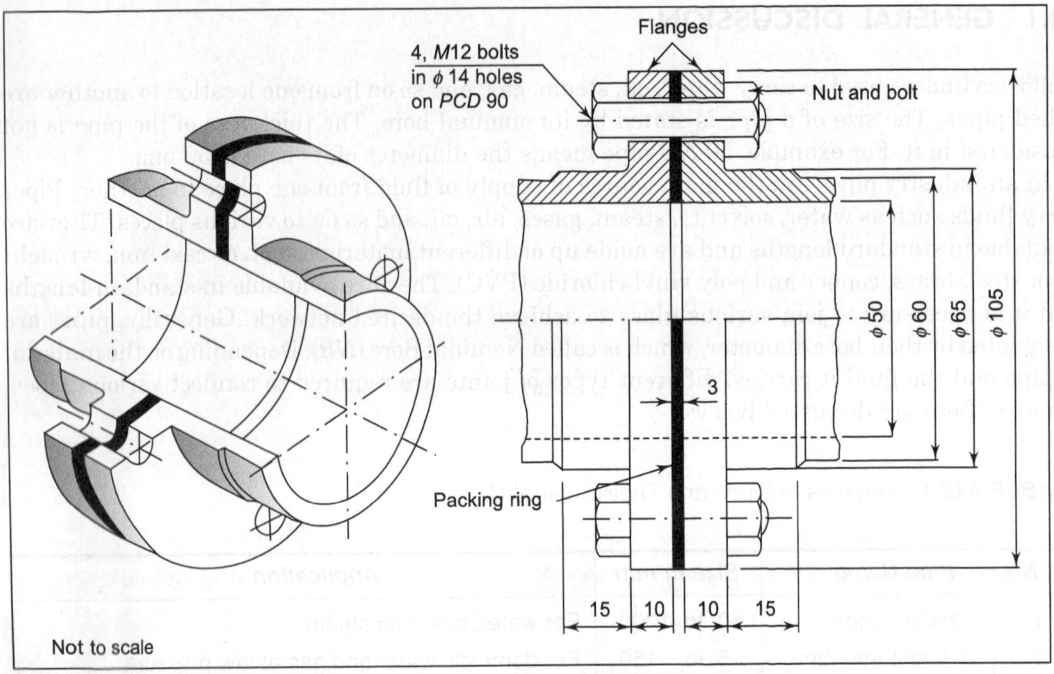

Fig. A10.1 Cast iron flanged joint

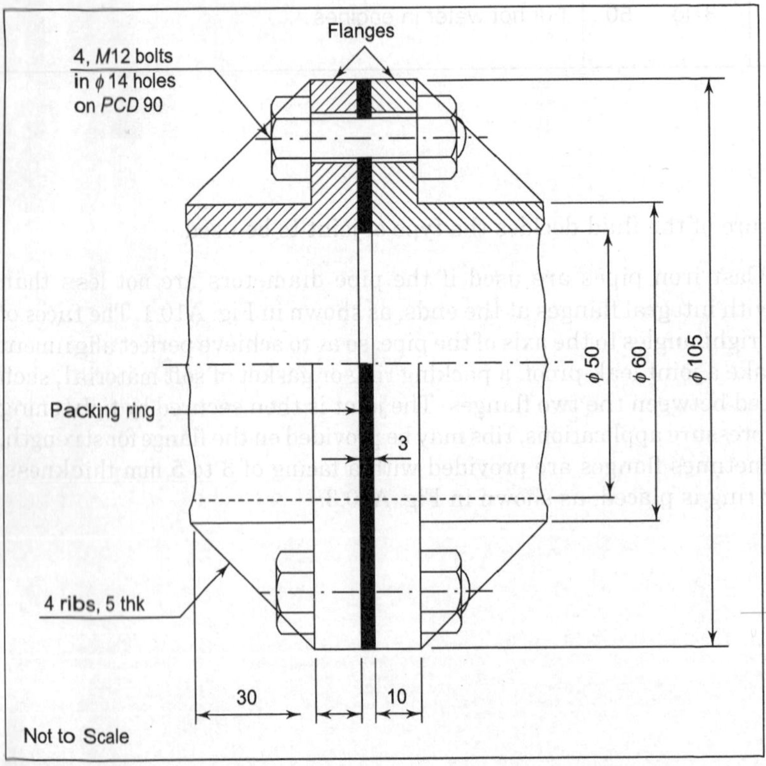

Fig. A10.2 Cast iron flanged joint for high pressure

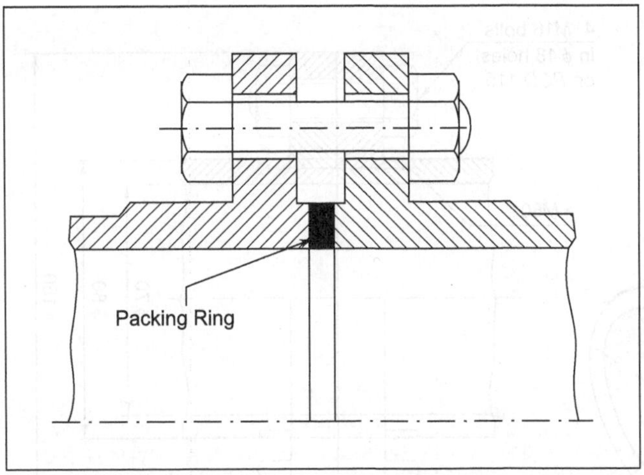

Fig. A10.3 Cast iron flanged joint with facings

Joints for wrought iron and steel pipes Wrought iron pipes are mainly used for domestic applications and are generally galvanised. These pipes have good strength if they are solid drawn and have in general replaced cast iron pipes. Wrought iron pipes can be butt welded and are available in small sizes from 12 mm to 100 mm internal diameters. These pipes can have flanges formed or welded on their ends, which can be fastened using bolts. Fig. A10.4 shows a wrought iron flange that has been formed integral with the pipes.

Another type of joint would be of the form shown in Fig. A10.5. Here flanges made of cast iron or steel are screwed on the ends of the threaded pipe and the flanges are then fastened using bolts.

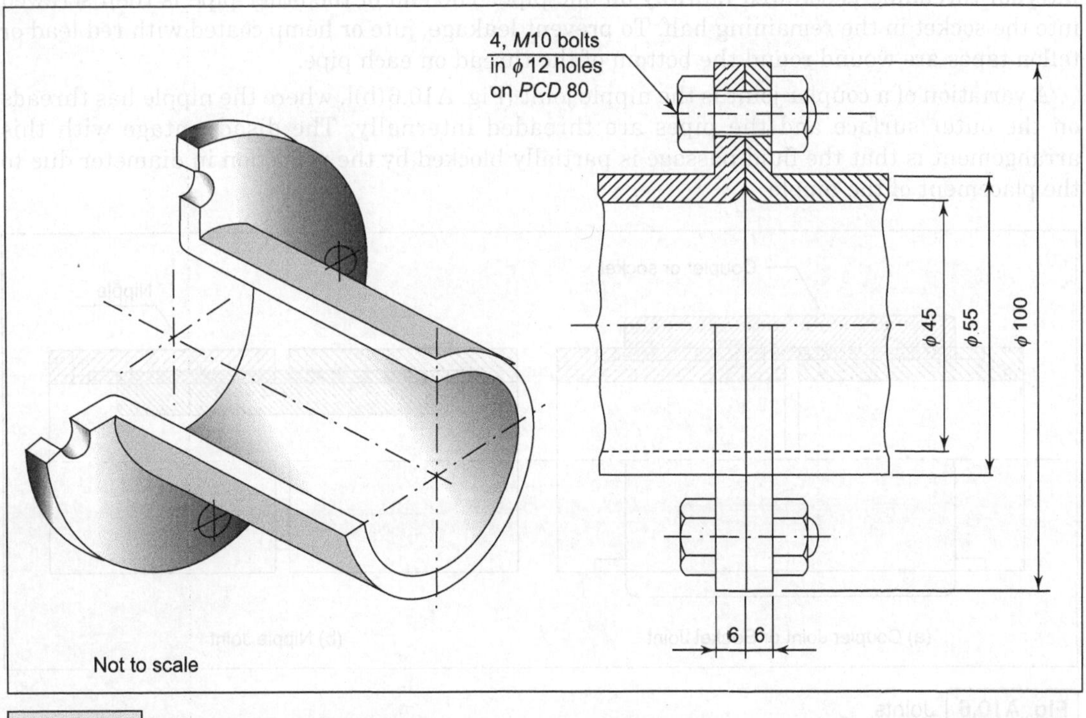

Fig. A10.4 Wrought iron flanged joint

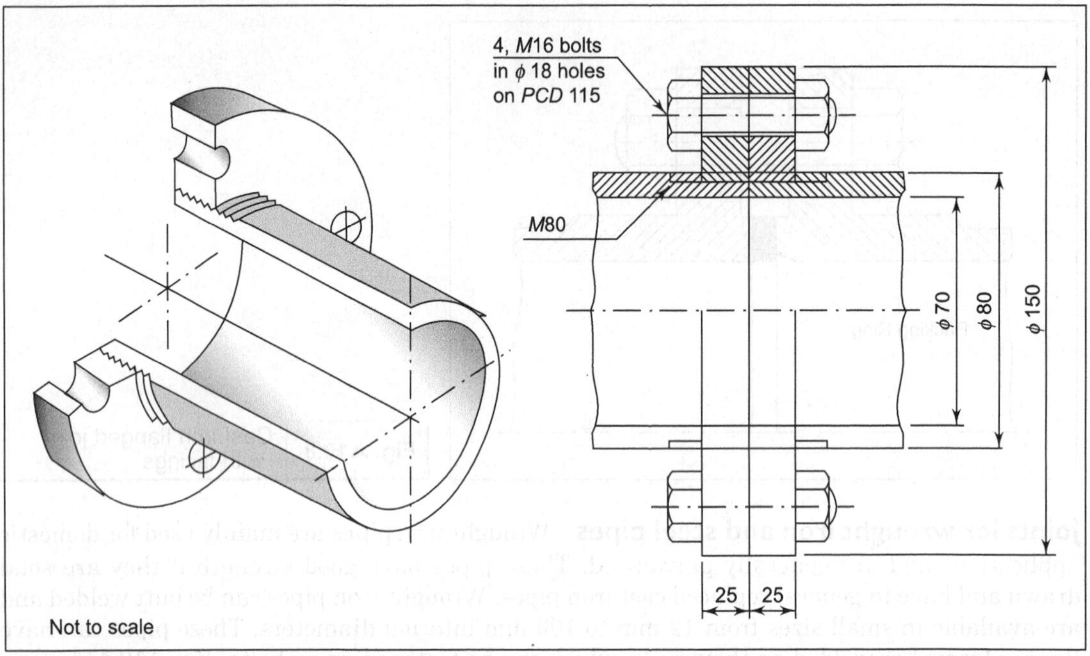

Fig. A10.5 Cast iron joint with screwed flanges

Steel pipes can also be connected to each other by a coupler or a socket joint, as shown in Fig. A10.6 (a). For this joint, the pipes are threaded on the outside. A socket or coupler with internal threading is screwed halfway on one pipe. The end of the other pipe is then screwed into the socket in the remaining half. To prevent leakage, jute or hemp coated with red lead or teflon tapes are wound round the bottom of the thread on each pipe.

A variation of a coupler joint is the nipple joint [Fig. A10.6 (b)], where the nipple has threads on the outer surface and the pipes are threaded internally. The disadvantage with this arrangement is that the fluid passage is partially blocked by the reduction in diameter due to the placement of the nipple.

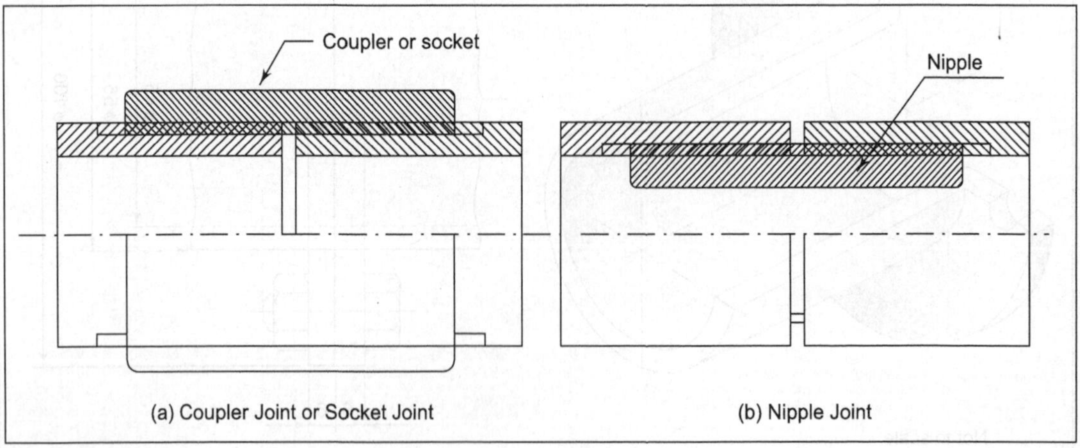

Fig. A10.6 Joints

Joint for copper pipes Copper joints are seamed longitudinally or are solid drawn. Separate flanges made of brass or gun metal are attached to the pipe ends by brazing as shown in Fig. A10.7.

Copper finds applications in heat exchangers, refrigeration coils, and steam pipes in marine engines.

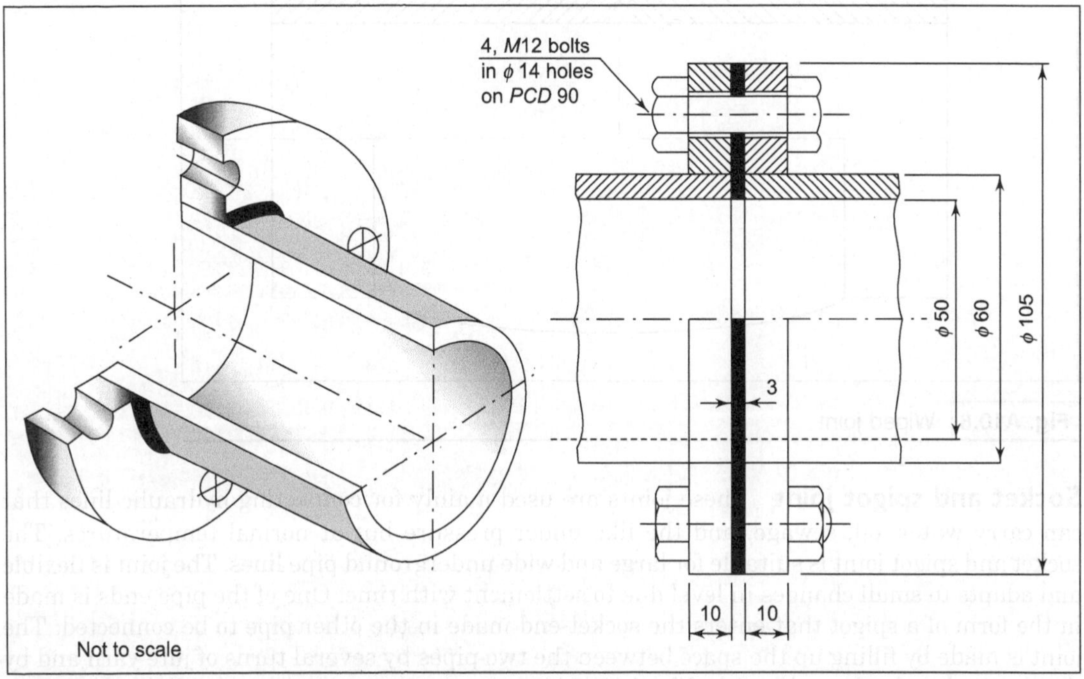

Fig. A10.7 Copper pipe with brazed flange joint

Joint for lead pipes Lead pipes are flexible and can be bent easily to form desired shapes. These pipes are used in low pressure lines. The most common type of a joint for connecting lead pipes is the wiped joint, as shown in Fig. A10.8.

In this joint, the end of one pipe is opened slightly wider. The other pipe is filed to a sharp edge. The ends are polished and joined together. After placing the pipes, one within the other, molten solder is poured over the joint in order to seal the pipes together. The hot solder is rapidly wiped with a piece of cloth smeared with grease. The solder adhers to the polished surface of the pipe and the outside surface of the joint is smoothened by wiping.

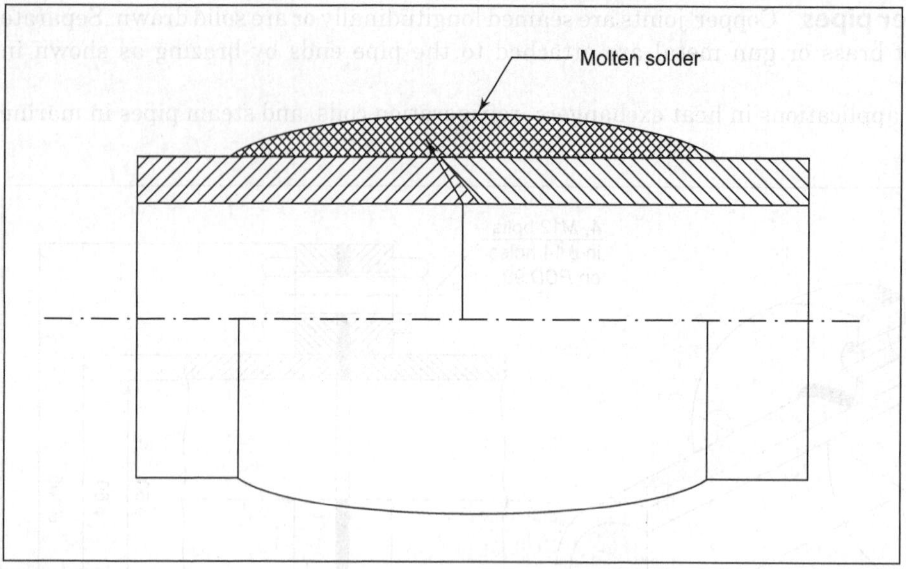

Fig. A10.8 Wiped joint

Socket and spigot joint These joints are used mainly for connecting hydraulic lines that can carry water, oil, sewage, and the like under pressure but at normal temperatures. The socket and spigot joint is suitable for large and wide underground pipe lines. The joint is flexible and adapts to small changes in level due to settlement with time. One of the pipe ends is made in the form of a spigot that enters the socket-end made in the other pipe to be connected. The joint is made by filling up the space between the two pipes by several turns of jute yarn and by pouring molten lead, as shown in Fig. A10.9.

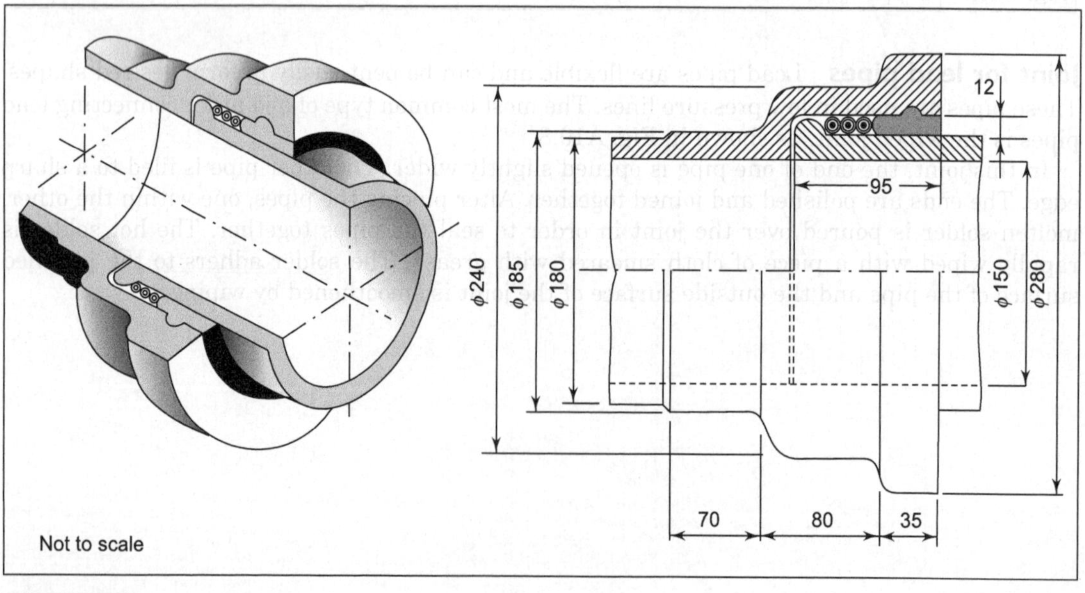

Fig. A10.9 Socket and spigot joint

Flanged hydraulic joint This form of joint is used for pipes connecting high pressure fuilds, where the pressure ratings vary from 6 to 9 N/mm^2. In this form of joint, oval flanges are cast at the end of the pipes. A spigot is provided in the centre of one flange while a machined socket is provided at the centre of the other flange, as shown in Fig. A10.10. This allows for proper alignment. A gasket made of rubber or canvas is inserted between the spigot and the socket to ensure that there is no leakage. The flanges are then fastened with bolts.

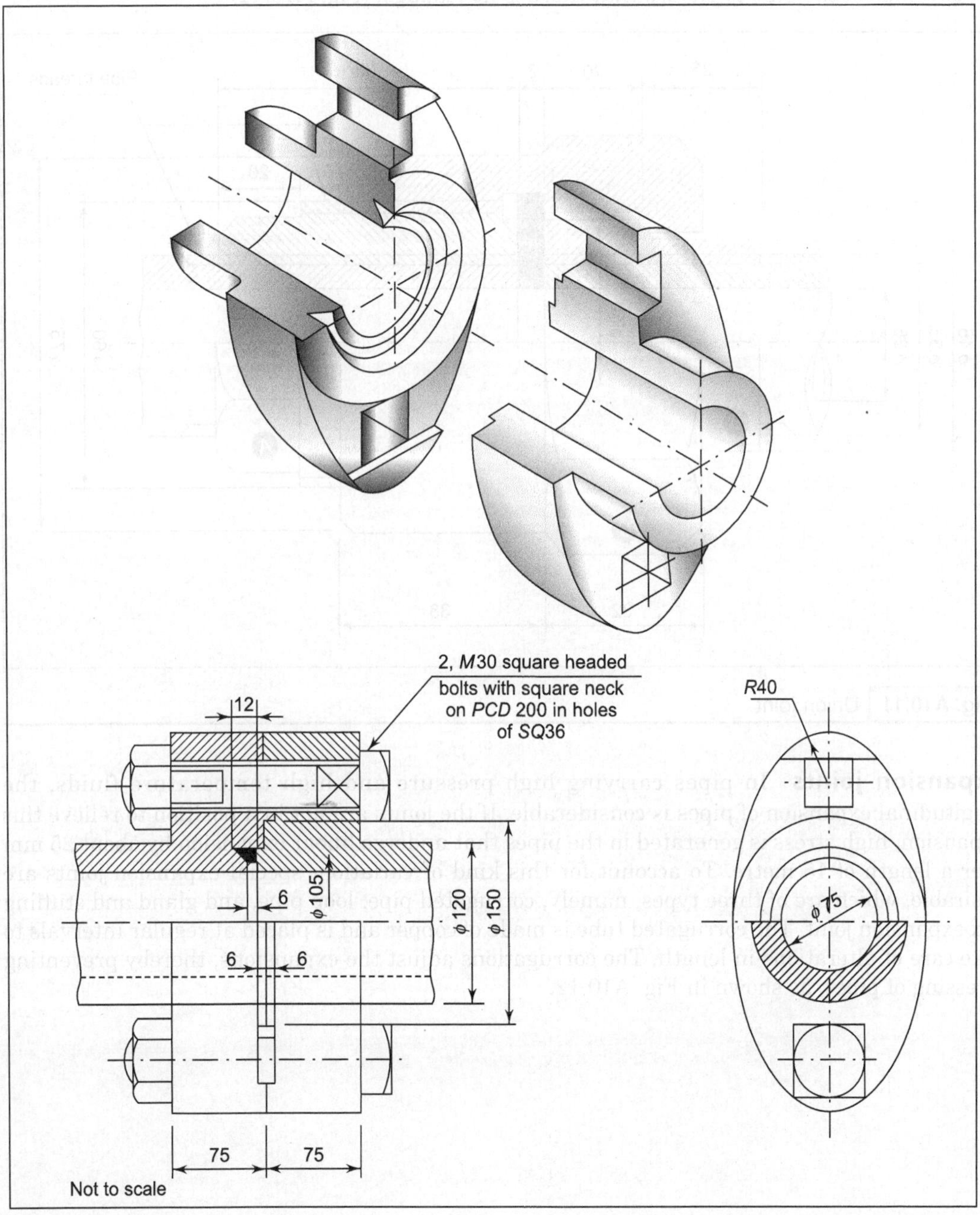

Fig. A10.10 Flanged hydraulic joint

Union joint This joint can be used for connecting small pipes. The joints are of use where connection and disconnection need to be achieved without disturbing the pipe layout.

This joint has a special form of nut *A*, which has inside and outside threading. The nut *A* is screwed on the end of one of the pipes. Another nut *B* is screwed on the end of the other pipe that is to be connected. The two nuts and pipes are then brought closer to each other by a coupler nut *C*, which screws on the outer threads of nut *A*. A packing ring is inserted between the two ends of the pipe to ensure that there is no leakage. (Fig. A10.11)

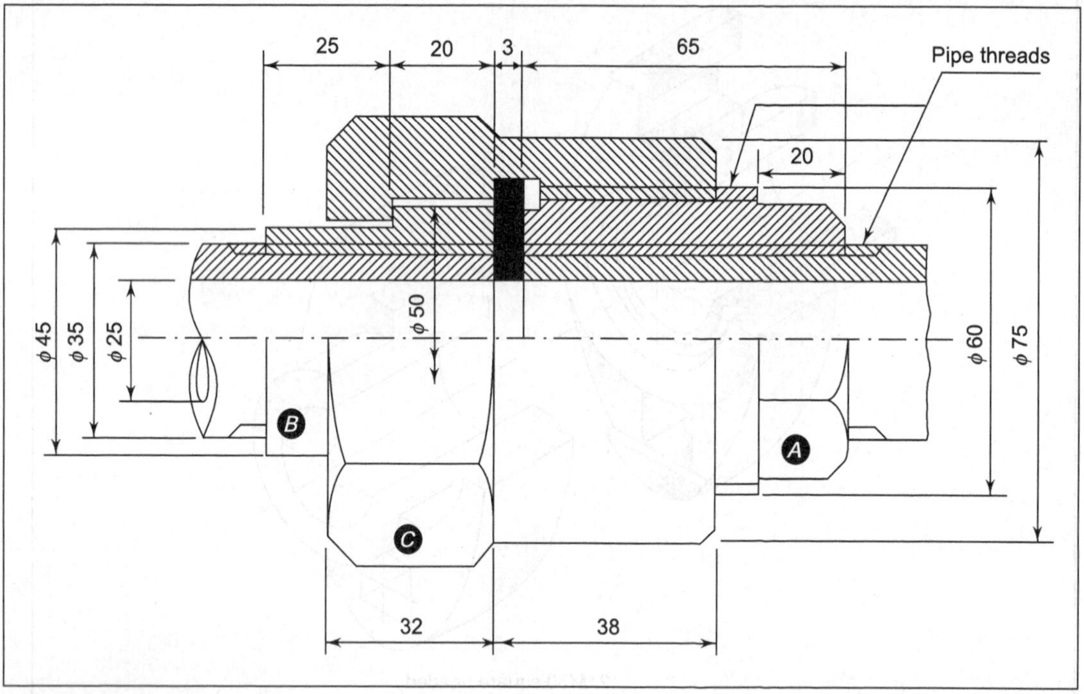

Fig. A10.11 Union joint

Expansion joints In pipes carrying high pressure and high temperature fluids, the longitudinal expansion of pipes is considerable. If the joints are not in a position to relieve this expansion, high stress is generated in the pipes that undergo linear expansion by about 25 mm over a length of 15 metre. To account for this kind of variation, special expansion joints are desirable, which are of three types, namely, corrugated pipe, loop pipe, and gland and stuffing box expansion joint. The corrugated tube is made of copper and is placed at regular intervals to take care of alterations in length. The corrugations adjust the expansions, thereby preventing stressing of pipes, as shown in Fig. A10.12.

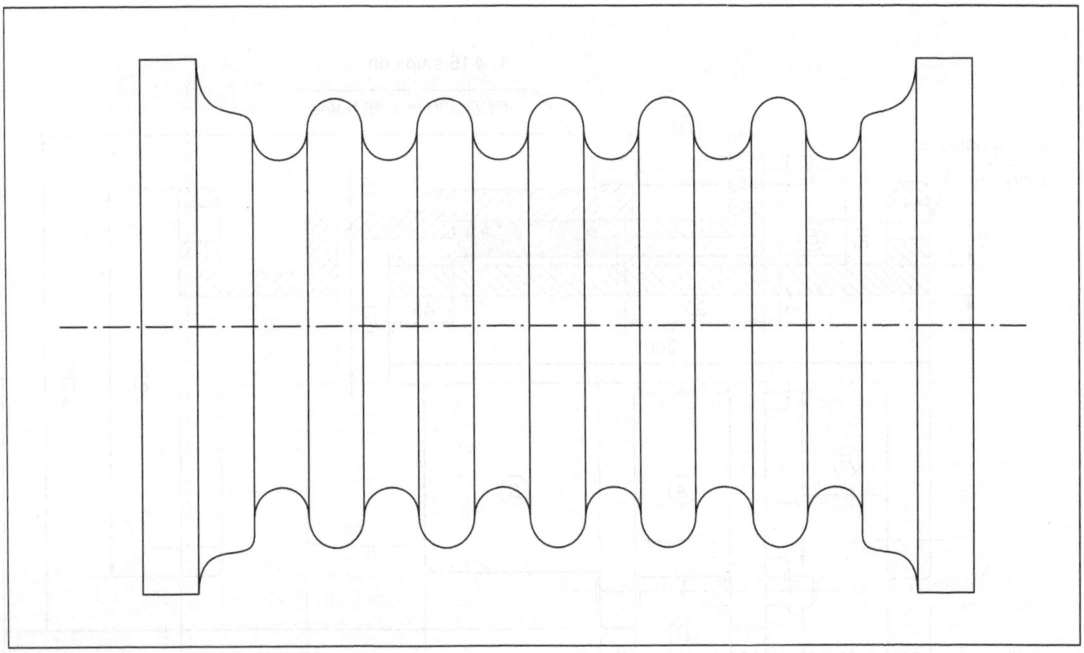

Fig. A10.12 Corrugated pipe joint

The loop pipe is also made of copper and it can be connected to pipes by flanges, as shown in Fig. A10.13. The radius R of the loop is taken as five times the diameter D of the pipe.

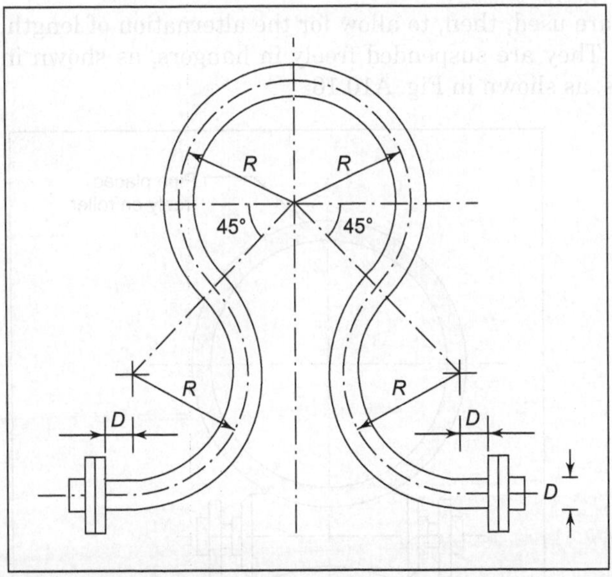

Fig. A10.13 Loop joint

Gland and stuffing box expansion joint consists of a pipe P, which is free to move in a stuffing box B, as shown in Fig. A10.14. Leakage is prevented by packing in the box with asbestos material C pressed against a gland A by means of studs D. Brass bushings fitted on brass liner E are fitted on pipe P to prevent corrosion. This joint gives better results and is used quite often.

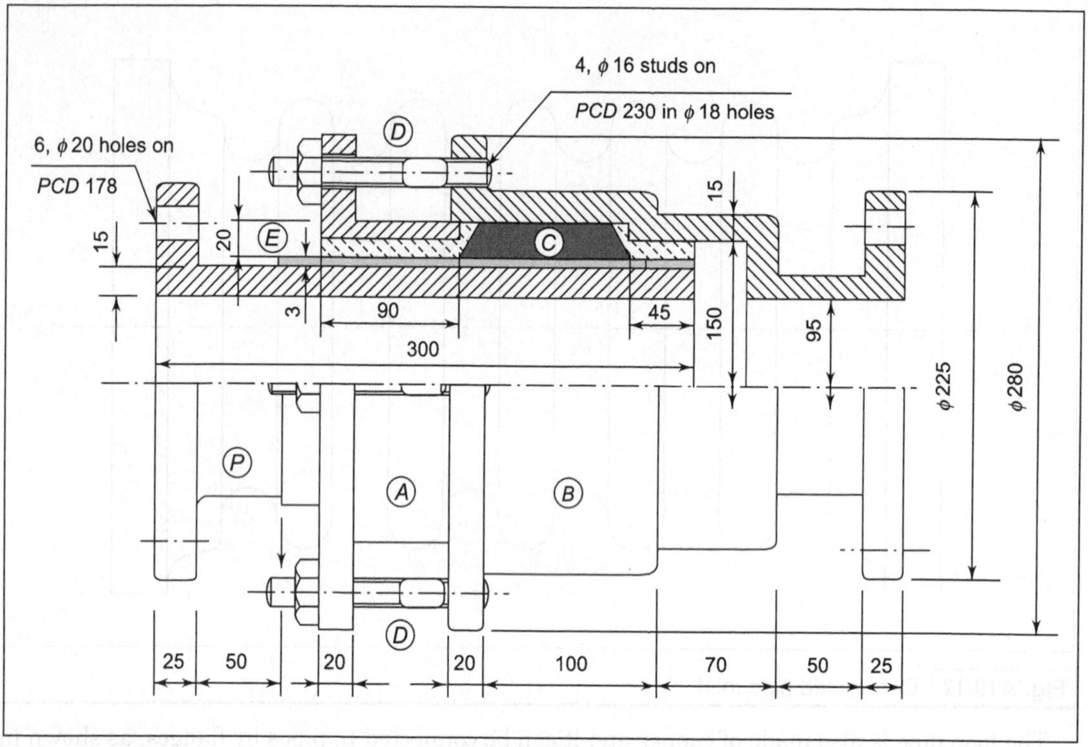

Fig. A10.14 Gland and stuffing box expansion joint

In any pipe layout if expansion joints are used, then, to allow for the alternation of length, the pipes should not be rigidly clamped. They are suspended freely in hangers, as shown in Fig. A10.15, or freely supported on rollers, as shown in Fig. A10.16.

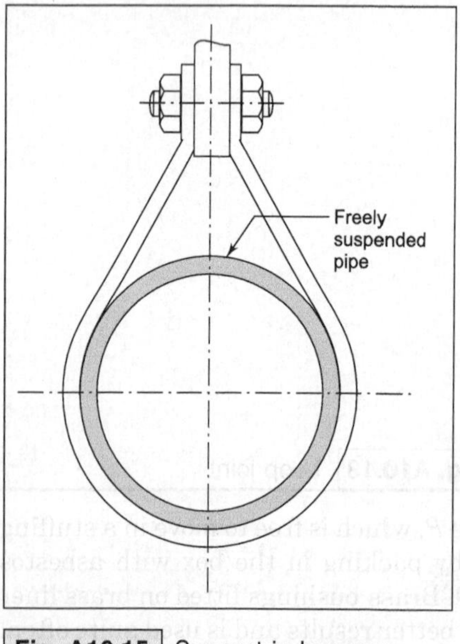

Fig. A10.15 Support in hangers

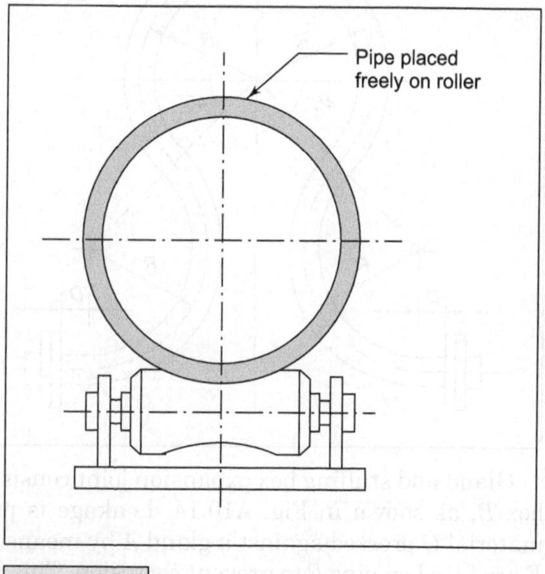

Fig. A10.16 Support on rollers

10.3 PIPE FITTINGS

To achieve the desired layout of pipes, various types of fittings are necessary. These fittings allow the designer to branch the pipelines, connect smaller pipes to larger pipes, or simply terminate the lines. Different types of pipe fittings such as bends, elbows, cross, and tee are available for smaller pipes. Plugs are used to terminate the open ends of a pipeline. These fittings are generally meant for wrought iron and steel pipes, as depicted in Fig. A10.17.

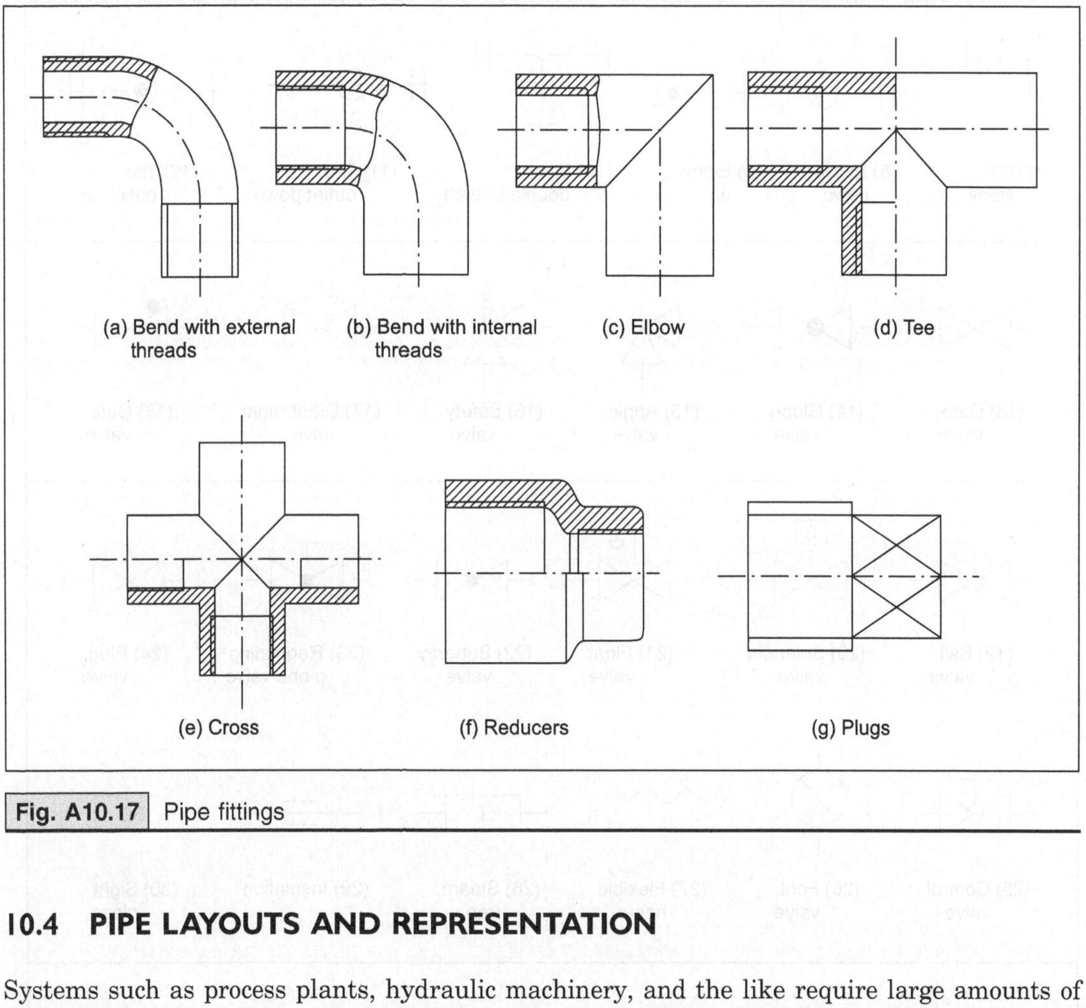

Fig. A10.17 Pipe fittings

10.4 PIPE LAYOUTS AND REPRESENTATION

Systems such as process plants, hydraulic machinery, and the like require large amounts of piping between interconnecting equipment. The representation of such piping details is called a piping layout. Making of piping drawings is a specialised job and requires good understanding of its principles. Piping drawings are generally of two types.

Process plant piping This involves the drawing of various pipes, their branches, and the layouts. The symbols representing the various fittings and valves are shown in Fig. A10.18.

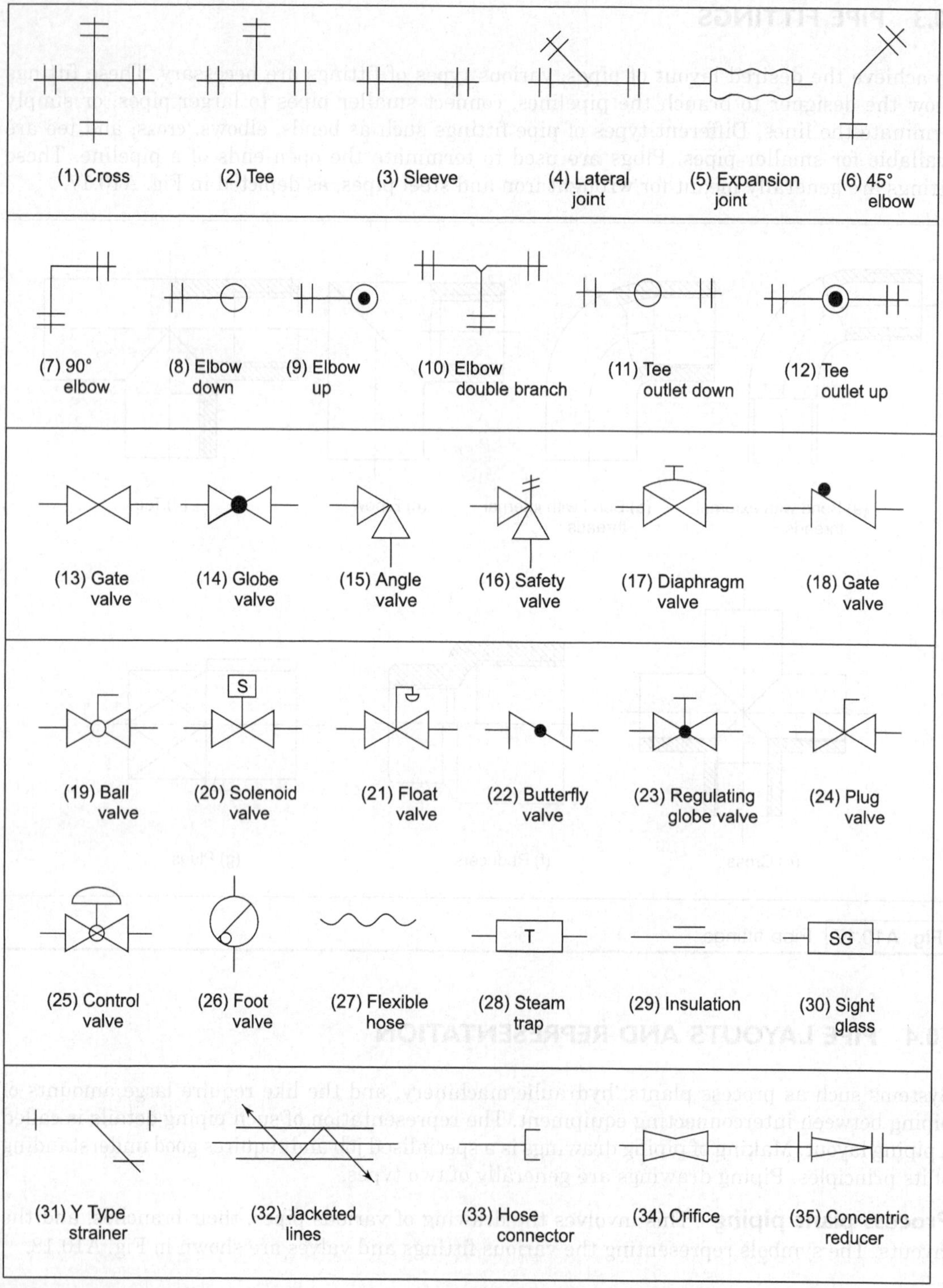

Fig. A10.18 Piping symbols

In drawing piping diagrams, the following rules apply:

- Vertical pipes — Projections drawn in the horizontal plane [Fig. A10.19(a)]
- Horizontal pipes — Projections drawn in the vertical plane [Fig. A10.19(b)]
- Not parallel to any coordinate plane pipes — Projections drawn in both the planes [Fig. A10.19(c)]
- Auxilliary projections — Projections drawn in the planes using thin continuous lines [Fig. A10.19(d)]

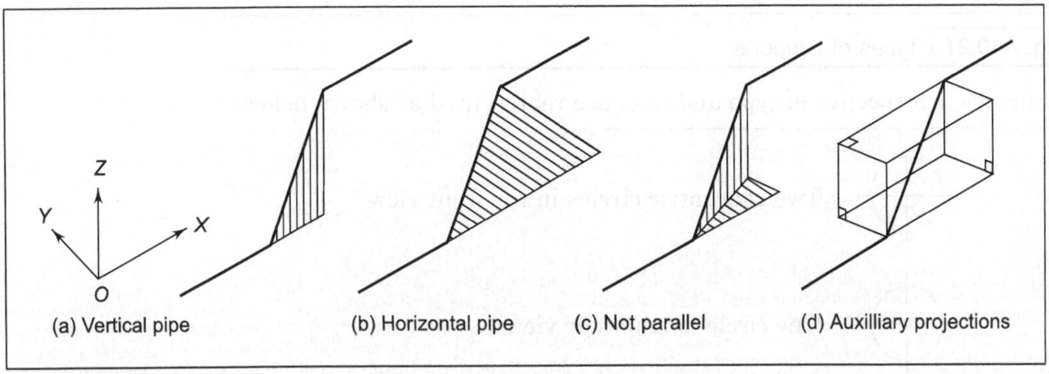

Fig. A10.19 Piping rules

Pipes are dimensioned by specifying the outer diameter (d) × thickness (t). Pipes having bends must be dimensioned from the central line or from the central line to the end of the pipe, as shown in Fig. A10.20(a). Radii and bends can be simplified and they can be representated as shown in Fig. A10.20(b). In certain cases, reductant dimensions are necessary for technical reasons. Such dimensions, which are also called auxilliary dimensions, should be shown in brackets, as shown in Fig. A10.20(c).

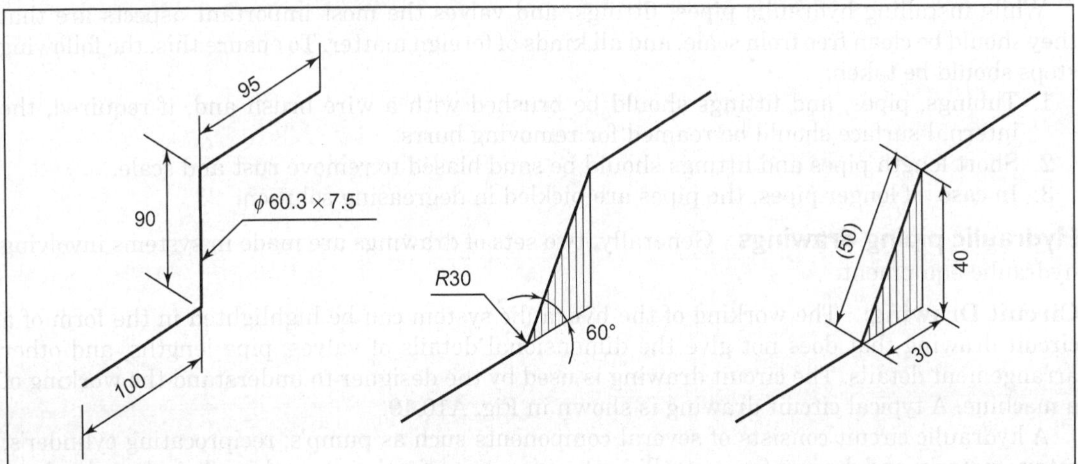

Fig. A10.20 Piping rules

Pipes, as mentioned earlier, are held in supports and hangers. The supports are represented as shown in Fig. A10.21. The hangers would be shown in a manner similar to supports, but in the reverse direction.

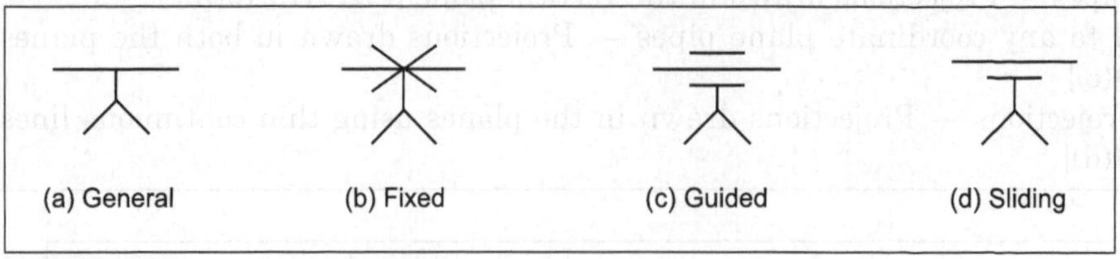

Fig. A10.21 Types of supports

Flanges, irrespective of type and size, are represented as shown below:

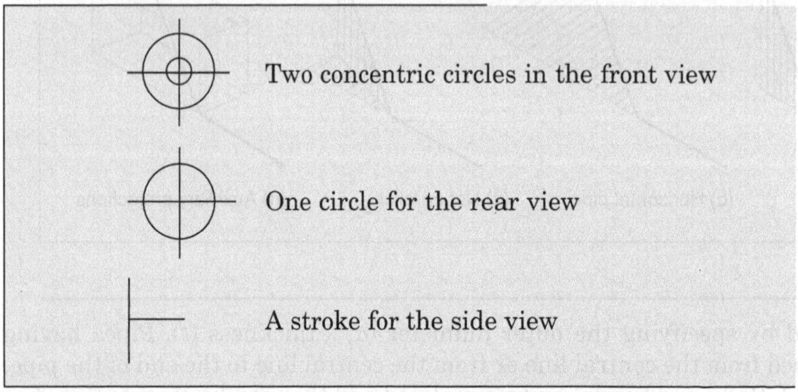

Nowadays in all machines where automatic or semiautomatic operations are desired, a hydraulic system is provided. The main components of a hydraulic system are a reservior, pipes, pipe fittings, and valves for working with fluids. Therefore, the study of pipes and pipe fittings is important.

While installing hydraulic pipes, fittings, and valves the most important aspects are that they should be clean free from scale, and all kinds of foreign matter. To ensure this, the following steps should be taken:

1. Tubings, pipes, and fittings should be brushed with a wire brush and, if required, the internal surface should be reamed for removing burrs.
2. Short length pipes and fittings should be sand blased to remove rust and scale.
3. In case of longer pipes, the pipes are pickled in degreasing solution.

Hydraulic piping drawings Generally, two sets of drawings are made in systems involving hydraulic equipment.

Circuit Drawing: The working of the hydraulic system can be highlighted in the form of a circuit drawing that does not give the dimensional details of valves, pipe lengths, and other arrangement details. The circuit drawing is used by the designer to understand the working of a machine. A typical circuit drawing is shown in Fig. A10.19.

A hydraulic circuit consists of several components such as pump's; reciprocating cylinder's; rotary motor's; and devices for controlling the operation of a circuit, such as flow control valves, pressure control valves, direction control valves, pressure relief valves, and check valves. Besides these, there are other accessories that enhance the performance of the system, such as accumulators, filters, coolers, heaters, seals, and so on. Some typical components of a hydraulic system and its conventional symbols of representation are shown in Table A10.2.

TABLE A10.2 *Components of Hydraulic System and their Conventional Symbols*

Pumps

Graphic Symbol		*Item*	*Description*
(a) 	(b) 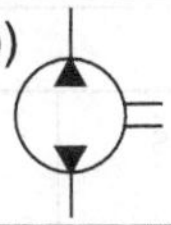	Fixed displacement pump	Unidirectional (a) Bidirectional (b)
(a) 	(b) 	Variable displacement pump	Unidirectional (a) Bidirectional (b)

Motors

Graphic Symbol		*Item*	*Description*
(a) 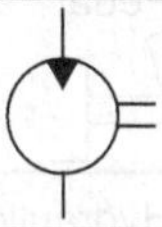	(b) 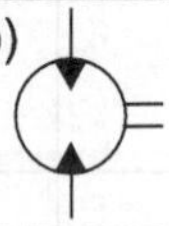	Fixed displacement motor	Unidirectional (a) Bidirectional (b)
(a) 	(b) 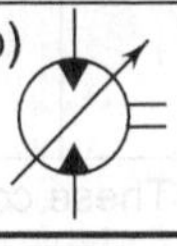	Variable displacement motor	Unidirectional (a) Bidirectional (b)

Valves

Pressure Relief Valve		*Sequence Valve*	
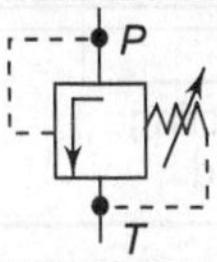	Relieves excess pressure build up by diverting fluid to the tank	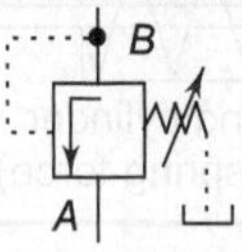	The valve is used in sequencing operations, that is, where a particular cycle is to follow the end of the first cycle
Pressure Reducing Valve		***Counterpressure Valve***	
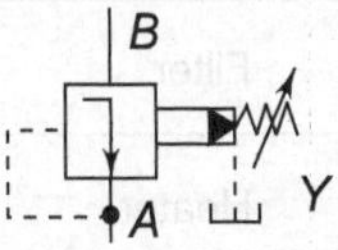	This valve is used to reduce pressure at any desired portion of the system	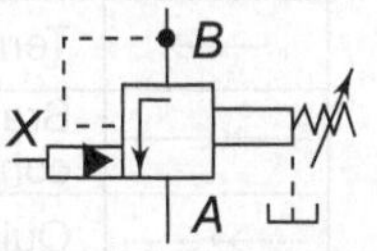	This valve is suitable in circuits where the presence of pressure is necessary to regulate the valve

Flow Control Valves

Variable Throttling Valve		*Compensated Flow Regulator*
Two Way	*With Check*	*Two Way*
	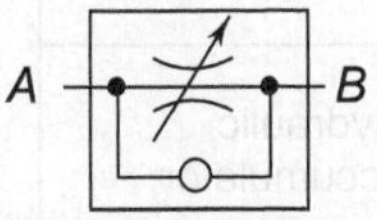	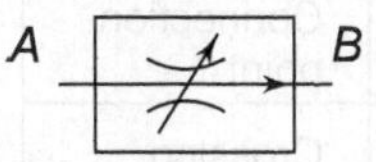

These are flow control valves that can be used to regulate feeds in a machine.

Directional Control Valves

Check Valves		
Standard 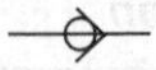	Calibrated 	Check valves are for ensuring unidirectional motion of a fluid. They can also be calibrated by having a spring loaded type of check valve.

Directional Valves			
2 ways-2 positions	3 ways-2 positions	4 ways-2 positions	4 ways-3 positions
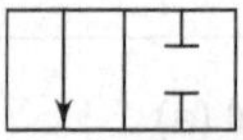	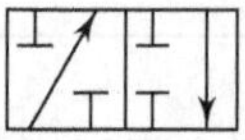	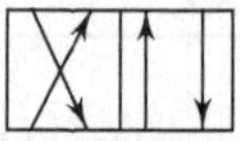	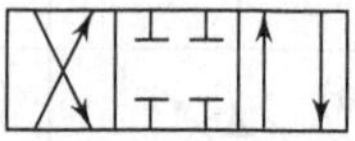

Directional control valves are those that can be manipulated to divert flow from one port to another. These valves become essential in applications that require feeding and retrieving. Depending on different positions and ports, different valves are used.

Controls for Directional Valves			
Mechanical 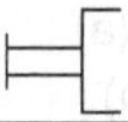	Pushbutton 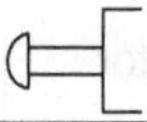	Lever 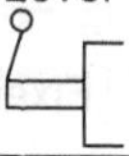	Pedal 
Spring 	Electric (solenoid) 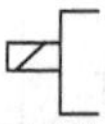	Pneumatic 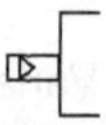	Hydraulic 

To operate the directional control valves, various controls are required. These controls can be of push-button type or any other mechanism that activates the valve.

Cylinders

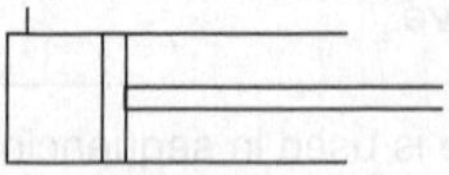	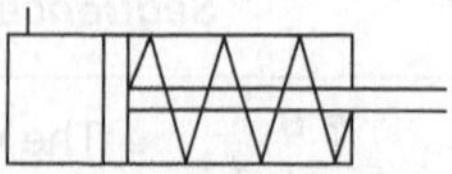	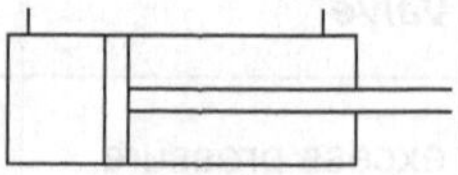	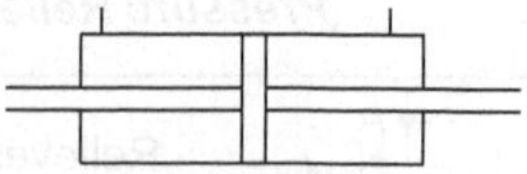
Single acting cylinder (return by external force)	Single acting cylinder (return by spring force)	Double acting single rod	Double acting double rod

Accessories and Fittings

Symbol	*Description*
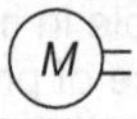	Electric motor
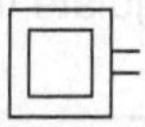	Engine
	Main line
	Pilot line
	Drain line
	Flexible house
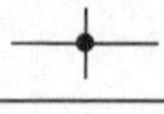	Connection point
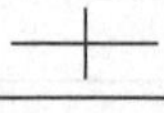	Crossing

Symbol	*Description*
	Termination
	Branching with connected piping
	Quick coupling
	Reservoir with pipe above level
	Reservoir with pipe under level
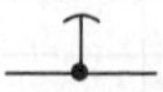	Air bleed
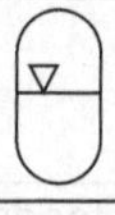	Hydraulic accumulator

Symbol	*Description*
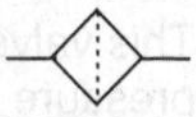	Filter
	Heater
	Cooler
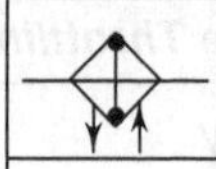	Liquid operated cooler
	Pressure gauge

General arrangement drawing (GA Drawing): The general arrangement layout drawing shows the actual arrangement of the various components of the system. Simplified outline representation of standard valves is sufficient to give the general idea of its shape and position. This drawing is necessary for manufacturing the system. The various symbols can be used, as in Fig. A10.18, to mark details on the layouts.

Review Questions

1. What is a pipe joint? Discuss the various types.
2. What is a pipe fitting? Discuss the various types.
3. State the difference between joints for CI pipes and joints for copper pipes.
4. Draw the various types of piping symbols.
5. Sketch pipe fittings for: (a) tee, (b) elbow, (c) expansion bend.
6. Sketch pipe joints for: (a) socket joint, (b) corrugated expansion joint.
7. State the applications of: (a) expansion joint, (b) hydraulic pipe joint.
8. Draw the general arrangement of fresh water piping arrangement in your house.

CHAPTER 11

Keys, Cotter Joints and Pin Joints

This chapter contains information about metal keys, and cotter and pin joints

After reading the chapter you will be able to

- distinguish between the different types of keys
- Identify the different types of joints
- identify cotter and knuckle joints

11.1 GENERAL DISCUSSION

Cotter and pin joints are temporary joints. A cotter joint is useful for joining two rods/shafts that are axial for the transmission of motion from one shaft to other. The pin joint has a little more flexibility as compared to a cotter joint and is used to join two rods/shafts that are either axial or at any angle in same plane after assembly. Keys are also used to join two metal parts like a bush and a shaft, a collar and a shaft, a coupling and a shaft, and so on. This also helps in transmitting power and motion from one part to another.

These temporary joints are employed as fasteners to transmit forces, usually in combinations of the shaft-hub type, to secure gears, pulleys, discs, and other similar parts to shafts or axles. Cotter joints and pin joints are used for rods in tension or compression. Cotter joint is a rigid joint and does not permit compression or any axial misalignment. The pin joint is a flexible joint that permits angular misalignment.

11.2 KEYS JOINTS

A key is used to connect two pieces, for example, a shaft and a pulley, in such a way that there is no relative rotational movement between them. It is subjected to shearing and torsional stresses and, hence, is always made of steel. Groove cut in the shaft as well as in the hub, to accomodate the key is called keyway (Fig. A11.1).

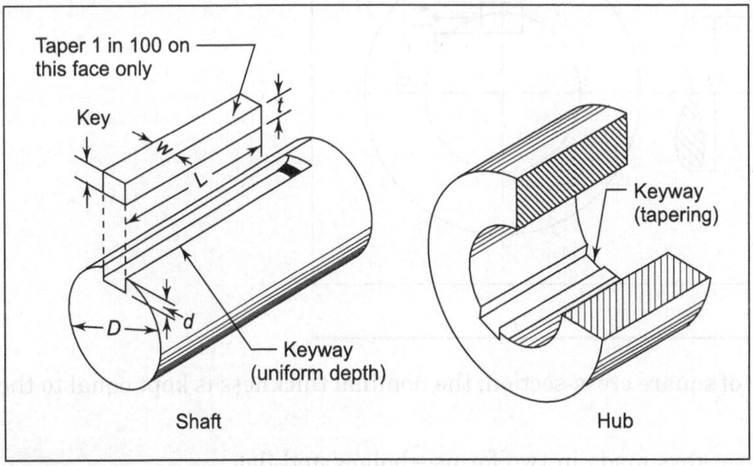

Fig. A11.1 Key and keyways

11.2.1 Types of Keys

A key is a metal piece inserted between a shaft and pulley or a shaft and bush or a shaft and coupling, and prevents relative motion between these components. As the shaft of any machine rotates, the key comes under torsional and shear stress. To sustain these stresses, the key is made up of alloy steel. The keyway is a groove provided on the shaft and inside the bore of a bush/pulley/shaft coupling that is of the same size as the key, to insert and assemble the key with these components.

Keys may be divided into two main types—taper keys and parallel keys.

Taper keys A taper key is uniform in width but tapered in thickness. When inserted in its position, it prevents relative rotational as well as axial movement between the two pieces. The standard taper of this key is 1 in 100. The thickness T at the large end is the nominal thickness of the key (Fig. A11.1).

Sunk taper key This is a standard form of key and may be either rectangular or square in cross section. It is sunk in the shaft to a depth of one-half its nominal thickness. The remaining portion fits in the keyway inside the hub of the mating piece. It is tightfit in both the pieces. The depth of the keyway is uniform inside the shaft, but tapering in the hub (Fig. A11.2).

Approximate proportions of a sunk taper key of rectangular cross-section are:

If	D is the diameter of the shaft, then	
	Width of key	$W = 0.25\,D + 2$ mm
	Nominal thickness	$T = 0.2\,D + 2$ mm

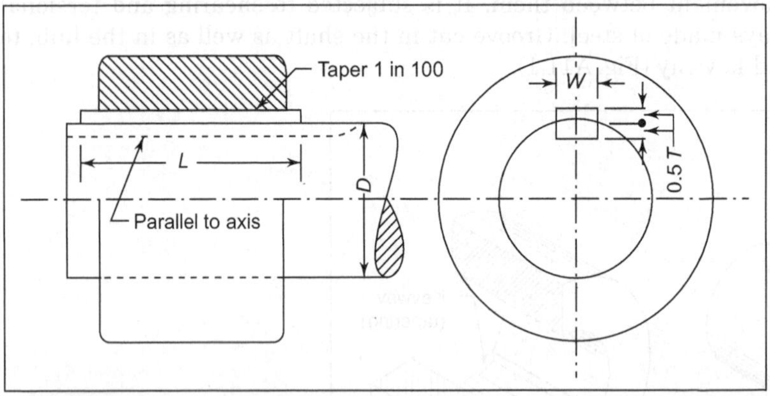

Fig. A11.2 Sunk taper key

In case of a sunk taper key of square cross-section, the nominal thickness is kept equal to the width of the key.

Saddle keys These are taper keys made in two forms—hollow and flat.

A hollow saddle key has its underside hollow in order to fit the curved surface of the shaft. The keyway is cut in the mating piece only. Relative rotation between the two pieces is only prevented by friction between the key and the shaft. This key is, therefore, used when the power to be transmitted is small.

Proportions of a saddle key are:

If	D is the diameter of the shaft, then,	
	Width of key	$W = 0.25\,D + 2$ mm
	Nominal thickness	$T = 0.08\,D + 1$ mm

A flat saddle key sits on a flat surface provided on the shaft. The resistance to splitting in case of this key is greater than that in case of a hollow key and, hence, greater power can be transmitted by its use (Fig. A11.3).

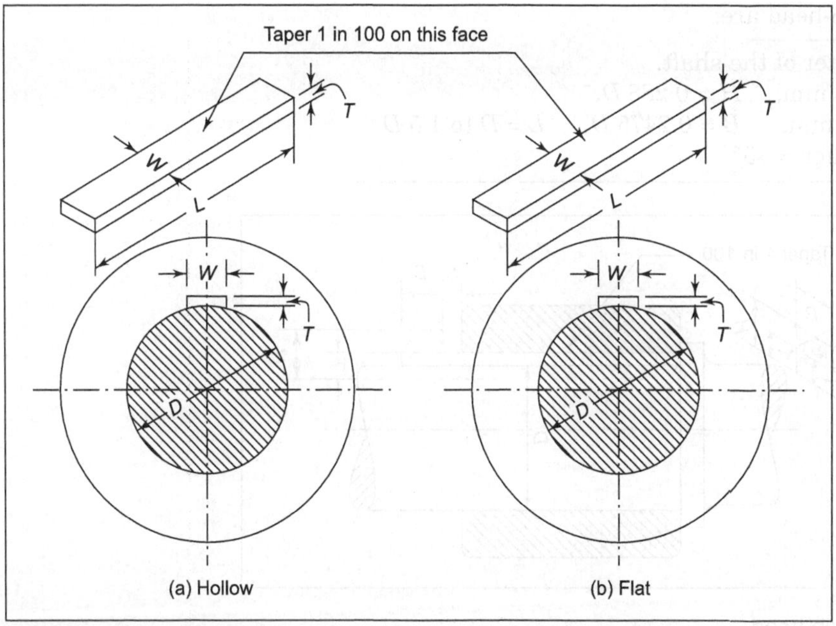

Fig. A11.3 Saddle keys

Round key or pin key This key has a circular cross-section and is driven into a hole drilled partly in the shaft and partly in the mating piece [Fig. A11.4(a)].

Proportions of round key are:

If D is the diameter of the shaft, then,
Diameter of the key $d = 0.2\,D$

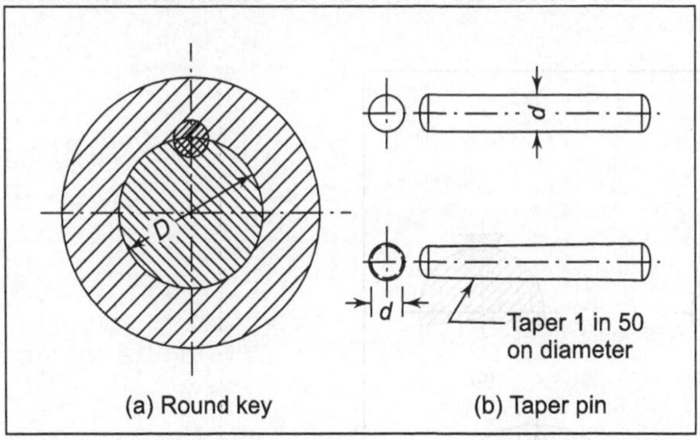

Fig. A11.4 Key and pin

Taper pin A taper key of circular cross section is called a taper pin. It has a standard taper of 1 in 50 on diameter. The diameter d at the larger end is the nominal diameter of the pin and is equal to 0.2 times the diameter of the shaft [Fig. A11.4(b)].

Gib-head A taper key is removed by hammering at its thinner end to facilitate its removal. The key with a gib-head is withdrawn by hammering on a wedge-shaped piece inserted in the space marked S (Fig. A11.5).

Proportions of a gib-head are:

Let D be the diameter of the shaft.

$W = 0.2\,D + 2$ mm, $H = 0.288\,D$,

$T = 0.2\,D + 2$ mm, $B = 0.2475\,D$, $L = D$ to $1.5\,D$

Angle of chamfer = 45°

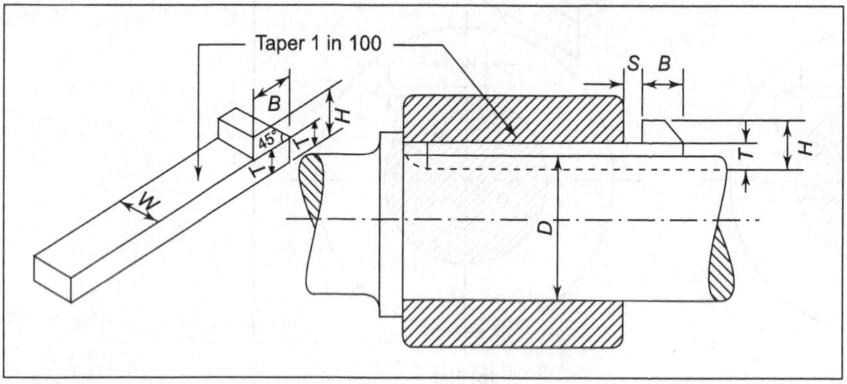

Fig. A11.5 Key with gib-head

Parallel or feather keys A feather key is a sunk key of uniform thickness. While preventing relative rotation it permits sliding or axial movement of the mating piece on the shaft. It may have a rectangular [Fig. A11.6(a)], square [Fig. A11.6(b)] or dove-tail [Fig. A11.6(c)] cross section. The ends of a feather key are sometimes rounded. The key may be secured to either the shaft or the mating piece. When secured to the shaft the keyway in the mating piece is made in such a way that the piece can slide freely on the shaft in the axial direction. Fig. A11.6(d) shows a feather key secured to a shaft by means of two cap-screws having countersunk heads. Feather keys secured to the mating piece, and free to slide in the keyways in the shaft, are shown in Fig. A11.7.

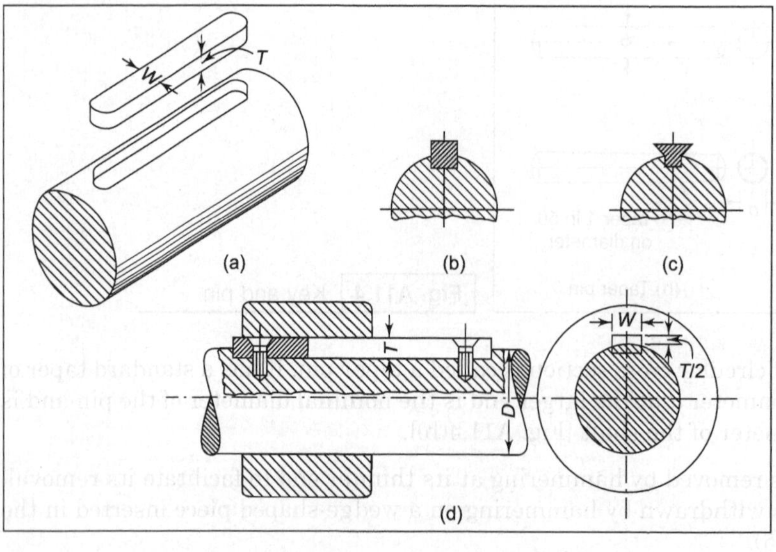

Fig. A11.6 Feather keys

The peg key in Fig. A11.7(a) has a peg (in the centre) that fits inside the hub. The single-headed feather key [Fig. A11.7(b)] is attached to the hub by means of a screw. The double-headed feather key is shown in [Fig. A11.7(c)].

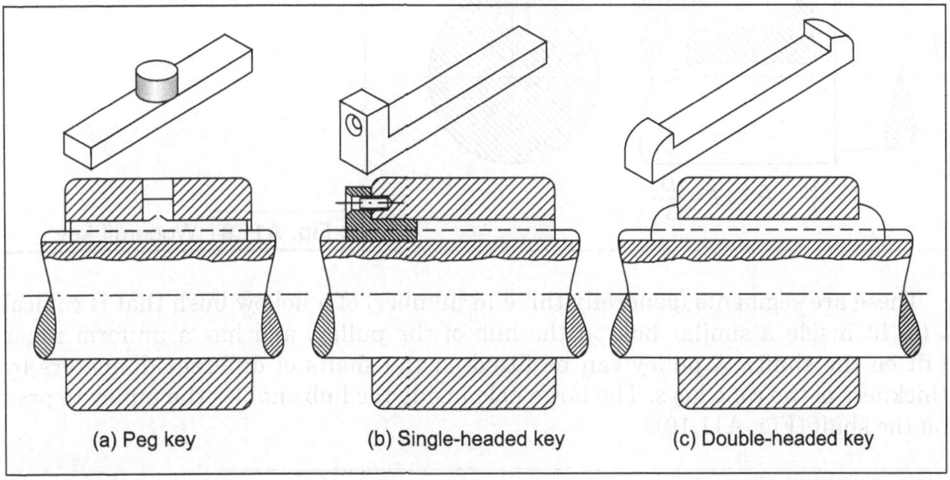

Fig. A11.7 Feather keys

Spline shafts These are used to withstand the stresses. Splines or feathers of uniform width are formed by cutting a number of equally spaced grooves of uniform depth on the surface of the shaft and parallel to the axis. The sliding piece is provided with a corresponding number of keyways to mate with the spline shaft. The spline may be of involute shape or straight sided. It is widely used in the gear-box of automobiles and machine tools (Fig. A11.8).

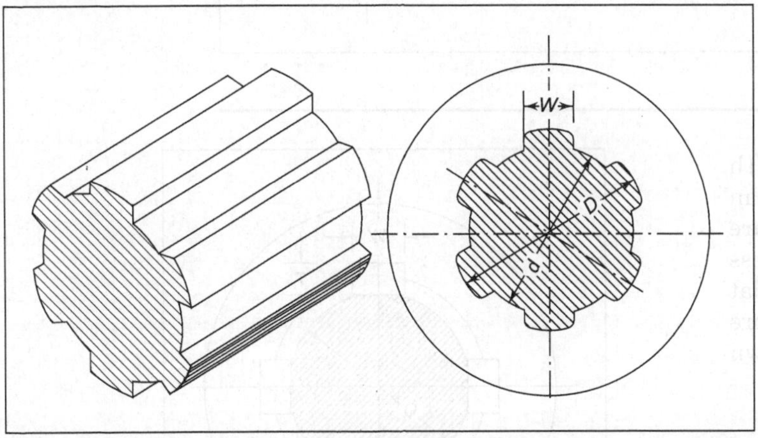

Fig. A11.8 6-spline shaft

Woodruff key This key is in the form of a segment of a circular disc of uniform thickness. It fits into a corresponding keyway cut in the shaft such that the flat portion projects outside. This projecting part fits in a keyway inside the mating piece, adjusting itself to any taper in it. The shaft is weakened due to the comparatively greater depth of the keyway cut in it. It is used in the feed gearbox of lathe machines (Fig. A11.9).

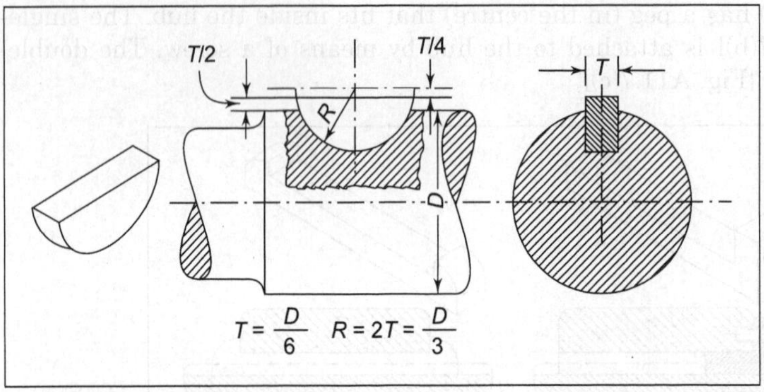

Fig. A11.9 Woodruff key

Cone keys These are segments (generally three in number) of a hollow bush that is conical from outside (to fit inside a similar hole in the hub of the pulley) and has a uniform inner diameter (to fit on the shaft). A pulley can be fitted on the shafts of different diameters by varying the thickness of the cone keys. The large hole inside the hub enables the pulley to pass over collars on the shaft (Fig. A11.10).

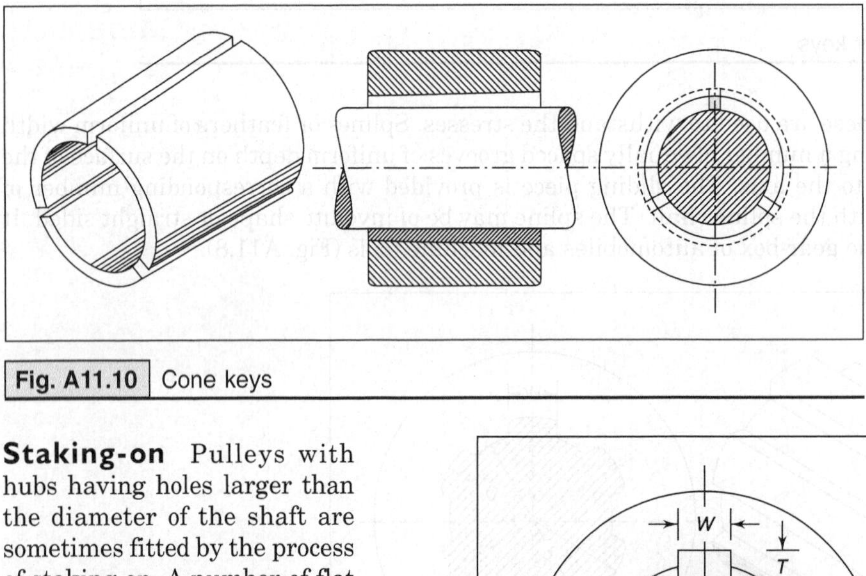

Fig. A11.10 Cone keys

Staking-on Pulleys with hubs having holes larger than the diameter of the shaft are sometimes fitted by the process of staking-on. A number of flat saddle keys with gib-heads are used for the purpose, as shown in Fig. A11.11.

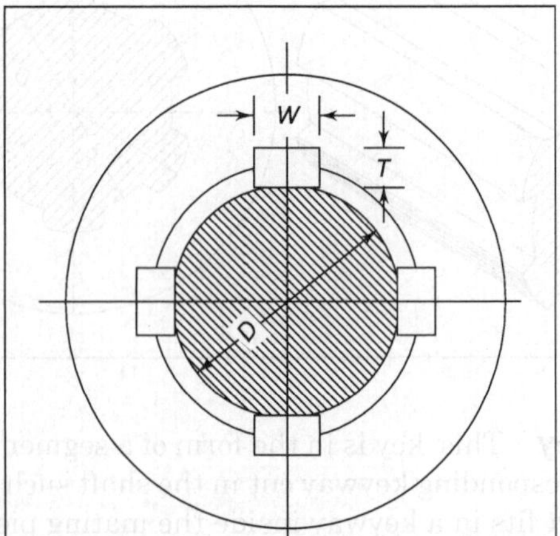

Fig. A11.11 Staking-on

11.3 COTTER JOINTS

A cotter is a flat wedge shaped piece of steel of rectangular cross-section which is used to rigidly connect two rods subjected to tensile forces and is inserted at right angles to the axes of the rods [Fig. A11.12(a)]. It is uniform in thickness but taper in width. The usual taper is 1 in 30. When a larger taper is adopted, a locking device becomes necessary to prevent the cotter from slackening or coming out.

In practice, the following proportions are generally adopted for a cotter.

Let D be the diameter of the rod.
Width of the cotter at the centre $W = 1.3\ D$
Thickness of the cotter $T = 0.31\ D$
Taper 1 in 30

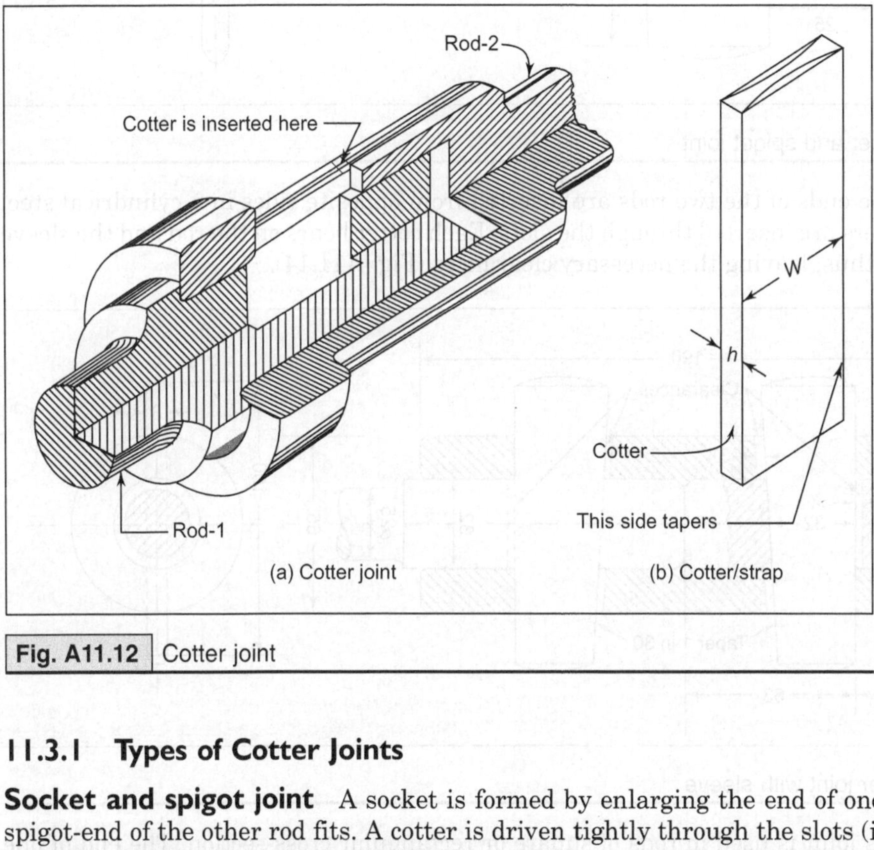

Fig. A11.12 Cotter joint

11.3.1 Types of Cotter Joints

Socket and spigot joint A socket is formed by enlarging the end of one rod in which the spigot-end of the other rod fits. A cotter is driven tightly through the slots (in the rods), which are kept slightly out of alignment [Fig. A11.12(b)].

The cotter comes in contact with the two rods on opposite sides, thus, leaving clearances on both sides. Clearances are absolutely necessary for the proper functioning of the cotter. The cotter tends to pull the slots in a line, thus, making the joint perfectly tight and rigid. Fig. A11.13 shows two views of such a joint.

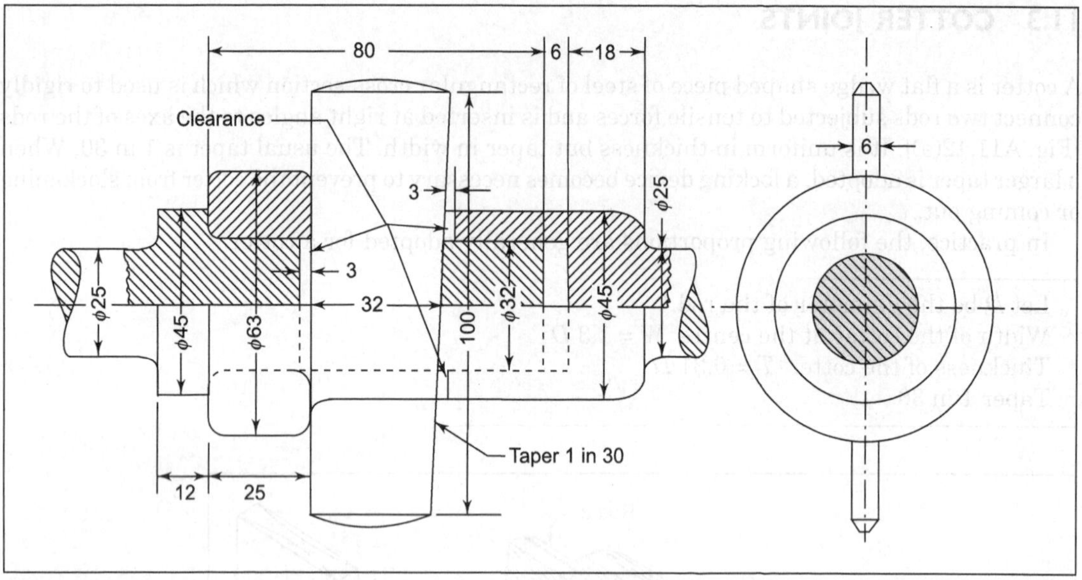

Fig. A11.13 Socket and spigot joint

Sleeve joint The ends of the two rods are inserted from opposite sides in a cylindrical steel box or sleeve. Cotters are inserted through the slots. Each cotter bears on the rod and the sleeve on opposite sides, thus, leaving the necessary clearances (Fig. A11.14).

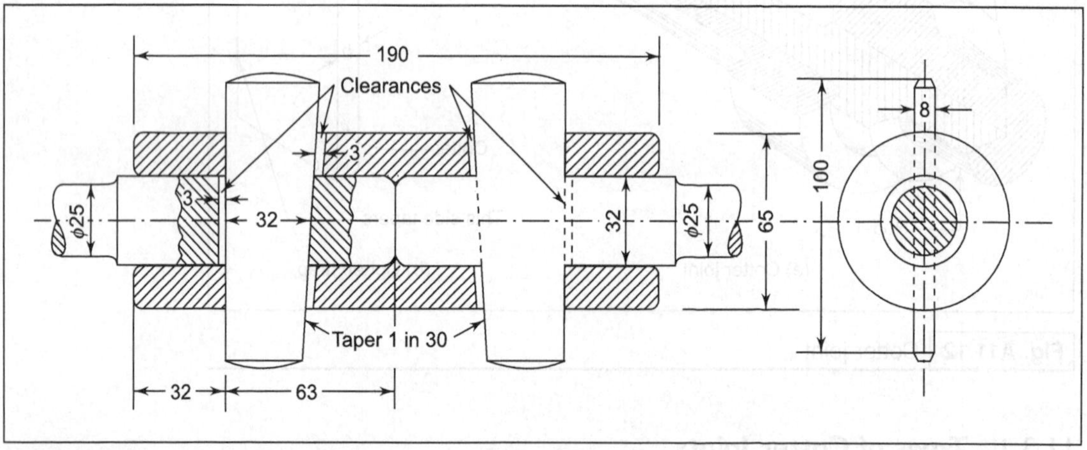

Fig. A11.14 Cotter joint with sleeve

Strap joint This joint is used in rods of square or rectangular cross-section. The end of one rod is made in the form of a strap in which the end of the other rod fits. A gib is used along with the cotter to make the joint. It has projections called gib-heads at its two ends. These act like hooks and prevent the strap-ends of the rod from opening wide. The depth and the width of the gib-heads are usually kept equal to the thickness of the cotter. The tapering sides of the gib and cotter mate with each other while their outer sides are parallel to each other and perpendicular to the common axis of the rods. Hence when a gib is used along with a cotter the opposite faces of the slots in the rods are parallel to each other.

When a cotter has a taper on both the sides, two gibs—one on each side—are used.

In Fig. A11.15 the cotter and the gib are illustrated in section to show the clearances accurately. In the orthographic half-sectional view (Fig. A11.16), they are shown according to the convention.

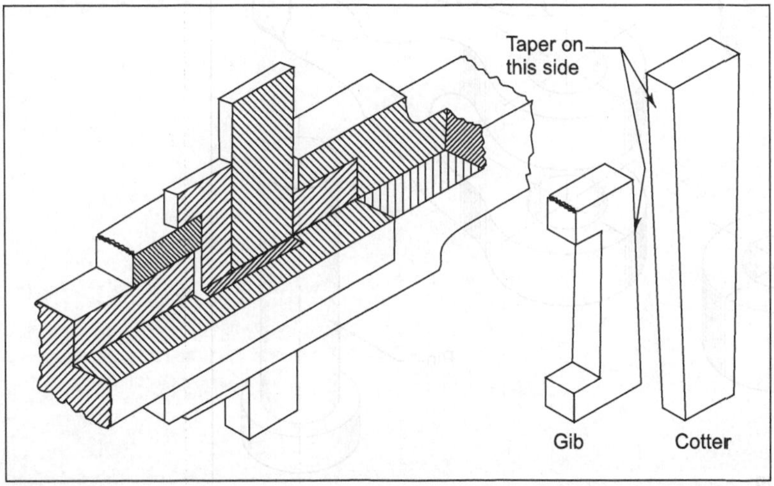

Fig. A11.15 Strap joint (Pictorial view)

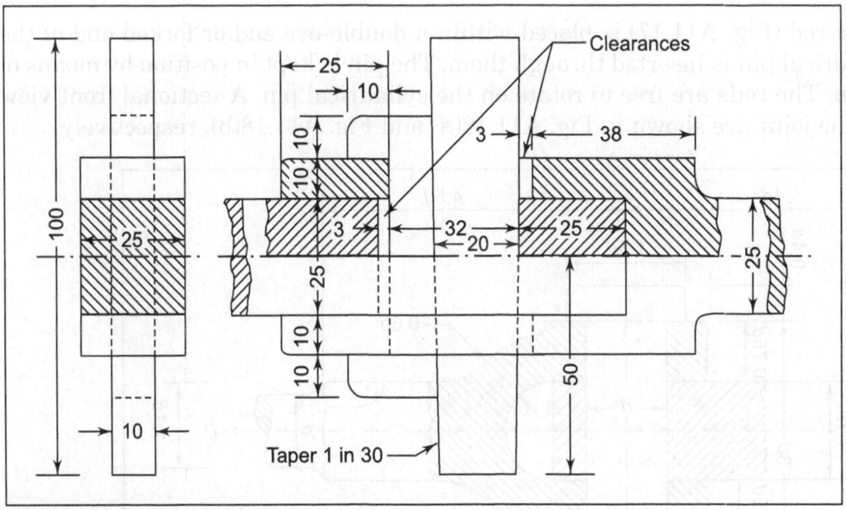

Fig. A11.16 Strap joint with gib and cotter

11.4 PIN JOINT OR KNUCKLE JOINT

This joint is used for rods that are in tension or in a compresed state and may not be in alignment with each other, but their axes intersect. The joint is commonly used where angular movement between the rods occurs and a reciprocating motion is to be converted into a rotary motion or vice-versa.

The pin joint is commonly used for joining the *D*-slide valve and electric rod of a steam engine, air brake of locomotives, and many other kinds of lever and rod connections.

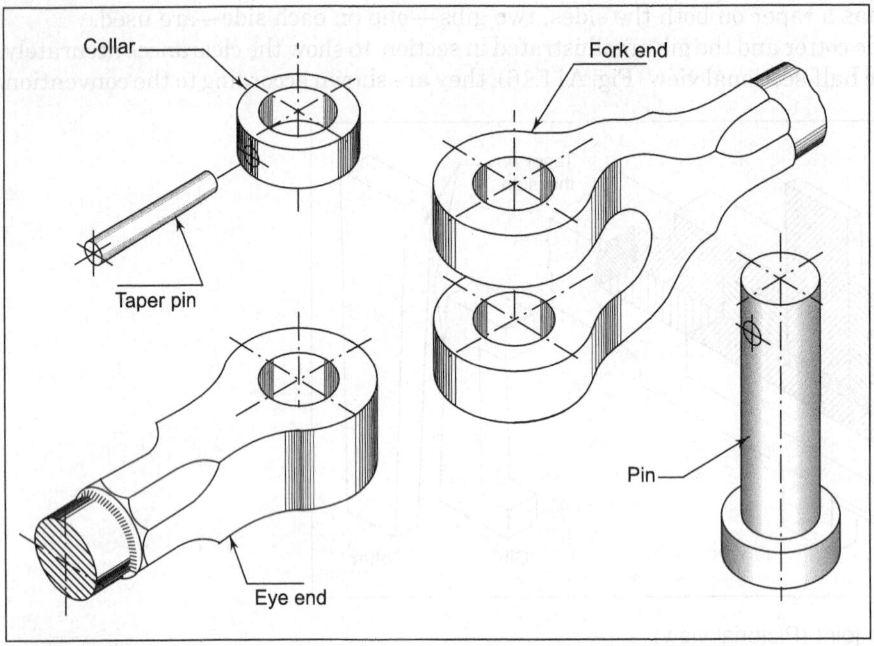

Fig. A11.17 Parts of knuckle joint

A single eye of one rod (Fig. A11.17) is placed within a double-eye end or forked end of the other rod and a cylindrical pin is inserted through them. The pin is kept in position by means of a collar and taper pin. The rods are free to rotate on the cylindrical pin. A sectional front view and the top view of the joint are shown in Fig. A11.18(a) and Fig. A11.18(b), respectively.

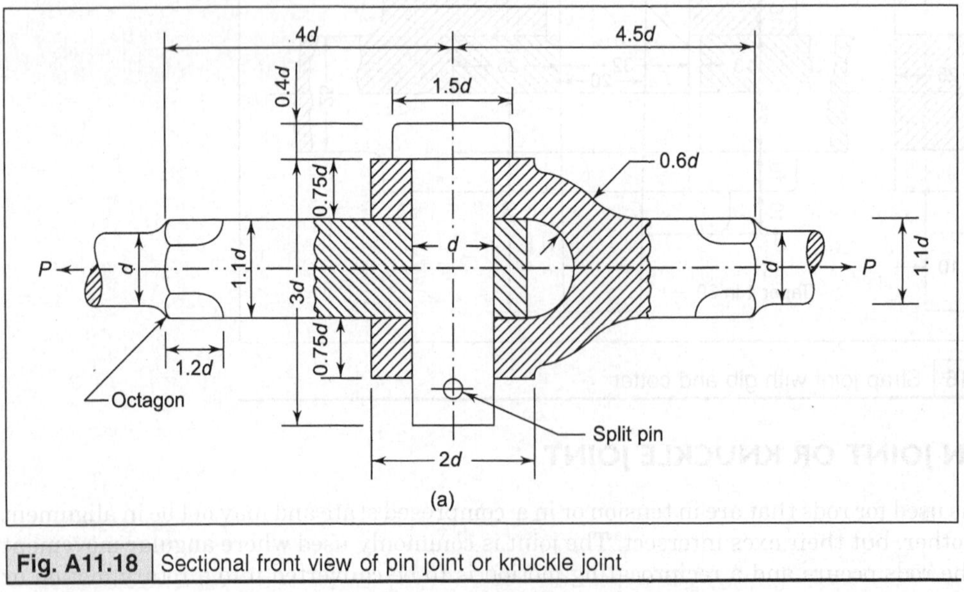

Fig. A11.18 Sectional front view of pin joint or knuckle joint

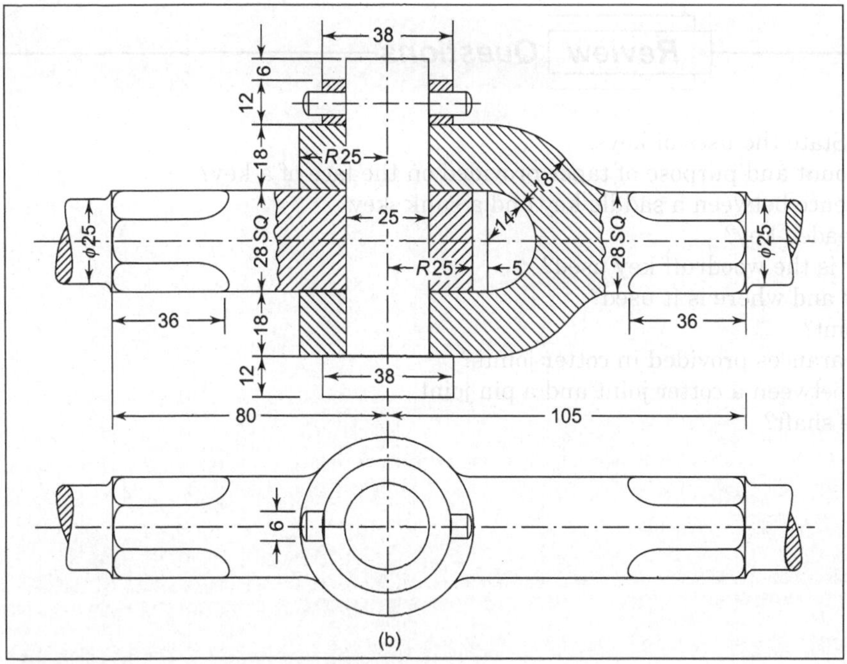

Fig. A11.18 Pin joint or knuckle joint

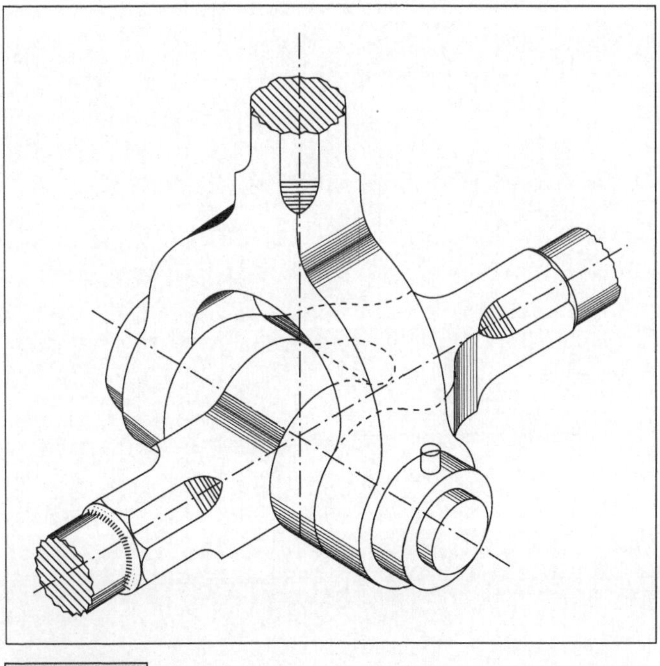

Fig. A11.19 Knuckle joint for three rods

Fig. A11.19 shows a view of a knuckle joint for three rods whose axes intersect at a common point.

Review Questions

1. What is a key? State the uses of keys.
2. What is the amount and purpose of taper provided on the face of a key?
3. Write the difference between a saddle key and a sunk key.
4. What is a gib headed key?
5. Where and why is the woodruff key used?
6. What is a cotter and where is it used?
7. What is a pin joint?
8. Why are the clearances provided in cotter joints?
9. Differerentiate between a cotter joint and a pin joint.
10. What is a spline shaft?

CHAPTER 12

Riveted Joints and Welded Joints

This chapter contains information about riveted and welded joints

After reading the chapter you will be able to

- differentiate between riveting and welding
- identify the types of riveted and welded joints
- draw symbols of welded joints

12.1 GENERAL DISCUSSION

Welding is a process by which permanent joints are made. It is process of joining two metal parts by fusing them. Rapid development in the welding techniques has replaced riveted joints. Since welding is used extensively, efforts are on to refine it.

Rivets are used to permanently fasten two or more plates or pieces of a metal. Such joints made with rivets are called riveted joints. Main applications include shipbuilding and for the construction of buildings, bridges, boilers, tanks, and so on. Plates once joined cannot be disconnected without chipping-off rivet-heads from one side of the joint.

Rivets are usually made of *C*-30; it is a low alloy steel with 0.3 per cent carbon content and is used in manufacturing rivets. In its initial form, [Fig. 12.1(a)] a rivet comprises of (1) a head, (2) a cylindrical body of shank and (3) a slightly tapered tail. A rivet is specified by the diameter of its shank. The length of the tail, out of which another head is formed, is kept about 1.25 times the diameter of the rivet.

12.2 RIVETS AND RIVETING

The process of forming another rivet-head after the rivet has been placed in the hole previously drilled or punched through the plates is called riveting. The diameter of this hole is kept slightly larger (about 1 mm to 1.5 mm) than the diameter of the rivet. Any burr formed at the edges of the hole due to drilling is removed by counter sinking. The rivet is made red-hot in a charcoal furnace and then inserted within the hole in the plates. The head of the rivet is held fast against the adjoining plate while the tail is hammered and another rivet-head is forged. The rough surface of the new head is smoothened by the use of a special tool having a cavity of the desired form of the head [Fig. A12.1(b)].

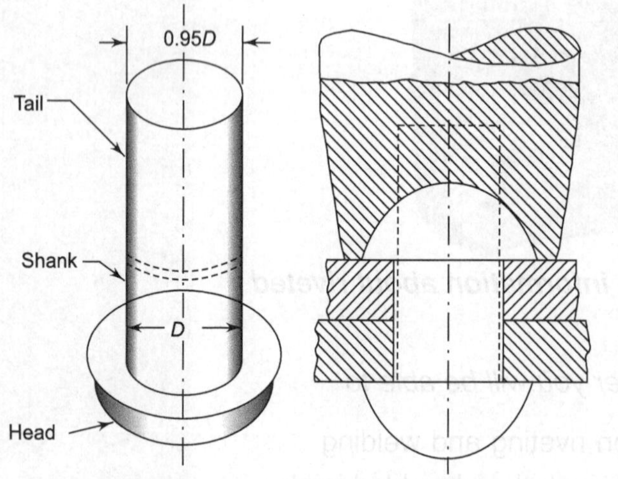

Fig. A12.1 Rivet and riveting

A hot rivet is easier to work on. It also binds the plates more closely together on cooling on account of contraction of the metal. Riveting is done in cold (i.e., without heating the rivet) in case of rivets of small sizes or when they are made of soft ductile metals such as copper, aluminium, and so on. Machine riveting is a common modern practice employed when the work is to be done fast or on a large scale. The rivet-head is formed under a steady force applied by means of hydraulic or pneumatic pressure. Due to the steady pressure, the hole in the plates is filled up more completely.

12.2.1 Caulking and Fullering

To avoid leakage through the joint, the plates are firmly forced together by caulking or fullering processes. The edges of the plates are hammered and driven-in by a caulking tool (Fig. A12.2) or fullering tool (Fig. A12.3). The caulking tool is in the shape of a blunt chisel. The thickness of the fullering tool is about the same as that of the plates. To facilitate these operations, the edges of the plates are usually machined to an angle of about 80° before joining them together. This angle is increased to about 85^0 after the fullering process.

Leakage through the hole is prevented by the caulking operation on the edge of the rivet-head (Fig. A12.2). Both these processes are generally performed with the aid of pneumatic power.

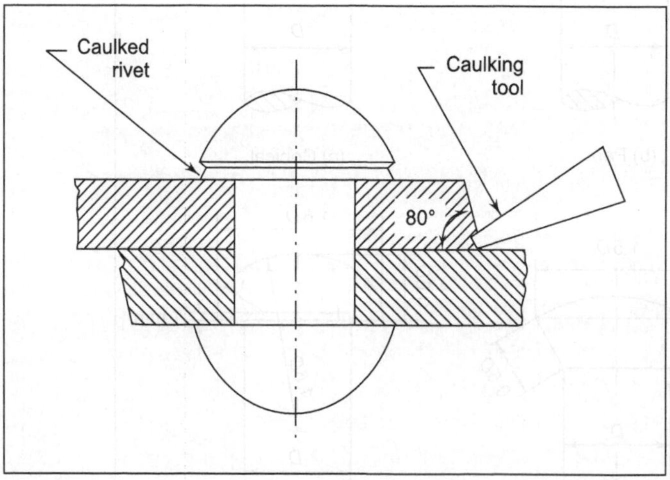

Fig. A12.2 Caulking

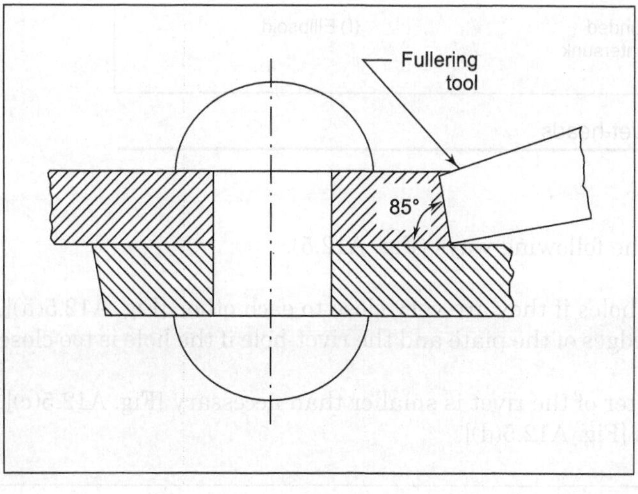

Fig. A12.3 Fullering

12.2.2 Forms and Proportions of Rivet-heads

Fig. A12.4 shows some of the various forms of rivet-heads used for general work. The proportions given are for the rivets of nominal diameters, between 12 mm and 50 mm. The snap or cup head is the most common form in use. The countersunk head is used when the surface of the plate is required to be free from projecting heads. Conical and ellipsoid heads are generally used in boilers.

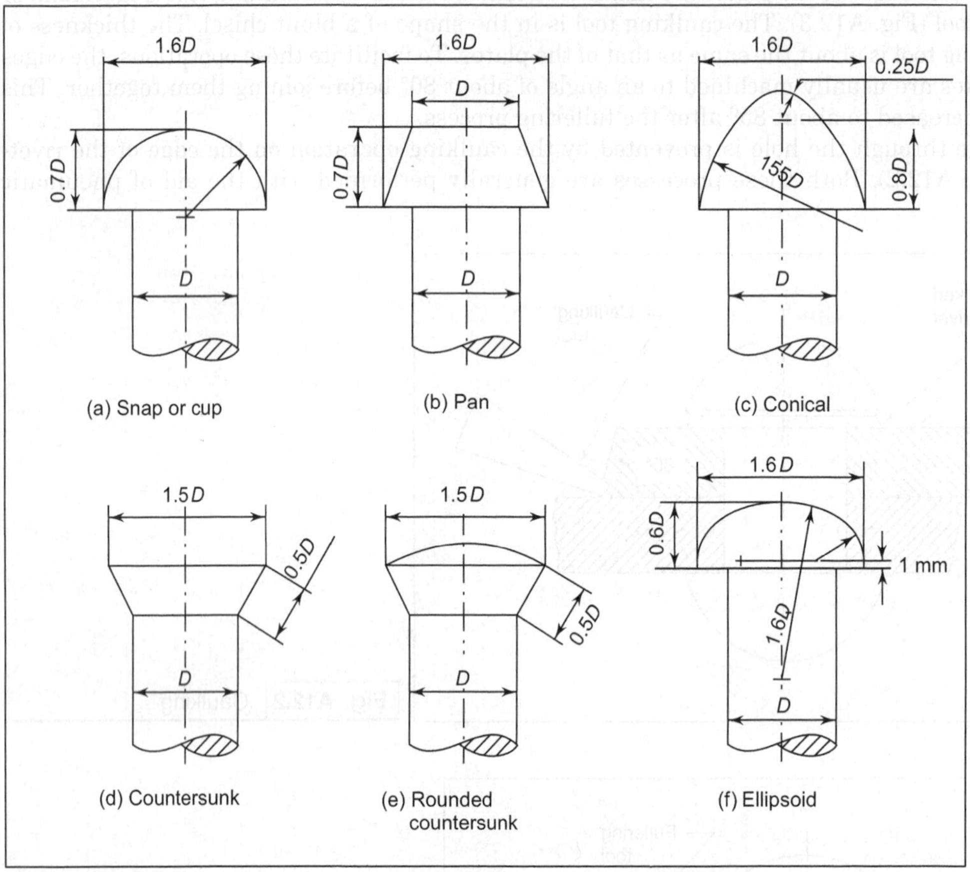

Fig. A12.4 Forms and proportions of rivet-heads

12.2.3 Failure of Riveted Joints

A riveted joint may fail in any one of the following ways (Fig. A12.5).

- Tearing of the plate between the holes if they are very close to each other [Fig. A12.5(a)].
- Tearing of the plate between the edges of the plate and the rivet-hole if the hole is too close to the edge [Fig. A12.5(b)].
- Shearing of the rivet if the diameter of the rivet is smaller than necessary [Fig. A12.5(c)].
- Crushing of the plate or the rivet [Fig. A12.5(d)].

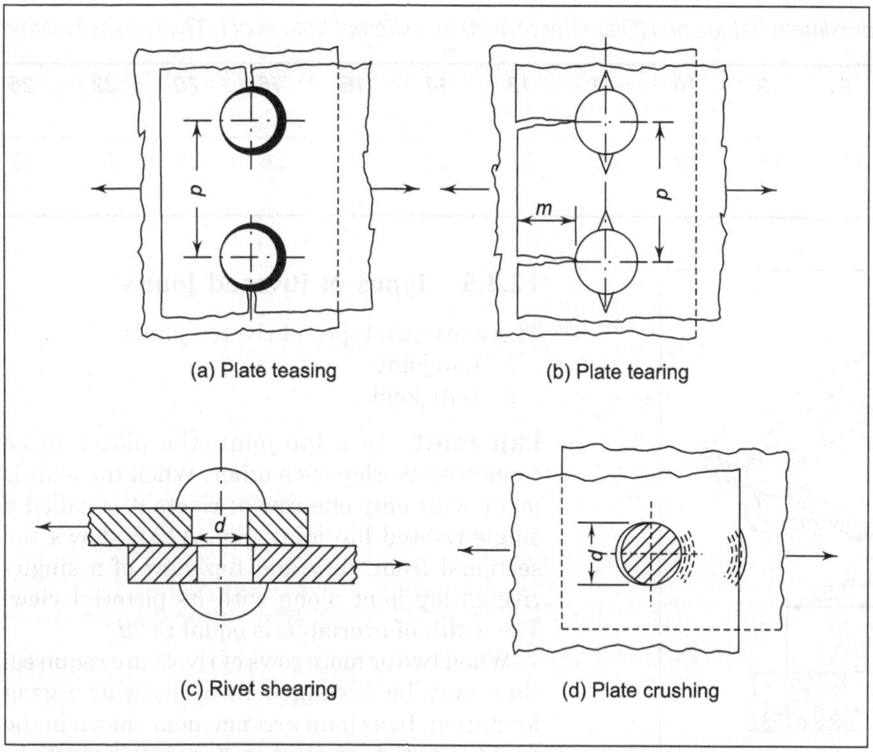

Fig. A12.5 Failure of riveted joints

12.2.4 Dimensions of Riveted Joint

To prevent failure, the joint should be carefully designed. For elementary work, suitable values of the rivet diameter, positions of holes, and so on, for a given thickness of the plates, may be obtained by using the following empirical formulae.

$d = 6\sqrt{t}$

$p = d$

$m = d$

where t is thickness of plates in mm

d is diameter of rivets

p is pitch, that is, the distance between centres of adjoining rivets in the same row (parallel to the edge of the plate)

m is equal to margin, that is, the distance between an edge of the plate and the nearest rivet-hole

As m 1= d, the distance of the centre line of the nearest row of rivets from the edge of the plate is equal to 1.5d.

TABLE A12.1 *Approximate Values of d (Diameter of rivet) for Different Values of t (Thickness of plates)*

Thickness of Plate, t (mm)	*8*	*9*	*10*	*11*	*12*	*14*	*16*	*18*	*20*	*22*	*25*
Diameter of rivet, *d* (mm)	17	18	19	20	21	22	24	26	27	28	30

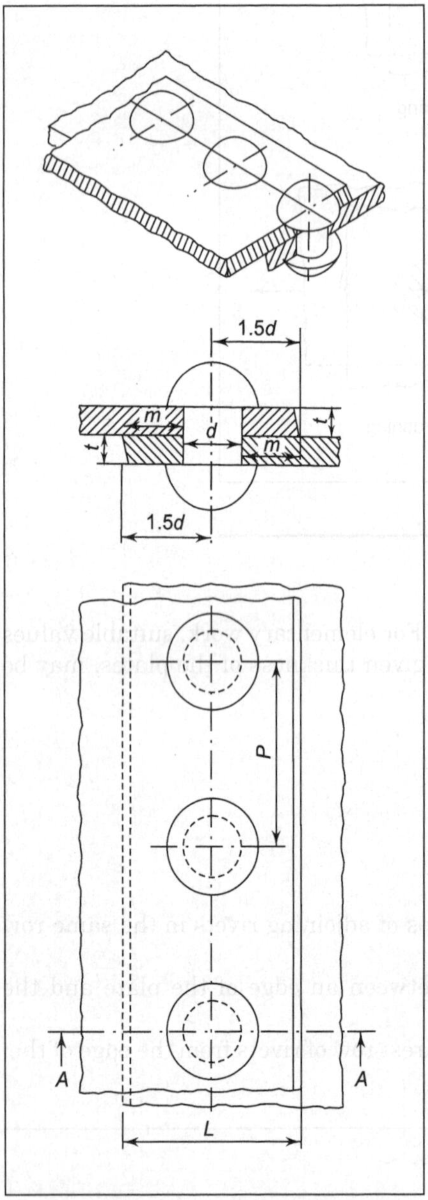

Fig. A12.6 Single-riveted lap joint

12.2.5 Types of Riveted Joints

There are two types of riveted joints

1. Lap joint
2. Butt joint

Lap joint In a lap joint, the plates to be connected overlap each other. When the joint is made with only one row of rivets it is called a single-riveted lap joint. Fig. A12.6 shows the sectional front view and top view of a single-riveted lap joint along with its pictorial view. The width of overlap L is equal to $3d$.

When two or more rows of rivets are required, they may be arranged in a chain or zigzag formation. In a chain arrangement shown in the double-riveted lap joint in Fig. A12.7, rivets in the adjoining rows are placed directly opposite to each other.

In a zigzag formation they are staggered, as shown in Fig. A12.8. The distance between the rows of rivets, called the row pitch p_r, should not be less than $0.6p$ for zigzag riveting and $0.8p$ for chain riveting.

In terms of d, the approximate values of p_r are $2d$ and $2d + 6$ mm respectively. In the zigzag arrangement, the distance between the centre of the nearest rivet in the adjoining row is called the diagonal pitch.

This pitch is often considered instead of the row pitch and its value is obtained by the formula

$$p_d = \frac{2p + d}{3}$$

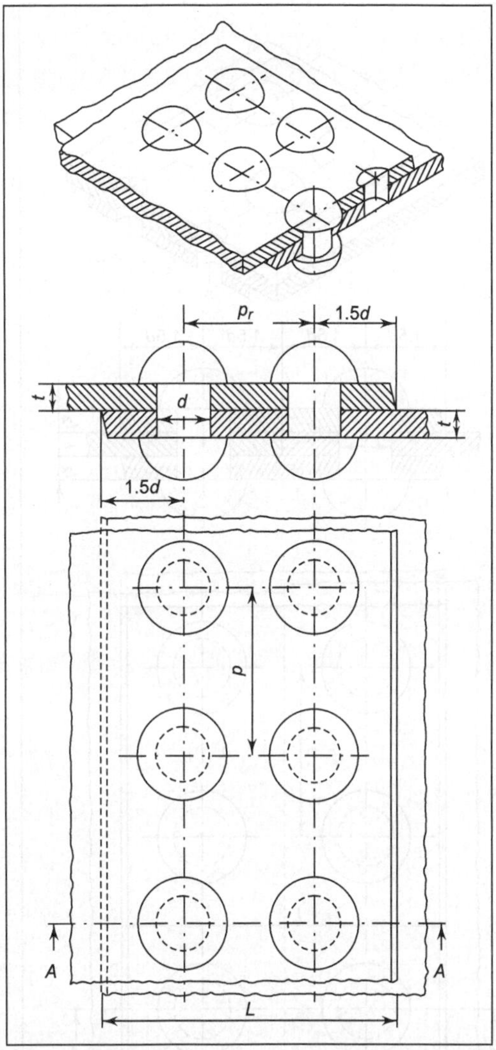

Fig. A12.7 Double-riveted (Chain) lap joint

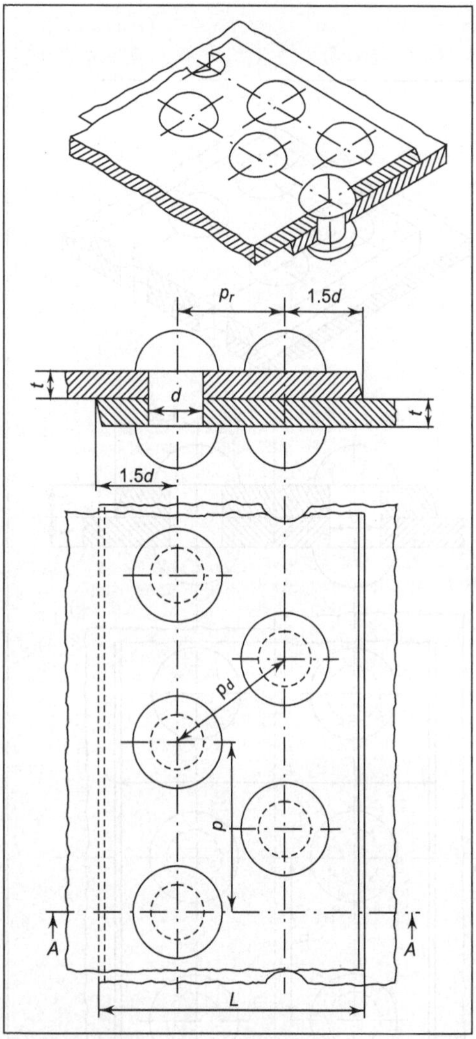

Fig. A12.8 Double-riveted (Zigzag) lap joint

Butt joint In a butt joint the edges of the plates to be connected butt against each other and the joint between them is covered by buttplates or butt-straps (also called cover plates or cover straps) on one or both sides. At least two of the rivets, one in each connected plate, are necessary to make the joint. When only one butt strap is used, its thickness is $t_1 = t$ to $1.125t$; when two straps are used, their thickness is $t_2 = 0.7t$ to $0.8t$, where t is the thickness of the plates to be connected.

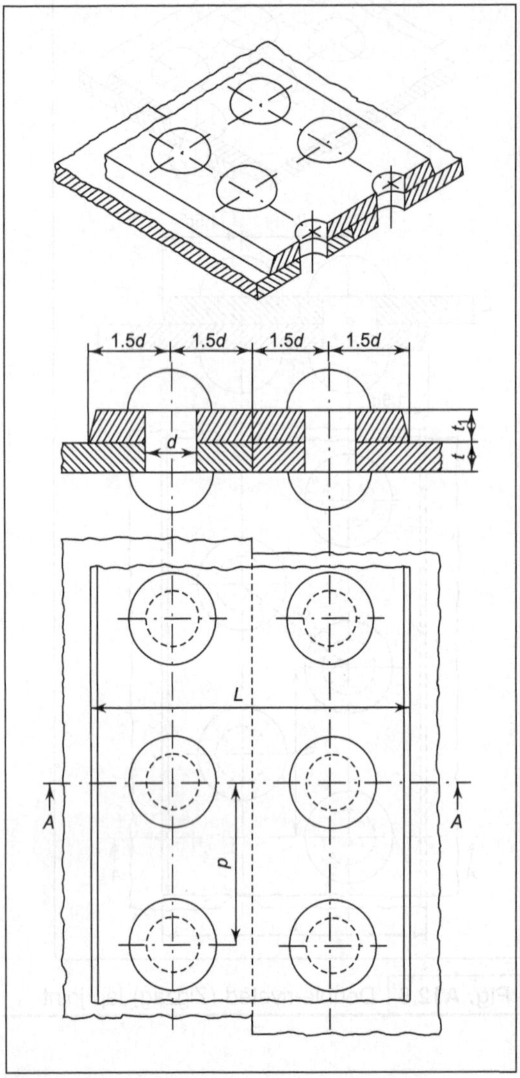

Fig. A12.9 Single-riveted (Single strap) butt joint

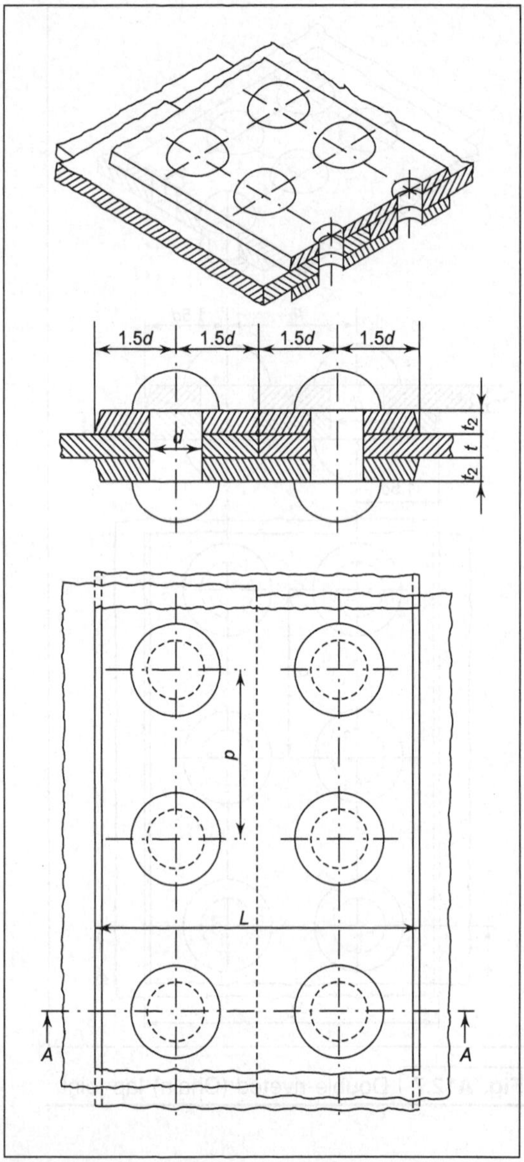

Fig. A12.10 Single-riveted (Double strap) butt joint

Fig. A12.9 shows two views of a single-riveted butt joint with one cover-strap. The same joint with two straps is shown in Fig. A12.10.

A double-riveted butt joint with double straps is illustrated in Fig. A12.11. Note that there are two rows of rivets in each of the main plates and that the rivets are arranged in zigzag formation.

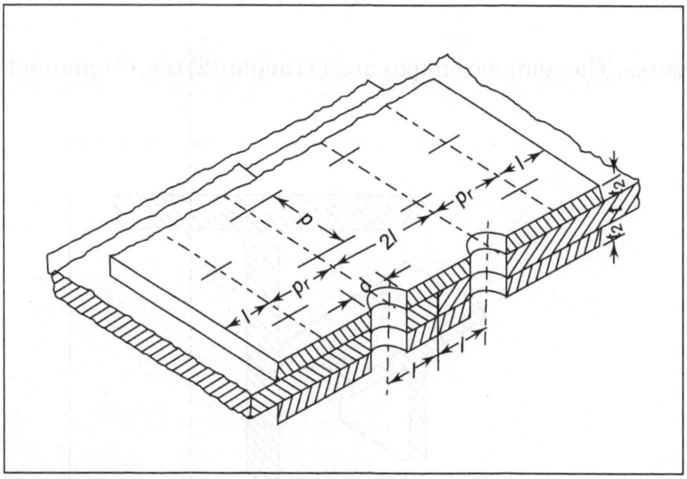

Fig. A12.11 Double-riveted butt joint

Two rings of a steamboiler shell are connected, as shown in Fig. A12.12. The circumferential joint is a double-riveted lap joint. A triple-riveted butt joint with two cover straps is used for longitudinal joints.

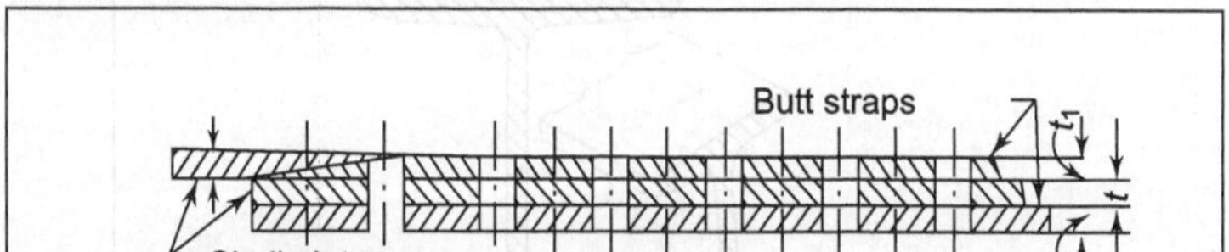

Fig. A12.12 Boiler-shell plate joints

12.2.6 Rolled-steel Sections

These are largely used in steel structures. The common shapes are: (1) angle, (2) tee, (3) channel, and (4) *H* or joist (Fig. A12.13).

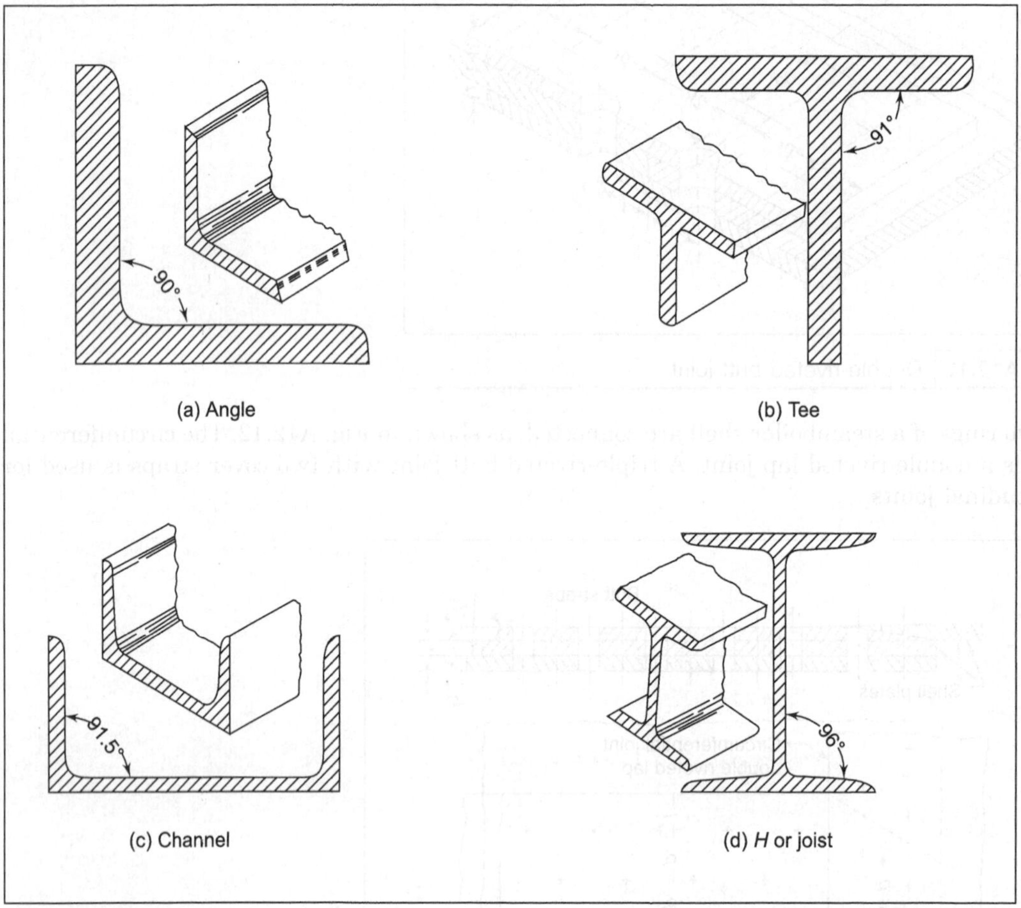

Fig. A12.13 Rolled-steel sections

Connection of plates at right angles Plates may be connected at right angles by flangining one of the plates. Fig. A12.14(a) shows a plate bent inside. The plate may also be bent outside, as shown in Fig. A12.14(b). The radius *R* of the inside curve of the bent plate should be less than twice its thickness. Another method in which an angle-section is used is shown in Fig. A12.14(c). The angle is often placed outside.

These methods of connection, as employed in boiler construction, are shown in Fig. A12.15. The front end-plate is connected to the shell-plate by a ring of anglesection placed on the outside. The back end-plate is bent inside and flanged to the shell plate. Note carefully how the flue plates are joined to each end plate.

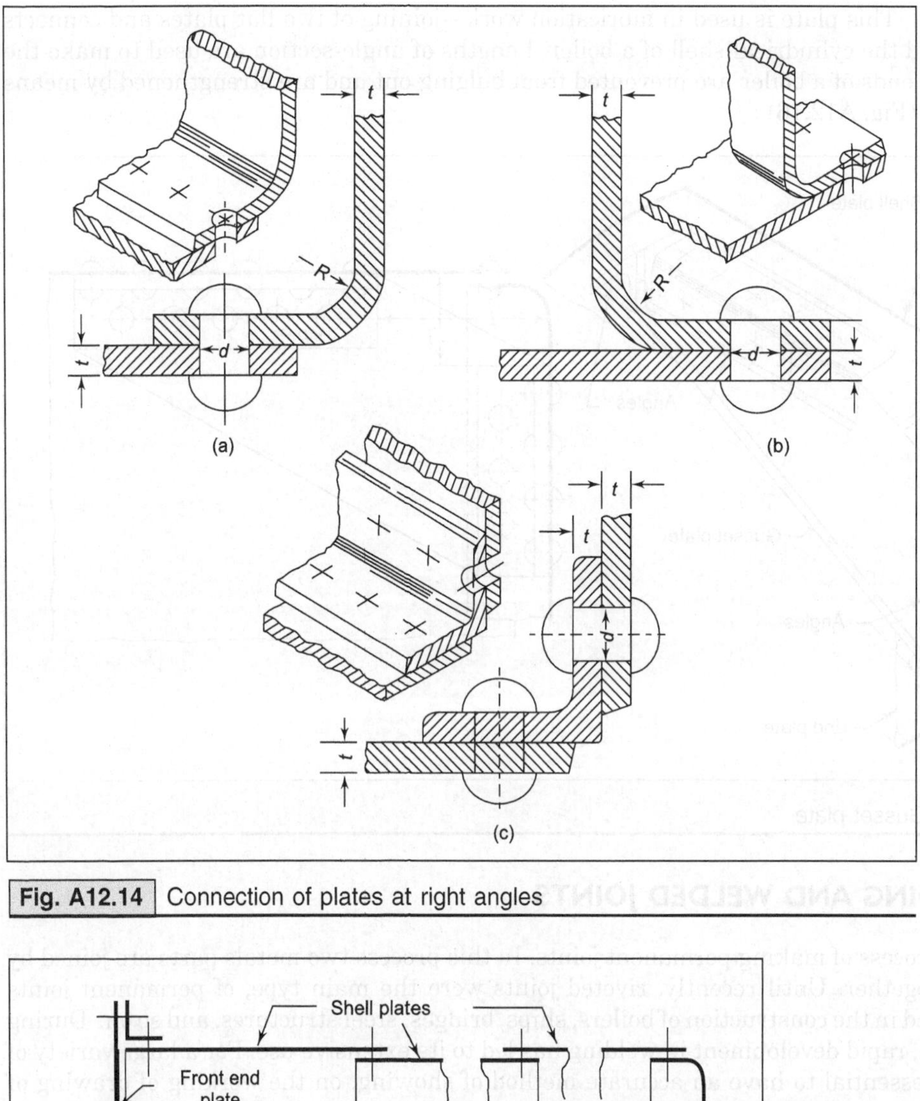

Fig. A12.14 Connection of plates at right angles

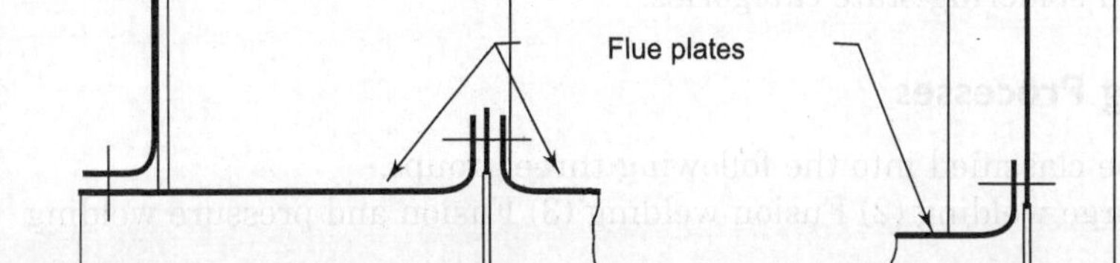

Fig. A12.15 Boiler joints

Gusset plate This plate is used in fabrication work—joining of two flat plates and connects the flat end and the cylindrical shell of a boiler. Lengths of angle-section are used to make the joints. The flat ends of a boiler are prevented from bulging out and are strengthened by means of these plates (Fig. A12.16)

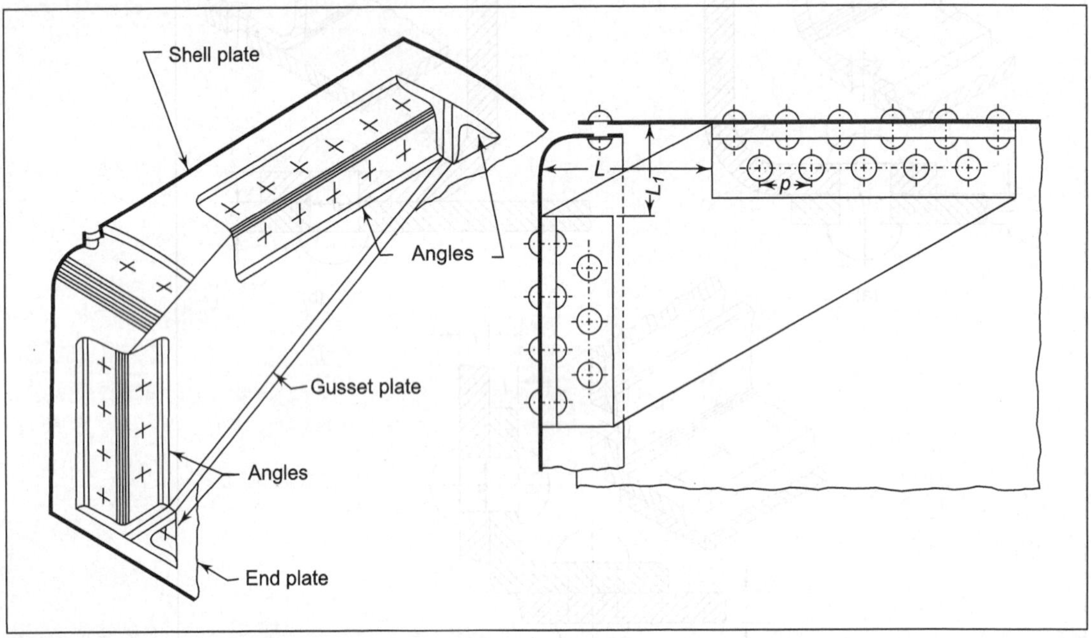

Fig. A12.16 Gusset plate

12.3 WELDING AND WELDED JOINTS

Welding is a process of making permanent joints. In this process two metals parts are joined by fusing them together. Until recently, riveted joints were the main type, of permanent joints extensively used in the construction of boilers, ships, bridges, steel structures, and so on. During the last decade, rapid development in welding has led to its extensive use. For a large variety of purposes it is essential to have an accurate method of showing, on the working of drawing of machines or structures, the types, sizes, and locations of welding desired by the machine designer.

Welding processes are divided into three basic categories: (1) fusion welding, (2) solid-state welding, and (3) brazing and soldering-state categories.

12.3.1 Types of Welding Processes

The welding processes can be classified into the following three groups.

(1) Pressure welding or forge welding (2) Fusion welding (3) Fusion and pressure welding

Pressure welding or forge welding: In this method two metal parts are heated at the joint upto plastic condition and then joined together by applying external mechanical pressure.

Fusion welding: It is a process of welding by local fusion with or without the use of a filler. The most commonly used fusion welding processes are gas welding and arc welding. This process constitutes a major category of welding; it comprises consumable or nonconsumable electrorode.

The welded joint undergoes important metallurgical and physical changes which, in turn, have a major efect on the properties and performance of the welded component or structure. In case of gas welding, the welding puddle is produced by the effect of a flame generated by a fuel gas (generally acetylene) and oxygen. The flame temperature varies from 2600°C to 3200°C, depending upon the fuel gases.

In case of arc fusion welding, the welding puddle is created by the effect of an arc. The arc burns between the melting electrode and the workpiece. The entrance of atmospheric gases into the arc and the welding puddle is prevented by the inert gases and slag produced by the electrode.

Fusion and pressure welding: In this method, a heavy current, of the order of 50,000A, is passed through the joint, causing the gases to melt. The welding is completely under external pressure. Spot welding, seam welding, flash butt welding, and pressure butt are the examples of pressure resistance welding.

Other methods of joining metal parts include:

Solid-state welding Whenever joining takes place without fusion; consequently, there is no liquid (molten) phase in the joint. The basic categories in this method are diffusion bonding and cold, ultrasonic, friction, resistance, and explosion welding. Diffusion bonding, combined with superplastic forming, has become an important manufacturing proces for complex shapes.

Brazing and **soldering** use filler metals and involve lower temperatures than welding; the heat required is supplied externally.

Adhesive bonding has been developed into an important technology because of its several advantages; it has unique applications requiring strength, sealing, insulation, vibration damping, and resistance to corrosion between dissimilar metals. Included in this category are electrically-conducting adhesives for surface-mount technologies. **Mechanical fastening** involves traditional methods of using various fasteners, bolts, nuts, and rivets. **Joining plastics** can be accomplished by adhesive bonding, fusion by various external or internal heat sources, and mechanical fastening.

The choice of a joining process depends on several factors: (a) the application, (b) the joint design, (c) the materials involved; and (d) the shapes of the components to be joined, their thicknesses, and their sizes. Other considerations are the location of the joint within the product, the number of individual components involved, the operator skill required, and equipment and labour costs.

12.3.2 Types of Welded Joints

There are four important types of welded joints, classified according to the positions of the workpiece being joined, as shown in Fig. A12.17.

Lap weld, butt weld, tee weld, and corner weld are all of forge welding processes. These are the oldest welding processes, and are carried out, in local blacksmith's shops. Here workpieces are heated in the furnace to a condition of plasticity on the surface to be joined. These parts are then power hammerrd or are joined together by pressing. The quality of weld depends on the amount of heating provided to the workpiece. If the ends to be joined are not heated enough, they will not stick together; if overheated, the metal burns and becomes brittle.

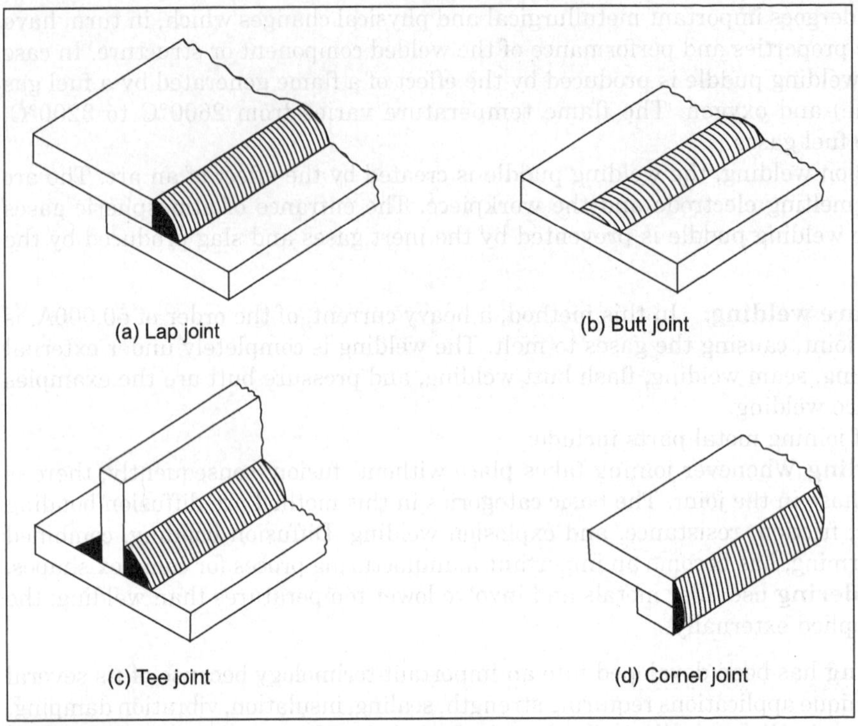

Fig. A12.17 Types of welded joints

Types of welds Arc and gas welds, which are of four types, are shown in Fig. A12.18.

Back weld: Here two metal pieces are ground, taken in the same plane and kept in line. This joint is only on one side of the plane. This is mainly used for closed vessels or cylinders.

Filled weld: This is a very common type of weld. Two metal pieces are kept at right angles to each other and with the help of electric arc welding, metal is filled in the gap of the two plates.

Plug weld: This is suitable where there is no space for joining two metal pieces with the help of fillet or back weld. Here, out of the two metal plates, one plate is drilled to a certain depth, the hole penetrates through the drilled plate, but is blind at the end. Welding material is filled in the drilled hole and the two plates are joined.

Square weld: Edges of the two plates that are to be joined are made parallel and perfectly square before welding, with help of grinding. This type of welding is useful where the projection of welding material is not allowed after joining two pieces.

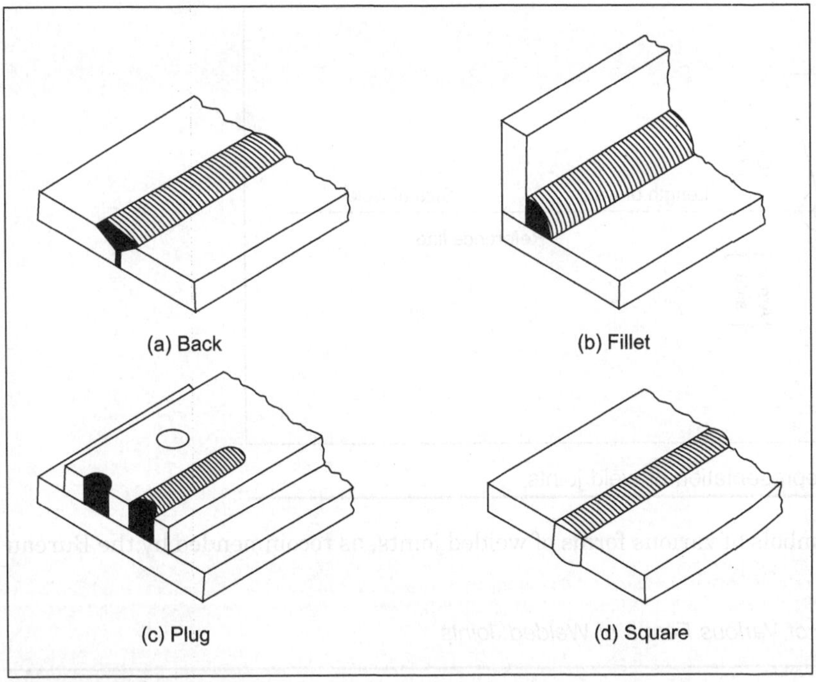

Fig. A12.18 Arc welds

Similarly, resistance welds are also of four types: (1) spot weld, (2) projection weld, (3) seam weld, (4) flash and upset welds. Symbolically, they are represented as shown in Fig. A12.19.

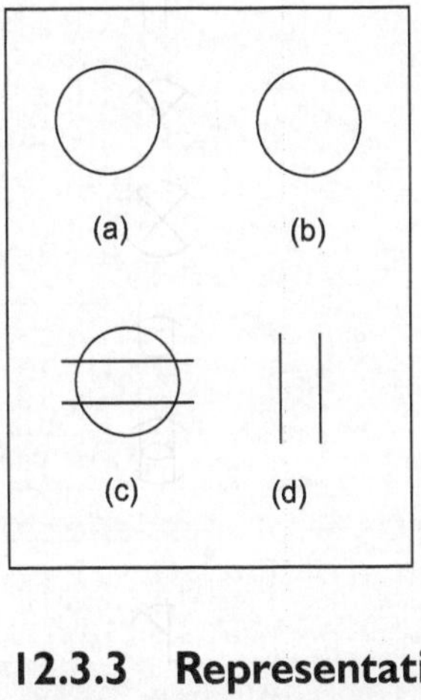

Fig. A12.19 Resistance weld symbols

12.3.3 Representation of Welded Joints

A welded joint is shown on a drawing by means of

1. a symbol that specifies the form of weld and
2. a bent arrow and a reference line indicating the location of the weld, as shown in Fig. A12.20.

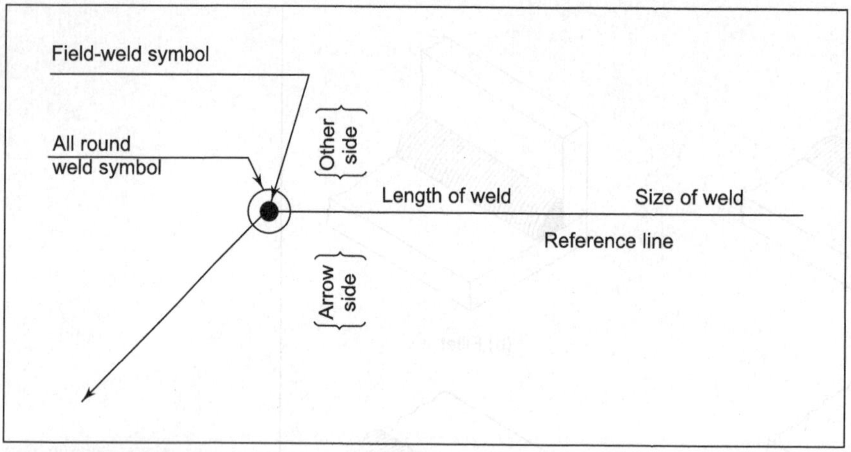

Fig. A12.20 Symbolic representation of weld joints

Table A12.2 shows symbols of various forms of welded joints, as recommended by the Bureau of Indian Standards.

TABLE A12.2 *Symbols of Various Forms of Welded Joints*

Form of Weld		*Symbol*
1. Fillet	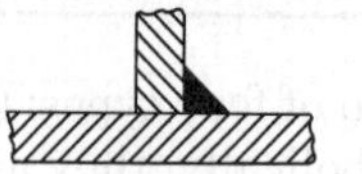	
2. Square butt		
3. Single-*V* butt		
4. Double-*V* butt		
5. Single-*U* butt		
6. Double-*U* butt		
7. Single-Bevel butt	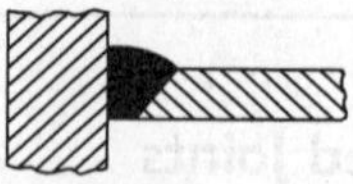	
8. Double-Bevel butt	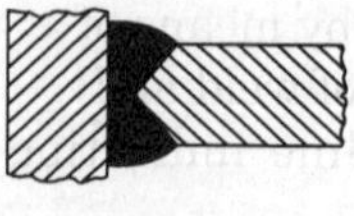	
9. Single-*J* butt		

Form of Weld		***Symbol***
10. Double-*J* butt		
11. Edge	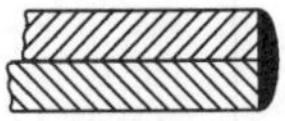	
12. Spot		

The method of showing the form and location of a fillet weld on a drawing is illustrated in Fig. A12.21.

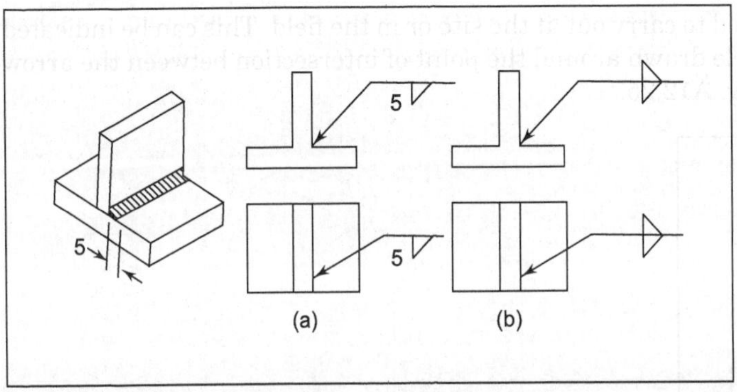

Fig. A12.21 Fillet weld

The symbol should be shown in only one of the views. When a weld is on the arrow side, the symbol should be inverted and placed below the reference line, as shown in Fig. 12.21(a). When a weld is to be on the other side, the symbol should be placed in its correct position, but over the reference line. For welds on both the sides, a symbol is placed above as well as below the reference line [Fig. A12.21(b)]. Note that in each case the vertical portion of the symbol is always kept on the lefthand side of the symbol. The size of weld is indicated on the leftside of the weld symbol.

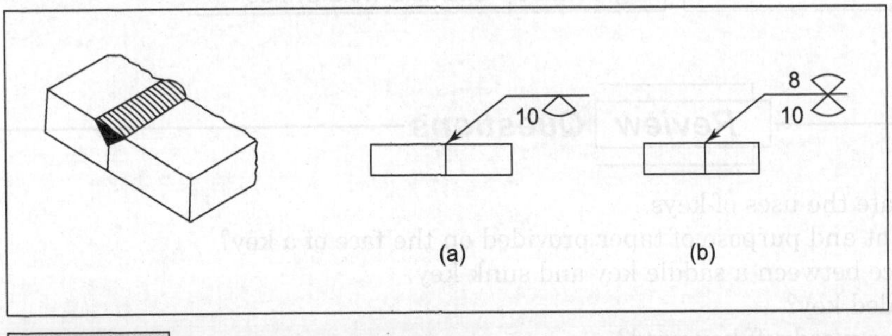

Fig. A12.22 Symbolic representation of weld

The depth of a partially penetrated *U*-butt or *V*-butt weld is indicated by means of a dimension placed to the left of the symbol. Figure A12.22(a) shows a *V*-butt weld partially penetrated on the top surface, while the dimensions of the same weld, partially penetrated on both the sides are shown in Fig. A12.22(b).

A butt-weld flush with the surface of the plate is indicated by placing a bar above the symbol as shown in Fig. A12.23. An allround weld is shown by means of a circle drawn around the point of intersection between the arrow and the reference lines, as shown in Fig. A12.24.

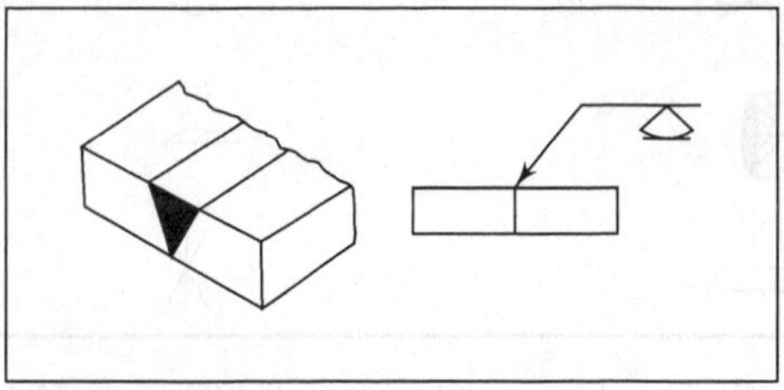

Fig. A12.23 Bult weld symbol

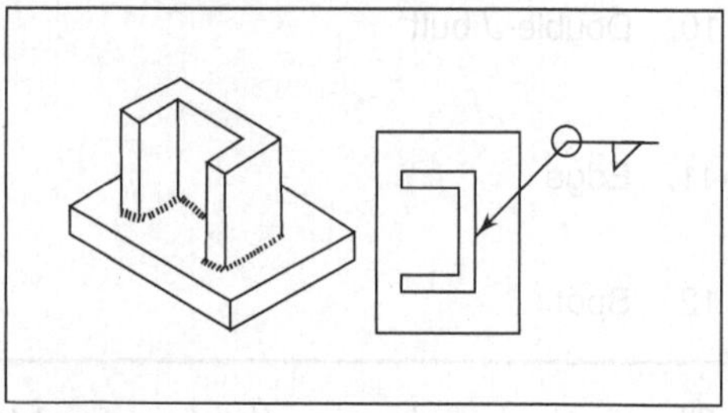

Fig. A12.24 All-round weld symbol

Sometimes the welding is required to carry out at the site or in the field. This can be indicated on a drawing by placing a filled circle drawn around the point of intersection between the arrow and reference lines as shown in Fig. A12.25.

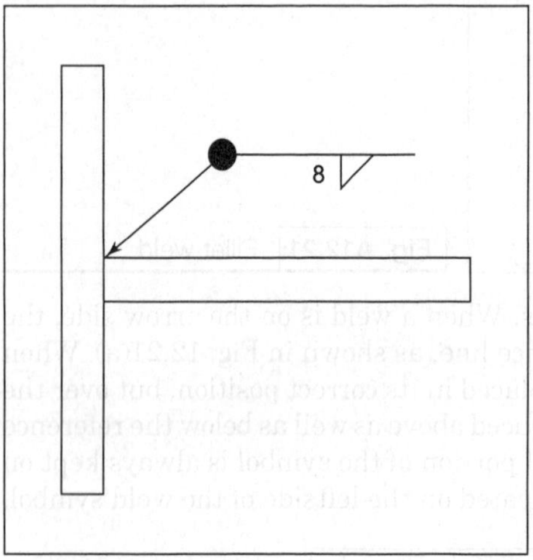

Fig. A12.25 Site/field weld symbol

Review Questions

1. What is a key? State the uses of keys.
2. What is the amount and purpose of taper provided on the face of a key?
3. State the difference between a saddle key and sunk key.
4. What is a gib-headed key?
5. Where and why is a woodruff key used?
6. What is a cotter and where it is used?
7. What is a pin joint?
8. Why are clearances provided in cotter joints?
9. Differentiate between a cotter joint and a pin joint.
10. What do you mean by spline shaft?

CHAPTER 13

Pulleys

This chapter contains information about pulleys

After reading the chapter you will be able to

- understand the application of pulleys
- indentify various types of pulleys
- understand assembly drawings of pulleys

13.1 PULLEYS

Pulleys and gears are used for transmission of power from one shaft to another by means of belts or through direct contact. They are made of castiron, wroughtiron, steel, wood, or plastic.

A pulley may be cast in one piece or may be builtup from separate parts. The three parts of a belt pulley are—hub or boss, the arms, and the rim as, shown in Fig. A13.1.

Fig. A13.2 shows three methods of mounting a pulley rigidly on a shaft, namely,

1. a key
2. cone keys
3. a set screw.

Generally three methods of mounting of pulleys are in use (Fig. A13.2).

1. Mounting of pulley with rectangular key
2. Mounting of pulley with cone key
3. Mounting of pulley with the help of a set screw

A pulley with the appropriate bore diameter is selected for the shaft on which it is to be mounted. The boring operation is normally carried out by a bosing machine. Keyway operation then follows on the slotting machine, on the outer surface of the shaft and inside the bore of the pulley. Thereafter, the pulley is assembled on the shaft with the help of a key. This proce-dure is common for all the three methods stated above. In case of method (b), a certain amount of gap is maintained inbetween the bore diameter of pulley and outside diameter of the shaft, and a cone key is pressed in this gap to assemble the pulley on the shaft.

In case of method (c), a small hole is drilled on the hub of the pulley and it is tapped to match the size of a standard setscrew. By screwing the setscrew, the pulley is assembled on the shaft.

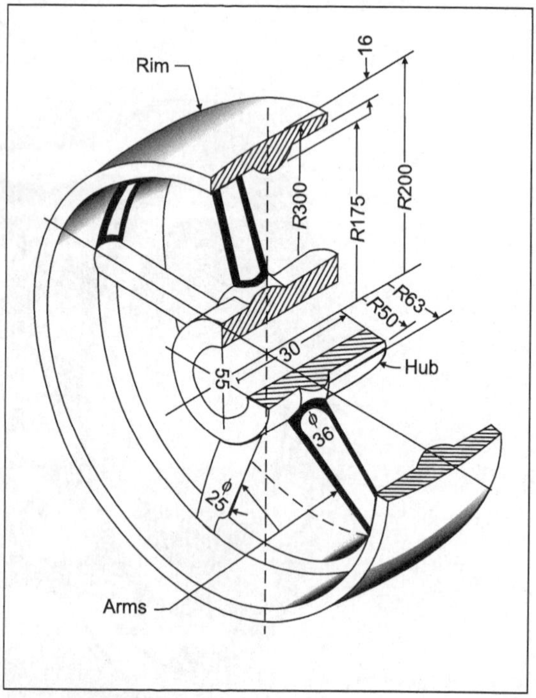

Fig. A13.1 Belt pulley with straight arms

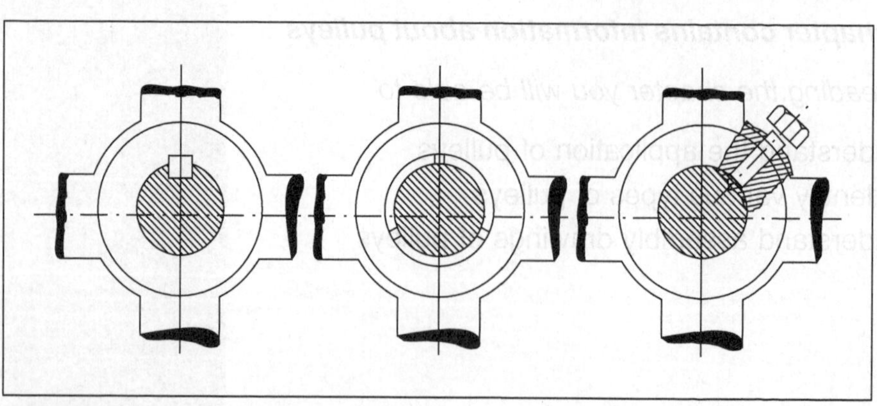

Fig. A13.2 Methods of fixing a pulley

13.2 TYPES OF PULLEYS

Pulleys are used to transmit motion from one component to the other component, mainly shafts. There are different types of pulleys. Some of them are discussed below.

1. *CI* belt pulley
2. Fast and loose pulley
3. Speed or stepped pulleys
4. Split pulleys
5. Built-up pulleys
6. Rope pulleys
7. *V*-belt pulleys

CI belt pulleys The rim of a cast iron belt pulley is generally made convex, that is, crowning or bulging outwards at the centre of its width. The curvature tends to keep the belt in the middle of the rim. When the rim is kept flat, it is usually provided with flanges at the sides to prevent the belt from slipping off. The arms of a pulley may be of circular or elliptical cross-section, but are larger at the hub than at the rim. The thickness of the arm of the elliptical shape is kept equal to about one-half of its width. During the process of casting a pulley, stresses are set up in the material on account of contraction while cooling. Due to these stresses, straight arms crack or at near the junction of arms with the rim. Curved arms conveniently yield to these stresses by straightening. Arms are therefore made curved, particularly in case of large size pulleys and are shaped such that they avoid sharp corners. Fig. A13.1 shows a pulley with straight arms, while those with curved arms is shown in Fig. A13.3.

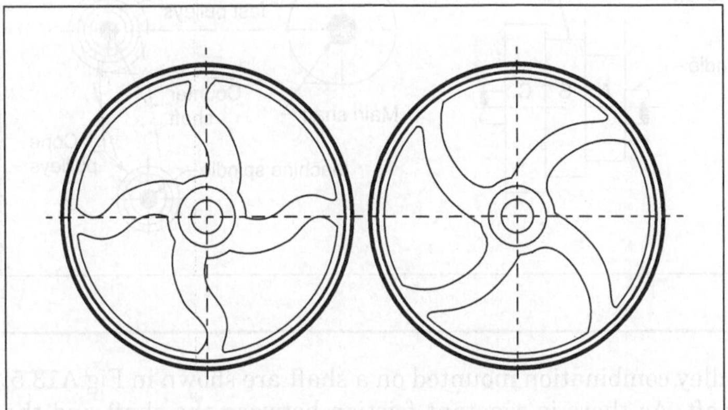

Fig. A13.3 Belt pulley with curved arms

Table A13.1 provides the diameters of pulleys to be used with flat belts as well as *V*-belts.

TABLE A13.1 *Types of Belts and Their Diameters*

Belt Name	***Diameter (mm)***
Only for *V*-belt	20, 22, 25, 28, 32, 36
Flat and *V*-belts	40, 45, 50, 56, 63, 71, 80, 90, 100, 112, 125, 140, 160, 180, 200, 224, 250, 280, 315, 355, 400, 450, 500, 560, 630, 710, 800, 900, 1000

Fast and loose pulleys A pulley rigidly attached to a shaft is called a fast pulley. A loose pulley is free to rotate on the shaft. A machine can easily be started or stopped whenever required by using a pair of fast and loose pulleys. This pair is mounted on a counter-shaft (Fig. A13.4), which is placed above or near the machine that is to be operated. When the driving belt from the main shaft is on the fast pulley, the countershaft is set in motion. If the belt is shifted from the fast pulley on to the loose pulley, the countershaft stops rotating. The shifting is done by means of a beltfork through which the belt passes and which is fixed to a striking bar. The bar and belt fork can be moved to the left and back again by pulling the cords or wires hanging down and they are within easy reach of the person operating the machine. As the machine spindle is driven by a beltdrive from the countershaft, it is started or stopped by pulling the respective cords.

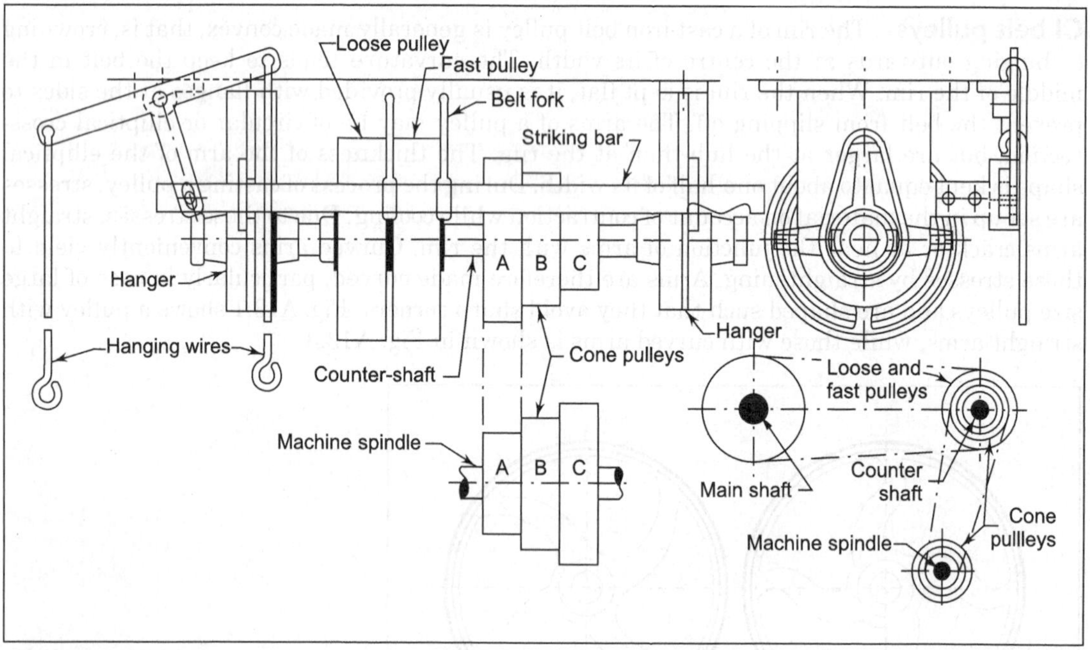

Fig. A13.4 Machine countershaft

Two views of a fast and loose pulley combination mounted on a shaft are shown in Fig.A13.5. The fast pulley is keyed to the shaft. As there is constant friction between the shaft and the loose pulley, the latter is generally lined with a gunmetal bush that can be replaced when worn out. A collar fixed on the shaft by a grub screw prevents the loose pulley from moving along the length of the shaft.

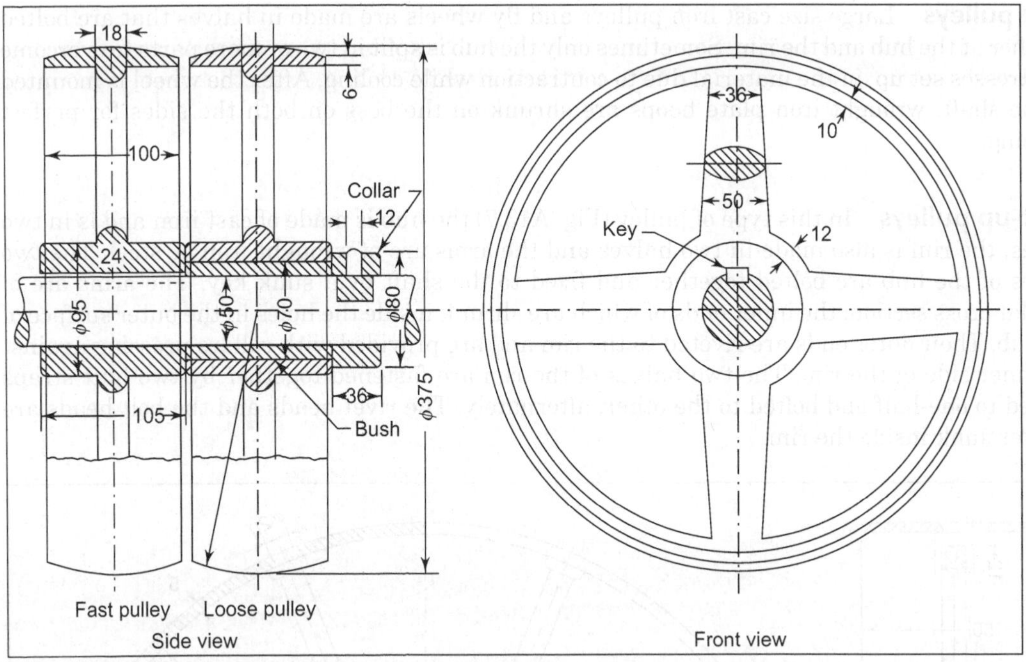

Fig. A13.5 Fast and loose pulleys

Speed cones or stepped pulleys These are made of cast iron and are secured to the shafts by means of keys of set screws (Fig. A13.5). A machine spindle can be driven at different speeds by fixing stepped pulleys on the spindle and on the countershaft in reverse positions, as shown in Fig. A13.4. The speed of the machine spindle will be (i) greater than, (ii) equal to, and (iii) less than that of the countershaft when the belt is in positions *AA*, *BB*, and *CC*, respectively.

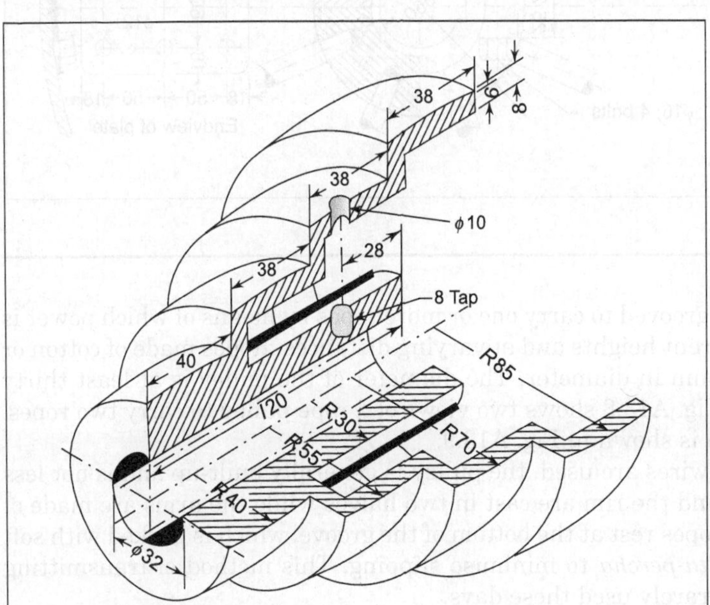

Fig. A13.6 Speed cone

Split pulleys Large size cast iron pulleys and fly-wheels are made in halves that are bolted together at the hub and the rim. Sometimes only the hub is split in two or more parts to overcome the stresses set up in the material due to contraction while cooling. After the wheel is mounted on the shaft, wrought iron plate hoops are shrunk on the boss on both the sides for perfect gripping.

Built-up pulleys In this type of pulley (Fig. A13.7) the hub is made of cast iron and is in two halves, the rim is also made in two halves and the arms are of wrought iron or steel. The two halves of the hub are bolted together and fixed to the shaft by a sunk key. The arms are of circular cross section, the inner ends of which are shrunk inside the holes in the outer surface of the hub. Their outer ends are riveted to the rim and are provided with collars pressing against the inner side of the rim. The two halves of the rim are fastened together by two butt straps riveted to one-half and bolted to the other, alternately. The rivet heads and the bolt heads are counter sunk inside the rim.

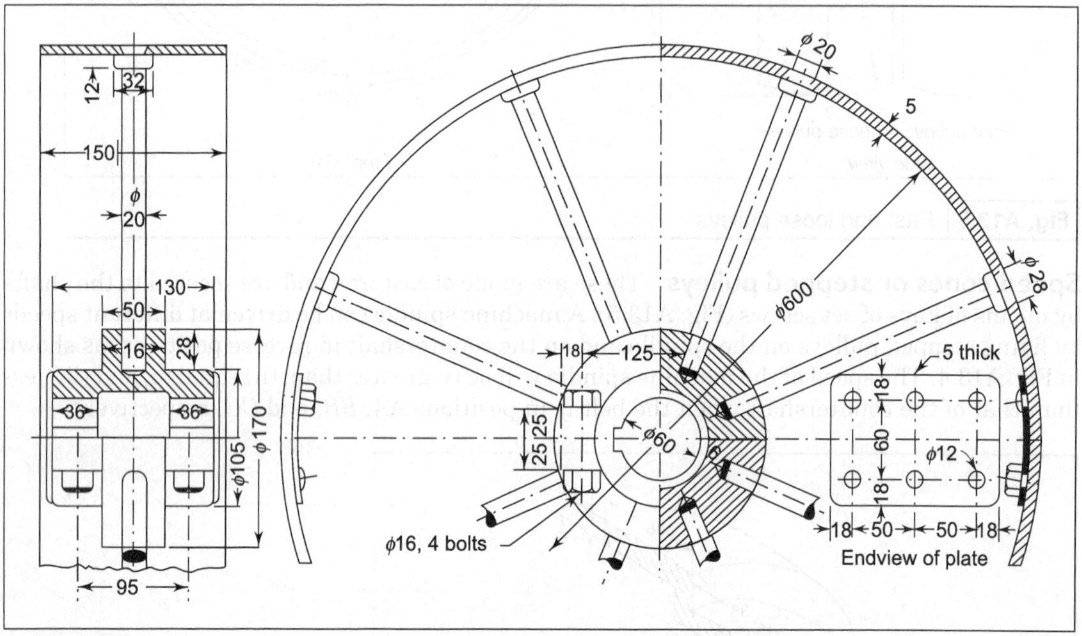

Fig. A13.7 Built-up pulley

Rope pulley Rope pulleys are grooved to carry one or more ropes by means of which power is transmitted to the shafts at different heights and at varying distances. Ropes made of cotton or hemp are usually 25 mm to 50 mm in diameter. The diameter of the pulley is at least thirty times the diameter of the rope. Fig. A13.8 shows two views of a rope pulley to carry two ropes. The view of the dimensioned rim is shown in Fig. A13.9.

When the ropes made of steel wires are used, the pulley is generally built-up and is not less than 2 m in diameter. The hub and the rim are cast in two halves while the arms are made of steel. Unlike cotton ropes, steel ropes rest at the bottom of the groove, which is packed with soft materials such as leather or *gutta-percha* to minimise slipping. This method of transmitting power by means of ropes is very rarely used these days.

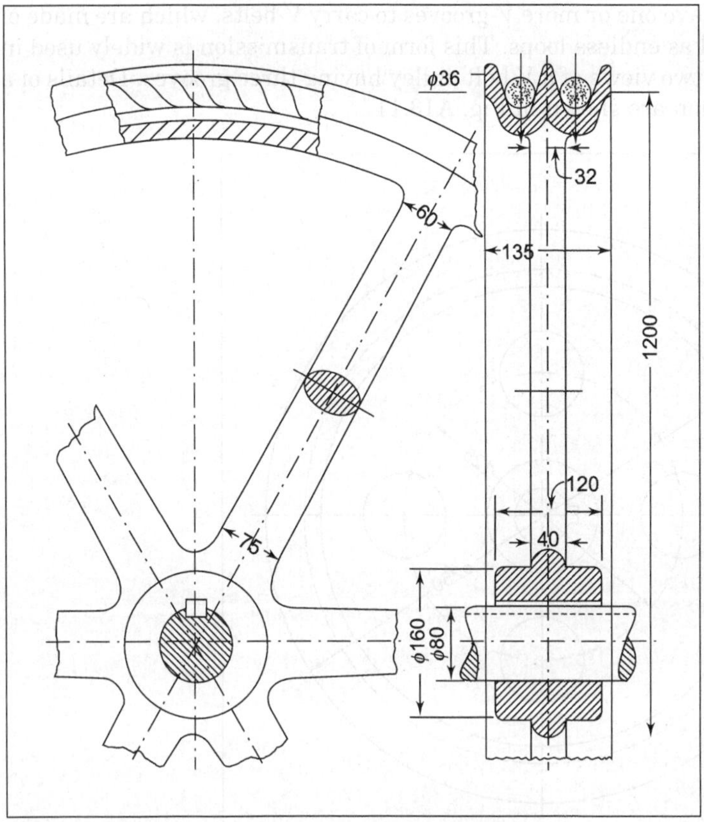

Fig. A13.8 Rope pulley

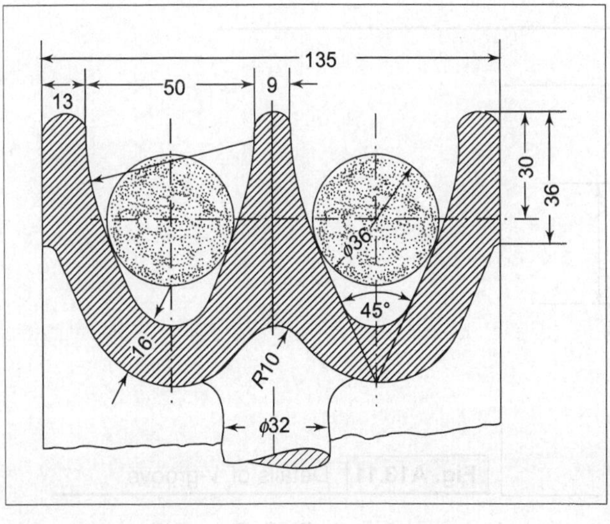

Fig. A13.9 Section of a rim

V-belt pulleys These pulleys have one or more *V*-grooves to carry *V*-belts, which are made of rubber and fibre and are moulded as endless loops. This form of transmission is widely used in modern times. Fig. A13.10 shows two views of a *V*-belt pulley having three grooves. Details of a *V*-groove along with the belt-section are shown in Fig. A13.11.

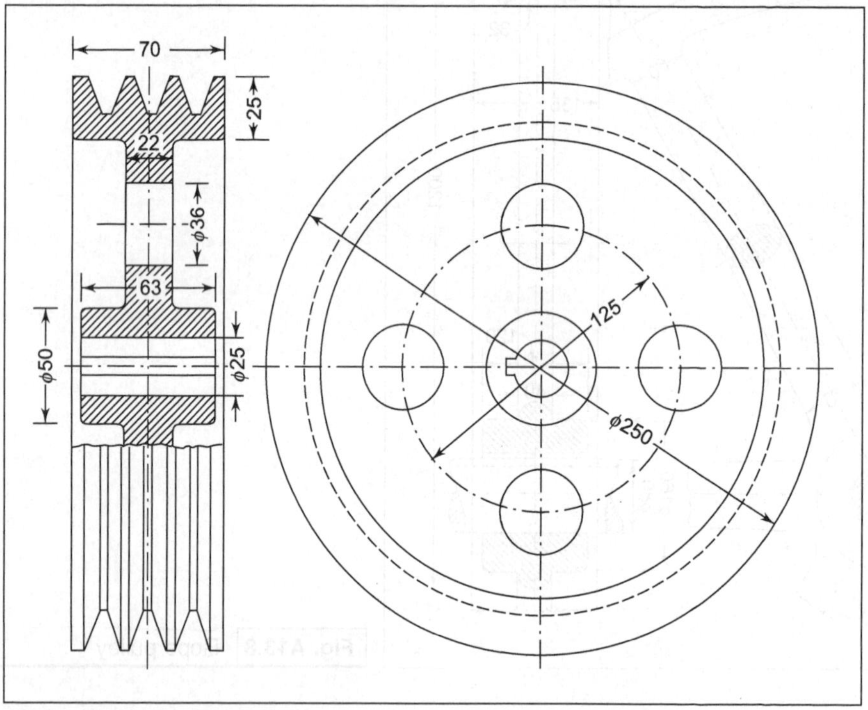

Fig. A13.10 *V*-belt pulley

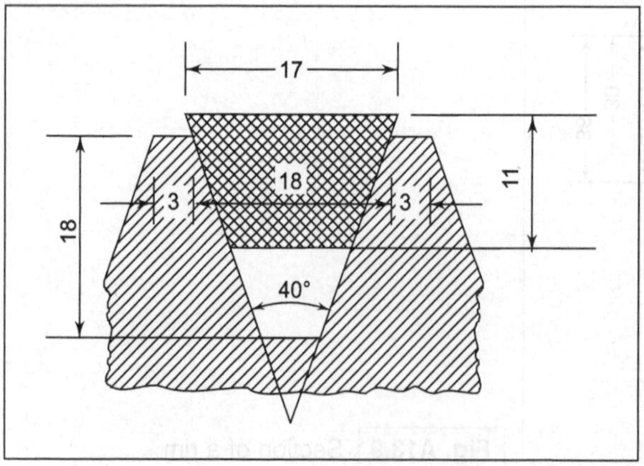

Fig. A13.11 Details of *V*-groove

V-belts are available in standard belt-sections, designated as *A*, *B*, *C*, *D*, and *E*.

Table A13.2 shows details of various cross sections of *V*-belts.

TABLE A13.2 *Details of Cross-section of V-belts*

Cross-section Symbol	Nominal Top Width (mm)	Nominal Thickness (mm)
A	13	8
B	17	11
C	22	14
D	32	19
E	38	23

A *V*-belt of cross section *A* and nominal inside length of 1212 mm shall be designated as *A*-1212 IS:2494 (1974).

The calculations of a *V*-belt drive are confined to (1) the selection of a standard profile and the length and (2) the number of belts are determined for the transmission of the, given power. The number of belts should not exceed 8 to 12. If it exceeds the limit then the next larger belt-section should be used.

Table A13.3 shows useful data for *V*-belt drives.

TABLE A13.3 *V-Belt Drive Data*

Width of belt (mm)	5	6	8	10	13	17	20
Height of belt (mm)	3	4	5	6	8	11	12.5
Length of belt (mm) from	150	212	296	420	585	832	1000
to	860	1262	1916	2820	4275	6332	9540

Review Questions

1. State the use of pulleys.
2. What is a pulley? Name the different parts of a pulley and also discuss the various types.
3. What is crowning? Where and why is it applied?
4. What are the various types of belt pulleys?
5. What do you mean by fast and loose pulleys?
6. What is the application of the rope drive?
7. What are the main advantages of stepped pulleys?
8. What is split pulley?
9. Identify ten applications where pulleys are used.

CHAPTER 14

Shaft Couplings

This chapter contains information on joining of shafts with couplings

After reading the chapter you will be able to

- differentiate between the various types of couplings and their uses
- draw coupling sketches
- select suitable couplings for different types of joints

14.1 COUPLINGS

In engineering, mechanical power is used to run a large number of machines. Transmiting power from one place to another is possible only with an assembly of shafts, pulleys, belts, and so on. Thus, shafts play an important role in the transmission of power. Shafts are available only in 6 to 10 meters lengths in the market, as it is difficult to manufacture and transport shafts of larger lengths. However, in mechanical engineering shafts of large lengths are required, so often, two or more shafts need to be joined axially using shaft couplings. In general, shaft couplings are assemblies of sleeves, flanges, pins, cotters, bolts and nuts, and so on and are used for transmitting rotary motion directly from one shaft to another. They may be classified as:

1. Fast or rigid couplings
2. Loose or disengaging couplings
3. Flexible couplings

14.2 FAST OR RIGID COUPLINGS

Rigid couplings are also called solid or forged flange couplings and are generally used in heavy power transmission systems like marine engines. Very often the flanges are forged solid with the shafts and are connected together by means of a number of taper bolts, without heads.

These couplings do not permit any relative rotation between two shafts. The various types of rigid couplings are described below.

Box or muff coupling It consists of a cast iron cylindrical muff in the middle of which ends of the two shafts butt against each other. A long sunk taper key is driven through, both, the shafts and the muff. Sometimes two separate keys are inserted from the opposite ends of the muff. The outside diameter and the length of the muff are, respectively, kept equal to about two times and three times the diameter of the shaft (Fig. A14.1).

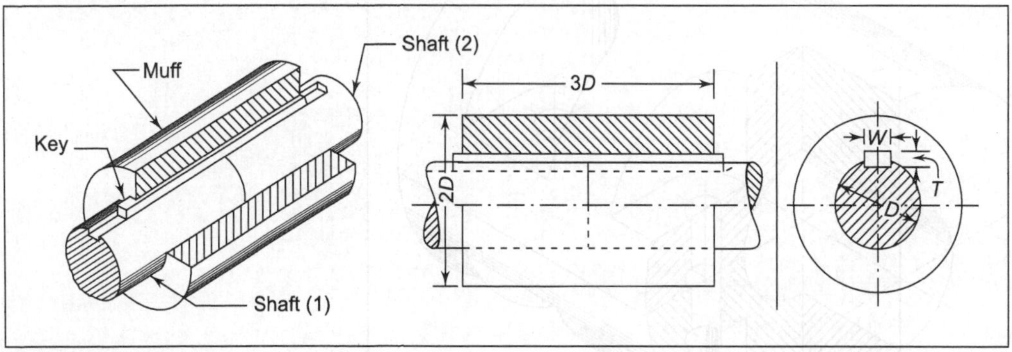

Fig. A14.1 Box or muff coupling

Half-lap couplings It is a muff coupling in which the ends of the shafts are made to overlap each other for a short length. The taper in the overlap prevents the shafts from separating, if pulled in opposite directions due to axial force. A hollow saddle key is used to connect the muff and the shafts. The diameter and the length of the muff are both equal to about twice the diameter of the shafts (Fig. A14.2).

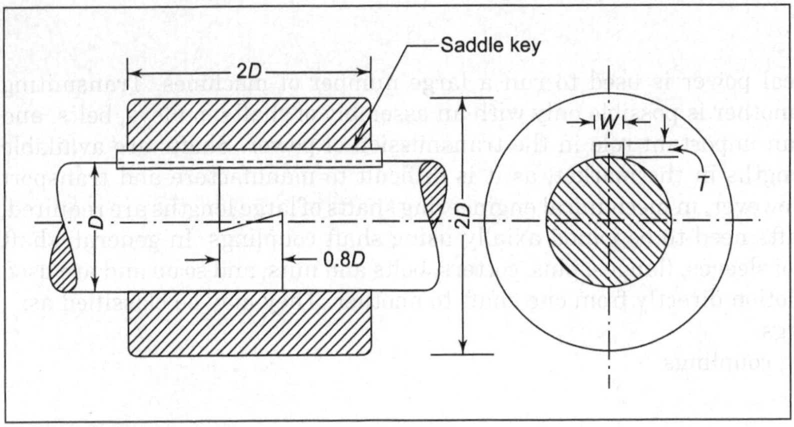

Fig. A14.2 Half-lap coupling

Slip-muff coupling The muff is made by two semi-cylindrical halves that are joined together by means of bolts and nuts. The halves are recessed to accomodate the necessary number of bolt-heads and nuts. During the boring operation, the two halves are separated by a thin shim or a strip of paper. The halves, when are drawn together by bolts, clip the shafts tightly. The coupling is done with the help of a feather key. Fig. A14.3(b) shows the pictorial view of a split muff partly in section [Fig. A14.3(a) and (b)].

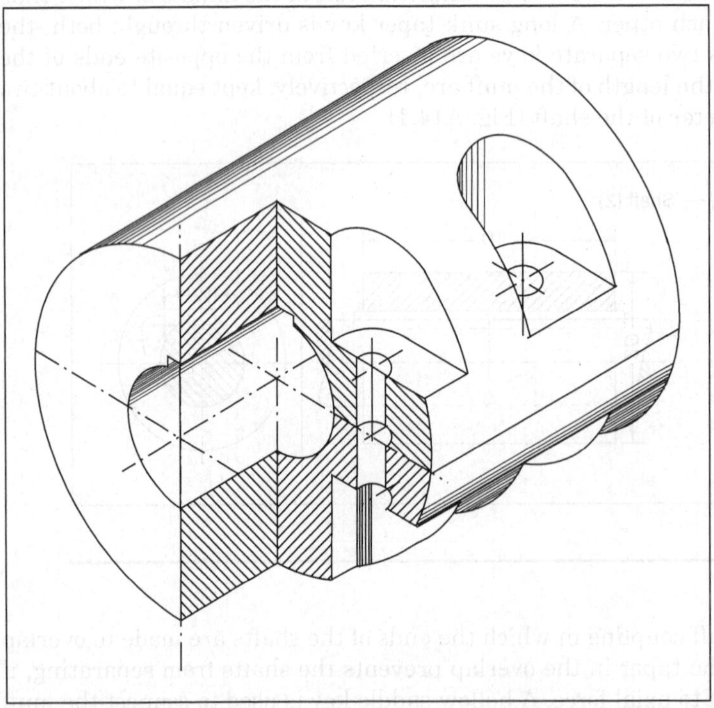

Fig. A14.3(a) Split-muff

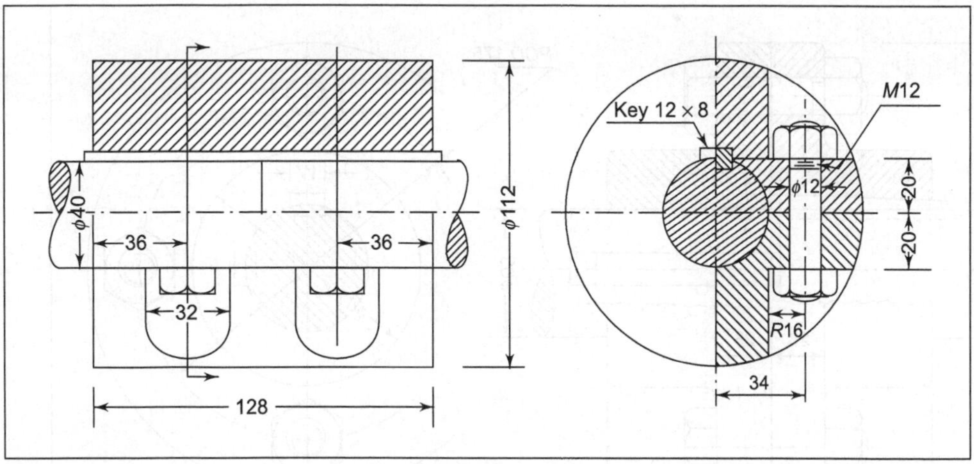

Fig. 14.3(b) Split-muff coupling

14.3 LOOSE OR DISENGAGING COUPLING

14.3.1 Flanged Couplings

[Fig. A14.4(a) and (b)]

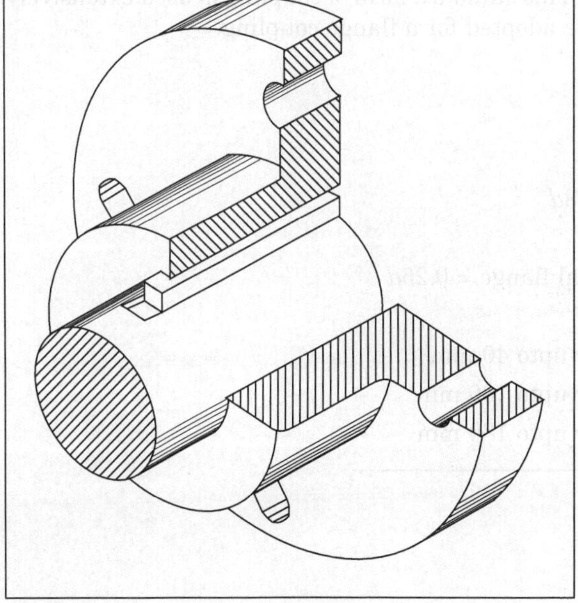

Fig. A14.4(a) Flanged couplings

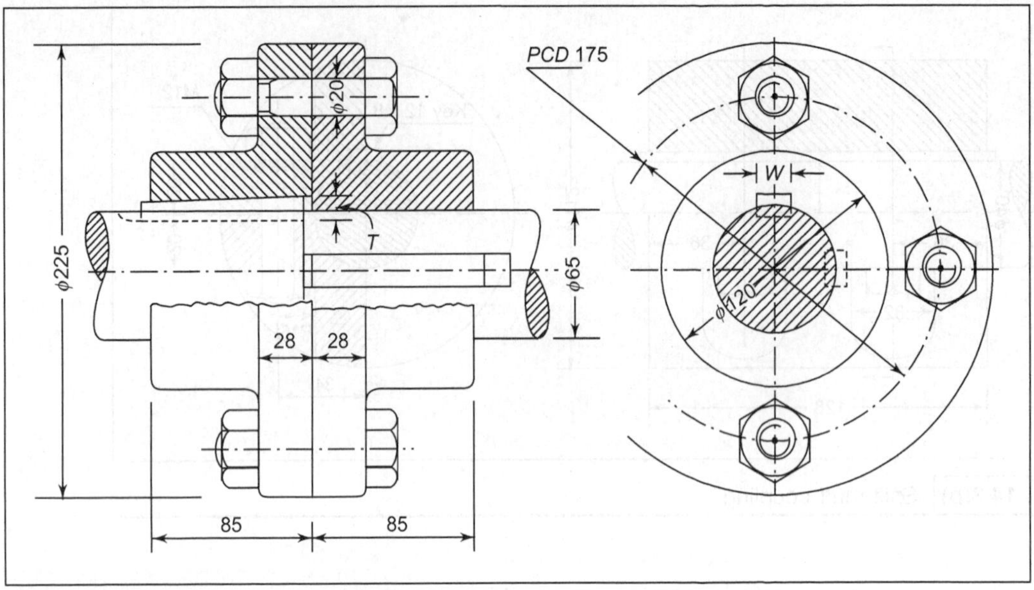

Fig. A14.4(b) Flanged couplings

It consists of two cast iron flanges that are keyed to the ends of the two shafts and fastened together by means of a number of tightfitting bolts. Sunk taper keys of rectangular or square cross-section are commonly used for this purpose. For ensuring correct alignment, one of the shafts is extended so that its end partly enters the flange keyed to the other shaft. The two shafts thus remain in a straight line. The shafts may be kept in alignment by the spigot and socket arrangement, as shown in Fig. A14.5. This standard form of coupling is used extensively.

The following standard proportions can be adopted for a flange coupling.

Let d be the diameter of a shaft

Outside diameter of the hub = $2d$

Pitch circle diameter of the bolts = 2.8 to $3d$

Thickness of the flange = $0.5d$

Thickness of the protective circumferential flange = $0.25d$

Length of the hub = $1.5d$

Number of bolts n = 3 for shaft diameter upto 40 mm

= 4 for shaft diameter upto 100 mm

= 6 for shaft diameter upto 180 mm

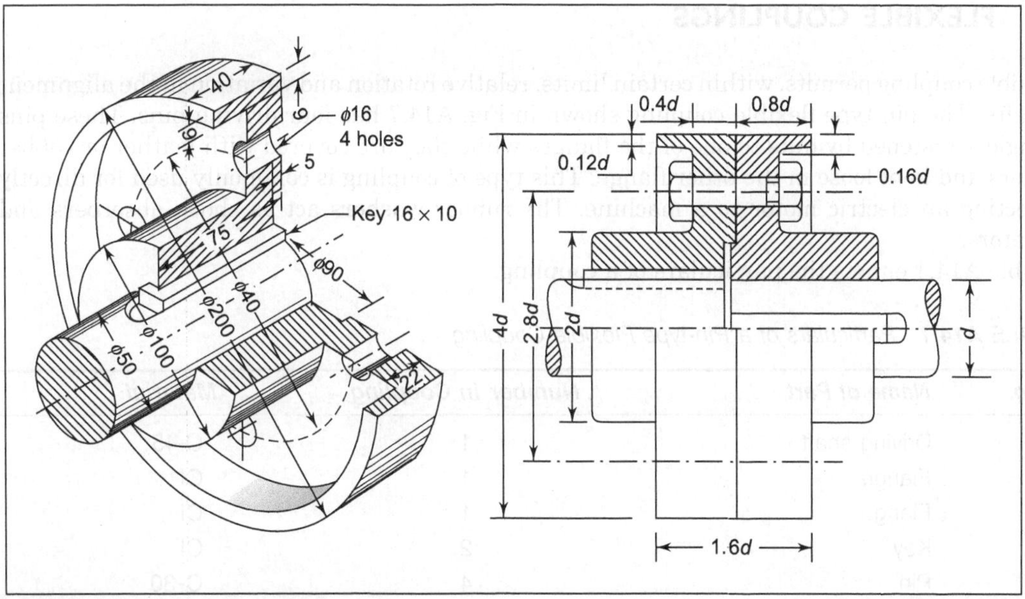

Fig. A14.5 Protected type flange coupling

Protected type flange coupling Fig. A14.5 shows a shurded or protected type of flanged coupling. Each flange is provided with a shroud or annular projection that shelters the boltheads or the nuts and prevents them from catching clothes of workmen. This type of coupling is sometimes used in belt pulleys. For correct alignment of the shafts, a spigot or projection is formed in the centre of one flange. This spigot is made to fit accurately into a corres-ponding socket or recess provided in the other flange.

Instead of providing a shroud, the flanges are very often made thicker and then counter-bored to accomodate bolt-heads and nuts for the same purpose.

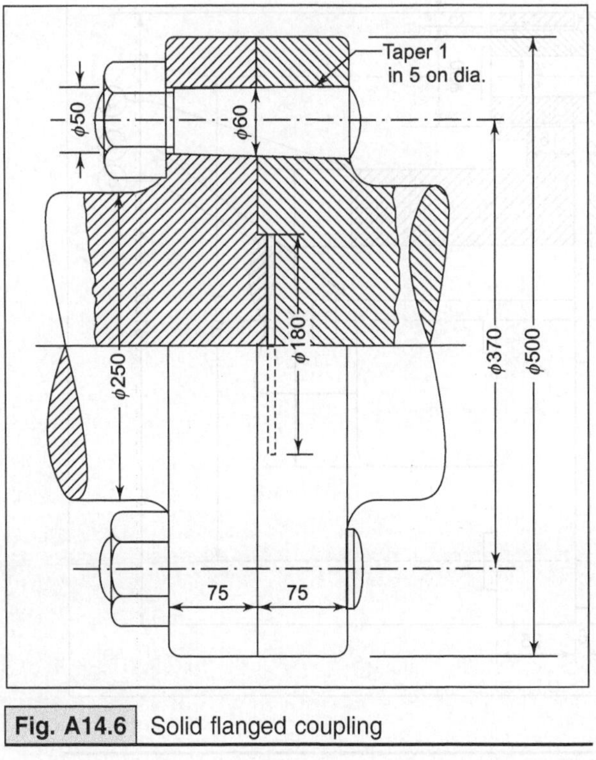

Fig. A14.6 Solid flanged coupling

Solid flanged coupling This type of coupling is used mainly for shafts of marine engines. The flanges are forged solid with the shafts and are connected together by means of a number of headless taper bolts (Fig. A14.6).

14.4 FLEXIBLE COUPLINGS

A flexible coupling permits, within certain limits, relative rotation and variation in the alignment of shafts. The pin-type flexible coupling shown in Fig. A14.7 has four driving pins. These pins are rigidly fastened by nuts to one of the flanges while they are covered with leather or rubber washers and kept loose in the other flange. This type of coupling is commonly used for directly connecting an electric motor to a machine. The rubber washers act as shock absorbers and insulators.

Table A14.1 enlists the particulars of a coupling.

TABLE A14.1 *Particulars of a Pin-type Flexible Coupling*

Sl. No.	*Name of Part*	*Number in Coupling*	*Material*
1	Driving shaft	1	C-45
2	Flange	1	CI
3	Flange	1	CI
4	Key	2	CI
5	Pin	4	C-30
6	Driven shaft	1	C-45
7	Rubber washers	5	Rubber

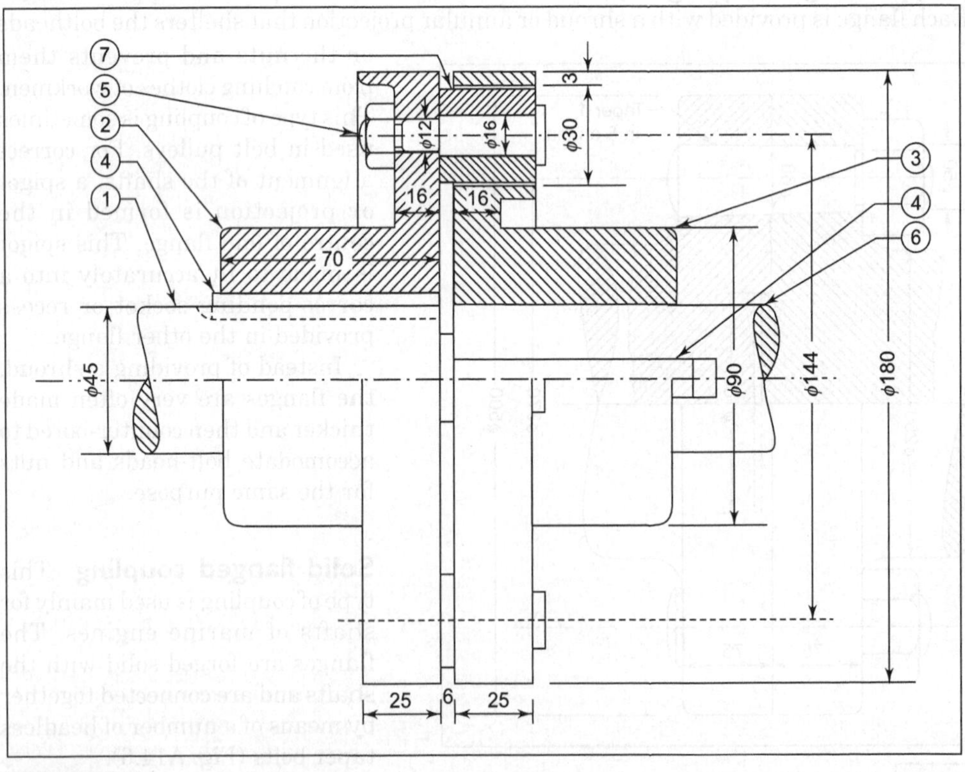

Fig. A14.7 Pin-type flexible coupling

Universal coupling or Hooke's joint This form of coupling is used to couple together two shafts whose axes intersect. Two similar forks are keyed on to the ends of the two shafts. These are pin jointed to a centre block, having two arms at right angles to one another. In this type of coupling the angle between the shafts may be varied even when they are in motion. Fig. A14.8(b) shows a sectional view of the coupling with all parts assembled together.

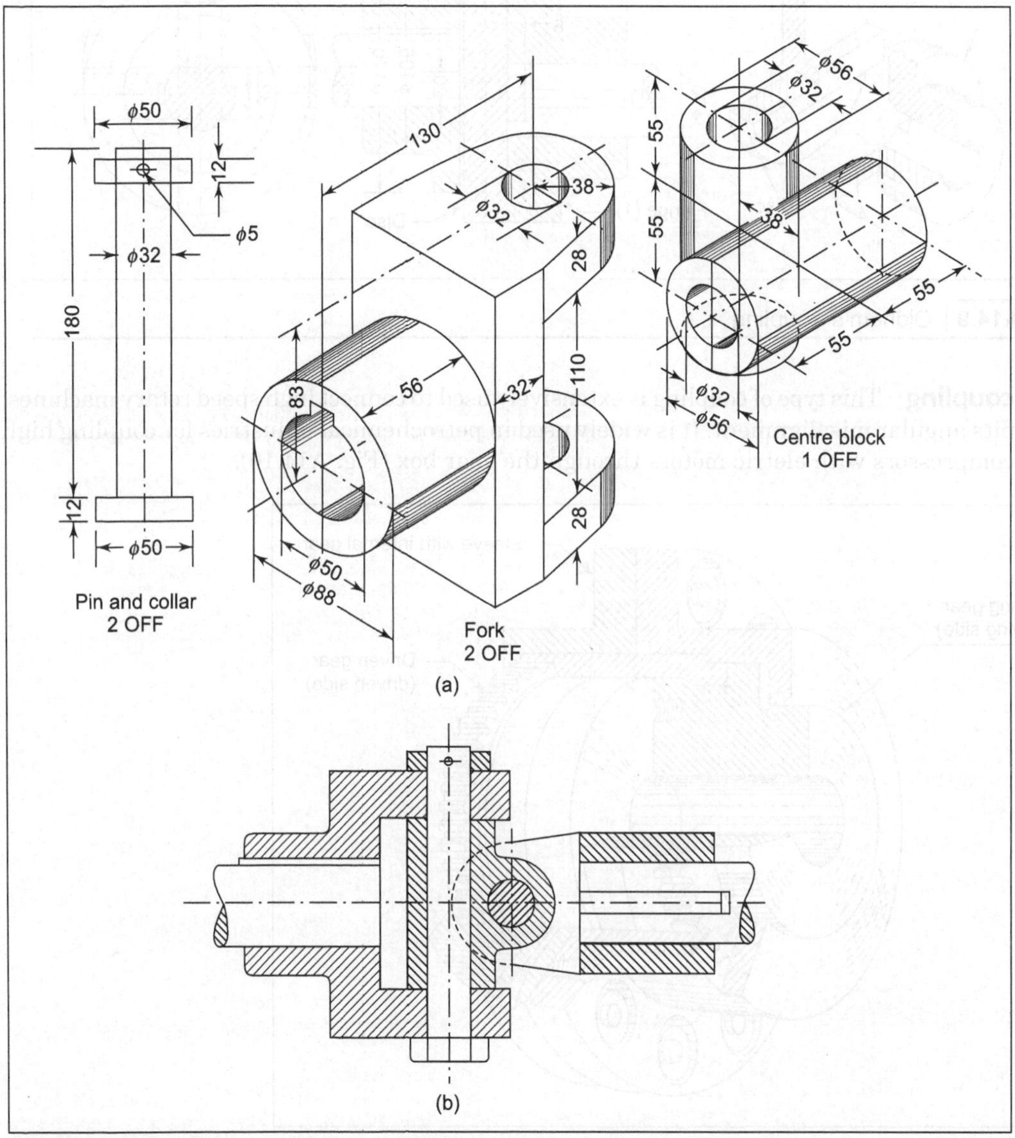

Fig. A14.8 Universal coupling

Oldham's coupling Shafts having their axes parallel, but not in alignment, are joined together by this type of coupling. Two flanges, each having a rectangular recess, are keyed one on each shaft. Another piece, in the form of a circular disc with two rectangular projecting parts on its opposite sides and at right angles to one another, is placed between the two flanges, as shown in Fig. A14.9.

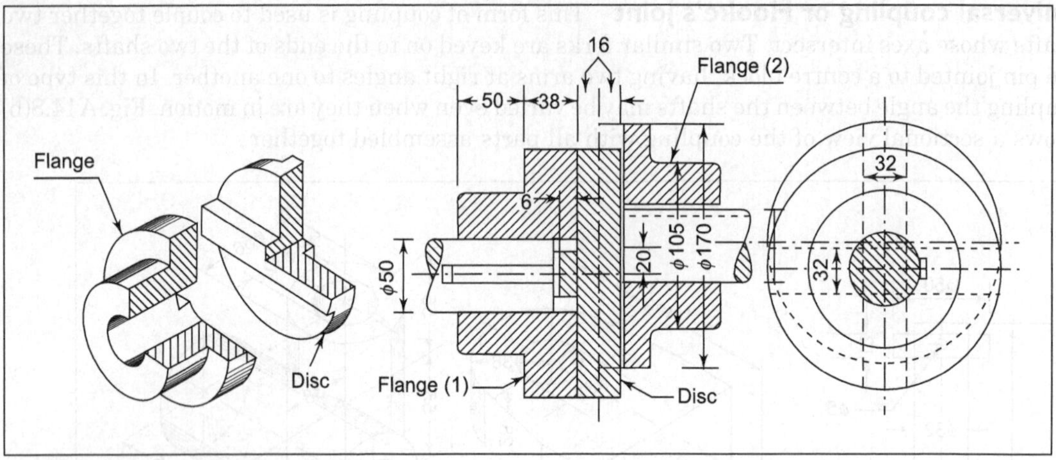

Fig. A14.9 Oldham's coupling

Gear coupling This type of coupling is extensively used to connect high speed rotary machines. It permits angular misalignment. It is widely used in petrochemical industries for coupling high speed compressors with eletric motors through the gear box (Fig. A14.10).

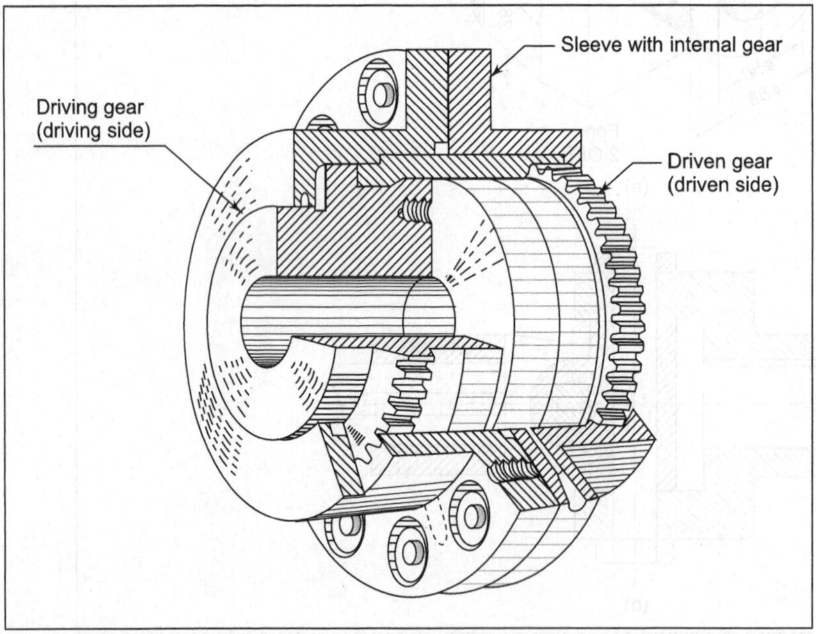

Fig. A14.10 Gear coupling

Review Questions

1. What do you mean by shaft couplings?
2. Name the different types of couplings.
3. What is the difference between rigid and flexible couplings?
4. What is the advantage of providing flanges in a protected flange coupling?
5. What are universal couplings?
6. Differentiate between box-muff coupling and half-lap muff coupling.
7. Differentiate between flanged and flexible coupling.
8. Differentiate between rigid and flexible coupling.
9. Differentiate between pin type and bush-pin type flanged coupling.
10. Identify five examples of the use of couplings.

CHAPTER 15

Spur Gears

This chapter contains information about spur gears

After reading the chapter you will be able to

- understand the use of gears in power transmission
- classify gears
- construct an involute gear tooth profile

15.1 GEARS

Gears are components used in the assembly of machines for transmission of power and motion. The gear drive provides motion and power transmission just like the belt drive, chain drive, and rope drive. Gears being solid metallic components, they provide fixed velocity ratio in assemblies. Velocity ratio is the ratio of speed of the driving pulley/shaft to the driven pulley/shaft. Hence, with a gear drive it is possible to achieve power and motion transmission with very low losses.

The two commonly used systems of generating tooth curves for gear tooth are the involute system and the cycloidal system. The curve most commonly used for gear tooth profiles is the involute of a curve.

An involute is a curve generated by a point on a straight edge as the straight edge is rolled on a cylinder.

A cycloid is the curve generated by a point on the circumference of a circle as the circle rolls on a straight line.

15.2 SPUR GEARS

Imagine two plain wheels *A* and *B* (Fig. A15.1) fixed rigidly to two parallel shafts and pressed tightly in contact with each other for rolling. If wheel *A* is rotated about its axis, wheel *B* will also rotate due to the friction between them. The rotary motion is thus transmitted from one shaft to another. The surfaces of two wheels will move at the same speed if there is no slipping. It is obvious that with an increase in the load to be transmitted, the wheels will begin to slip over each other. To prevent slipping, grooves may be cut on the cylindrical surfaces of the wheels and the projections can be placed between them. These grooves and projections form the teeth and the wheels accompanied with teeth becomes toothed wheels or gears (Fig. A15.2).

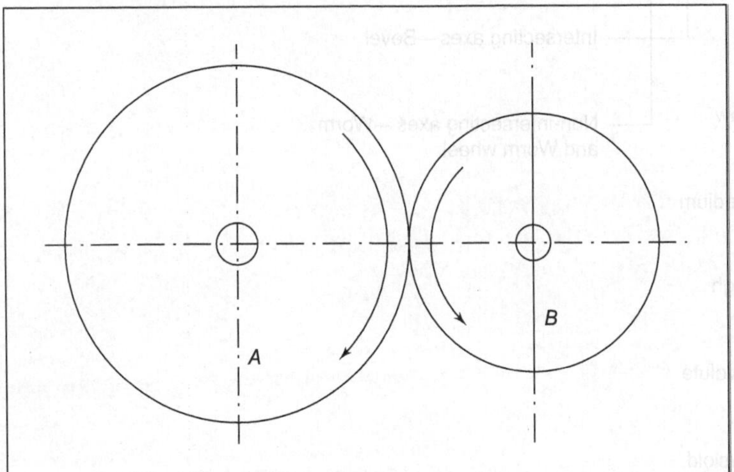

Fig. A15.1 Plain wheels

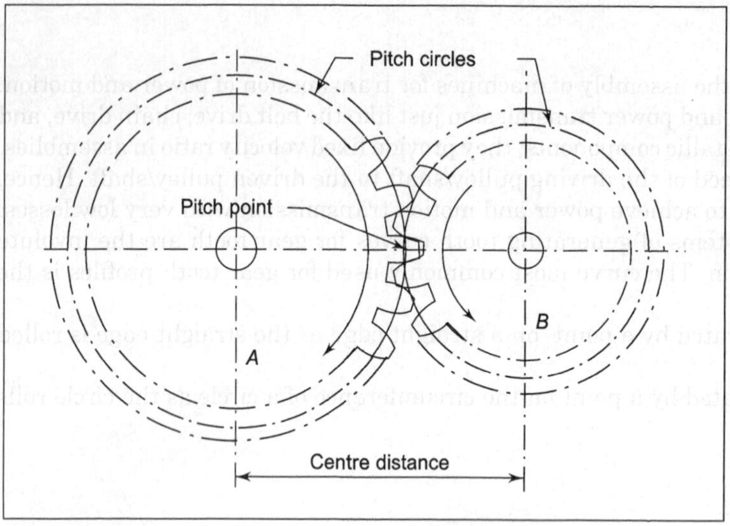

Fig. A15.2 Toothed wheels

Toothed gears can be classified according to:

1. the mutual position of the shafts
2. the relative motion of the shafts
3. the forms of teeth

The detailed classification of the toothed gears is depicted below.

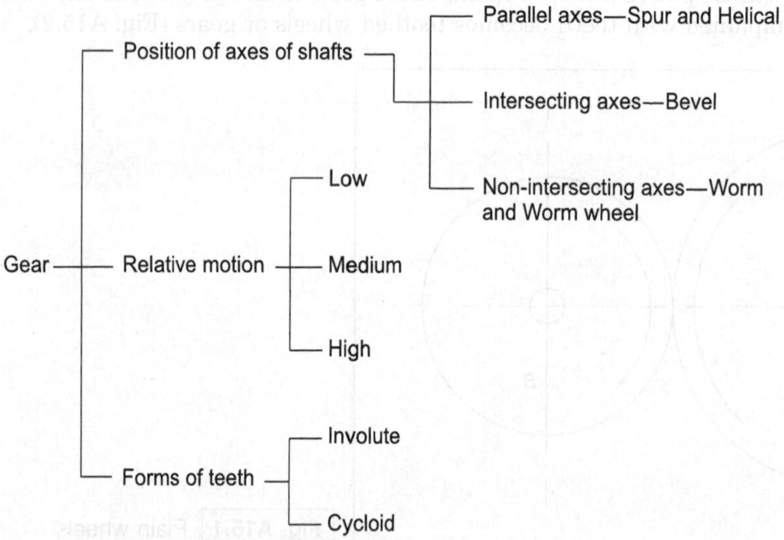

Gears that transmit motion between two parallel shafts are called spur gears. Gears that transmit power between shafts in the same plane, but whose axes would intersect if produced, are called bevel gears. The angle between shafts is usually a right angle.

Terminology

Pitch circle: In every pair of gears, in mesh, the two circles representing the two plain wheels in contact are always assumed to exist. Each of these circles is called a pitch circle. Its diameter is the 'pitch circle diameter'.

Pitch surface: The cylindrical surface represented by the pitch circle is the pitch surface.

Pitch point: It is the point of contact between the pitch circles of two gears in mesh. It lies on the line joining their centres.

Centre distance: It is the point of contact between the centres of a pair of mating gears and is equal to the sum of the radii of the pitch circles of the two gears (Fig. A15.2).

Tooth face: It is the side surface of the tooth above the pitch circle, perpendicular to the plane of the gear.

Tooth flank: It is the side surface of the tooth below the pitch circle, perpendicular to the plane of the gear.

Crest of tooth: It is the outside surface of the tooth perpendicular to the plane of the gear.

Root of tooth: It is the junction of the tooth with the material at the bottom of the tooth space.

Tooth thickness: It is the thickness of the tooth measured along the pitch circle (Fig. A15.3)

Addendum: It is the radial height of a tooth above the pitch circle.

Dedendum: It is the radial depth of a tooth below the pitch circle.

Clearance: It is the difference between the addendum and dedendum.

Whole depth: It is the sum of the addendum and dedendum of a tooth.

Working depth: It is the distance by which a tooth extends into the space of the mating gear. It is equal to the whole depth minus the clearance or twice the addendum.

Addendum circle: It is a circle that contains the crest of the teeth. Its diameter is called the outside or blank diameter.

Dedendum circle: It is a circle that contains the bottom of the tooth space. It is also called a root circle. Its diameter is the root diameter.

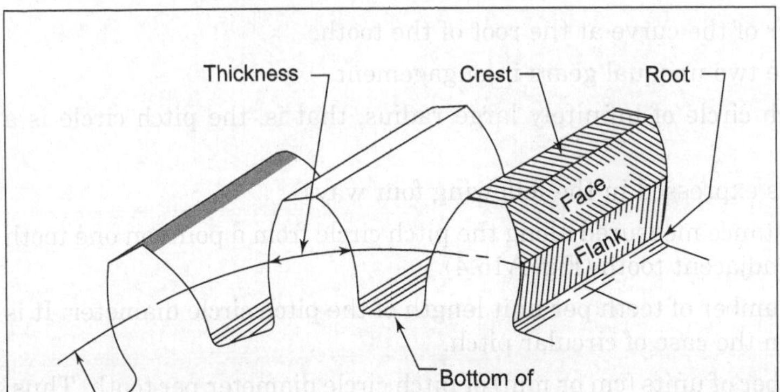

Fig. A15.3 Involute gear tooth nomenclature

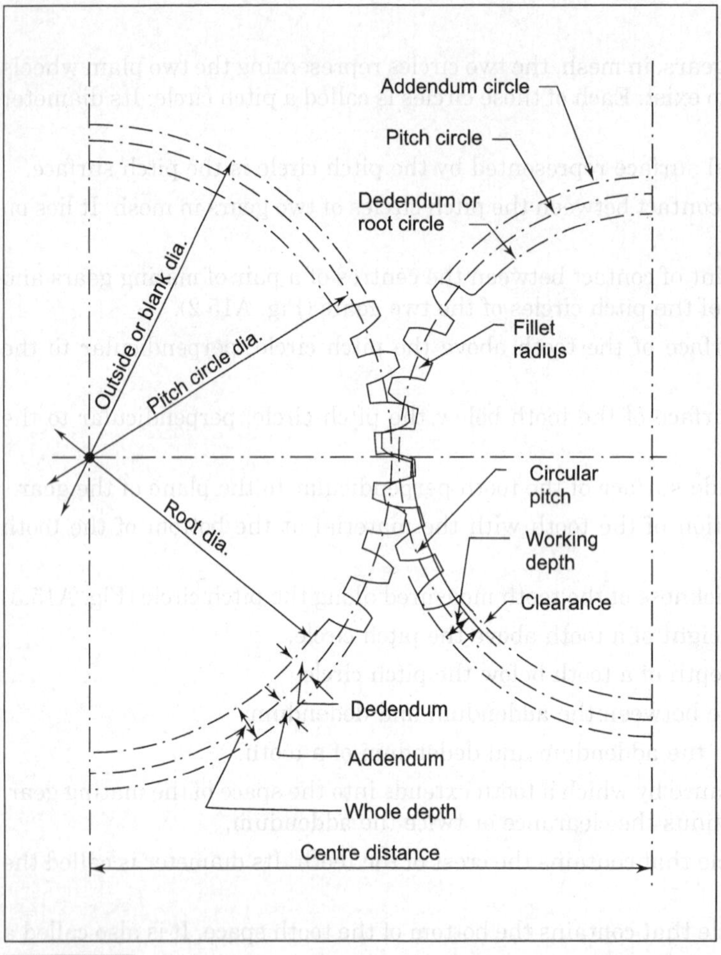

Fig. A15.4 Involute gear tooth nomenclature

Fillet radius: It is the smaller of the curve at the root of the tooth.

Pinion: It is the smaller of the two unequal gears in engagement.

Rack: It is a gear with a pitch circle of infinitely large radius, that is, the pitch circle is a straight line.

Pitch: The pitch of the teeth is expressed in the following four ways.

Circular pitch (CP) It is the distance measured along the pitch circle from a point on one tooth to a corresponding point on the adjacent tooth (Fig. A15.4).

Diameter pitch (DP) It is the number of teeth per unit length of the pitch circle diameter. It is a ratio and not a dimension as in the case of circular pitch.

Module pitch (m) It is the number of units (cm or mm) of pitch circle diameter per tooth. Thus, it is the reciprocal of diameter pitch and is obtained by dividing the pitch circle diameter by the number of teeth. Its value is generally expressed in the metric system.

Chordal pitch It is the shortest distance or the length of the chord of the pitch circle between a point on one tooth and a corresponding point on the adjacent tooth, both on the pitch circle (Fig. A15.4).

Relationship between the Pitches

$$\text{Circular pitch} = \frac{\text{Pitch circle diameter} \times \pi}{\text{Number of teeth}}$$

$$CP = \frac{PCD \times \pi}{N} = \pi m$$

$$\text{Diameter pitch} = \frac{\text{Number of teeth}}{\text{Pitch circle diameter}}$$

$$DP = \frac{N}{PCD} = \frac{1}{m}$$

that is,

$$CP \times DP = \frac{PCD \times \pi}{N} \times \frac{N}{PCD}$$

$$= \pi = \text{constant}$$

$$CP = \frac{\pi}{DP} \text{ and } DP = \frac{\pi}{CP}$$

$$\text{Module pitch } m = \frac{\text{Pitch circle diameter}}{\text{Number of teeth}}$$

$$m = \frac{PCD}{N} = \frac{1}{DP}$$

$$CP = \frac{\pi}{DP} = \pi \times \frac{1}{DP} = \pi \times m$$

15.3 TOOTH PROPORTIONS

The proportions of a standard gear tooth are given in terms of a module (m) in Table A15.1.

TABLE A15.1 *Proportions of a Standard Gear Tooth*

Sl. No.	Description	14.5° Full Depth Involute	20° Full Depth Involute
1	Addendum	m	m
2	Dedendum	1.157 m	1.157 m
3	Whole depth	2.157 m	2.157 m
4	Clearance	0.157 m	0.157 m
5	Tooth thickness	1.157 m	1.157 m
6	Circular pitch	π m	π m

To keep the velocity ratio constant, teeth profiles of mating gears should be shaped such that the common normal to the profile at the point of contact always passes through the pitch point. This is possible when teeth have profiles of either involute or cycloidal form.

The module of gears can be selected from Table A15.2 published by the BIS. Students are advised to consult IS 2535:1978 (revised 19191) for further details.

TABLE A15.2 Standard Module in mm

1,	1.125,	1.25,	1.35,	1.5,	1.75,	2,	2.25,	2.5,	2.75,	3,	3.25,	3.5,	3.75,	4,	4.5,
5,	5.5,	6,	6.5,	7,	8,	9,	10,	11,	12,	14,	16,	18,	20		

15.4 INVOLUTE SPUR GEARS

In modern times the involute form of teeth profile is exclusively used mainly because it is easier to manufacture.

Involute is a curve generated by a point on a straight line which rolls around the circumference of a circle. This curve forms the profile or outline of teeth of involute gears.

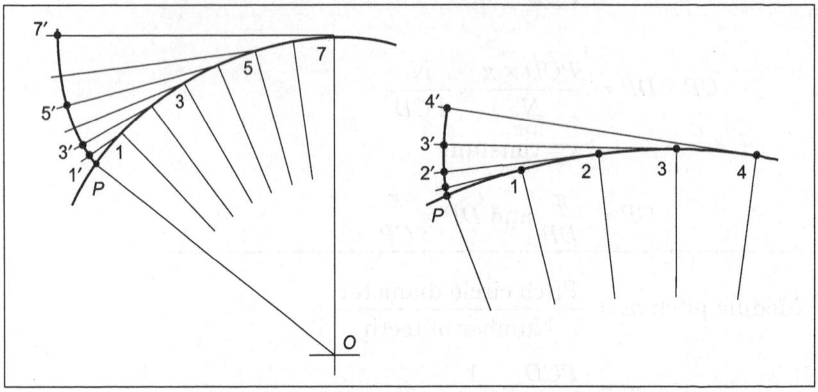

Fig. A15.5 Involutes

Fig. A15.5 shows an involute generated by a point P on each of the two circles of different diameters. The lines 1′1, 2′2, 3′3 and so on are tangential to the circle and are equal in length to the arc lengths $P1$, $P2$, $P3$, and so on. They are involutes. It can be seen that in the case of circles of large diameter, the involute is straighter than in the case of circles with smaller diameter. Hence, when the circle is of an infinitely large diameter (as in case of a rack), the involute is a straight line.

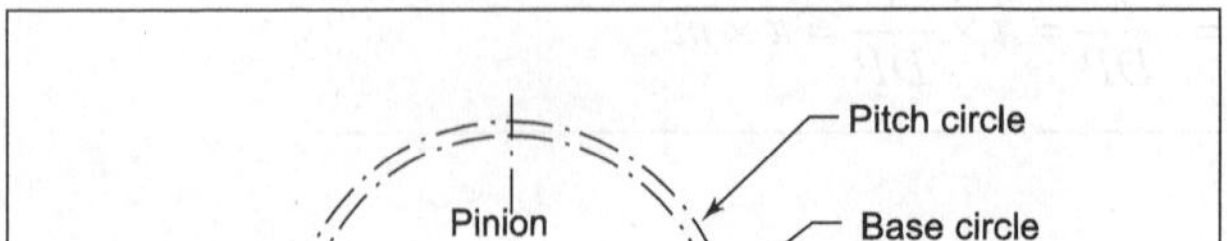

Fig. A15.6 Construction of base circles

Base circle: The circle from which the involute curve for the tooth profile is generated is called the base circle. It is concentric with the pitch circle (Fig. A15.6).

Line of action: In involute teeth, the path of contact is a straight line that passes through the pitch point.

This line is called the line of action. It is the common tangent to the two base circles of the involutes that form teeth profiles of the mating gears (Fig. A15.6).

Pressure angle or angle of obliquity: It is the angle between the line of action and the common tangent to the pitch circles at the pitch point. Its value varies between 14.5° and 22°. It is usually taken as 20° (Fig. A15.6).

15.5 CONSTRUCTION OF BASE CIRCLES

Draw the base circle for two mating gears of different pitch circle diameters, assuming the pressure angle to be θ (Fig. A15.6).

With centres O and O'(centre distance apart), draw two circles touching each other at a point P, which is the pitch point. It lies on the line joining O with O'. Through P, draw a line TT' as a common tangent to the two pitch circles. It will be perpendicular to OO'. Again, through P draw a line LL' making an angle θ with TT'. LL' is the line of action. Through O and O' draw lines OE and OE', both perpendicular to LL'. Each of these lines will make an angle θ with line OO'. With O and O'as centres, and OE and OE', respectively, as radii, draw two circles, which are the required base circles. Note that the line of action LL' is tangential to these circles.

Example A15.1

Draw the profile of an involute teeth for a gear having 24 teeth and module equal to 8.33 mm, assuming a pressure angle of 20°.

Calculation

Pitch circle diameter	= Number of teeth × module
	= 24 × 8.33 = 20 cm = 200 mm
Circular pitch	= $\pi m = \pi \times 8.33$ = 26.16 mm
Addendum	= m = 8.33 mm
Clearance	= 0.157 m = 0.157 × 8.33 = 1.308
Addendum circle diameter	= PCD + 2 × addendum
	= 200 + 2 × 8.33 = 216.66 mm
Clearance	= 0.157 m = 1.308 mm
Dedendum	= Addendum + Clearance
	= 8.33 + 1.309 = 9.639 mm
Dedendum circle diameter	= PCD – 2 × dedendum
	= 200 – 2 × 9.639
	= 180.722 mm
Tooth thickness	= 1.57 m = 13.08 mm

Construction (Fig. A15.7)

With centre O, draw a pitch circle of 200 mm diameter. At any point O on it, draw a tangent TT'. Through O, draw a line OE, inclined at π to OP (it will be perpendicular to LL'). With centre O and radius OE, draw the base circle. Draw the addendum and dedendum circles of diameters 216.66 mm and 180.72 mm.

Construct an involute curve from this base circle, as shown at X. Trace out this curve and part AB of the base circle on a piece of tracing paper, as shown at Y.

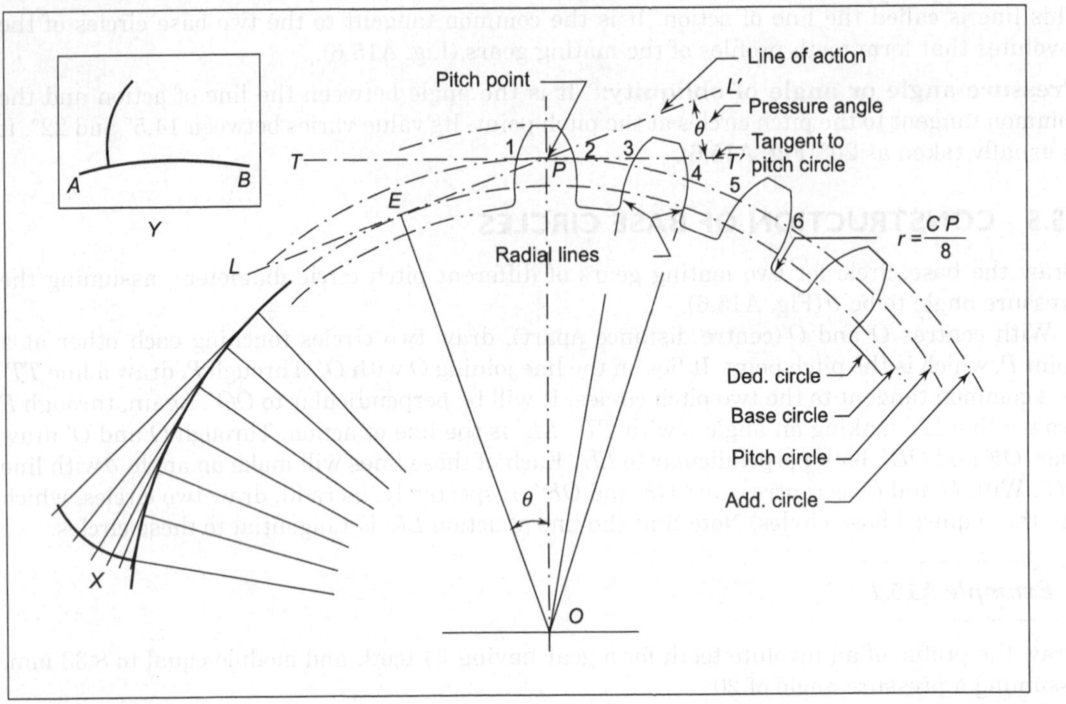

Fig. A15.7 Construction of involute teeth profile

On the pitch circle mark points 1, 2, 3, and so on $\frac{\text{distance}}{2} = \frac{CP}{2}$ = 26.18 = 13.09 mm apart. Place the tracing paper on the base circle in such a manner that the arc *AB* coincides with the base circle, while the curve passes through, say, point 1. Prick a few points on the curve between the addendum circle and the base circle. Join these points by means of a french curve. Complete one side of the tooth profile by drawing a radial line below the base circle and then join it with the bottom of the tooth space by a fillet of radius $r = \frac{CP}{8} = \frac{26.18}{8}$ = 3.272 mm.

Reverse the tracing paper and plot the curve through point 2 in the same manner, thus completing a tooth profile. Repeat construction for each tooth.

Example A15.2

Draw the profile of involute teeth for a gear having 25 teeth and a module pitch equal to 10 mm, assuming a pressure angle of 20°.

Calculation

$$PCD = m \times N = 10 \times 25 = 250 \text{ mm}$$

$$CP = \times m = 3.14 \times 10 = 31.4 \text{ mm}$$

$$\text{Addendum} = \frac{CP}{\pi} = m = 10 \text{ mm}$$

$$\text{Addendum circle diameter} = PCD + 2 \times \text{addendum}$$

$$= 250 + 2 \times 10 = 270 \text{ mm}$$

$$\text{Clearance} = \frac{CP}{20} = \frac{31.4}{20} = 1.57 \text{ mm}$$

$$\text{Dedendum} = \text{Addendum} + \text{Clearance}$$
$$= 10 + 1.57 = 11.57 \text{ mm}$$
$$\text{Dedendum circle diameter} = PCD - 2 \times \text{dedendum}$$
$$= 250 - 2 \times 11.57$$
$$= 250 - 23.14 = 226.86 \text{ mm}$$

$$\text{Tooth thickness} = \frac{CP}{2} = \frac{31.4}{2} = 15.7 \text{ mm}$$

Approximate construction

1. For gears of 30 teeth and above (Fig. A15.8): With centre O, draw the pitch circle, addendum circle, and dedendum circle. The diameters may be calculated from the given data and from the tooth proportions, as shown in Example 2. Mark a point P on the pitch circle. With OP as diameter, that is, with radius equal to $\frac{PCD}{4}$, draw a circle. With P as centre, radius equal to $\frac{PCD}{8}$, draw an arc cutting the semi-circle at a point Q. With O as centre, draw a circle passing through Q. This is then the circle on which the centre of the arcs for teeth profiles will lie. The radius for these arcs is $R = \frac{PCD}{8}$. To draw the arcs, mark points 1, 2, 3, and so on on the pitch circle at a distance of $\frac{CP}{2}$ apart from each other. With

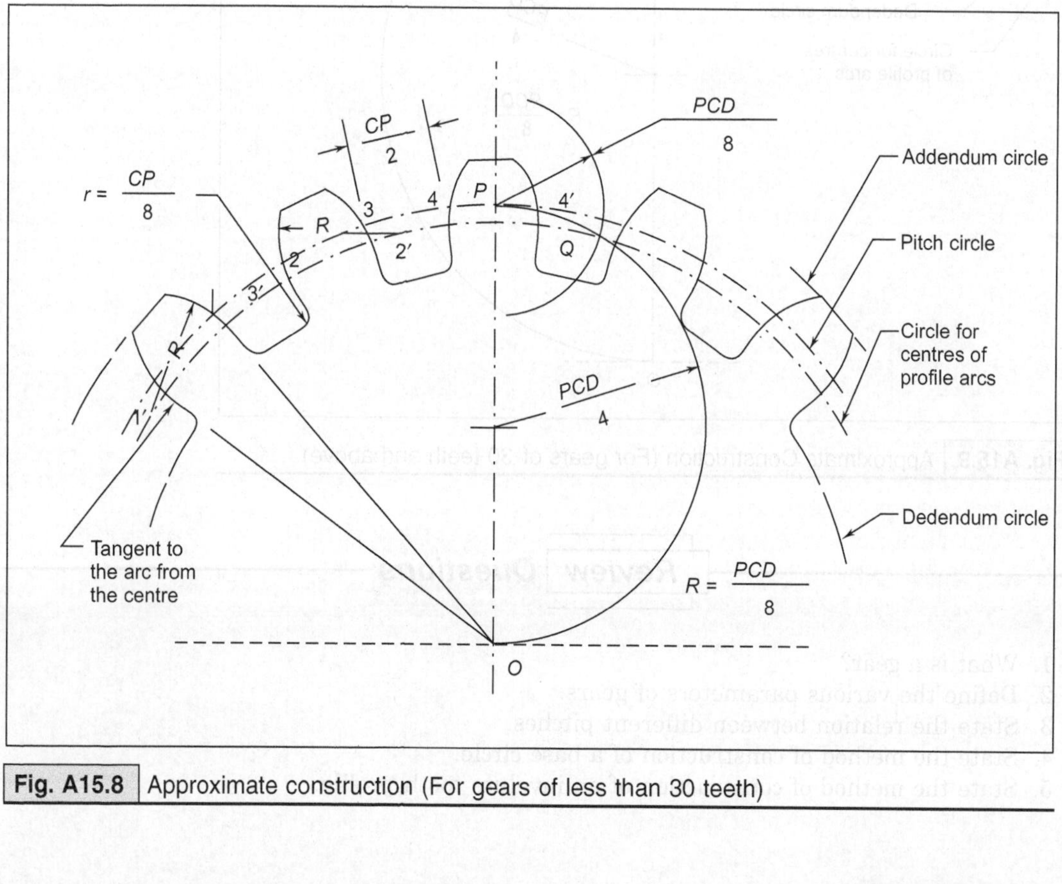

Fig. A15.8 Approximate construction (For gears of less than 30 teeth)

each of these points as centre and radius = $\frac{PCD}{8}$, locate the centres for these arcs 1′, 2′, 3′, and so on. (on the circle for centres). With 1′, 2′, 3′, and so on as centres and radius R, draw the profile arcs. Join each arc with the bottom of the tooth space by a fillet of radius $r = \frac{CP}{8}$.

2. For gears of less than 30 teeth (Fig. A15.9): Determine the centres 1′, 2′, and so on. Draw the arcs as described in 1 above. From O, draw lines tangential to these arcs. Join each of these lines with the bottom of the tooth space by a fillet of radius $r = \frac{CP}{8}$.

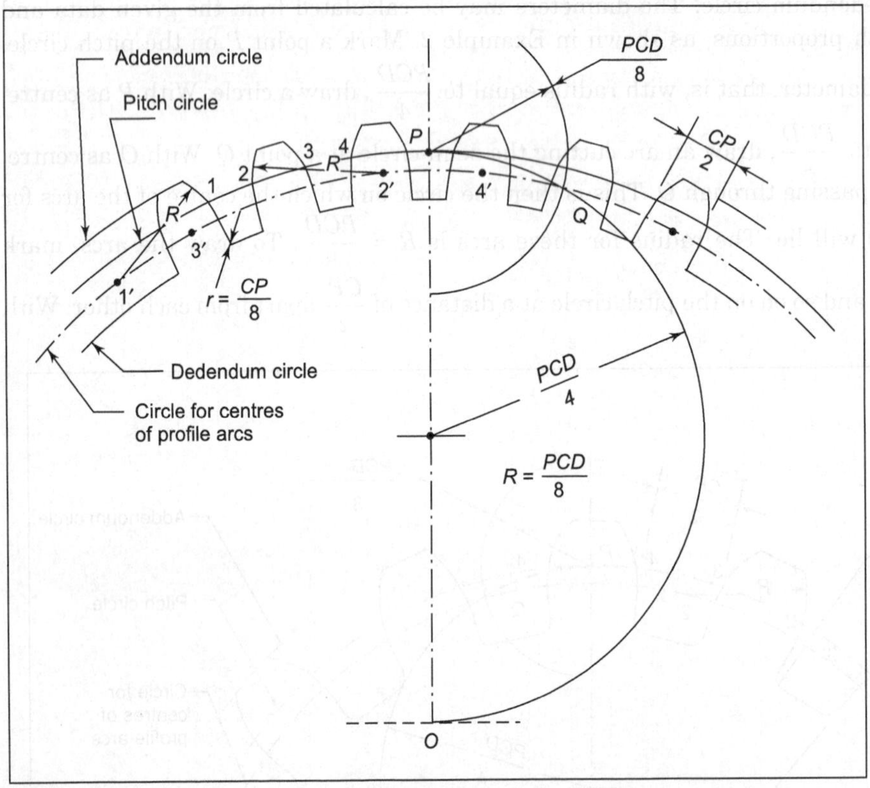

Fig. A15.9 Approximate Construction (For gears of 30 teeth and above)

Review Questions

1. What is a gear?
2. Define the various parameters of gears.
3. State the relation between different pitches.
4. State the method of construction of a base circle.
5. State the method of construction of an involute tooth profile.

CHAPTER 16

Introduction to Computer Aided Drafting

This chapter contains information about computer aided drafting

After reading this chapter you will be able to

- understand the advantage, CAD software has over conventional drafting
- understand the basic operations that can be done on AutoCAD

16.1 COMPUTER AIDED DRAFTING (CAD)

With the increasing use of computers most of the companies and offices are replacing their drafting boards with Computer Aided Drafting Systems, which have a tremendous advantage over conventional drafting. In the process of manufacturing a product, the first step is to make a design or sketch on paper. Designing is a complex process, and requires a large amount of calculations. Hence, the process is lengthy. Computers simplify the process to a great extent. Different drafting softwares are available in the market, which enable semi-automatic dimensioning, tolerancing, hatching, sectional views, rotated views, scaling, enlargement of details, and so on. They help in increasing productivity and accuracy in the drafting process. Drawing is the last step in the design process. It provides a clear illustration of the product to the manufacturer. The designer uses imagination and creativity to develop new products and communicates with the manufacturer with the help of drawings. Hence, drawings are important in the manufacturing environment. Sound knowledge of drawing in its various forms is very important for people in the enginering profession.

Free hand sketching with the pencil is the first step in drafting. The final engineering drawing would be made with a pencil and later traced with ink on a tracing paper for the purpose of storage or reproduction. Various instruments are used in drafting the different geometrical shapes, in order to make precise and neat drawings. Complete drawings are checked for accuracy and are corrected or even redrawn if there are errors, as it is important to maintain standards. Hence, the conventional drawing system is enormously repetitive. This can be reduced or even eliminated by using CAD.

16.2 ADVANTAGES OF COMPUTER AIDED DESIGN AND DRAFTING

The advantages of CAD are as follows:

1. High productivity with reducing lead time
2. Accurate designing
3. Standardisation of design
4. Modifications in designs are easy and quick
5. Automatic database creation assists in the creation of manufacturing documentation
6. Better control over the complete project
7. Better communication and presentation of design
8. Optimised solutions as a result of detailed engineering analysis

Design drafting software like AutoCad can be customised by using the programming language AutoLISP, which makes it possible to achieve the above advantages.

16.3 CAD SOFTWARE

Generally graphics designing software must provide the user a variety of functions for creating and manipulating the picture. The fundamental requirements are:

- The software should be user friendly and easy to understand
- The software should operate in a consistent and predictable manner
- The software should be capable of implementing variety of design problems
- The software should entail low maintenance cost
- The sofware should be capable of wide usability

AutoCAD, VersaCAD, CADKey, DesignCAD are some examples of drafting systems that are avilable in the market. Of these, AutoCAD is the most popular and low budget CAD software.

16.4 WORKING WITH AUTOCAD

Listed below are the recommended hardware requirements for running AutoCAD on your system.

- Pentium 133 or higher (or compatible processor)
- 64 MB RAM (32 MB minimum)
- Windows® 98, Windows 95, or Windows NT® 4.0 operating system
- 1024×768 VGA video display (800 × 600 VGA video display minimum)
- Windows video display driver
- 130 MB free harddisk space and 64 MB swap space
- Pointing device (mouse or digitiser with Wintab driver)
- 4X CD-ROM drive
- IBM-compatible parallel port
- Serial port (for digitisers and some plotters)
- Printer or plotter
- Modem (connection to Internet, not required)
- Sound card for multimedia learning
- TCP/IP or IPX support (required only for multi-user or floating license configurations)

AutoCAD commands operate by using command tools, toolbars, and a series of prompts. The prompts appear in the command line box and ask for a selection or numerical input so that a command sequence can be completed. The line command will be used to demonstrate the various inputs and prompt sequences that are typical of AutoCAD commands.

Most of the commands are contained in the Draw and Modify toolbars (Fig. A16.1). When a command is activated, an Active Assitance Dialog Box appears on the screen. The box may be closed by clicking the close button.

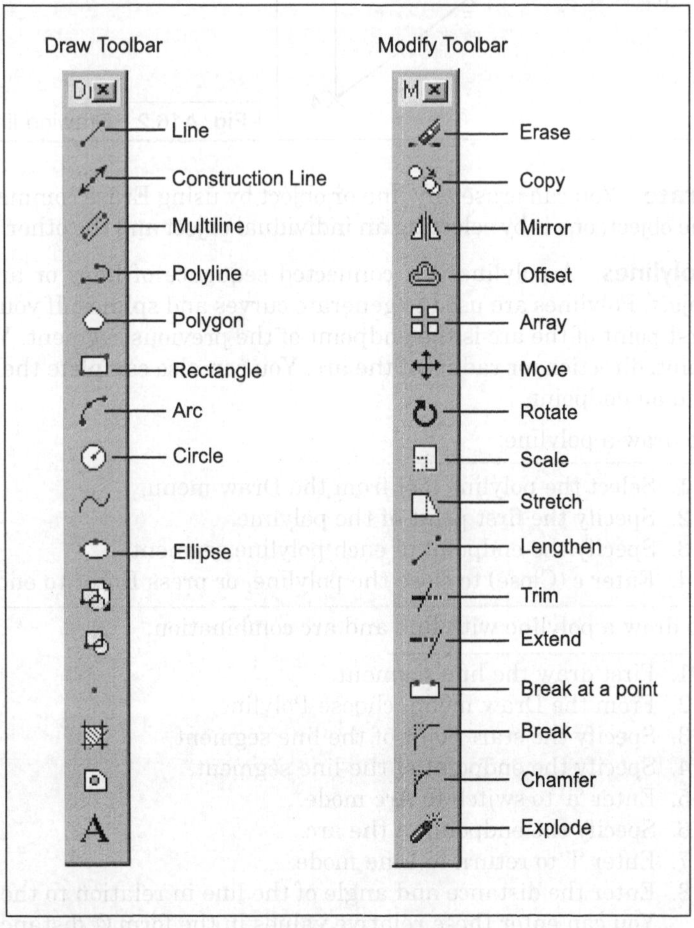

Fig. A16.1 Draw and modify tool bar in Auto CAD

Line The line command allows you to draw straight lines (Fig. A16.2). You can specify the desired endpoints.

To draw a line:

1. From the Draw menu choose Line.
2. Specify the start point (1).
3. Specify the endpoint (2).
4. Specify the endpoints of the successive segments (3, 4, 5, 6).
5. Press Enter to complete the line.

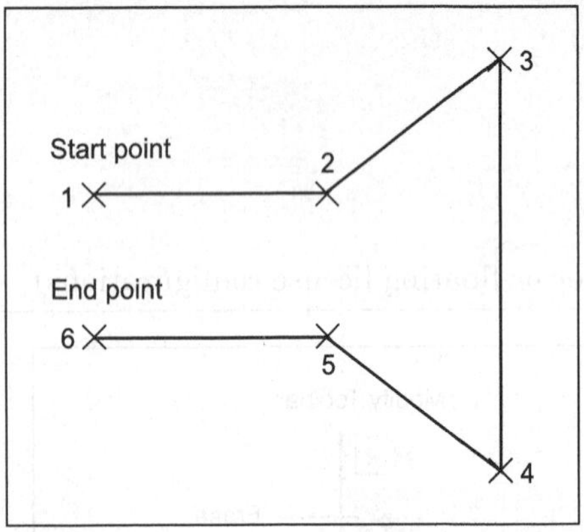

Fig. A16.2 Drawing lines in AutoCAD

Erase You can erase any line or object by using Erase command. There are two ways to erase the object, one is by selecting an individual object and the other is by window selection of objects.

Polylines A polyline is a connected sequence of lines or arc segments created as a single object. Polylines are used to generate curves and splines. If you want to draw arc segments, the first point of the arc is the endpoint of the previous segment. You can specify the angle, centre point, direction, or radius of the arc. You can also complete the arc by specifying a second point and an endpoint.

To draw a polyline:

1. Select the polyline tool from the Draw menu.
2. Specify the first point of the polyline.
3. Specify the endpoint of each polyline segment.
4. Enter c (Close) to close the polyline, or press Enter to end the command.

To draw a polyline with line and arc combination:

1. First draw the line segment.
2. From the Draw menu, choose Polyline.
3. Specify the start point of the line segment.
4. Specify the endpoint of the line segment.
5. Enter 'a' to switch to Arc mode.
6. Specify the endpoint of the arc.
7. Enter 'l' to return to Line mode.
8. Enter the distance and angle of the line in relation to the endpoint of the arc. You can enter these relative values in the form @ distance < angle.
9. Press Enter to end the polyline.

Circle The circle tool is on the Draw toolbar. Circles can be defined in several ways. The default method is to specify the centre point and radius. One can also specify the centre point and diameter; or diameter alone with two or three points on the diameter. One can also create a circle by tangents to two existing objects or arcs and a radius value.

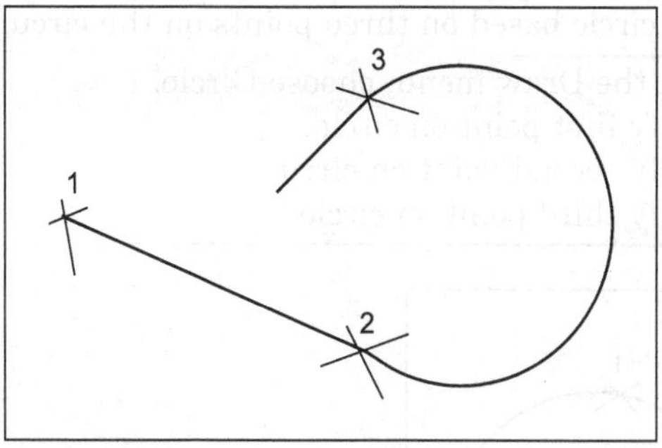

Fig. A16.3 Drawing a circle using AutoCAD

To draw a circle by specifying a centre point and radius (Fig. A16.4):

1. From the Draw menu, choose Circle: Centre, Radius.
2. Specify the centre point.
3. Specify the radius.

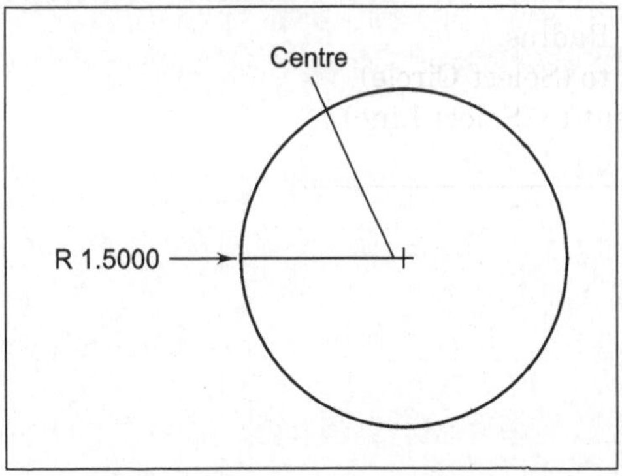

Fig. A16.4 Drawing a circle using centre point and radius

To draw a circle by specifying a centre point and diameter (Fig. A16.5):

1. From the Draw menu, choose Circle: Center, Diameter
2. Specify the centre point
3. Specify the diameter

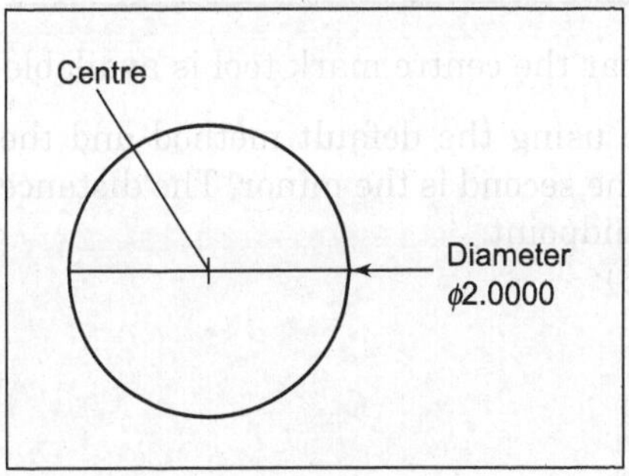

Fig. A16.5 Drawing a circle in the centre point and diameter

To draw a circle based on three points on the circumference (Fig. A16.6):

1. From the Draw menu, choose Circle.
2. Specify first point on circle.
3. Specify second point on circle.
4. Specify third point on circle.

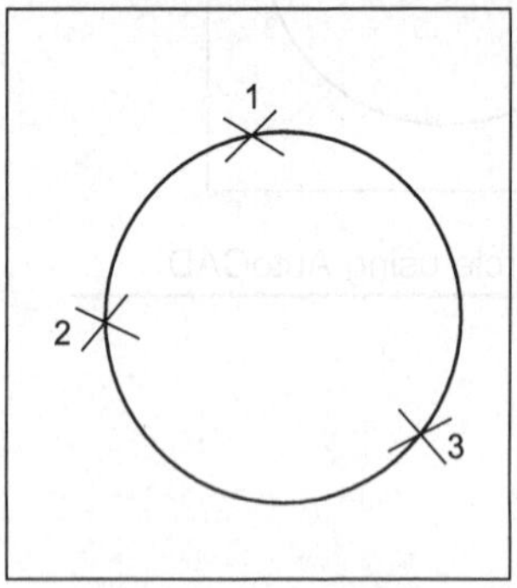

Fig. A16.6 Drawing a circle in 3 points on the circumference

To create a circle tangent to existing objects (Fig. A16.7):

1. From the Draw menu, choose Circle: Tan, Tan, Radius.
2. Select the first object to draw the circle tangent to (Select Circle).
3. Select the second object to draw the circle tangent to (Select Line).
4. Specify the radius of the circle.

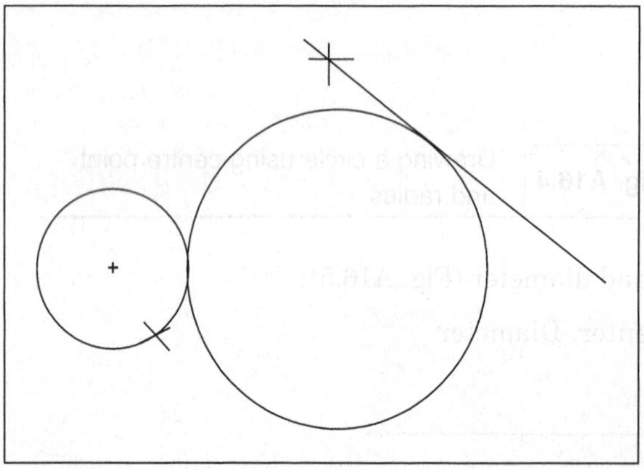

Fig. A16.7 Drawing a circle tangent to an existing object

To mark the centre of the circle, on the Dimension toolbar the centre mark tool is available

Ellipse In the following procedure, you draw an ellipse using the default method and the pointing device. Here, the first axis is the major axis, and the second is the minor. The distance increases as you drag the pointing device away from the midpoint.
To draw a ellipse using endpoints and distance (Fig. A16.8):

1. From the Draw menu, choose Ellipse: Axis, End.
2. Specify the first endpoint of the first axis.
3. Specify the second endpoint of the first axis.
4. Drag the pointing device away from the midpoint of the first axis and click to specify the distance.

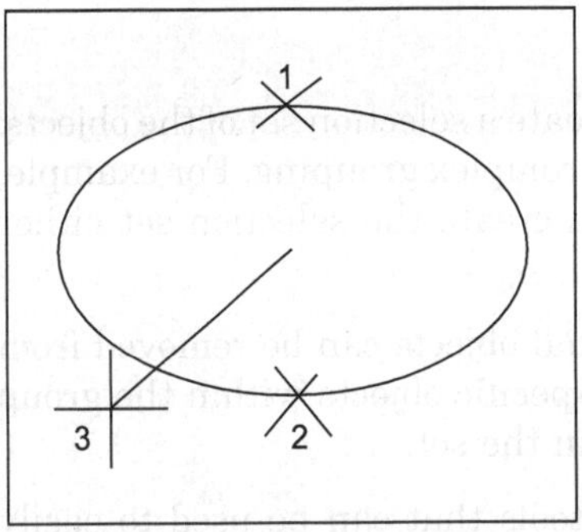

Fig. A16.8 Drawing an ellipse on AutoCAD

Rectangle To draw rectangle choose rectangle (Fig. A16.9):

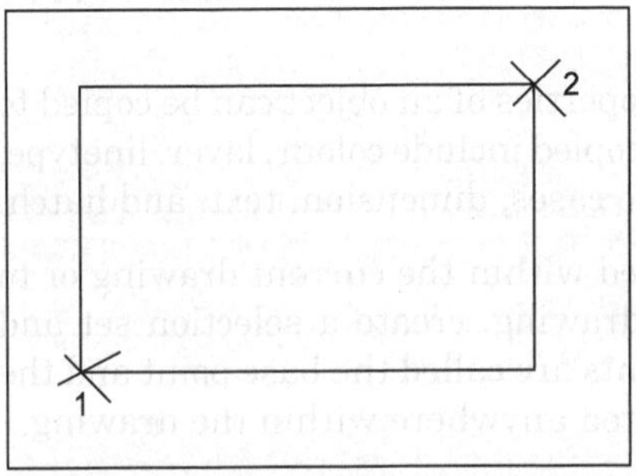

Fig. A16.9 Drawing a rectangle on AutoCAD

1. Specify first corner point or [Chamfer/Elevation/Fillet/Thickness/Width]: Enter an option or specify a point (1)
2. Specify other corner point: Specify a point (2)

Polygon A polygon is a closed figure bounded by straight lines. To draw a polygon (Fig. A16.10):

1. Command line: Polygon
2. Enter number of sides: Enter a value between 3 and 1024 or press Enter.
3. Specify centre of polygon or (Edge): Specify a point (1) or Enter e.

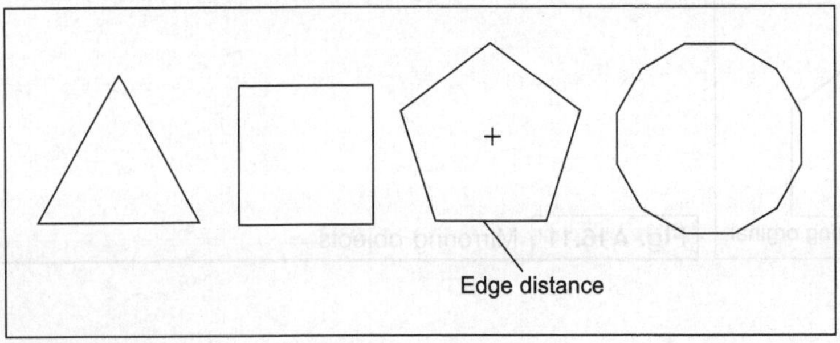

Fig. A16.10 Drawing polygons on AutoCAD

16.4.1 Editing Methods

Selecting objects Before you can edit objects you need to create a selection set of the objects. A selection set can consist of a single object, or it can be a more complex grouping. For example, a set of objects of a certain colour on a certain layer. You can create the selection set either before or after you choose an editing command.

Removing objects After a selection set is created, individual objects can be removed from it. An entire group of grouped objects, can be select and then specific objects within the group are removed leaving only the objects that need to be retained in the set.

Editing object properties AutoCAD provides two main tools that can be used to easily edit object properties such as layers, colours, linetypes, and lineweights.

Object properties toolbar This toolbar provides options for viewing or changing the object properties that are common to all objects, including layers and layer properties, colours, linetypes, lineweights, and plot style.

Matching properties of other objects Some or all properties of an object can be copied to one or more objects using Matchprop. Properties that can be copied include colour, layer, linetype, linetype scale, lineweight, thickness, plot style, and in some cases, dimension, text, and hatch.

Copying objects Single or multiple objects can be copied within the current drawing or to other drawings or applications. To copy objects within a drawing, create a selection set and specify a start point and an endpoint for the copy. These points are called the base point and the second point of displacement, respectively, and can be located anywhere within the drawing.

Mirroring objects Objects can be mirrored around a mirror line, which is defined by two points, as shown in Fig. 16.11. The original objects can be deleted or retained. Mirroring works in any plane parallel to the *XY* plane of the current UCS (User Co-ordinate System). Although a viewport object can be mirrored in paper space, doing so has no effect on its model space view or model space objects.

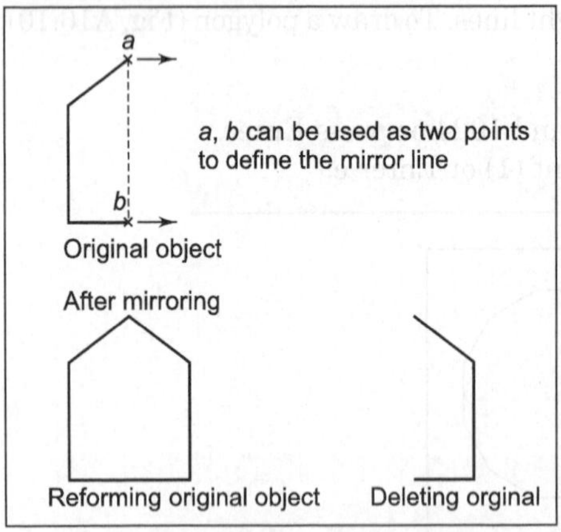

Fig. A16.11 Mirroring objects

16.4.2 Hatching

AutoCAD supplies a solid fill and more than 50 industry-standard hatch patterns that can be used to differentiate the components of objects or represent object materials. AutoCAD supplies 14 hatch patterns that conform to the ISO (International Standards Organisation) standards.

To use a predefined hatch pattern:

Command line: Bhatch

1. From the Draw Menu, choose Hatch.
2. In the Boundary Hatch and Fill dialog box, choose Pick Points or Select Objects.
3. Specify an internal point or select an object.
4. In the Boundary Hatch and Fill dialog box, Hatch tab, select Predefined in the Type box.
5. Specify the angle and spacing of the hatch pattern.
6. From the Pattern box, select a pattern.
7. Choose OK.

Create User-Defined Hatch Patterns:

Command line: Bhatch

In addition to using predefined hatch patterns, a simple hatch pattern can be defined based on the current line type. The pattern can be defined by changing the angle and spacing of the hatch lines.

To create a user-defined hatch pattern:

1. Specify the line type for the user-defined hatch pattern by making a particular line type current.
2. From the Draw Menu, choose Hatch.
3. In the Boundary Hatch and Fill dialog box, choose Pick Points or Select objects.
4. Specify an internal point or select an object.
5. In the Boundary Hatch and Fill dialog box, Hatch tab, select User-Defined in the Type box.
6. Specify the angle and spacing of the hatch pattern.
7. To use intersecting lines in the pattern, select Double.
8. Choose OK.

16.4.3 3*D* Modelling

AutoCAD supports three types of 3*D* modeling: Wire frame, Surface, and Solid. Each type has its own creation and editing techniques (Fig. A6.12).

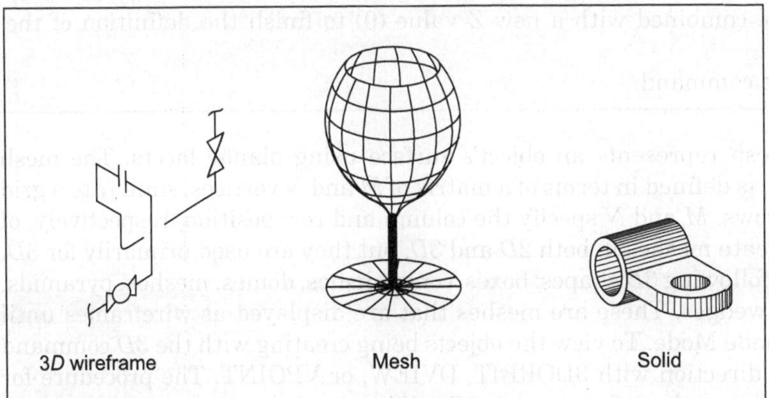

Fig. A16.12 3*D* molding on AutoCAD

A wireframe model is a skeletal ascription of a $3D$ object. There are no surfaces in a wire frame model; it consists only of points, lines, and curves that describe the edges of the object. With AutoCAD, wire frame models can be created by positioning $2D$ (planar) objects anywhere in $3D$ space. AutoCAD also provides some $3D$ wireframe objects, such as $3D$ polylines (that can only have a Continuous line type) and splines. Because each object that makes up a wireframe model must be independently drawn and positioned, this type of modeling can be very time consuming.

Surface modeling is more sophisticated than wireframe modeling in that it defines not only the edges of a $3D$ object but also its surfaces. The AutoCAD surface modeler defines faceted surfaces using a polygonal mesh. Because the faces of the mesh are planar, the mesh can only approximate curved surfaces. With Mechanical Desktop®, we can create true curved surfaces. To differentiate these two types of surfaces, AutoCAD calls faceted surfaces 'meshes'.

Solid modeling is the easiest type of $3D$ modeling. With the AutoCAD solid modeler, $3D$ objects can be made by creating basic $3D$ shapes like boxes, cones, cylinders, spheres, edges, and tori (donuts). These shapes can then be combined to create more complex solids by joining or subtracting them, or finding their intersecting (overlapping) volume. Solids can also be created by sweeping a $2D$ object along a path or revolving it about an axis.

Methods for creating wire frame models With AutoCAD, wire frame models can be created by positioning any $2D$ planar object anywhere in $3D$ space, using several methods:

- Entering $3D$ coordinates. You enter coordinates that define the X, Y, and Z location of the object.
- Setting the default construction plane (XY plane) on which you will draw the object by defining a UCS.
- Moving or copying the object to its proper $3D$ location after you create it.

To drop a perpendicular line from a $3D$ point down to the XY plane:

1. From the Draw menu, choose Line.
2. Use an object snap to specify a point on an object that is not on the XY plane of the UCS. This defines the first point of the line.
3. Enter XY and then enter @ at the Of prompt. This operation extracts the X and Y coordinate values from the first point.
4. Enter 0 to specify the Z value. Using coordinate filters, the X and Y values extracted from the first point and are combined with a new Z value (0) to finish the definition of the second point.
5. Press Enter to end the command.

Creating surfaces A mesh represents an object's surface using planar facets. The mesh density, or number of facets, is defined in terms of a matrix of M and N vertices, similar to a grid consisting of columns and rows. M and N specify the column and row position, respectively, of any given vertex. We can create meshes in both $2D$ and $3D$, but they are used primarily for $3D$.

$3D$ command creates the following $3D$ shapes: boxes, cones, dishes, domes, meshes, pyramids, spheres, tori (donuts), and wedges. These are meshes that are displayed as wireframes until you use Hide, Render, or Shade Mode. To view the objects being creating with the $3D$ command more clearly, set a viewing direction with 3DORBIT, DVIEW, or VPOINT. The procedure for creating $3D$ shapes are similar to those for creating $3D$ solids.

Creating 3*D* solids

Creating a solid box: The Box command is used to create a solid box. The base of the box is always parallel to the *XY* plane of the current UCS.

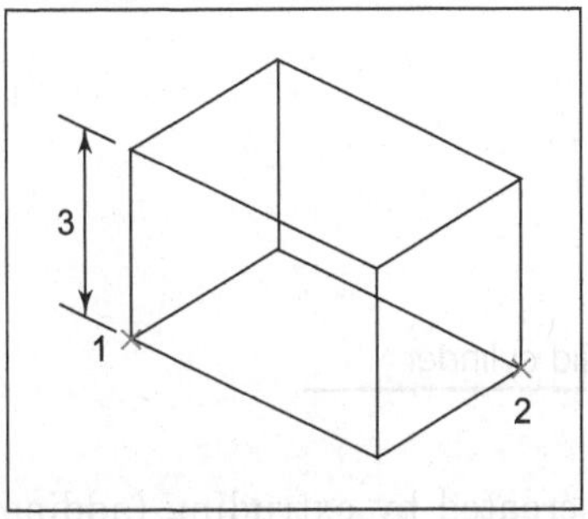

Fig. A16.13 Creating a solid box

The Rectang or Pline command creates a rectangle or closed polyline from which a box can be created using Extrude. The 3*D* command creates a box shape defined by surfaces only (Fig. A16.13).

Creating a solid cone: Command Cone is used to create a solid cone defined by a circular or an elliptical base tapering to a point perpendicular to its base. By default, the cone's base lies on the *XY* plane of the current UCS. The height, which can be positive or negative, is parallel to the *Z*-axis. The apex determines the height and orientation of the cone.

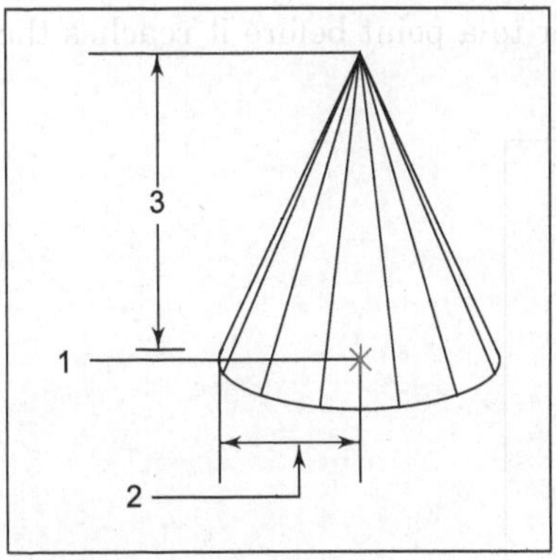

Fig. A16.14 Creating a solid cone

To create a truncated cone or a cone that requires a specific angle to define its sides, draw a 2*D* circle and then use Extrude to taper the circle at an angle along the *Z*-axis. To complete the truncation, a box has to be cut off from the tip of the cone with the Subtract command. Circle creates a circle from which a cone can be created using Extrude with its Taper option. The 3*D* command creates a conical shape defined by surfaces only (Fig. A16.14).

Creating a solid cylinder: Command Cylinder is used to create a solid cylinder with a circular or an elliptical base. The base of the cylinder lies on the *XY* plane of the current UCS. For the construction of a cylinder with special details, such as grooves along its sides, create a 2*D* profile of its base with a closed Pline and use Extrude to define its height along the *Z*-axis. Circle creates a circle from which a cylinder can be created using Extrude.

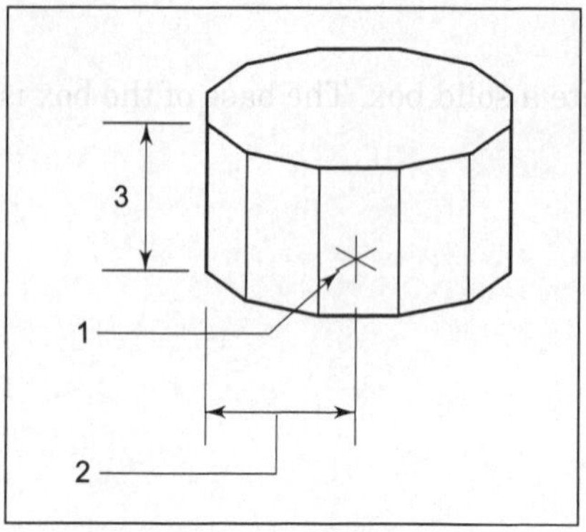

Fig. A16.15 Creating a solid cylinder

Creating an extruded solid: With Extrude, solids can be created by extruding (adding thickness to) selected objects. Closed objects such as polylines, polygons, rectangles, circles, ellipses, closed splines, donuts, and regions can be extruded. Objects can be extruded along a path, or by specifying a height value and a tapered angle.

Use Extrude to create a solid from a common profile of an object, such as a gear or sprocket. Extrude is particularly useful for objects that contain fillets, chamfers, and other details that might otherwise be difficult to reproduce, except in a profile. If a profile is created using lines or arcs, use the Join option of Pedit to convert them to a single polyline object or make them into a region before Extrude.

Tapering the extrusion is useful specifically for parts that need their sides defined along an angle, such as a mold used to create metal products in a foundry. Avoid using extremely large tapered angles. If the angle is too large, the profile can taper to a point before it reaches the specified height.

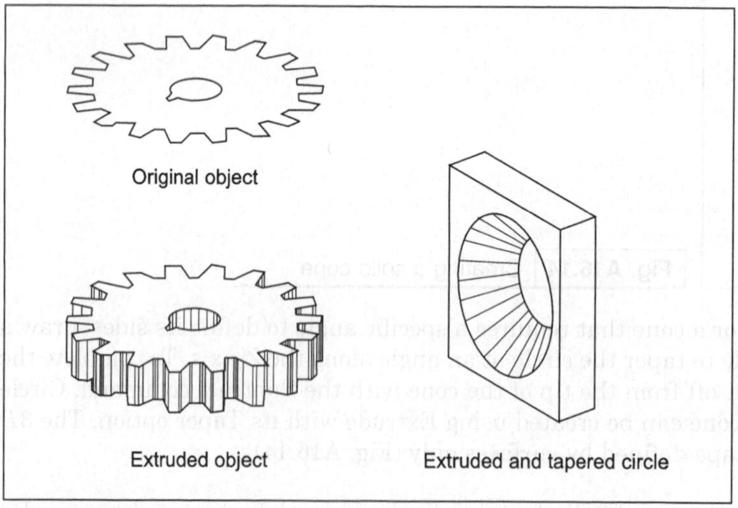

Fig. A16.16 Creating an extruded solid

Creating a composite solid: With Union, the total volume of two or more solids or two or more regions can be combined into a composite object (Fig. A16.17).

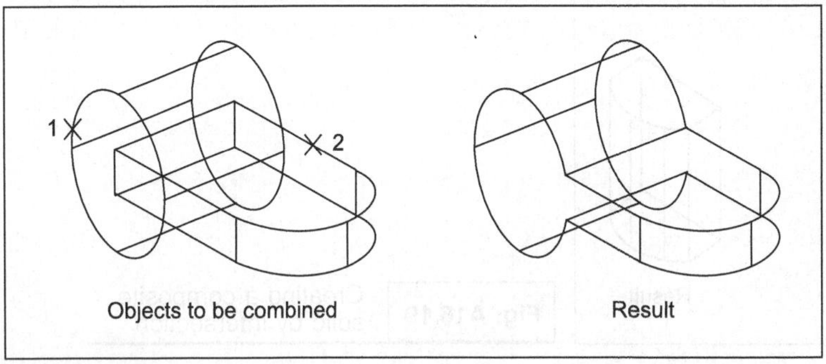

Fig. A16.17 Creating a composite solid by contiaining two objects

With Subtract, the common area of one set of solids can be removed from another. For example, Substract can be used to add holes to a mechanical part by subtracting cylinders from the object (Fig. A16.18).

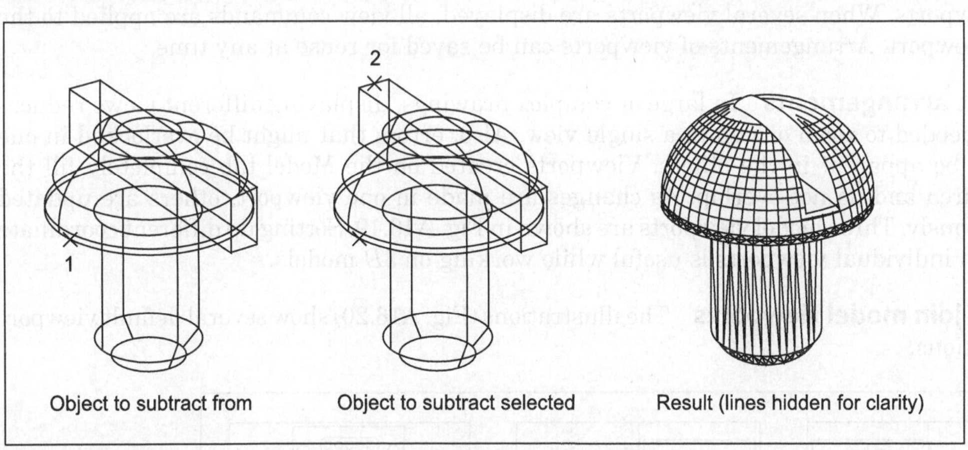

Fig. A16.18 Creating a composite solid by subtraction

With Intersect, a composite solid can be created from the common volume of two or more overlapping solids. Intersect removes the nonoverlapping portions and creates a composite solid from the common volume (Fig. A16.9).

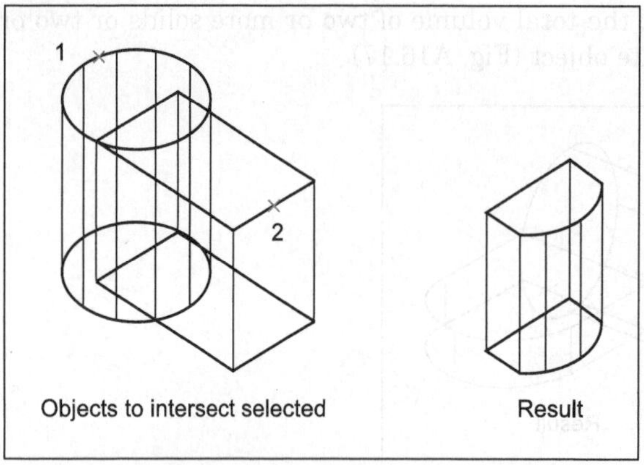

Fig. A16.19 Creating a composite solid by intersection

Interfere performs the same operation as Intersect, but Interfere keeps the original two objects.

16.4.4 Display Multiple Views (Viewports)

To see several views at the same time, the drawing area can be split into separate viewing areas called viewports. When several viewports are displayed, all view commands are applied to the current viewport. Arrangements of viewports can be saved for reuse at any time.

Viewport arrangements In large or complex drawings, displaying different views reduces the time needed to zoom or pan in a single view. Also, errors that might be overlooked in one view may be apparent in the others. Viewports created on the Model tab completely fill the drawing area and do not overlap. As changes are made in one viewport, others are updated simultaneously. Three model viewports are shown in Fig. A16.19. Setting up different coordinate systems in individual viewports is useful while working on *3D* models.

Split and join model viewports The illustrations (Fig. A16.20) show several default viewport configurations.

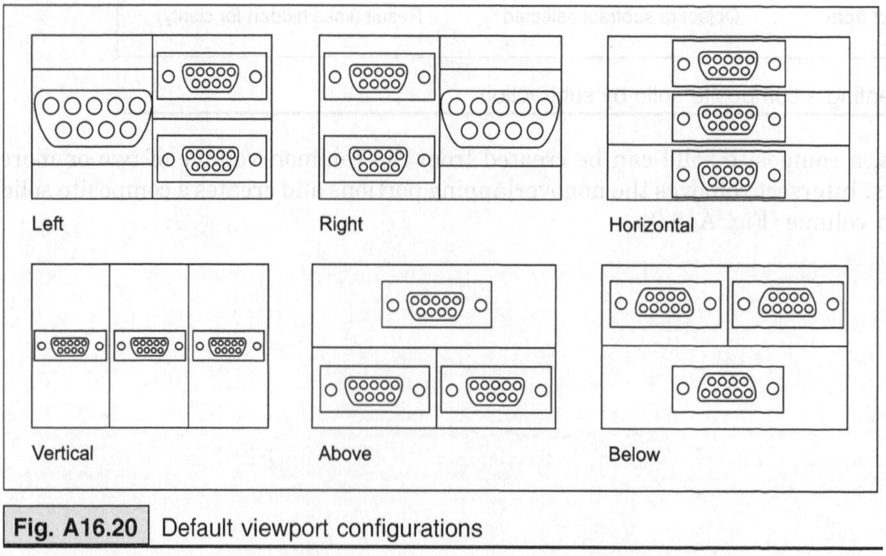

Fig. A16.20 Default viewport configurations

Command line: Vports

To subdivide a viewport on the Model Tab:

1. Click inside the Viewport to create more than one viewport.
2. From the View menu, choose Viewports and then either 2 Viewports, 3 Viewports, or 4 Viewports to indicate how many viewports should be created.
3. At the Next prompt, specify the arrangement of the new viewports.

Selection of the current viewport: In AutoCAD, only one viewport can be actuated at a time. If there are multiple viewports, one of them is the current viewport. When a viewport is current, the cursor is displayed as crosshairs rather than an arrow and the viewport boundary is highlighted. The current viewport can be changed at any time, except when a viewing command is in progress.

To make a viewport the current viewport, click inside it or press CTRL +R to cycle through the existing viewports.

Specifying 3*D* Views When working in 3*D*, it is often required to display several different views so that the 3*D* effect can easily be verified. One of the viewpoints is often an isometric view to reduce the number of visually overlapping objects. From a selected viewpoint, it is possible to create new objects, edit existing objects, or generate a hidden line or shaded view.

Viewing a parallel projection in 3*D* The parallel projection of a 3*D* model can be viewed from any point in model space.

To determine the point or angle in model space, the following methods can be followed.

- Choose a preset 3*D* view from a toolbar
- Enter a coordinate or angles that represent the viewing location in 3*D*
- Change to a view of the *XY* plane of the current UCS, a saved UCS, or the WCS

Choosing preset 3*D* views A quick way to set a view is to choose one of the predefined 3*D* views, that is, selection of predefined standard orthographic and isometric views by name or description. These views represent commonly used options: Top, Bottom, Front, Left, Right, and Back. In addition, the views can be set from isometric options: *SW* (southwest) Isometric, *SE* (southeast) Isometric, *NE* (northeast) Isometric, and *NW* (northwest) Isometric (Fig. A16.21).

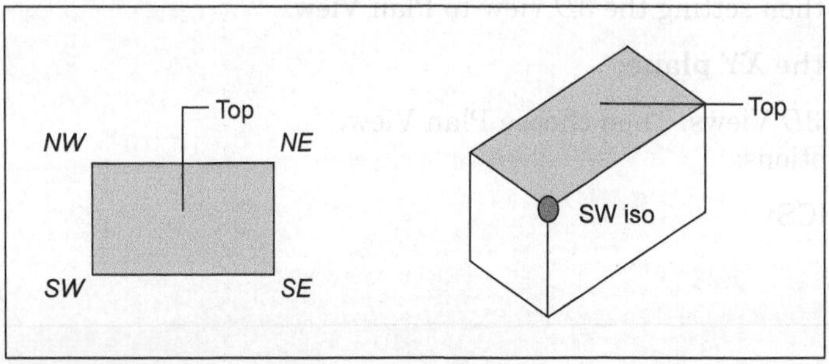

Fig. A16.21 Choosing preset 3*D* view

To use a preset 3*D* view: Choose 3*D* Views from the View menu. Then choose a preset view (Top, Bottom, Left, and so on).

Defining a 3*D* view with coordinate values or angles The viewing direction can also be defined by entering the coordinate values of a point or the measures of two angles of rotation. This point represents the position in 3*D* space as the view of the model while looking towards the origin (0,0,0). Viewpoint coordinate values are relative to the world coordinate system.

To set a view with a viewpoint coordinate:

1. From the View menu, choose 3*D* Views □ Viewpoint.
2. Click inside the compass to specify the viewpoint. The selected viewpoint is used to view the drawing in the direction of 0,0,0.

Command line: Vpoint

To set a view with two angles of rotation:

1. At the Command prompt, enter vpoint.
2. Enter '*r*' (rotate) to specify a new direction using two angles.
3. Enter an angle in the *XY* plane measured from the positive *X* axis.
4. Enter an angle from the *XY* plane, which represents the position while viewing the model in the direction of 0,0,0.

Command line: Vpoint

To set standard views with Vpoint:

1. On the command line, enter vpoint.
2. Enter a coordinate according to the viewpoint, as per the need:
   - Enter 0, 1, 0 for a top view.
   - Enter 0, 0, 1 for a front view.
   - Enter 1, 0, 0 for a right side view.
   - Enter 1, 1, 1 for an isometric view. This view is identical to one with a right-out of 45 degrees and a top-out of 35.267 degrees.

Command line: Vpoint

Change to view of *XY* plane In AutoCAD, a plan view is a view aimed toward the origin (0, 0, 0) from a point on the positive *Z*-axis. This results in a view of the *XY* plane. The view and coordinate system can be restored, that is, the default for most AutoCAD drawings can be set by the UCS orientation to World and then setting the 3*D* view to Plan View.

To change the current view to the *XY* plane:

1. From the View menu, choose 3*D* Views. Then choose Plan View.
2. Choose one of the following options:
   - Current (for the current UCS)
   - World (for the WCS)
   - Named (for a saved UCS)

16.4.5 Isometric Drawings

Isometric drawings simulate a 3*D* object from a particular viewpoint by aligning along three major axes.

If the snap angle is 0, the axes of the isometric planes are 30 degrees, 90 degrees, and 150 degrees. After setting the snap style to isometric, work can be performed on any of three planes, each with an associated pair of axes:

- Left—Aligns snap and grid along 90 and 150 degree axes.
- Top—Aligns snap and grid along 30 and 150 degree axes.
- Right—Aligns snap and grid along 30 and 90 degree axes.

Choosing one of the three isometric planes causes Ortho and the crosshairs to be aligned along the corresponding isometric axes. For example, when Ortho is on, the points specified align along the simulated plane where the drawing is developing. Therefore, after the top plane is drawn, switch to the left plane to draw another side, and to the right plane to complete the drawing. The Isometric Snap/Grid mode helps to create 2*D* isometric images that represent 3*D* objects. By setting the Isometric Snap/Grid, objects can be easily aligned along one of three isometric planes. However, isometric drawing appears to be 3*D*, it is actually a 2*D* representation.

To turn on an isometric plane:

1. From the Tools menu, choose Drafting Settings.
2. In the Drafting Settings dialog box, on the Snap and Grid tab under Snap Type and Style, select Isometric Snap.
3. Choose OK.

Alternatively it can also be done in a cycle through the three isometric planes, by pressing F5.

Command line: Dsettings to draw an isometric circle:

1. From the Tools menu, choose Drafting Settings.
2. In the Drafting Settings dialog box, Snap and Grid tab, select Isometric Snap.
3. Choose OK.
4. From the Draw menu, choose Ellipse. Then choose Axis, End.
5. Enter i (isocircle).
6. Specify the centre of the circle.
7. Specify the radius or diameter of the circle.

Command line: Ellipse

16.4.6 Modify Objects

With AutoCAD, the size, shape, and location of objects can easily be modified. This can be done by either entering a command first and then selecting the objects to modify, or by selecting the objects first and then entering a command to modify them. Important modifying commands are as follows:

Removing objects Objects can be removed from the drawing by several methods, including:

- Erasing them with Erase.
- Cutting them to the Clipboard.
- Pressing Delete.

To erase an object:

1. From the Modify menu, choose Erase.
2. At the Select Objects prompt, use a selection method to select the objects to be erased or enter one of the following options:
   - Enter 'L' (Last) to erase the last object drawn.
   - Enter 'p' (Previous) to erase the last selection set.

- Enter 'all' to erase all objects from the drawing.
- Enter '?' to see a list of all selection methods.

3. Press Enter to end the command.

Modify toolbar

Command line: Erase

To restore the last erased object:

- At the Command prompt, enter oops.

The last objects that were removed by Erase, Block, or Wblock are restored.

Command line: Oops

To cut objects to the Clipboard:

1. Select the objects to be cut.
2. From the Edit menu, choose Cut.

The objects are available to be pasted into other Windows applications.

Command line: Cutclip

Rotating objects Objects can be rotated around a specified point. To determine the angle of rotation, enter an angle value or specify a second point. Entering a positive angle value rotates the objects counterclockwise or clockwise, depending on the Direction Control setting in the Drawing Units dialog box. The plane of rotation and the direction of the zero angle depend on the orientation of the user coordinate system.

Rotate an object by a specified angle: This is done by choosing a base point and a relative or absolute rotation angle. Specify a relative angle to rotate the object from its current orientation around the base point by that angle. Specify an absolute angle to rotate the object from the current angle to a new absolute angle.

It can also be done by specifying a relative angle using one of two methods:

- Enter a rotation angle value from 0 to 360 degrees.
- Drag the object around the base point and specify a second point. Turning on Ortho and Polar Tracking, or using object snaps for the second point, is often useful with this method.

To rotate an object using a reference angle:

1. From the Modify menu, choose Rotate.
2. Select the object to rotate
3. Specify the base point for the rotation and the endpoint of the object you are aligning with.
4. Specify the Angle of rotation.

Modify toolbar

Command line: Rotate

Move objects Objects can be moved without changing their orientation or size. This is done by using coordinates and object snaps.

In this example (Fig. A16.22), the block is moved representing a window. After choosing Move, select the object to be moved (1) Specify the base point for the move (2) followed by the point of displacement (3). The object is moved from point 2 to point 3.

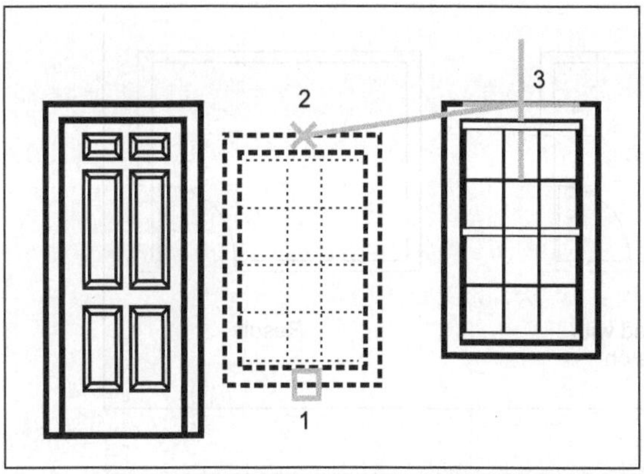

Fig. A16.22 Moving objects on AutoCAD

It can also be performed by using a relative distance by entering coordinate values for the first point and pressing Enter for the second point of displacement. This instructs AutoCAD to use the coordinate values as a relative displacement rather than a base point. The objects selected are moved to a new location determined by the relative coordinate values that have been entered.

To move an object using a displacement:

1. From the Modify menu, choose Move.
2. Select the object to move.
3. Enter the displacement in the form of a Cartesian, polar, cylindrical, or spherical coordinate value. Do not include the @ sign, because a relative coordinate is assumed.
4. At the prompt for the second point of displacement, press Enter.

By entering only one set of coordinate values, AutoCAD is instructed to use it as a relative displacement rather than a base point. The selected objects are moved to a new location determined by the relative coordinate values that have been entered.

Modify toolbar

Command line: Move

Move to move an object using two points:

1. From the Modify menu, choose Move.
2. Select the objects to move.
3. Specify a base point for the move.
4. Specify a second point, the point of displacement.

The selected objects are moved to a new location determined by the distance and direction between the first and the second points.

Modify toolbar

Command line: Move

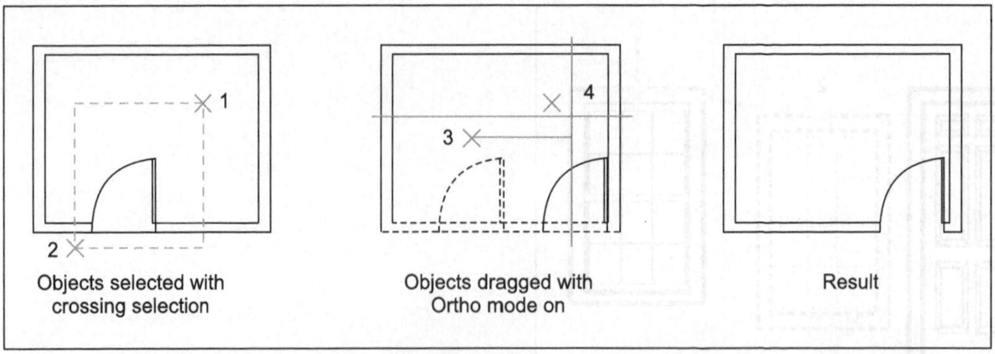

Fig. A16.23 To move an object using two points

In Fig. A16.23, notice that the door itself is entirely within the selection area and thus moves to the new location. The wall lines, on the other hand, merely cross the selection area. Only the endpoints that lie within the selection area move. Therefore, the wall lines stretch in accordance with the movement of the door. To move objects a specified distance, direct distance entry, can be used together with Ortho and Polar Tracking.

Change the size and shape of objects There are several methods for adjusting the lengths of existing objects, relative to other objects both symmetrically and asymetrically.

Trimming or extending objects: Objects can be shortened or lengthening the meet the edges of other objects. This means creating an object such as a, first and then later adjusting it to fit exactly between other objects. Objects that are selected as cutting edges or boundary edges are not required to intersect the object being trimmed. It is possible to trim or extend an object to a projected edge or to an extrapolated intersection; that is, where the objects would intersect if they were extended. If a boundary is not specified and Enter is pressed at the Select Objects prompt, all objects become potential boundaries. This is called implied selection.

Trimming objects: Cutting edges can be lines, arcs, circles, polylines, ellipses, splines, xlines, regions, blocks, and rays. They can also be layout viewport objects in paper space. In this example, the intersection of two walls is cleaned up smoothly by trimming. An object can be one of the cutting edges and one of the objects being trimmed. For example (Fig. A16.24), the circle is a cutting edge for the construction lines and is therefore, also trimmed.

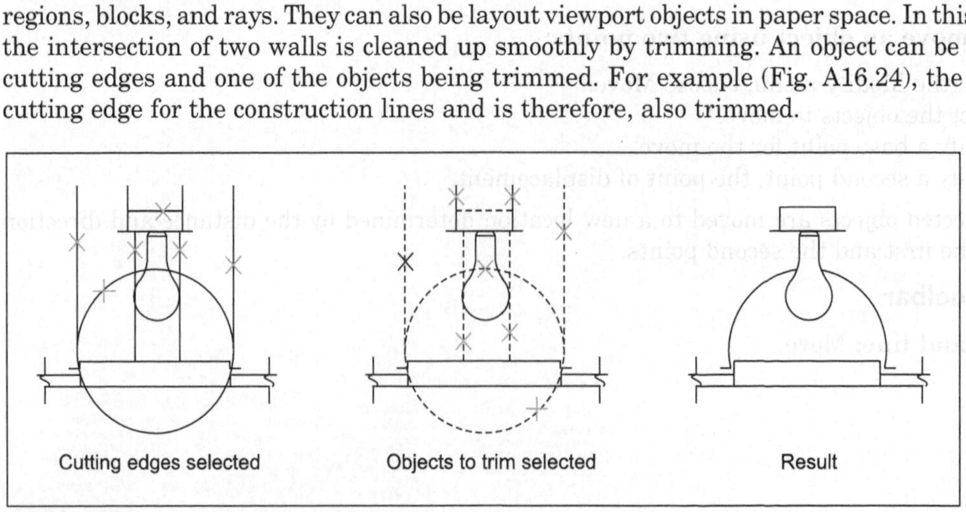

Fig. A16.24 Trimming objects on AutoCAD

To trim an object:

1. From the Modify menu, choose Trim.
2. Select the objects to serve as cutting edges.
   To select all objects in the drawing as without selecting any objects.
3. Select the objects to trim.

Modify toolbar

Command line: Trim

Extending objects: Extending operates the same way as trimming. In the example (Fig. A16.25), the lines are extended precisely to a circle, which is the boundary edge.

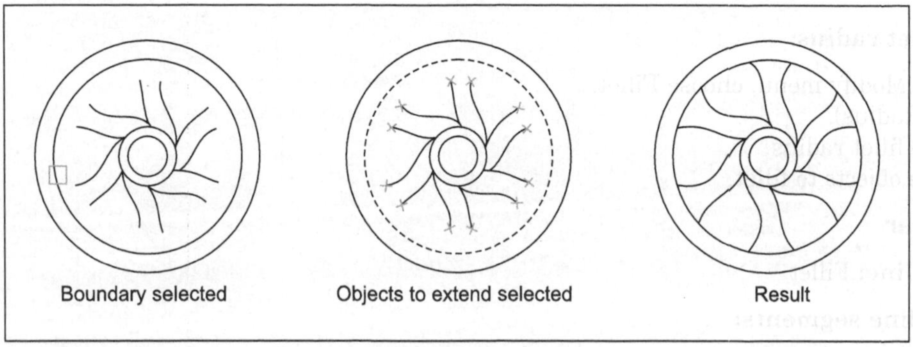

Fig. A16.25 Extending objects on AutoCAD

To extend an object:

1. From the Modify menu, choose Extend.
2. Select the objects to serve as boundary edges.
   To select all objects in the drawing as potential boundary edges, press Enter without selecting any objects.
3. Select the objects to extend.

Modify toolbar

Command line: Extend

Creating fillets, chamfers, or breaks in objects It is possible to change objects to meet at rounded or flattened corners. Gaps can also created in objects.

Creating fillets: Filleting connects two objects with a smoothly fitted arc of a specified radius. An inside corner is called a fillet and an outside corner is called a round; both can be created using Fillet in AutoCAD (Fig. A16.26).

Seting the fillet radius: The fillet radius is the radius of the arc that connects filleted objects. Changing the fillet radius affects subsequent fillets. If the fillet radius is set to 0, filleted objects are trimmed or extended until they intersect, but no arc is created.

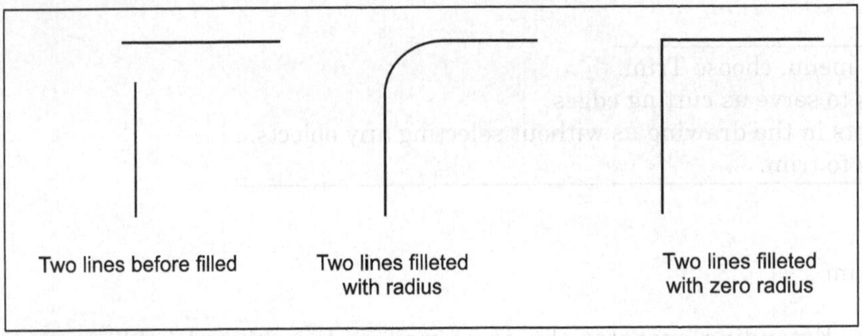

Fig. A16.26 Creating fillet on AutoCAD

To set the fillet radius:

1. From the Modify menu, choose Fillet.
2. Enter *r* (Radius).
3. Enter the fillet radius.
4. Select the objects to fillet.

Modify toolbar

Command line: Fillet

To fillet two line segments:

1. From the Modify menu, choose Fillet.
2. Select the first line.
3. Select the second line.

Creating Chamfers: Using Chamfer is a fast way of creating a line between two nonparallel lines (Fig. A16.27). It is usually used to represent a beveled edge on a corner. Chamfer can also be used to bevel all corners of a polyline. In the distance method, specify the amount that each line should be trimmed or extended. In the angle method, specify the length of the chamfer and the angle it forms with the first line. Chamfered objects can be retained as they were before the chamfer, or trimmed, or extended to the chamfer line.

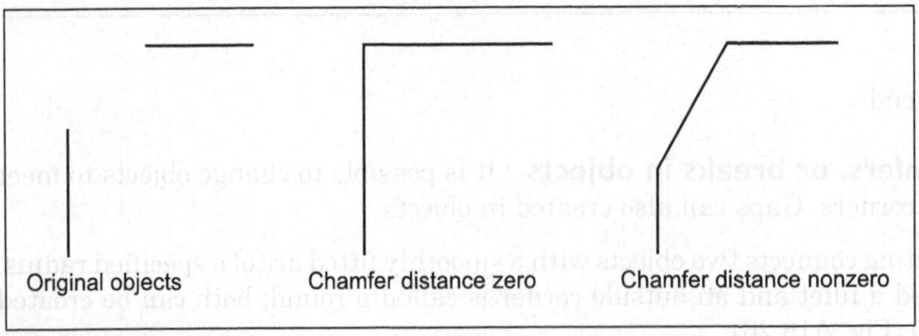

Fig. A16.27 Chamfering

To set chamfer distances:

1. From the Modify menu, choose Chamfer.
2. Enter *d* (distances).
3. Enter the first chamfer distance.
4. Enter the second chamfer distance.
5. Select the lines for chamfering.

Modify toolbar

Command line: Chamfer

To chamfer two nonparallel line segments:

1. From the Modify menu, choose Chamfer.
2. Select the first line.
3. Select the second line.

Create Breaks: Using Break is a convenient way to create a gap in an object, resulting in two objects with a space between them (Fig. A16.28). Break is often used to create space for block or text insertion.

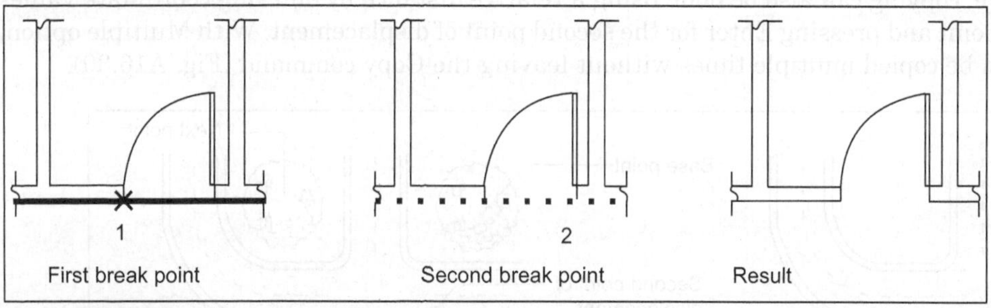

Fig. A16.28 Creating breaks in the AutoCAD

When breaking an object, either

- Select the object at the first break point and then specify a second break point
- Select the entire object and then specify two break points

To break an object:

1. From the Modify menu, choose Break.
2. Select the object to break.

   By default, the point at which the selected object is, is the first break point. To choose a different pair of break points, enter *f* (First) and specify the first break point.
3. Specify the second break point.

Copy, Offset, or mirror objects Duplicates of objects can be created in drawing that are either identical or similar to selected objects.

Copy objects: Copying is the creation of duplicates at a specified distance from the original objects. This is performed in the same way as in Move.

In Fig. A16.29, the distance and direction are specified by two points, point (1) and point (2), called the base point and the second point of displacement, respectively. These points can be located anywhere within the drawing.

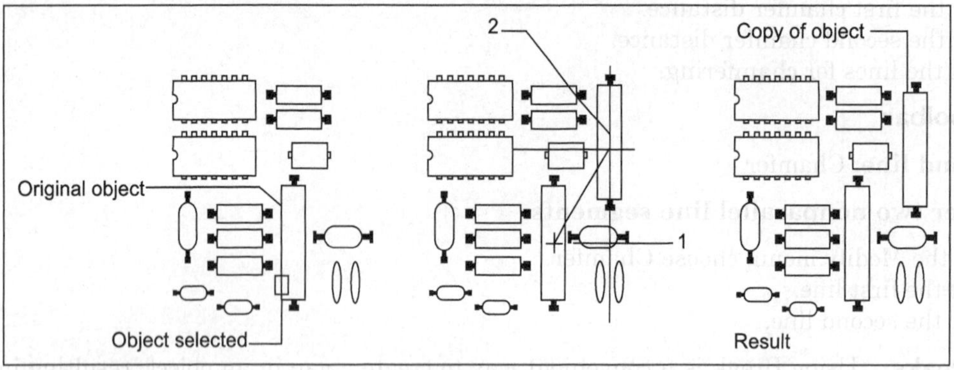

Fig. A16.29 Copying an object

As in Move, copying can also be done using a relative distance by entering coordinate values for the first point and pressing Enter for the second point of displacement. With Multiple option, an object can be copied multiple times without leaving the Copy command (Fig. A16.30).

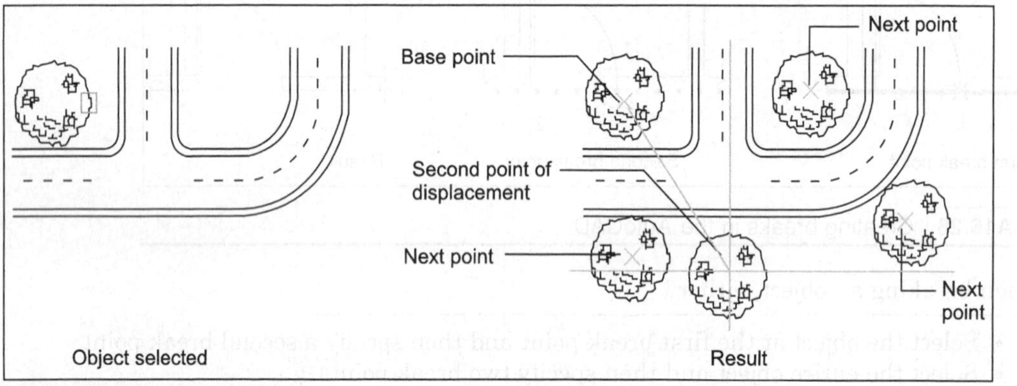

Fig. A16.30 Multiple copying

To copy an object multiple times:

1. From the Modify menu, choose Copy.
2. Select the objects to copy.
3. Enter m (multiple).
4. Specify the base point.
5. Specify the second point of displacement.
6. Specify the next point of displacement. Continue inserting copies, or press Enter to end the command.

Modify toolbar

Command line: Copy

Creating an array of objects Copies of objects can be created in a rectangular or polar (circular) pattern called an array. For rectangular arrays, control the number of rows and columns and the distance between each. For polar arrays, control the number of copies of the object and whether the copies are rotated. To create many regularly spaced objects, arraying is faster than copying (Fig. A16.31).

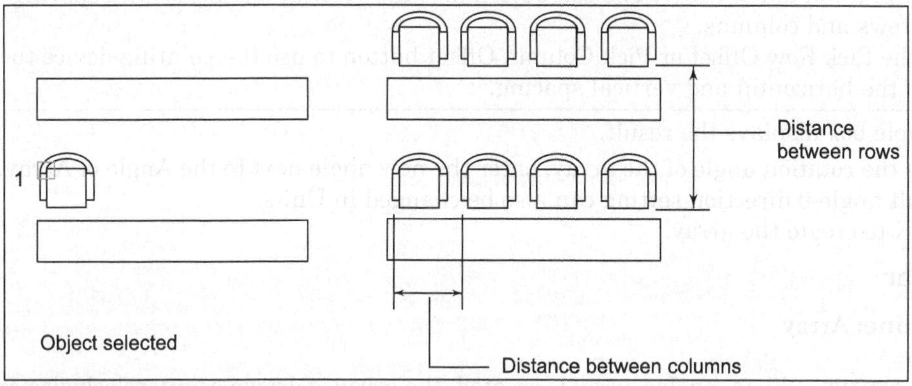

Fig. A16.31 Creating an array

Creating rectangular arrays AutoCAD builds a rectangular array along a baseline defined by the current snap rotation angle. This angle is zero by default, so the rows and columns of a rectangular array are orthogonal with respect to the *X* and *Y* axes. The default angle 0 direction setting can be changed is the Units command (Fig. A16.32).

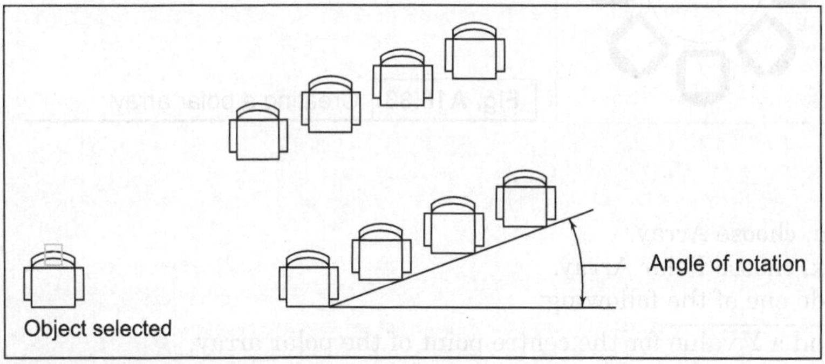

Fig. A16.32 Creating a rectangular array

To creating a rectangular array:

1. From the Modify menu, choose Array.
2. In the Array dialog box, choose Rectangular Array.
3. Choose Select Objects.

The Array dialog box closes and AutoCAD prompts for object selection.

4. Select the objects to be arrayed and press Enter.
5. In the Rows and Columns boxes, enter the number of rows and columns in the array.
6. Specify the horizontal and vertical spacing (offsets) between objects by using one of the following methods:

- In the Row Offset and Column Offset boxes, enter the distance between rows and between columns. Adding a plus sign (+) or a minus sign (–) determines direction.
- Click the Pick Both Offsets button to use the pointing device to specify the diagonal corners of a cell in the array. The cell determines the vertical and horizontal spacing of the rows and columns.
- Click the Pick Row Offset or Pick Column Offset button to use the pointing device to specify the horizontal and vertical spacing.

The example box displays the result.

7. To change the rotation angle of the array, enter the new angle next to the Angle of Array.
8. The default angle 0 direction setting can also be changed in Units.
9. Choose OK to create the array.

Modify toolbar

Command line: Array

Create polar arrays When a polar array is created, the array is drawn counterclockwise or clockwise, depending on whether a positive or a negative value has been entered for the angle that needs to be to filled (Fig. A16.33).

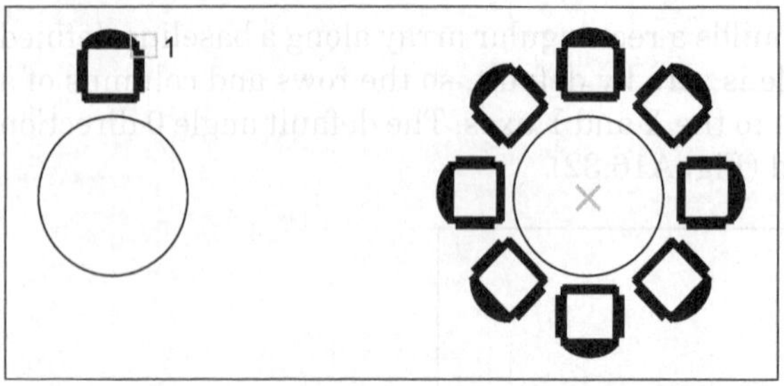

Fig. A16.33 Creating a polar array

To create a polar array:

1. From the Modify menu, choose Array.
2. In the Array dialog box, choose Polar Array.
3. Next to Centre Point, do one of the following:

- Enter an *X* value and a *Y* value for the centre point of the polar array.
- Click the Pick Centre Point button. The Array dialog box closes and AutoCAD prompts for object selection. Use the pointing device to specify the centre point of the polar array.

4. Choose Select Objects.

The Array dialog box closes and AutoCAD prompts for object selection.

5. Select the objects to be arrayed.
6. In the Method box, select one of the following methods:

- Total Number of Items and Angle to Fill
- Total Number of Items and Angle between Items
- Angle to Fill and Angle between Items

7. Enter the number of items (including the original object), if available.
8. Use one of the following methods:

- Enter the angle to fill and angle between items, if available. Angle to Fill specifies the distance to fill around the circumference of the array. Angle between Items specifies the distance between each item.
- Click the Pick Angle to fill button and the Pick Angle between Items button and use the pointing device to specify the angle that needs to be filled and the angle between items.

The Fig. A16.33 displays the result.

9. Set any of the following options:

- To rotate the objects as they are arrayed as illustrated in Fig. A16.33, select Rotate Items As Copied.
- To specify the *X*, *Y* base point, choose More, clear the Set to Object's Default option and enter values in the *X* and *Y* boxes, or click the Pick Base Point button and use the pointing device to specify the point.

10. Choose OK to create the array.

Modify toolbar

Command line: Array

Offset an object Offsetting creates a new object whose shape parallels the shape of a selected bject. Offsetting a circle or an arc creates a larger or smaller circle or arc, depending on which side is specified for the offset (Fig. A16.34).

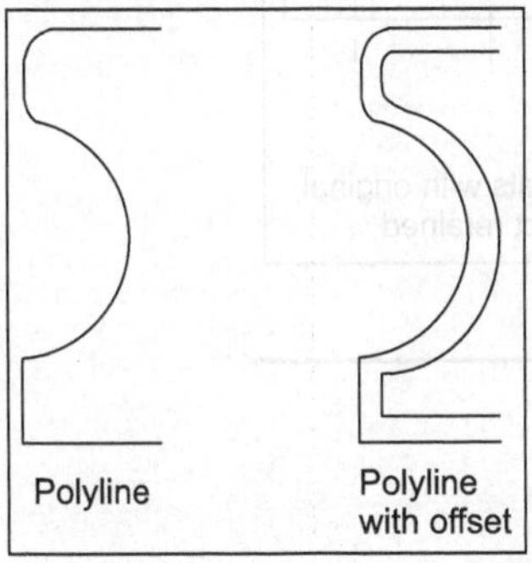

Fig. A16.34 Offset an object

A highly effective drawing technique is to offset objects and then trim or extend their ends.

To offset an object by specifying a distance:

1. From the Modify menu, choose Offset.
2. Specify the offset distance.

Enter a value or use the pointing device.

3. Select the object to offset.
4. Specify a point on the side where the new objects need to be placed.
5. Select another object to offset, or press Enter to end the command.

Modify toolbar

Command line: Offset

To offset an object through a point:

1. From the Modify menu, choose Offset.
2. Enter *t* (Through).
3. Select the object to offset.
4. Specify the through point.
5. Select another object to offset, or press Enter to end the command.

Modify toolbar

Command line: Offset

Mirror objects Mirroring creates a mirror image of objects. It is useful for creating symmetrical objects because half the object can be quickly drawn and then mirrored instead of drawing the entire object. Flip objects about an axis, called a mirror line, to create a mirror image. To specify this temporary mirror line, enter two points. Choose whether to delete or retain the original objects. Mirroring works in any plane parallel to the *XY* plane of the current UCS (Fig. A16.35).

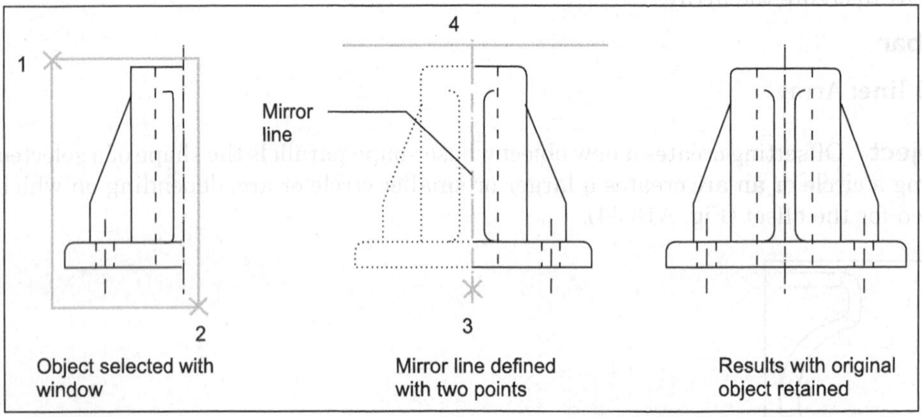

Fig. A16.35 Mirroring

To mirror objects:

1. From the Modify menu, choose Mirror.
2. Select the objects to mirror.
3. Specify the first point of the mirror line.
4. Specify the second point.
5. Press Enter to retain the original objects, or enter **y** to delete them.

Modify toolbar

Command line: Mirror

16.4.4 Dimensioning

Dimensioning is the process of adding measurement annotations to a drawing. AutoCAD® provides many ways to dimension objects and many ways to format dimensions. Dimensions for a wide variety of object shapes can be created in many different orientations. Dimension styles can also be created to format dimensions quickly and ensure that dimensions in drawings conform to industry or project standards.

Dimensions show the measurements of objects, the distances or angles between objects, or the distance of a feature from an origin that is specified. AutoCAD provides three basic types of dimensioning: linear, radial, and angular. Dimensions can be horizontal, vertical, aligned, rotated, ordinate, baseline, or continued. Some simple examples are shown in Fig. A16.35.

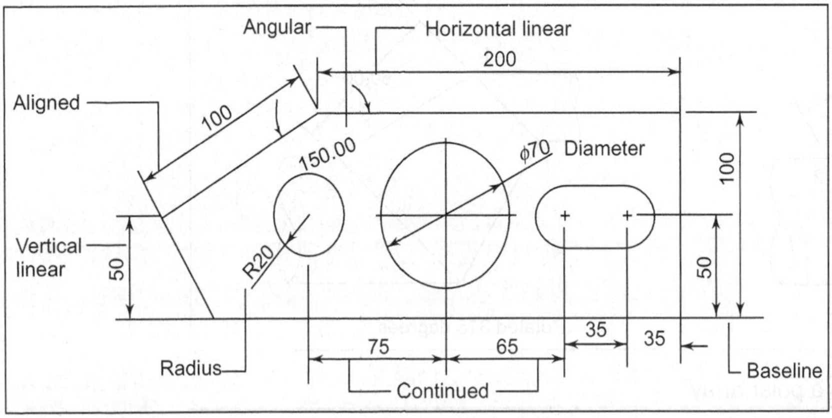

Fig. A16.36 Dimensioning drawings on AutoCAD

Creating linear dimensions Linear dimensions are maked with horizontal, vertical, and aligned dimension lines. These linear dimensions can also be stacked or can be created end to end.

Linear dimensions can be horizontal, vertical, or aligned. With aligned dimensions, the dimension line is parallel to the line (imaginary or real) between the extension line origins. Baseline (or parallel) and continued (or chain) dimensions are a series of consecutive dimensions that are based on a linear dimension. In all four illustrations in Fig. A16.37, the extension line origins are designated explicitly at 1 and 2, respectively. The dimension line location is specified at 3.

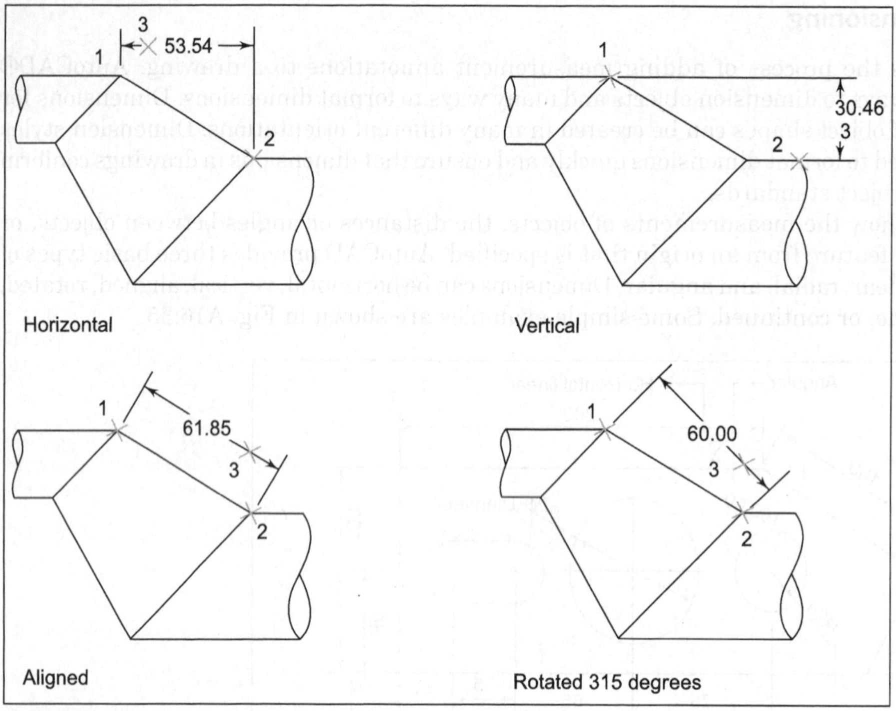

Fig. A16.37 Creating a polar array

Creating horizontal and vertical dimensions AutoCAD automatically applies a horizontal or vertical dimension according to the extension line origins that are specified or the location from where an object is selected.

To create a horizontal or vertical dimension:

1. From the Dimension menu, choose Linear.
2. Press Enter to select the object to dimension, or specify the first and second extension line origins.
3. Before specifying the dimension line location, override the dimension direction and edit the text, the text angle, or the dimension line angle:

- To rotate the extension lines, enter **r** (Rotated). Then enter the dimension line angle.
- To edit the text, enter **m** (Mtext). In the Multiline Text Editor, revise the text and choose OK. Editing within or overwriting the brackets (<>) changes or removes the dimension value calculated by AutoCAD. Adding text before or after the brackets appends text before or after the dimension value.
- To rotate the text, enter **a** (Angle). Then enter the text angle.

4. Specify the dimension line location.

Dimension toolbar

Command line: Dimlinear

Creating aligned dimensions In aligned dimensions, the dimension line is parallel to the extension line origins. The illustration shows two examples of aligned dimensioning. (1) the object is selected, and (2) the location of the aligned dimension is specified. The extension lines are drawn automatically.

To create an aligned dimension:

1. On the Dimension menu, click Aligned.
2. Other steps are similar as in linear dimensioning.

Dimension toolbar

Command line: Dimaligned

Creating baseline and continued dimensions Baseline dimensions are multiple dimensions measured from the same baseline. Continued dimensions are multiple dimensions placed end to end. A linear, aligned, or angular dimension must be created before baseline dimensioning or continued dimensions. Both baseline and continued dimensions are measured from the previous extension line unless another point is specified as the point of origin.

To create a baseline linear dimension:

1. From the Dimension menu, choose Baseline.
   By default, the origin of the last linear dimension created is used as the first extension line for the new baseline dimension. AutoCAD prompts for the second dimension line.
2. Use an object snap to select the second extension line origin, or press ENTER to select any dimension as the base dimension. AutoCAD automatically places the second dimension line at a distance specified by the Baseline Spacing option in the Dimension Style Manager, Lines and Arrows tab.
3. Use an object snap to specify the next extension line origin.
4. Continue to select extension line origins as required.
5. Press Enter twice to end the command.

Dimension toolbar

Command line: Dimbaseline

To create a continued linear dimension:

1. From the Dimension menu, choose Continue. AutoCAD uses the origin of the second extension line of the existing dimension as the first extension line origin.
2. Use object snaps to specify additional extension line origins.
3. Press Enter twice to end the command.

Dimension toolbar

Command line: Dimcontinue

To create a diameter dimension:

1. From the Dimension menu, choose Diameter.
2. Select the arc or circle to be dimensioned.
3. Enter options as needed:

- To edit the dimension text content, enter **t** (Text) or **m** (Mtext). Editing within or overwriting the brackets (<>) changes or removes the dimension value calculated by AutoCAD. Adding text before or after the brackets appends text before or after the dimension value.
- To change the dimension text angle, enter **a** (Angle).

4. Specify the leader line location.

Dimension toolbar

Command line: Dimdiameter

To create a radius dimension:

1. From the Dimension menu, choose Radius.
2. Select the arc or circle to dimension.
3. Enter options as needed:

- To edit the dimension text content, enter **t** (Text) or **m** (Mtext). Editing within or overwriting the brackets (<>) changes or removes the dimension value calculated by AutoCAD. Adding text before or after the brackets appends text before or after the dimension value.
- To edit the dimension text angle, enter **a** (Angle).

4. Specify the leader line location.

Dimension toolbar

Command line: Dimradius

Creating angular dimensions Angular dimensions measure the angle between two lines or three points. To measure the angle between two radii of a circle, select the circle and specify the angle endpoints. With other objects, select the objects and then specify the dimension location.

To create an angular dimension:

1. From the Dimension menu, choose Angular.
2. Use one of the following methods:

- To dimension a circle, select the circle at the first endpoint of the angle and then specify the second endpoint of the angle.
- To dimension any other object, select the first line, and then the second line.

3. Enter options as required:

- To edit the dimension text content, enter **t** (Text) or **m** (Mtext). Editing within or overwriting the brackets (<>) changes or removes the dimension value calculated by AutoCAD. Adding text before or after the brackets appends text before or after the dimension value.
- To edit the dimension text angle, enter **a** (Angle).

4. Specify the dimension line arc location.

Dimension toolbar

Command line: Dimangular

Dimension styles The appearance of dimensions can be controlled by changing settings. The Dimension Style Manager allows you to modify the following:

- Extension lines, dimension lines, arrowheads, centre marks or lines, and the offsets between them.
- The positioning of the parts of the dimension in relation to one another and the orientation of the dimension text.
- The content and appearance of the dimension text.

Dimension toolbar > style The following window will appear

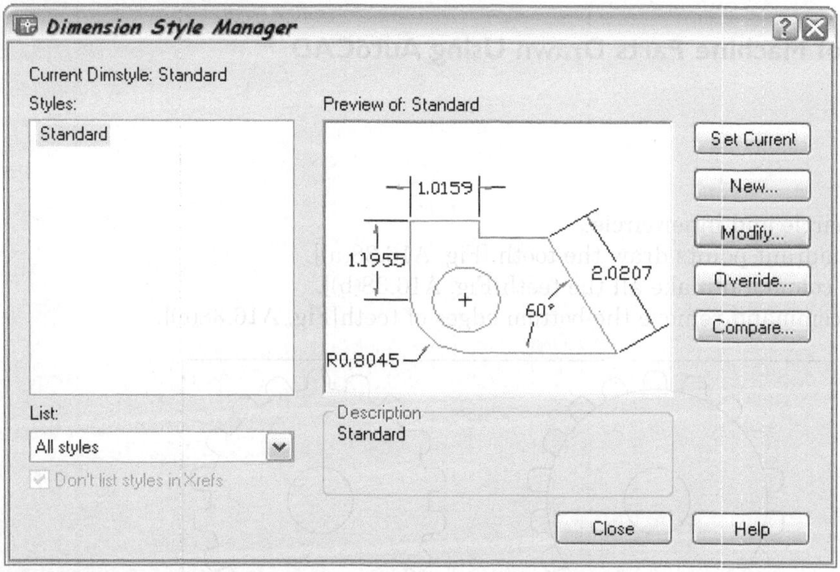

In the window click Modify to edit existing dimension appearance.

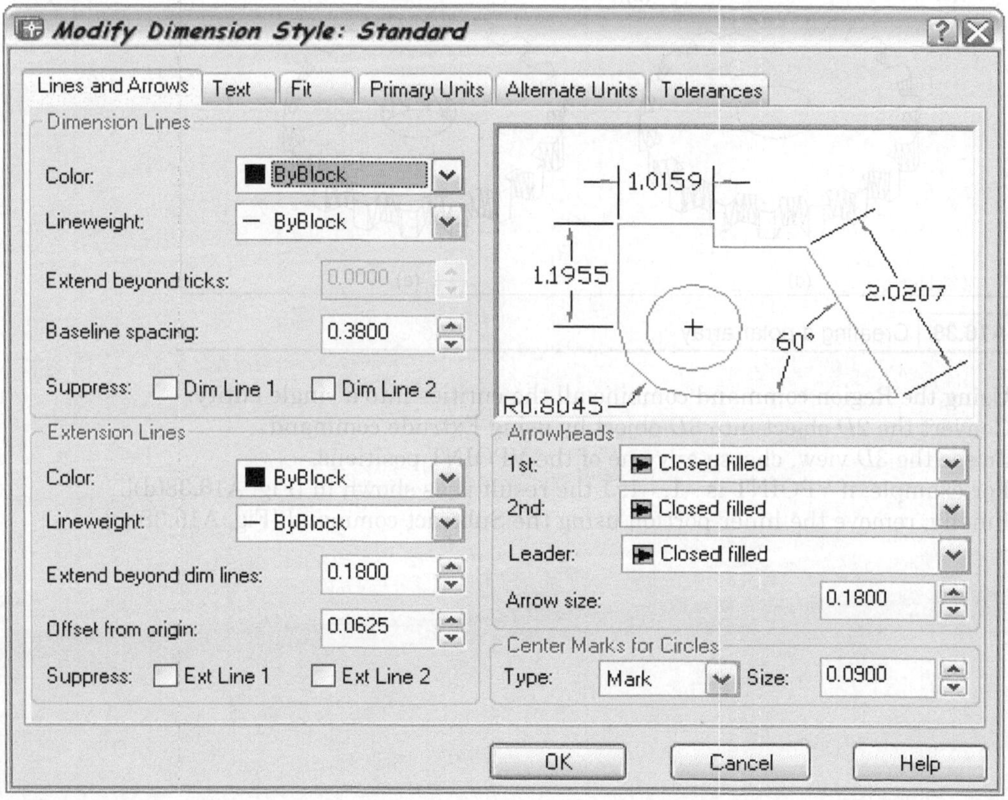

In the above window all modifications for various parameters stated earlier can be entered as per requirement.

16.4.5 Example of Machine Parts Drawn Using AutoCAD

GEAR

To draw a gear:

1. Draw the base circle and inner circle.
2. At one of the quadrant points draw the tooth.[Fig. A16.38(a)].
3. By using Array command make all the teeth[Fig. A16.38(b)].
4. By using Trim command remove the bottom edges of teeth[Fig. A16.38(c)].

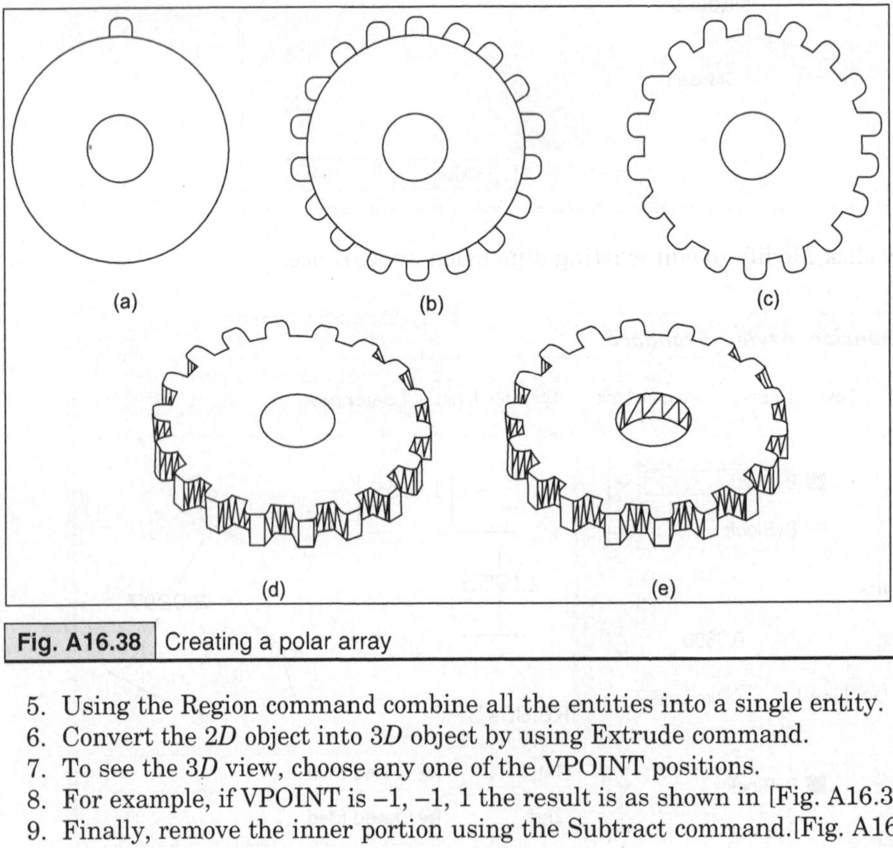

Fig. A16.38 Creating a polar array

5. Using the Region command combine all the entities into a single entity.
6. Convert the *2D* object into *3D* object by using Extrude command.
7. To see the *3D* view, choose any one of the VPOINT positions.
8. For example, if VPOINT is –1, –1, 1 the result is as shown in [Fig. A16.38(d)].
9. Finally, remove the inner portion using the Subtract command.[Fig. A16.38(e)].

PART B

Assembly Drawings

INTRODUCTION TO ASSEMBLY DRAWING

A drawing that displays the parts of a machine or a machine unit assembled in their relative working positions is known as an assembly drawing. An assembly drawing should be such that it should satisfy: (1) Manufacturing requirements (2) Operational requirements (3) Maintenance requirements

Types of Assembly Drawings

Assembly drawings are classified according to their use.

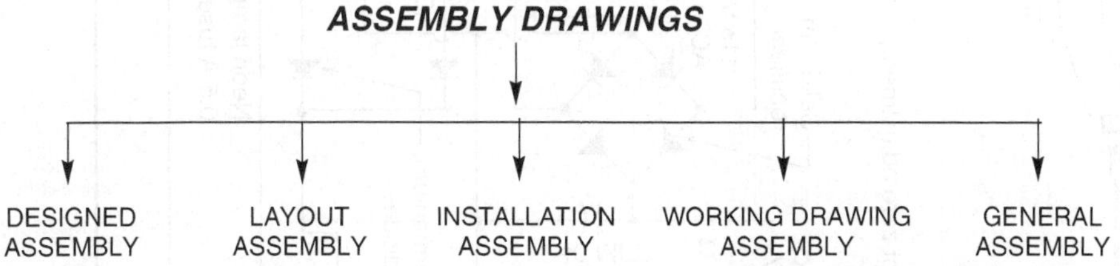

Types *K*, *KG*, and *KR* (2- and 3-phase)
Frames 203 to 326 Inclusive

Belt Drive, Two Sleeve or Ball Bearings
For Direct Drive Omit Base and Pulley

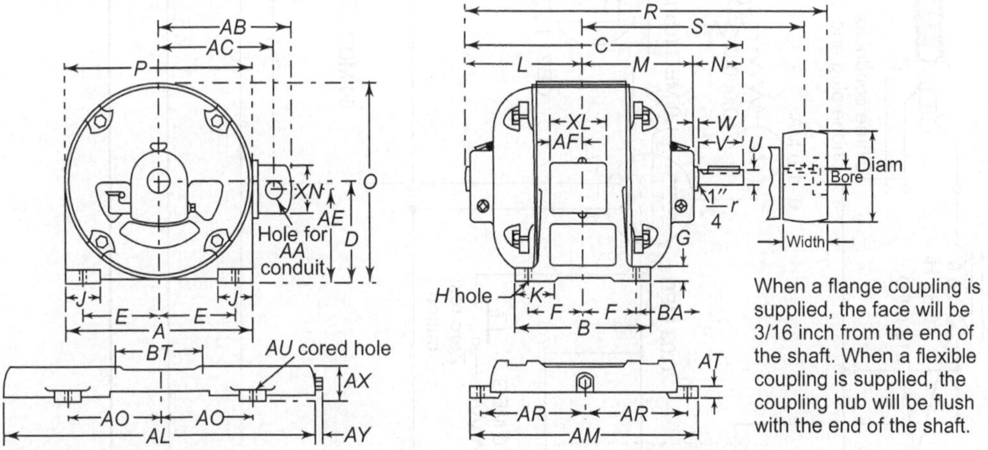

Providing mounting restrictions permit conduit boxes to be placed so that entrance can be made upward, downward, or from either side.

BASE AND PULLEY FURNISHED ONLY WHEN CALLED FOR ON REQUISITION

Frame, Base and Pulley No.	Approx net wt in LB			Dimensions in Inches																			
				Pulley			Belt Width	Keyway		Key Length	A	B	C	D	E	F	G	H	J	K	L	M	
	Motor only	Base	Pulley	Dia.	Width Over-all	Bore		Width	Depth														
203	67	19	1¼	3	3		2½	3/16	3/32	1⅜	9¾	6⅞	14	5 edded is re- of an	4	2 2/4	11/16	1 13/22	1¾	1 15/16	5⅞	5 2/4	
204	74	20	1¼	3	3		2½	3/16	3/32	1⅜	9¾	7⅞	15	5	4	3¼	11/16	1 13/22	1¾	1 15/16	6⅜	6¼	

Fig. B1 Installation assembly

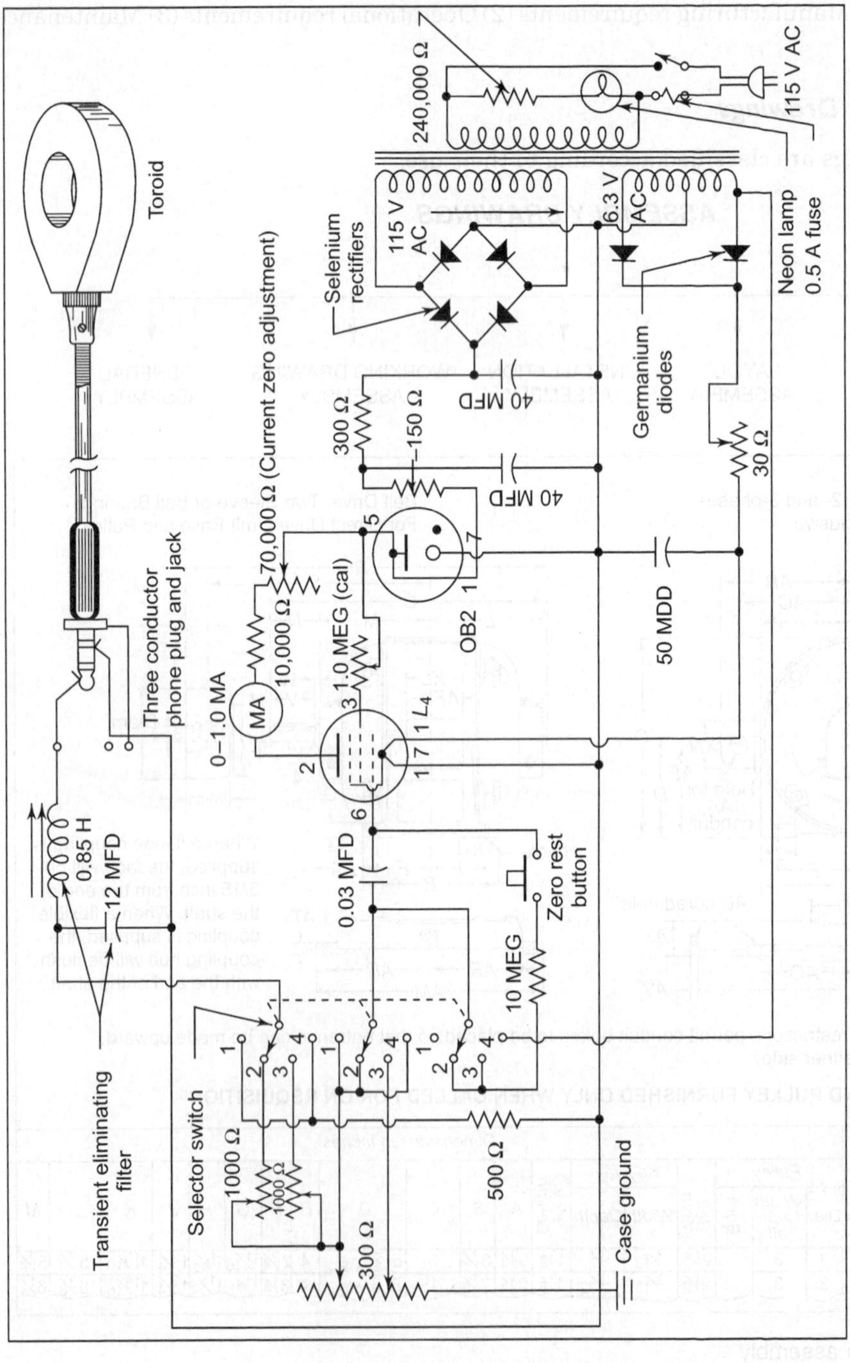

Fig. B2 Layout assembly

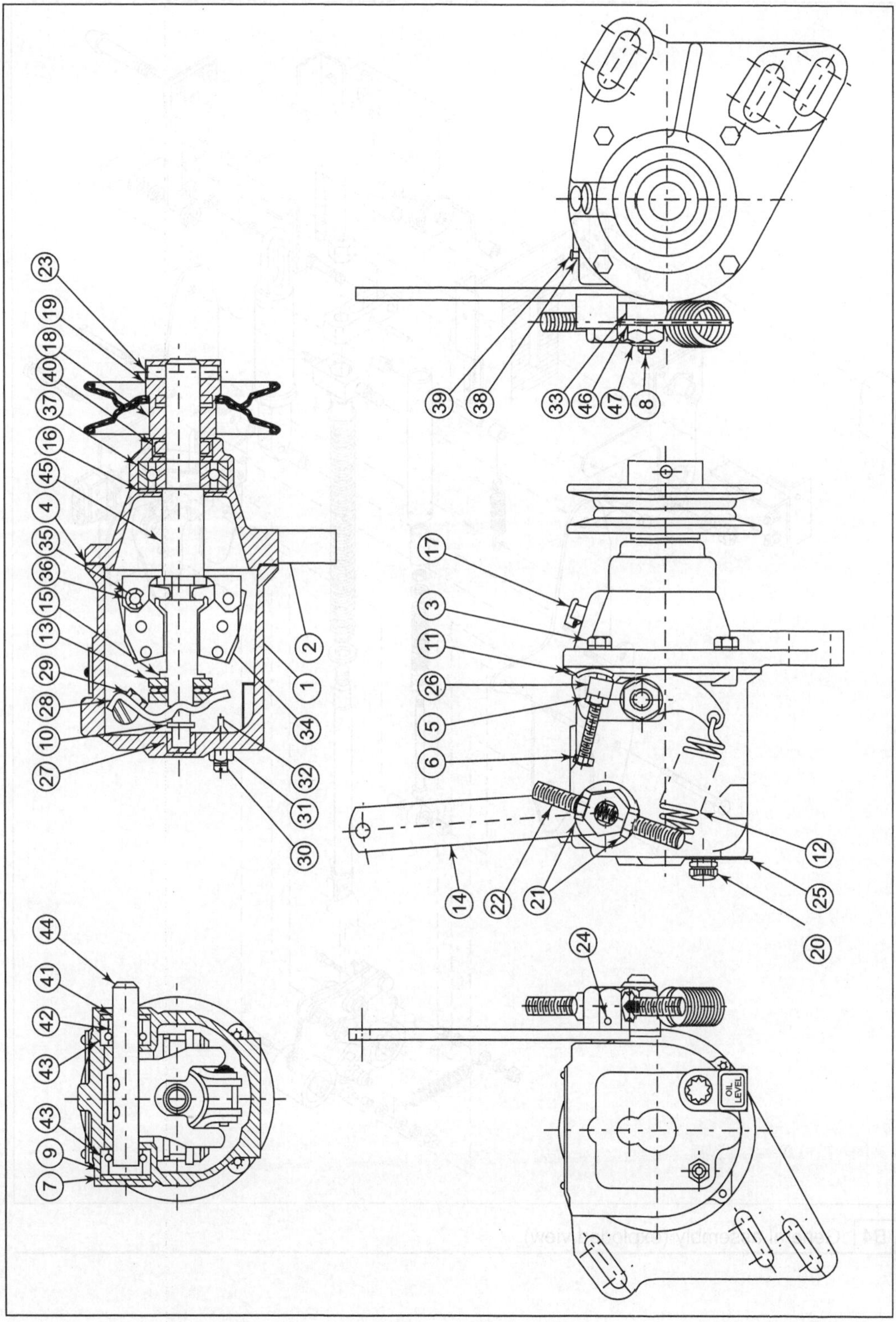

Fig. B3 Designed assembly

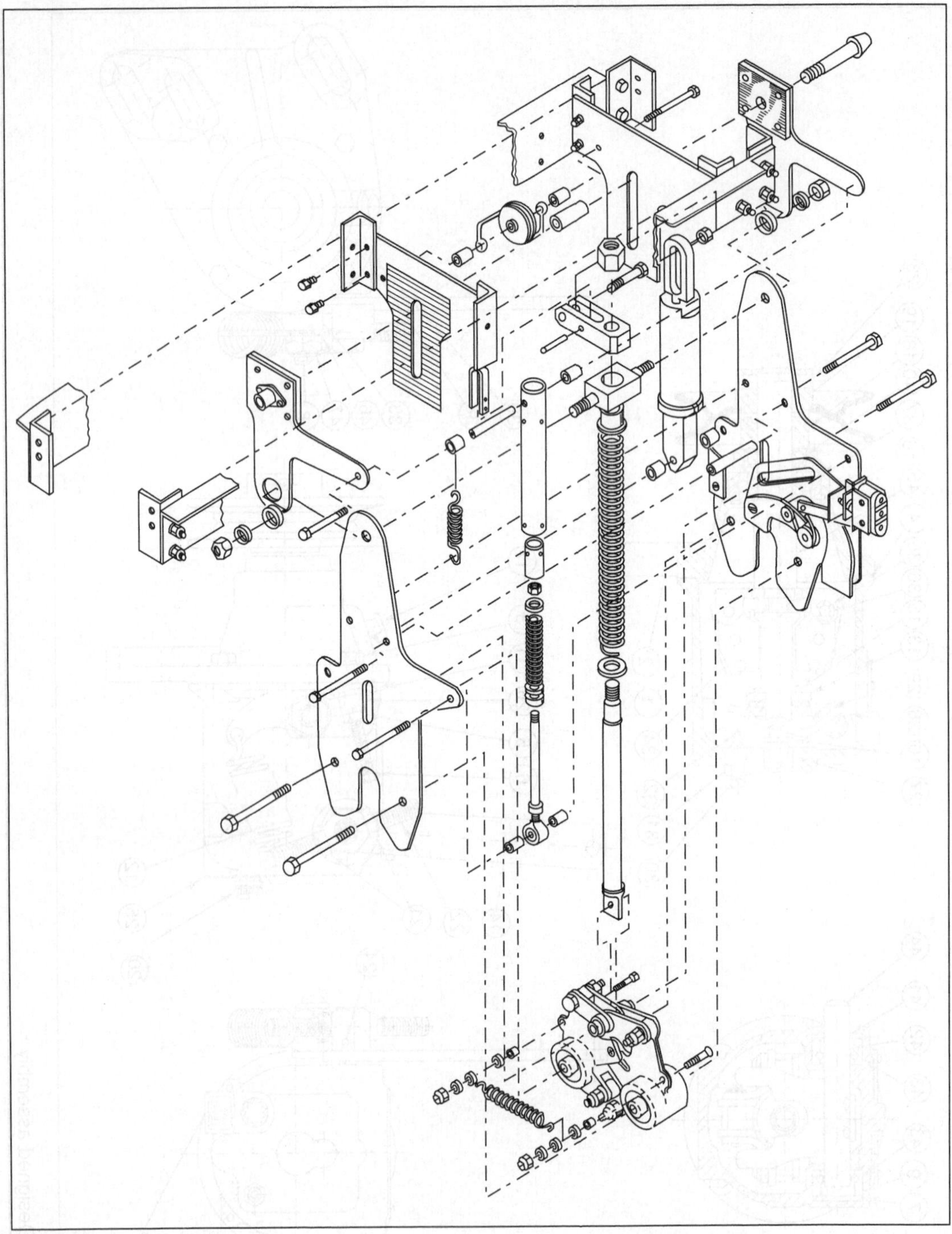

Fig. B4 General assembly (exploded view)

1. Designed assembly: An assembly drawing prepared at the design stage and on a large scale, mode is called a designed assembly drawing. (Fig. B3)
2. Layout assembly: An assembly drawing showing how various parts are assembled, with their basic proportions. (Fig. B2)
3. Installation assembly: An installation assembly is prepared for the installation or erection of a machine. It is also sometimes known as an outline assembly. (Fig. B1)
4. Working drawing assembly: This is a complete set of working drawings of a machine, comprising of detailed drawings, giving all necessary information for the production of individual parts and assembly drawings showing the location of each part. An assembly drawing should be ready before detailed drawings are accepted as finished and blue prints are made.
5. General assembly: Generally assembly comprises of detailed drawing of individual parts and includes sub-assembly and assembly drawings of the machine (Fig. B4)

Accepted norms to be observed for assembly drawings

1. Selection of views: The main or important view, which is usually in section, should show all the individual parts and their relative locations. Additional views are shown only if they add relevant information.
2. Sectioning: The parts should be sectioned according to the requirements (i.e., half-section or partial section) to show important assembly details.
3. Dotted lines: Dotted lines should be omitted from assembly drawings when a proper section is illustrated. However, if, in a drawing, a part of an object is illustrated as a half-section, the remaining portion of the drawing will have dotted lines to represent details that would normally be hidden in the view.
4. Dimensions: The overall dimensions and centre-to-centre distances, showing the relationship of parts to the machine as a whole, are sometimes shown. Detailed dimensions are given on working assembly drawings when detailed drawings are not prepared.
5. Bill of materials: Each part of the machine is identified on assembly drawings by the leader line and number, which are used in detailed drawings and in the bill of materials. The height of the number may be approximately 5 mm and it may be in 9 mm diameter. Leader lines are drawn radially touching the respective parts. The bill of materials also shows the following:
   (a) Number of parts
   (b) Material of parts required for one unit
   (c) Standard norm for standard components
   (d) Scale
   (e) Method of projection
   (f) Shop processes
   (g) Name of the company
   (h) Designed by, drawing by, and checked by
   (i) Any special remark

   A table containing the above information must be prepared.

Sequence of preparing an assembly drawing

1. Study functional requirements of each component and their inter-relationships. Learn the actual working of a machine.

2. Carefully study the views of each component in the detailed drawing and decide the relating location of each part for the proper functioning of the machine.
3. Decide the mating dimensions between two components that are required to be assembled.
4. Prepare free-hand sketch of the main view or an important view (generally front elevation). Add additional views, if necessary.
5. Select a suitable scale for the entire assembly drawing.
6. Lay out the views of the assembly drawing so that it becomes easier to understand.
7. Prepare the bill of materials.
8. Label each component by the leader line and number it.
9. Show overall dimensions.
10. Draw the section lines according to convention.
11. Use proper grade of pencil to make the drawing fair. Maintain uniform thickness of lines.
12. Show required fits and tolerance between the two mating components.

Assembly Drawing 1

Screwjack

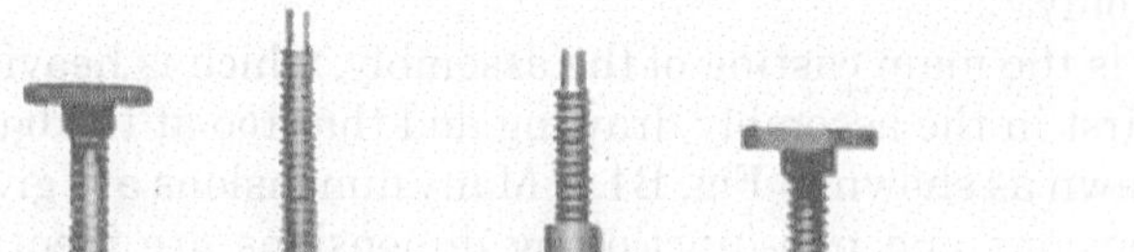

This section contains information about screwjack

After reading the section you will be able to

- draw the assembled views of a screwjack
- understand its functions and uses

Screwjack

Question

Draw the following assembled views of a screwjack to a suitable scale.

1. Sectional front view
2. Side view (Looking from out)
3. Top view (Looking from out)

Also include a bill of material.

Given below is a description of the screwjack and how to prepare an assembly drawing of this simple machine. In practical life, assemblies are known by some unique name from which we get an indication about the use and the set-up of the assembly. In this case, screwjack is the name of the assembly. This name gives us information about the main parts of the assembly. One of the main parts is the jack and the other is the screw. First of all, try to identify these two parts from the given part drawing. The drawing or question related to the assembly will either be an isometric projection or an orthographic projection. The given question is solved in an orthographic projection only.

Part 1 of the drawing is the main casting of the assembly, which is heavier than other parts. This part is dealt with first in the assembly drawing and that too at the bottom. The elevation of the main casting is drawn as shown in Fig. B1.1. Many dimensions are given to this part from which functioning dimensions and non-functioning dimensions are identified. Study of only functioning dimensions is necessary from the point of view of assembly drawings. Non-functional dimensions give information about the shope and profile of the parts. In the main body of the screwjack part 1, all dimensions are important for the manufacturing of the casting but an internal ϕ of 50 mm is more important from assembly point of view. So ϕ 50 mm of part 1 and 50 mm external diamension of part 2, which has internal threading will engage together and part 2 is filled in part 1. Therefore, the focus has to be on the functioning dimensions of part 2 and 3. Internal square threads of part 2 of the size *M*38 × 5 mm pitch will engage with the external square threads of part 3 of the same size. Thus, the assembly of part 3 with 1 will provide the name given to the assembly, that is, screwjack. On studying part 3, main functioning dimension sq. thd. *RH M*38′ is found engage with the main casting 1 and for 22′ × 10 mm, which has another part, whose female dimension is 22 mm, that has to be found. This size is available in part 4, that is, a cup, with ϕ 38′ ϕ 22′ × 10 mm depth as functioning dimensions. So female ϕ 22 mm will engage with male ϕ 22 of part 3 and thus, the cup with the main screw is obtained.

Another main functioning dimension of the main screw is the internal threading *TAP M*12. This will engage with the external threading *M*12 of set screw in part 6. But before tightening the set screw in part 6 to part 3, the main screw through part 4, that is, the cup, will have to be accommodated with part 5. Thus, the assembly is now ready to function. But how will it function? For that, tommy bar (part 7) is provided. Tommy, the bar, has male functioning dimension of ϕ 12 mm, which will engage with ϕ 12 mm female dimension of the hole drilled on the head of the main screw. With the leverage of this tommy bar and applying force at one end of tommy bar, the main screw can be lowered down or lifted up within permissible limits. Part 1 is the main body and is stationary, hence, it is made up of *CI.* Part 2, nut, will engage with the main body. As it is in contact with the main screw, friction will take place between part 2 and 3. But as part 3 is main and heavy part of the assembly and can be easily machined on lathe machine, it is made up of *MS* whereas nut is made up non-ferrous metal (gun metal). If one wants to replace or recorrect the functioning of the assembly, replacement of the nut is more economical and rapid than the replacement of the main screw. The cup is made up of cast steel, the density of which is more as compared to cast iron. As direct load comes on the cup, cast steel is selected. The remaining parts are made up of *MS*.

Prepare a bill of materials (table) as shown in the answer for every assembly drawing. Once the assembly is complete, check that it is functioning properly.

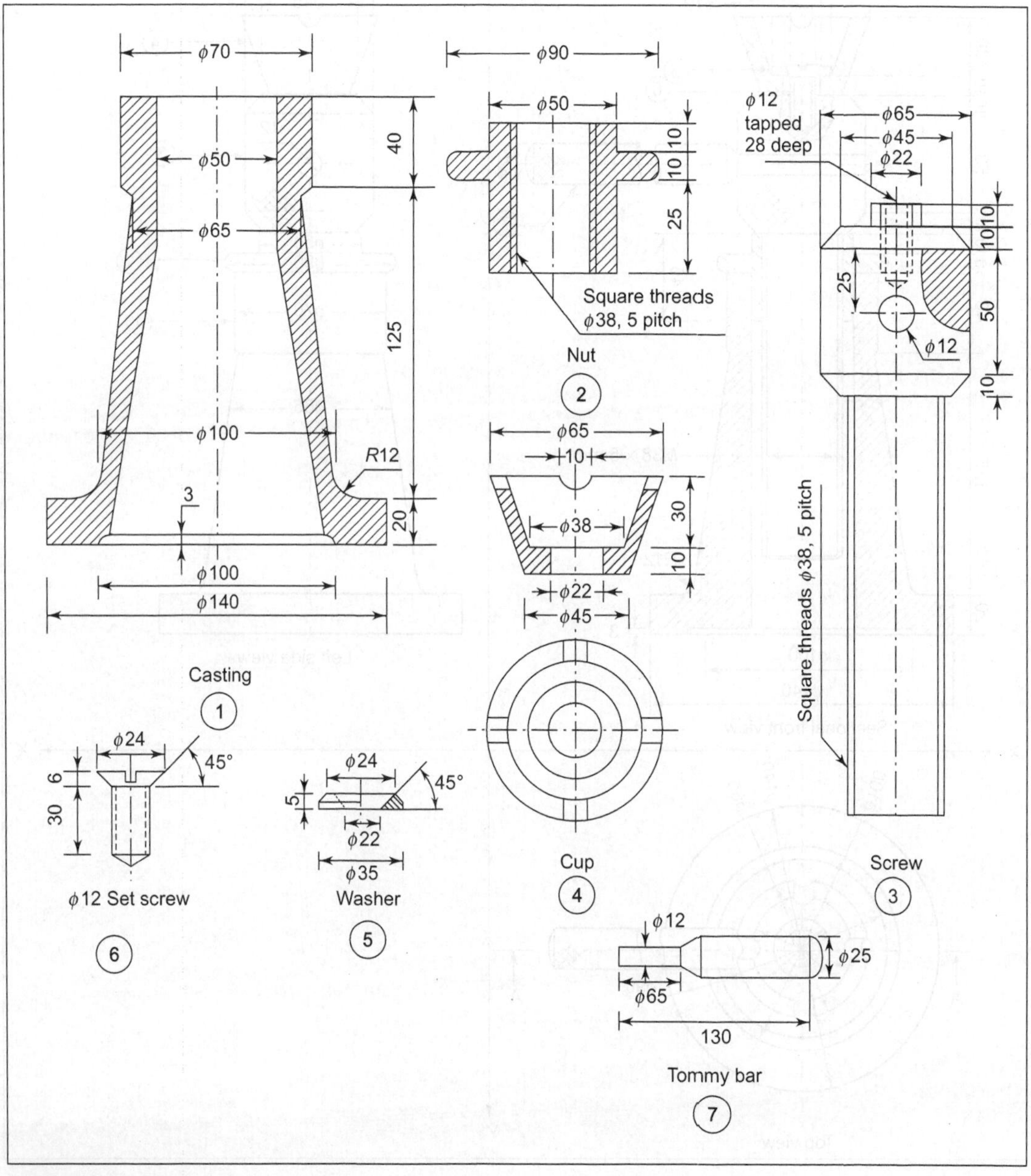

Fig. B1.1 Details of a screwjack

Note: Dimensions which are not given directiy can be assumed and taken in propovtionote with other dimensions of the part as in given dimensions are non-functional dimensions.

Answer

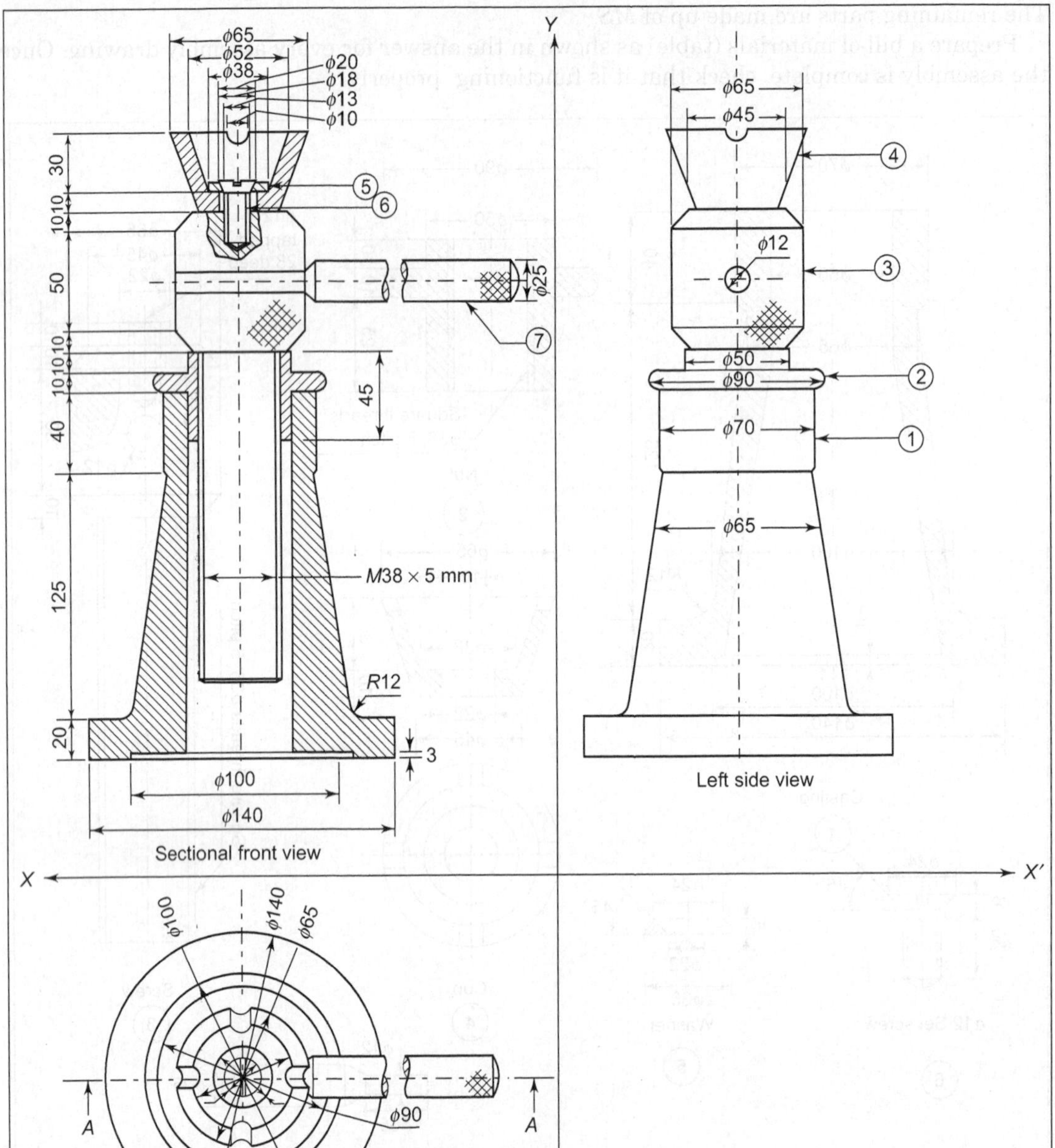

Fig. B1.2 Screwjack

The enlarged view of Fig. B1.2 is shown as under:

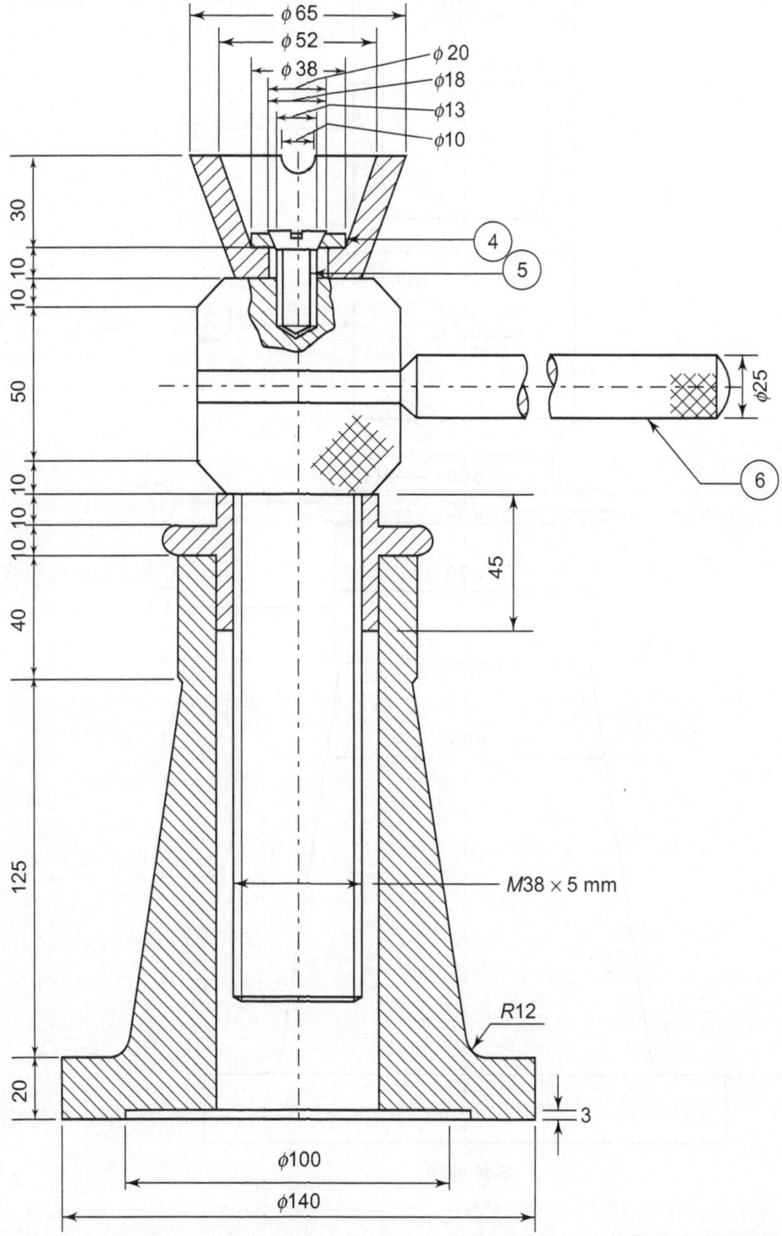

Sectional front view

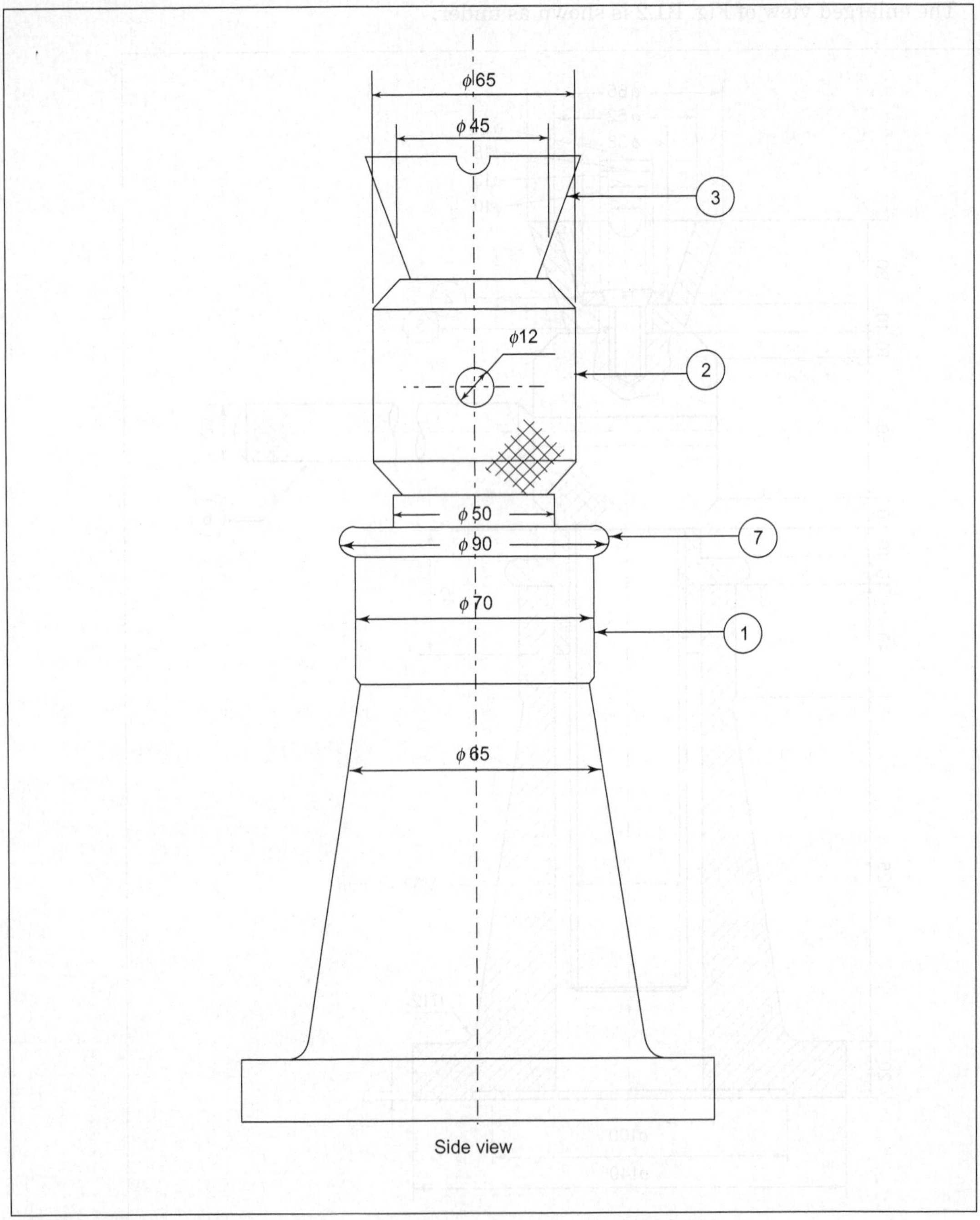

Fig. B1.2b Screwjack

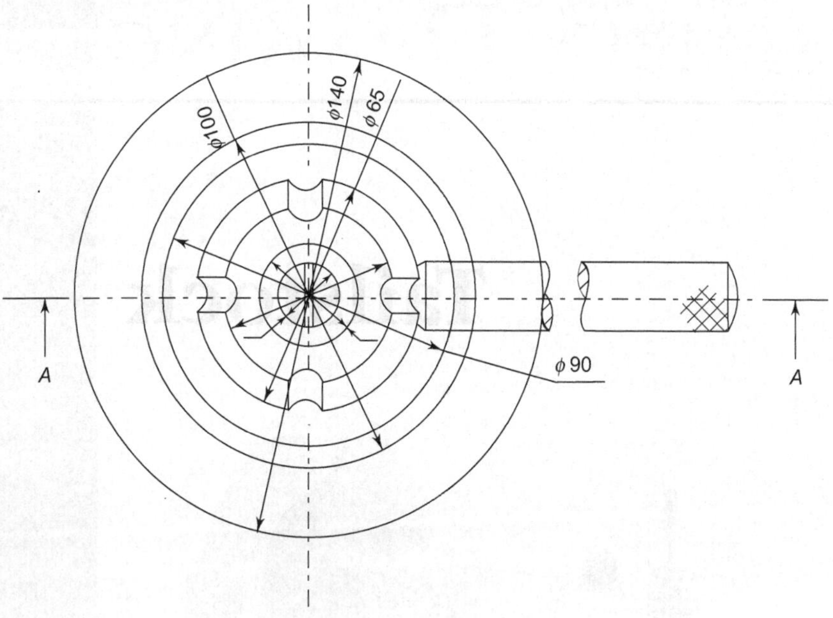

Top view

Bill of Material

Sl. No.	Part	Quantity	Material
1	Casting	1	CI
2	Nut	1	Gun material
3	Screw	1	MS
4	Cup	1	Cast steel
5	Washer	1	M S
6	Set screw	1	M S
7	Tommy bar	1	M S

As sectional front view, side view and top view are interrelated so while placing independent views, ensure that scale applied for enlargement of front view must be applicable to side view and top view.

ASSEMBLY DRAWING 2

Tailstock

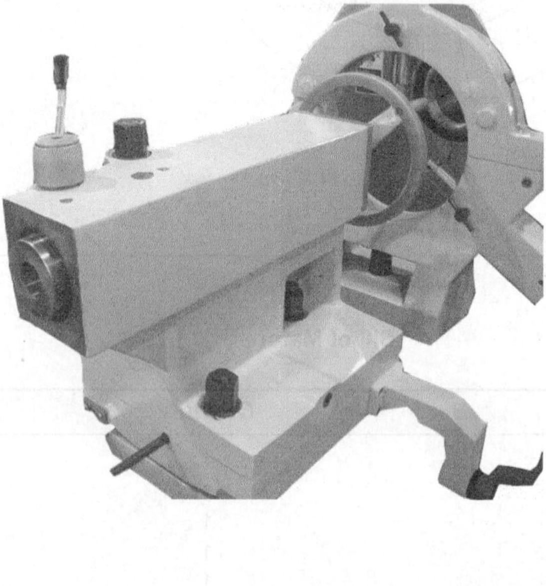

This section contains information about tailstock

After reading the section you will be able to

- draw assembled views of the tailstock
- understand its functions and uses

Tailstock

Question

Draw the following assembled views of tailstock.

1. Sectional elevation
2. End view

Also include a bill of material.

Details of tailstock are shown in Fig. B2.1.

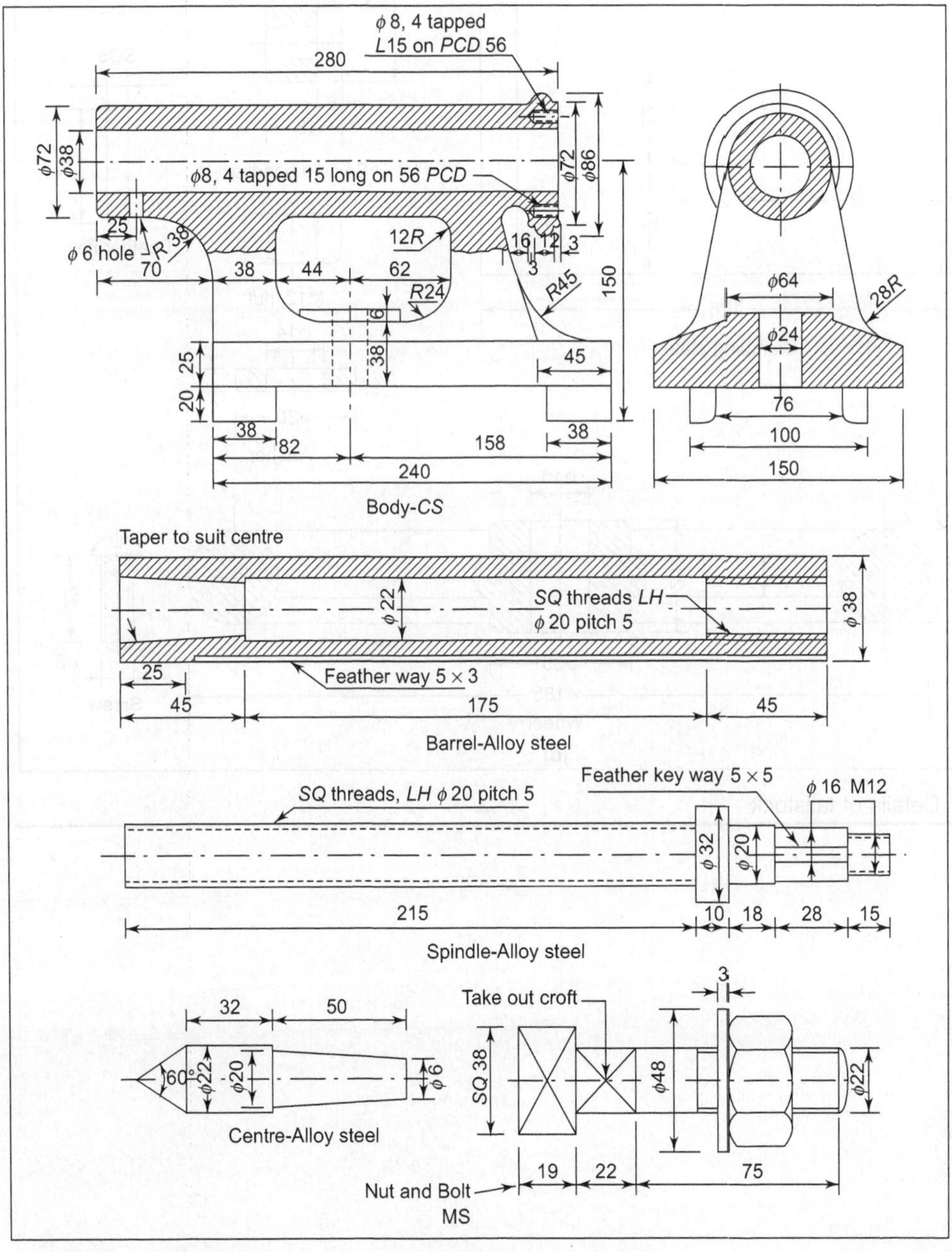

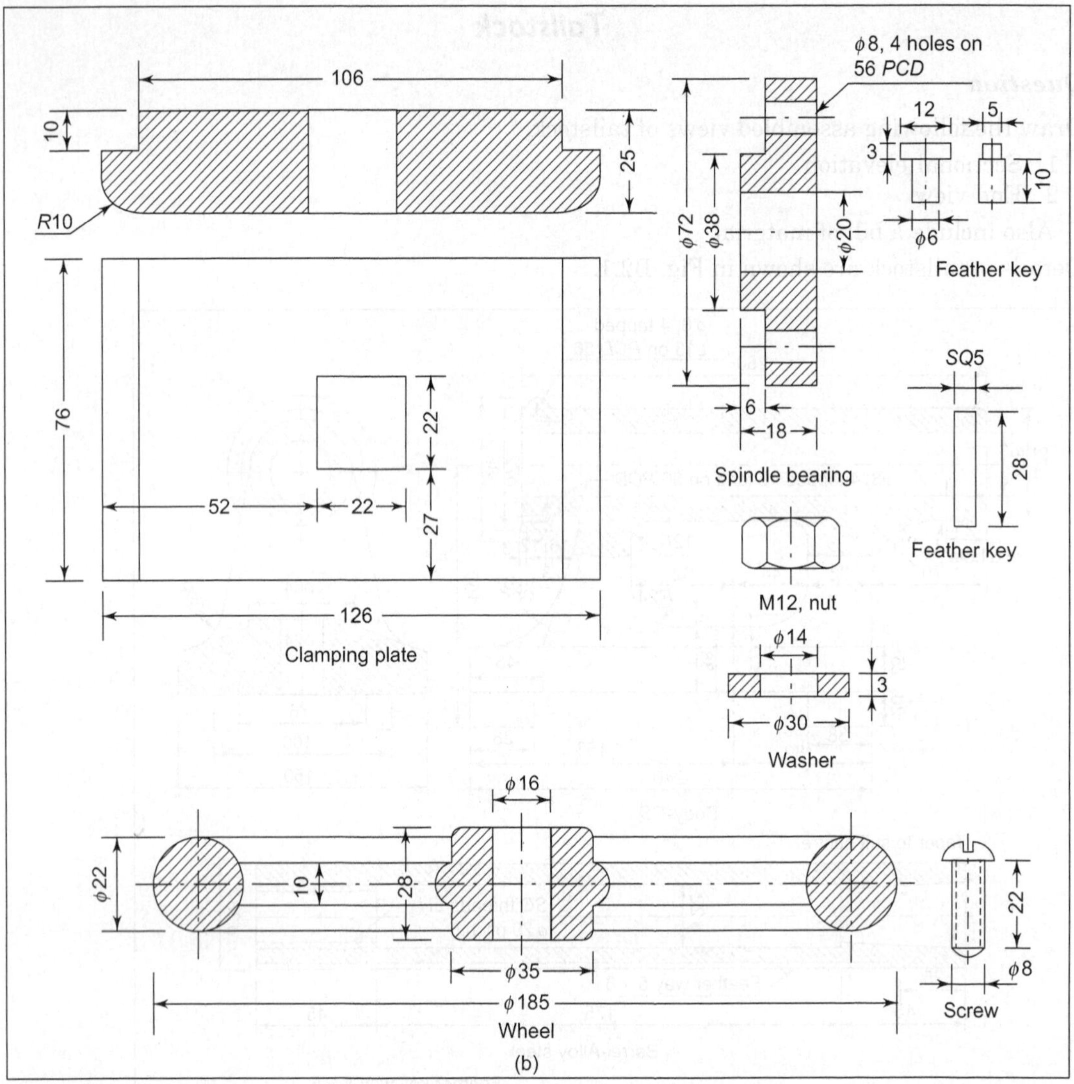

Fig. B2.1 Details of tailstock

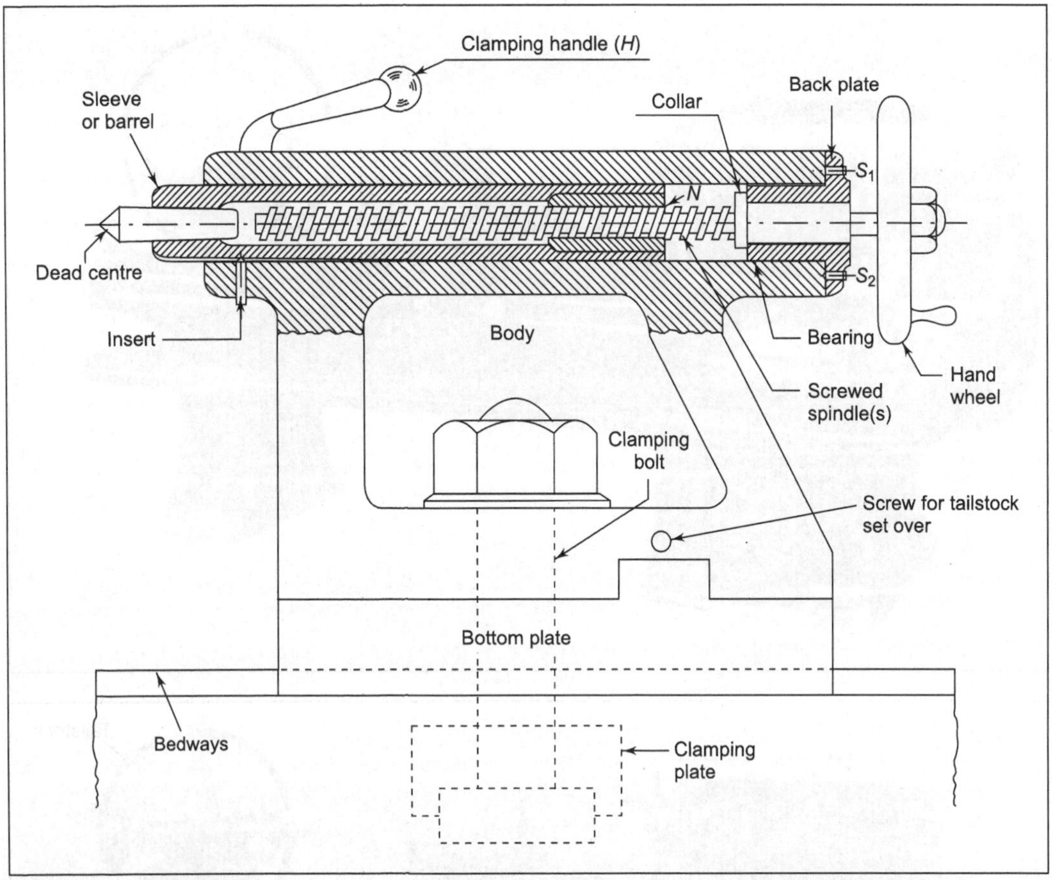

Fig. B2.2 Tailstock

Introduction

Tailstock is also called the loose head stock or puppet head. It is mounted on the bed of the lathe such that it is capable of sliding along the latter, maintaining its alignment with the head stock. On common types of medium and small lathes, it is moved along the bed by hand, whereas in heavier types of lathes, it is moved by means of a hand wheel through a pinion that meshes with the rack provided on the front of the lathe bed. The main function of a tailstock is to provide bearing and support to the job that is being worked between the centres. To enable this, the tailstock is made to possess a number of parts that collectively help in its successful functioning.

Uses

Tailstock has two main uses, namely,

1. It supports the other end of the work when the work/job is being machined between the centres.
2. It holds the tool for performing operations such as drilling, reaming, tapping, and so on. It is used for supporting and feeding drills, reamers, and so on when it is necessary to use for drilling work held in the chuck.

(b)

Fig. B2.3 Centre lathe machine with tailstock

Structure and Function

To accommodate different lengths of work pieces, the body of tailstock can be adjusted by sliding it to the desired position, where it can be clamped by bolts and plates. The upper casting of the body can be moved towards or away from the lathe chuck by way of adjusting the screws to the tailstock. Sometimes for taper turning operation realignment of the tailstock is required.

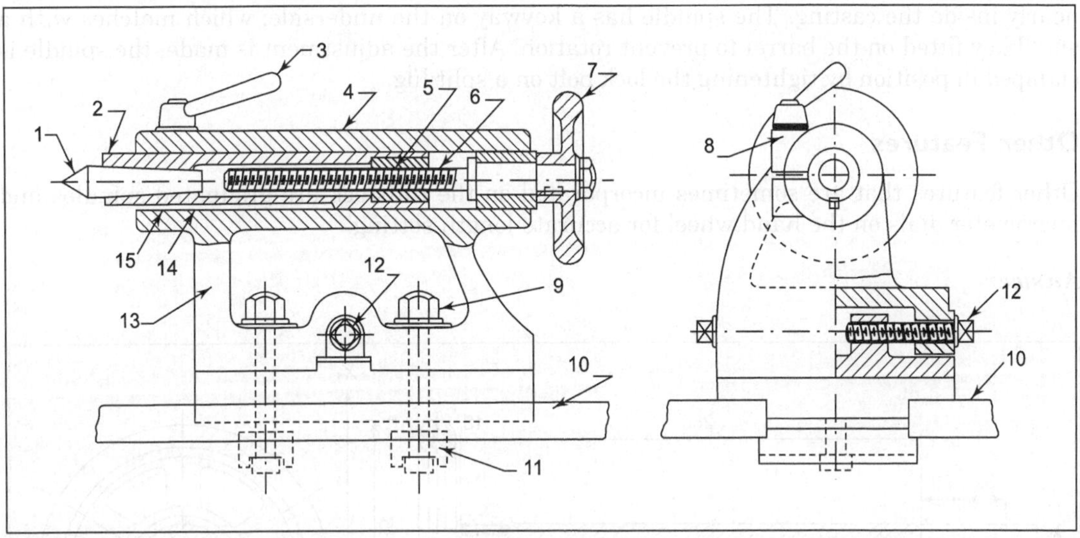

Fig. B2.4 Assembly of tailstock

The different parts of a tailstock are illustrated in Fig. B2.4.

1. Dead centre
2. Spindle
3. Spindle clamp
4. Barret
5. Bush
6. Square threaded screw
7. Hand wheel
8. Split lug
9. Tailstock clamping bolt
10. Lathe bed ways
11. Clamping plate
12. Setover screws
13. Body
14. Keyway
15. Key

The body of the tailstock is bored to act as the barrel that carries the tailstock spindle, which moves in and out of the barrel by means of a screw when tailstock hand wheel is turned. The front of the spindle has a taper hole into which a dead centre or other tools fit. The screw thread is left handed so that the clockwise rotation of the hand wheel causes the spindle to advance while the anticlockwise rotation causes the spindle to be drawn inwards and ultimately the end of the screw strikes the back of the dead centre or any tool that is fitted into the hole. To remove tools from the spindle, it is necessary to back upon the hand wheel until the spindle end is nearly inside the casting. The spindle has a keyway on the underside, which matches with a small key fitted on the barrel to prevent rotation. After the adjustment is made, the spindle is clamped in position by tightening the lock bolt on a split lug.

Other Features

Other features that are sometimes incorporated on the tailstock are graduated spindles and micrometer dials on the hand wheel for accurate length setting.

Answer

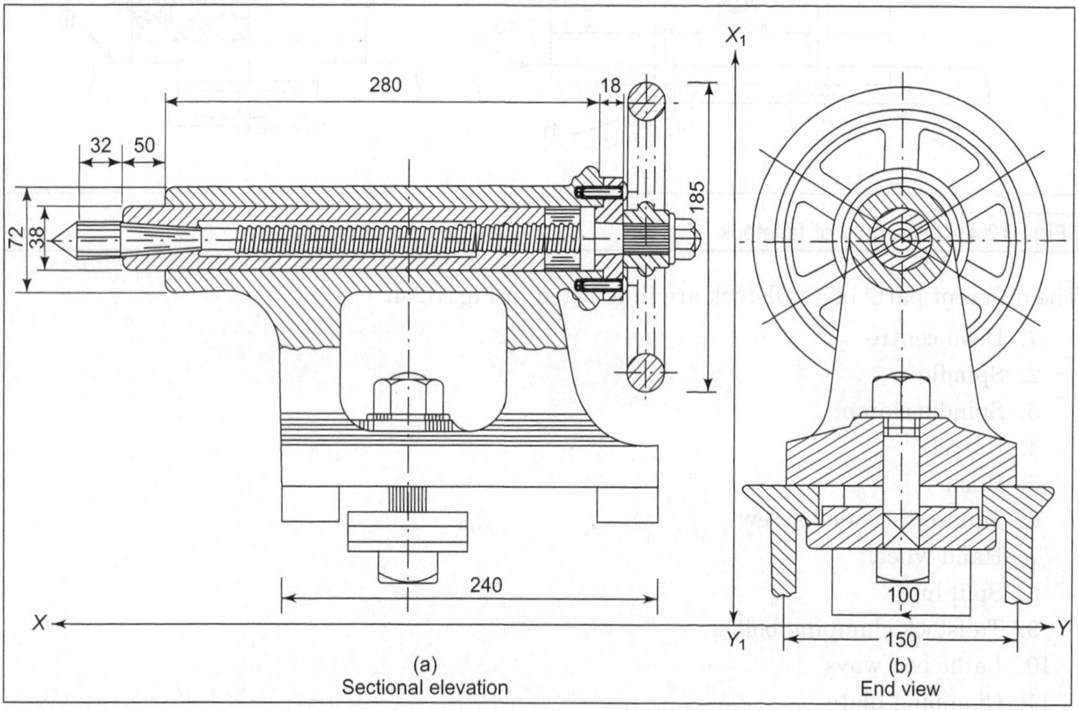

Fig. B2.5 Assembly of tailstock

The enlarged view of Fig. B2.5 is shown as under:

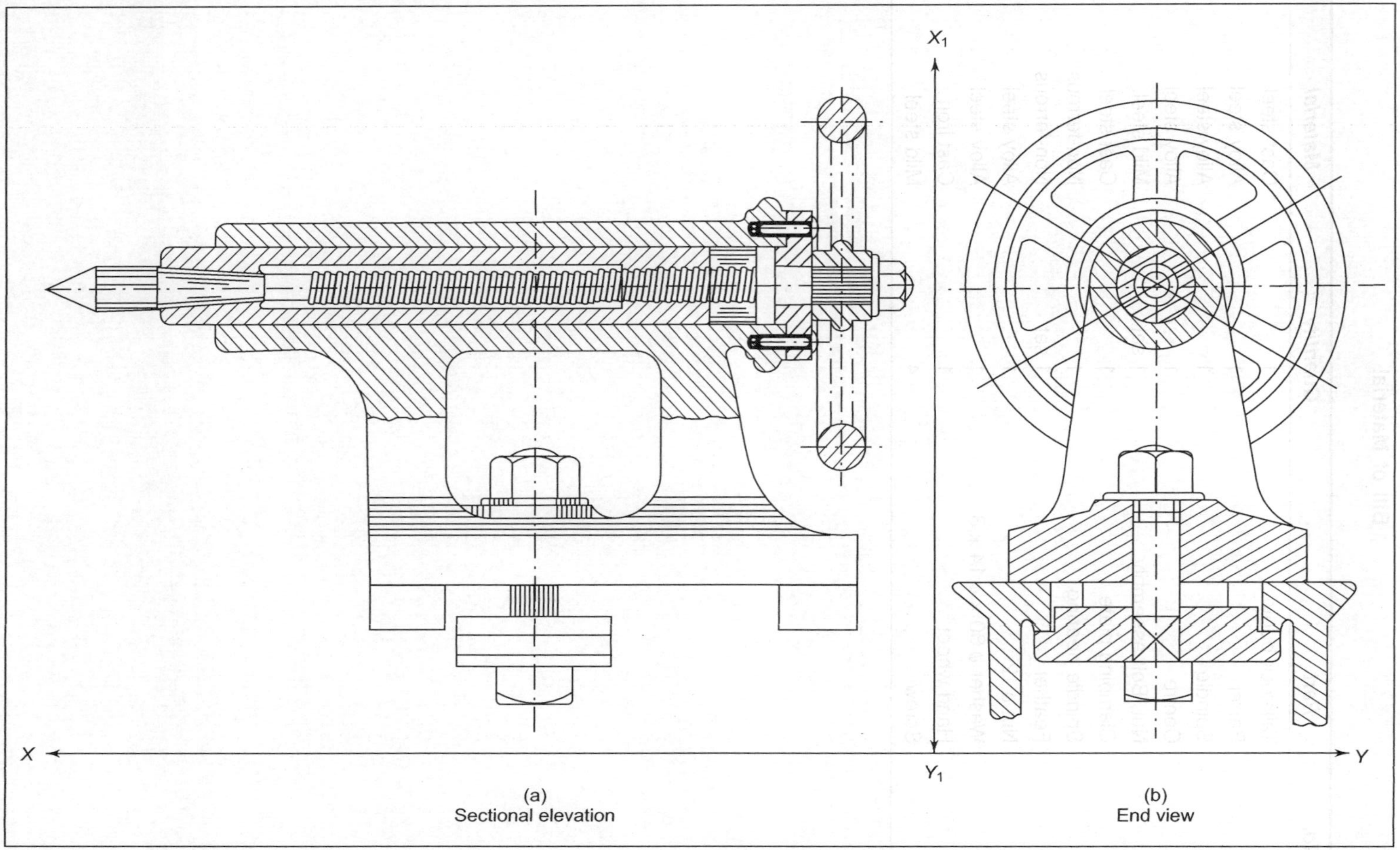

Fig. B2.5 Assembly of tailstock

Bill of Material

Sl. No.	Part	Quantity	Material
1	Tailstock body	1	Cast steel
2	Barrel	1	Alloy steel
3	Spindle	1	Alloy steel
4	Centre	1	Alloy steel
5	Nut, Bolt assembly	1 set	Mild steel
6	Clamping plate	1	Cast steel
7	Spindle bearing	1	Non-ferrous
8	Feather	1 set	Non-ferrous
9	Nut ϕ 12	1	Alloy steel
10	Washer ϕ 30 × ϕ 14 × 3	1	Alloy steel
11	Hand wheel	1	Cast iron
12	Screw	4	Mild steel

ASSEMBLY DRAWING 3

Crane Hook

This section contains information about the crane hook

After reading the section you will be able to

- identify the different types of hooks
- understand the steps in making assembly drawings

Crane Hook

Question

Draw the following assembled views of crane hook.

1. Sectional front view through the axis of the swivel block showing all assembled parts
2. Side view
3. Sectional top view through the axis of the swivel block.

Also include a bill of material.

Details of a crane hook are shown in Fig. B3.1.

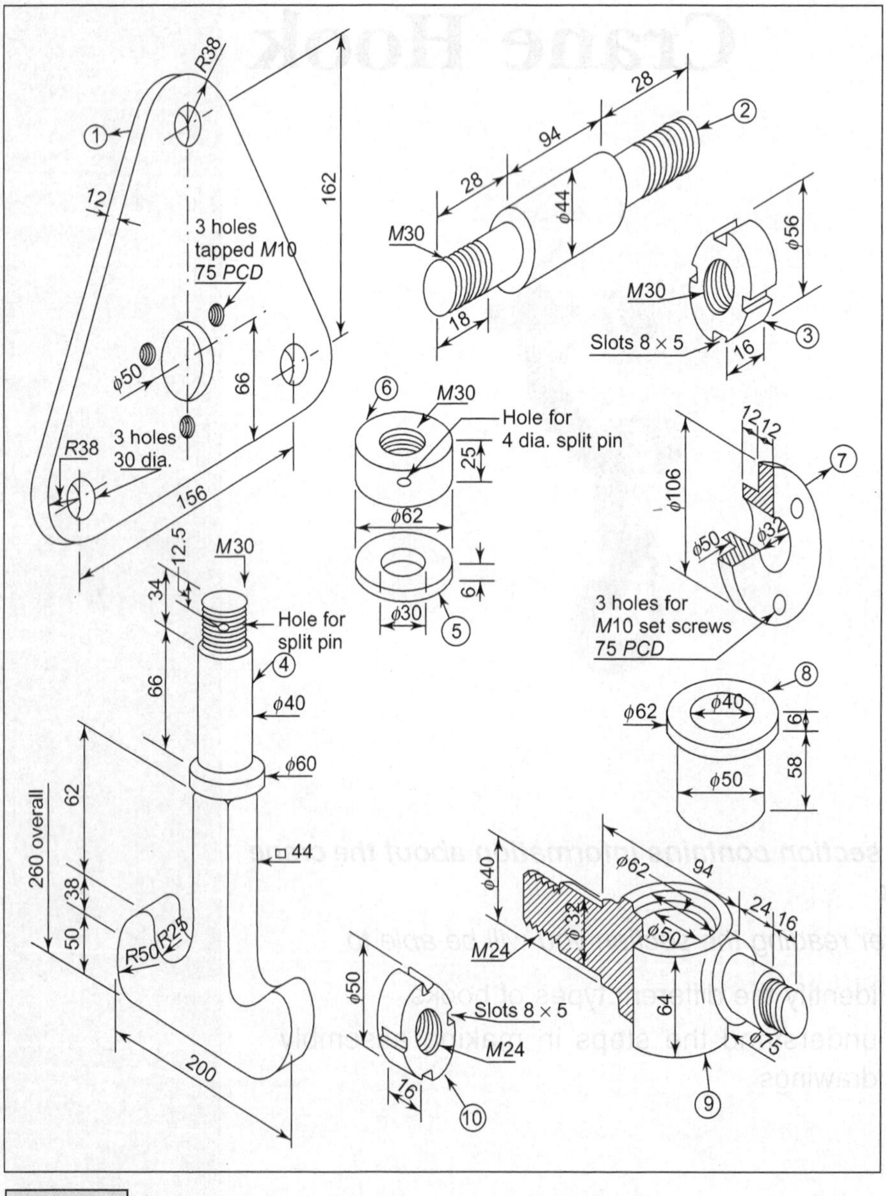

Fig. B3.1 Crane hook

Introduction

Hooks are mainly used for hanging loads. There are different varities of hooks in the engineering field, some of which are given below. Hooks are used alongwith cranes for lifting heavy loads. Cranes are used either to shift heavy loads from one place to other or for lifting heavy loads of machinery for reparing purposes. The cross-section of the hook at the centre is elliptical to provide more strength against the load to be lifted. The crane alongwith a hook can be called as lifting machine and is used for procuring more output, in terms of lifting heavy loads, by applying minimum effort.

Hooks are generally used to lift heavy loads in industrial applications.

Types of Hooks

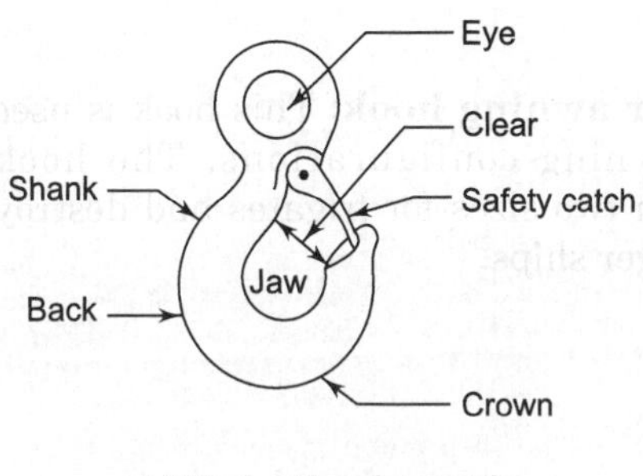

(a) Hook with safety catch

Hook with safety catch: This type of hook is commonly used in the British Royal Navy. It has a safety catch in the form of a spring operated tongue, which dispenses the requirement for mousing. The hook is designed for use in conjunction with an appropriate size shackle, although in certain applications it is acceptable to spring a thimble through the eye and splice the cordage or wire round the thimble. Cordage must not be spliced directly to the eye.

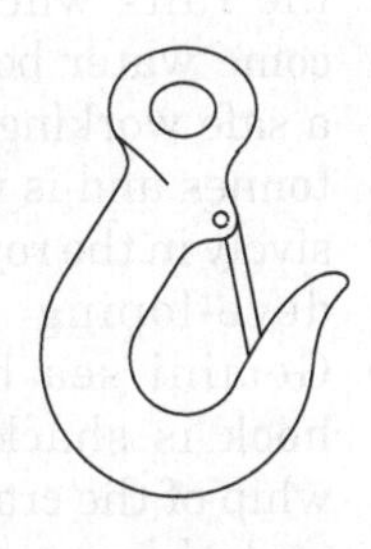
(b) Spring hooks

Spring hook: The spring hook is used for picking-up ropes and a boat's bad weather pendants. This hook also has a safety catch to prevent accidental unhooking. It is used in conjunction with a pin and pallet shackle and has a machined surface on either side of the shackle bolt hole to facilitate a snug fit between the shackle and the hook.

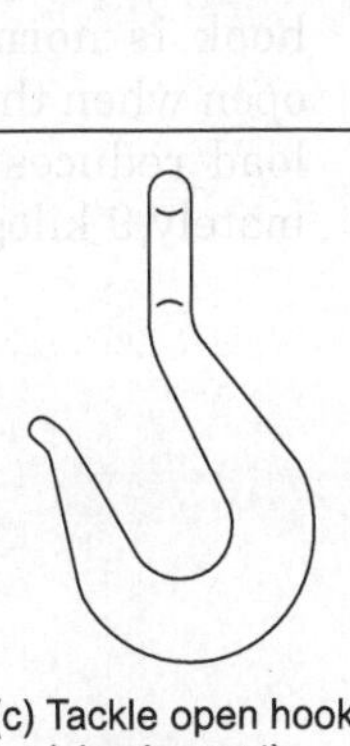
(c) Tackle open hook (obsolescent)

Tackle hook: It is incorporated into the eye of the metal and the wooden tackle block's and is gradually being replaced by the hook's hoist, with a safety catch, which can be shackled to a SRBF (spring hook for picking up ropes and boats foul weather) block. The tackle hook is no longer available at naval stores.

(d) Swivel spring hook

Swivel-spring hook: These hooks are fitted with a swivel that prevents any twist in the whip being transmitted to the hook. They are specified for certain mine sweeping and storing tasks.

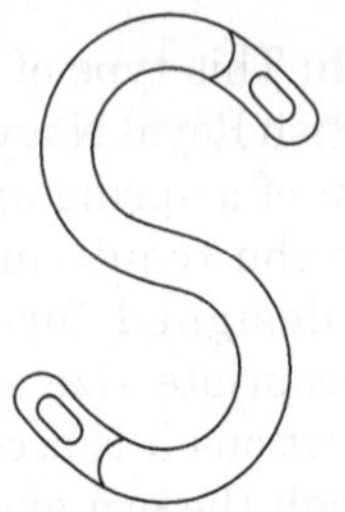

(e) 'S' hook

'S' Hook or awning hook: This hook is used in certain awning configurations. The hook is available in two sizes for frigates and destroyers and for larger ships.

RFD automatic release hook: This hook is a development on the davit release hook, designed originally for merchant fleets to provide a method of automatically releasing davit launched life rafts when they become water borne. It has a safe working load of 2.5 tonnes and is used extensively in the royal navy for deve-loping RIB and Gemini sea boats. The hook is shackled to the whip of the crane or davit and the master link of boat's lifting sling. The hook is normally set to open when the suspended load reduces to approximately 9 kilograms.

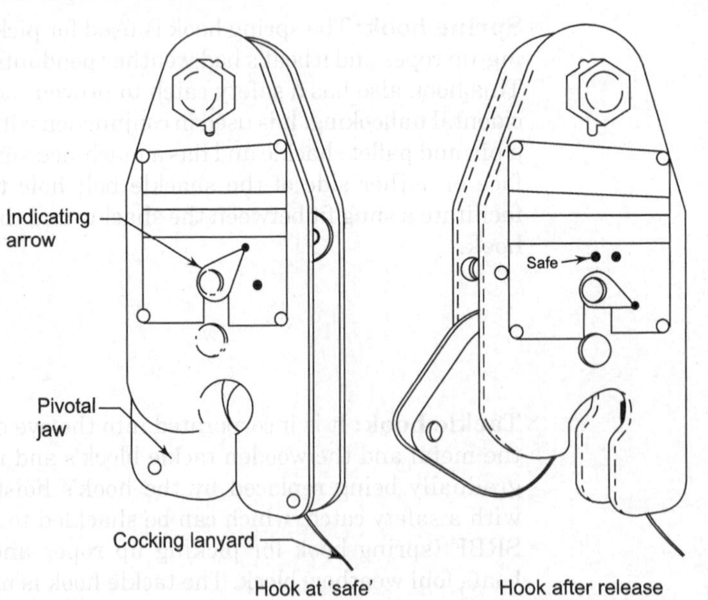

(f) RFD automatic hook release

Fig. B3.2 Types of hooks

Steps for Assembly Drawing

Little assistance is required when the crane hook is assembled. It is a fairly common piece of equipment on ships as well as on industrial shop floors.

The side plates are held together by three 44 mm diameter studs, one of which is shown in Fig. B3.1. The swivel block fits into the two bushes, which are secured to the side plates by three *M*10 set screws in their flanges. The hook is free to revolve in a 40 mm diameter bush fitted in the swivel block and a thrust washer is fitted between the *M*30 nut on the screwed end of the hook and the top of the bush.

Leaving sufficient room for the sectional elevation, commence this exercise by drawing in the triangular shaped side plate in the end view and mark off the centres for the three securing studs, together with the centre of the swivel block. The circular flange of the swivel block bush, three *M*10 set screws securing this bush, and four slotted nuts could also be drawn.

For the depth of side plate in the sectional elevation, project from the end view and draw these 12 mm thick and 94 mm apart inside. Draw in the vertical centre line on this view and again project across for the centres of the 44 mm diameter stud (only the top stud will be seen). The horizontal centre line of the swivel block is also obtained by projecting from the end elevation. Build up the swivel block around the centre line, showing part of it in section, in order to show the vertical bush. The hook can now be drawn in, noting that there is a clearance of a 2 mm between the 60 mm diameter collar and the bottom of the swivel block. Finish off this view by adding the slotted nuts in section, the thrust washer, and *M*10 sets screw in each bush flange.

The plan view taken through the horizontal centre line of the swivel block is now added and the widths of the items in this view are obtained by projecting from the elevation immediately above it. The portion of the side plate seen in section in this view is found from the end view, by scaling off the distance from the vertical centre line to the point where the horizontal centre line through the swivel block intersects the sides of the triangular side plate. The bottom two 44 mm diameter studs will also be seen in this view.

Answer

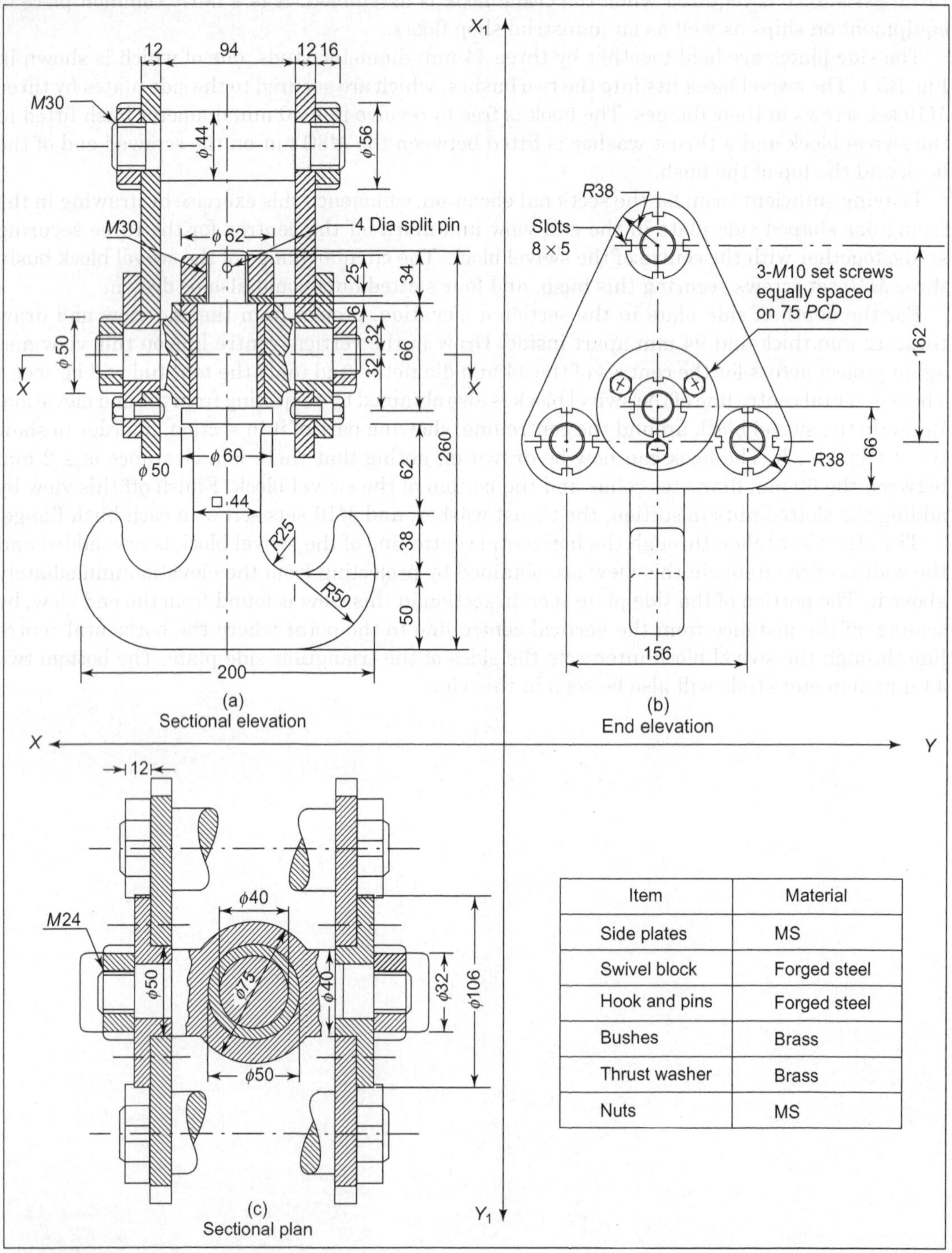

Item	Material
Side plates	MS
Swivel block	Forged steel
Hook and pins	Forged steel
Bushes	Brass
Thrust washer	Brass
Nuts	MS

Fig. B3.3 Assembly of Crane Hook

The enlarged view of Fig. B3.3 is shown as under:

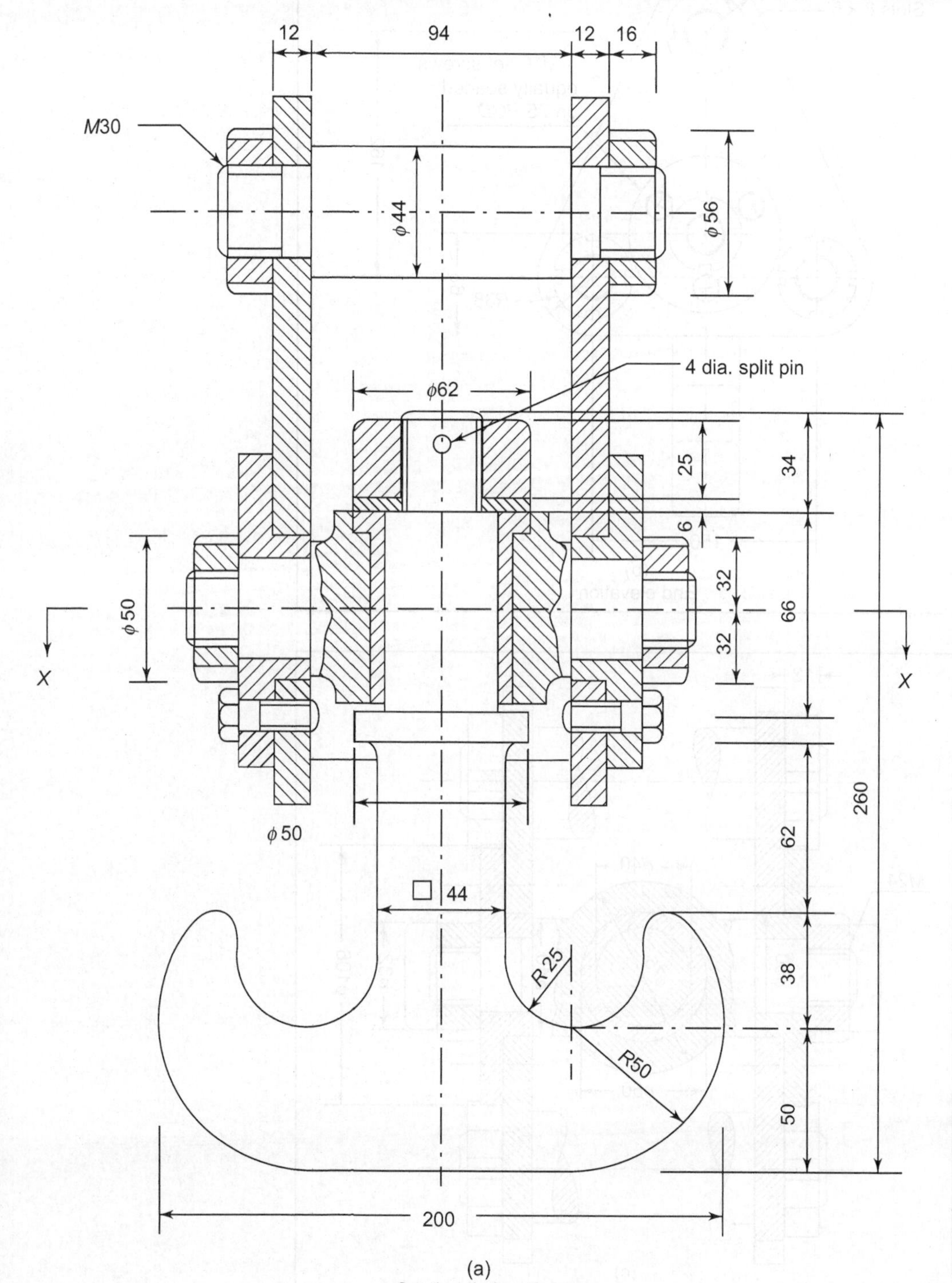

(a)
Sectional elevation

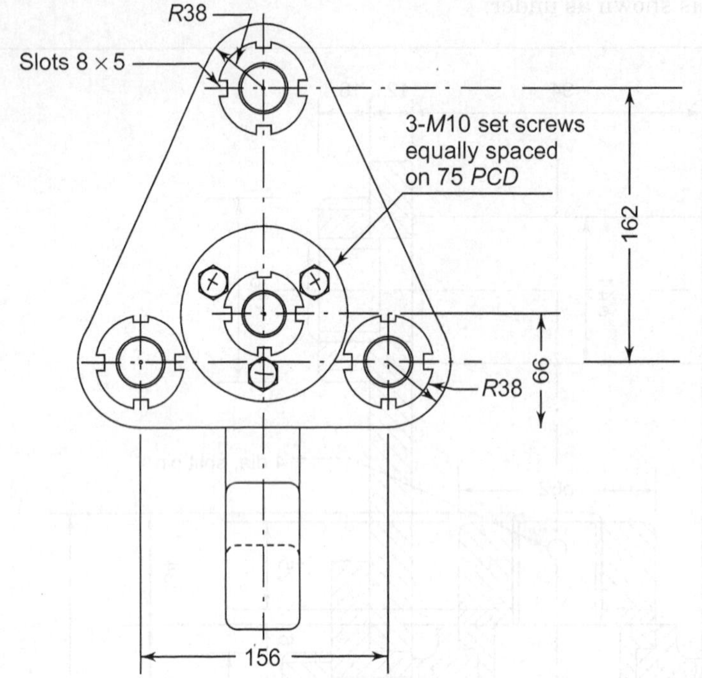

(b)
End elevation

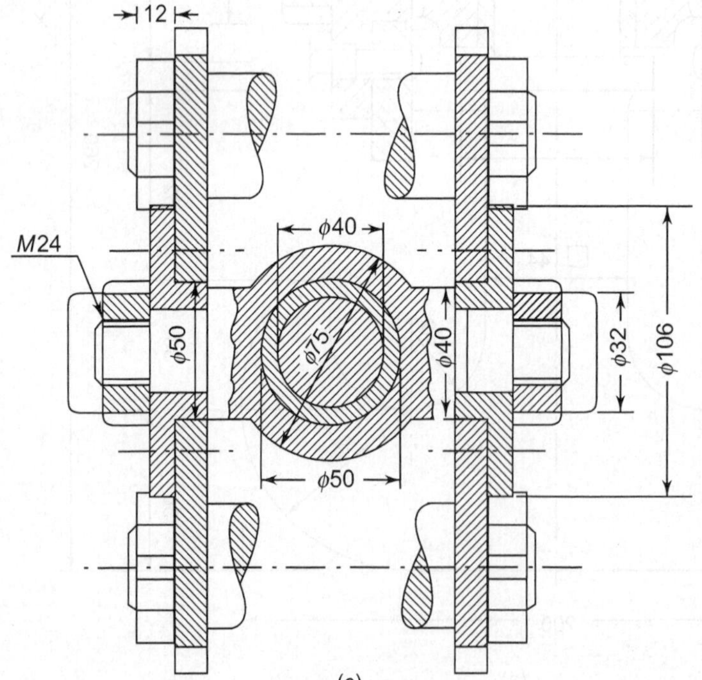

(c)
Sectional plan

Bill of Material

Sl. No.	*Part*	*Quantity*	*Material*
1	Side plates	2	MS
2	Pin	3	Forged steel
3	Slotted nut	6	MS
4	Hook	1	Forged steel
5	Thrust washer	1	Brass
6	Round nut	1	MS
7	Bush	2	Brass
8	Bush	1	Brass
9	Swivel block	1	Forged steel
10	Slotted nut	2	MS

Assembly Drawing 4

Gear Pump

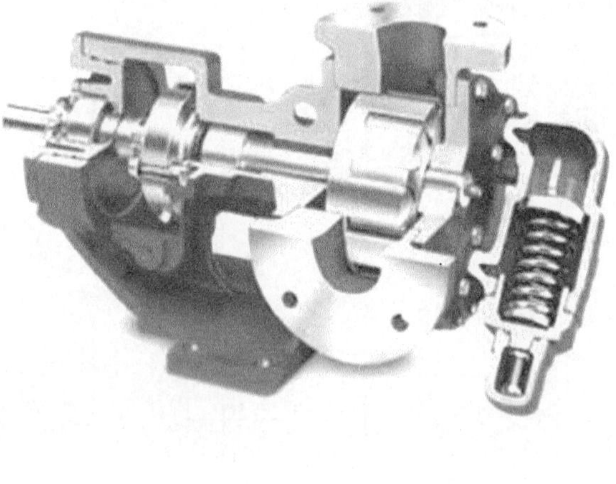

This section contains information on the different types of gear pumps, namely, lobe pump, internal gear pump, and gerotor pump

After reading the section you will be able to

- Understand the principle of gear pumps
- draw their assembled views

Gear Pump

Question

Draw the following assembled views of gear pump

1. Sectional elevation through longitudinal axis of driving shaft showing an assembled pump.
2. End view, one half of outside view looking on gland with coupling removed, the other half, a section through the pump casing, looking at the ends of the pinions.

Also include a bill of material.

The details of gear pumps are shown in Fig. B4.1.

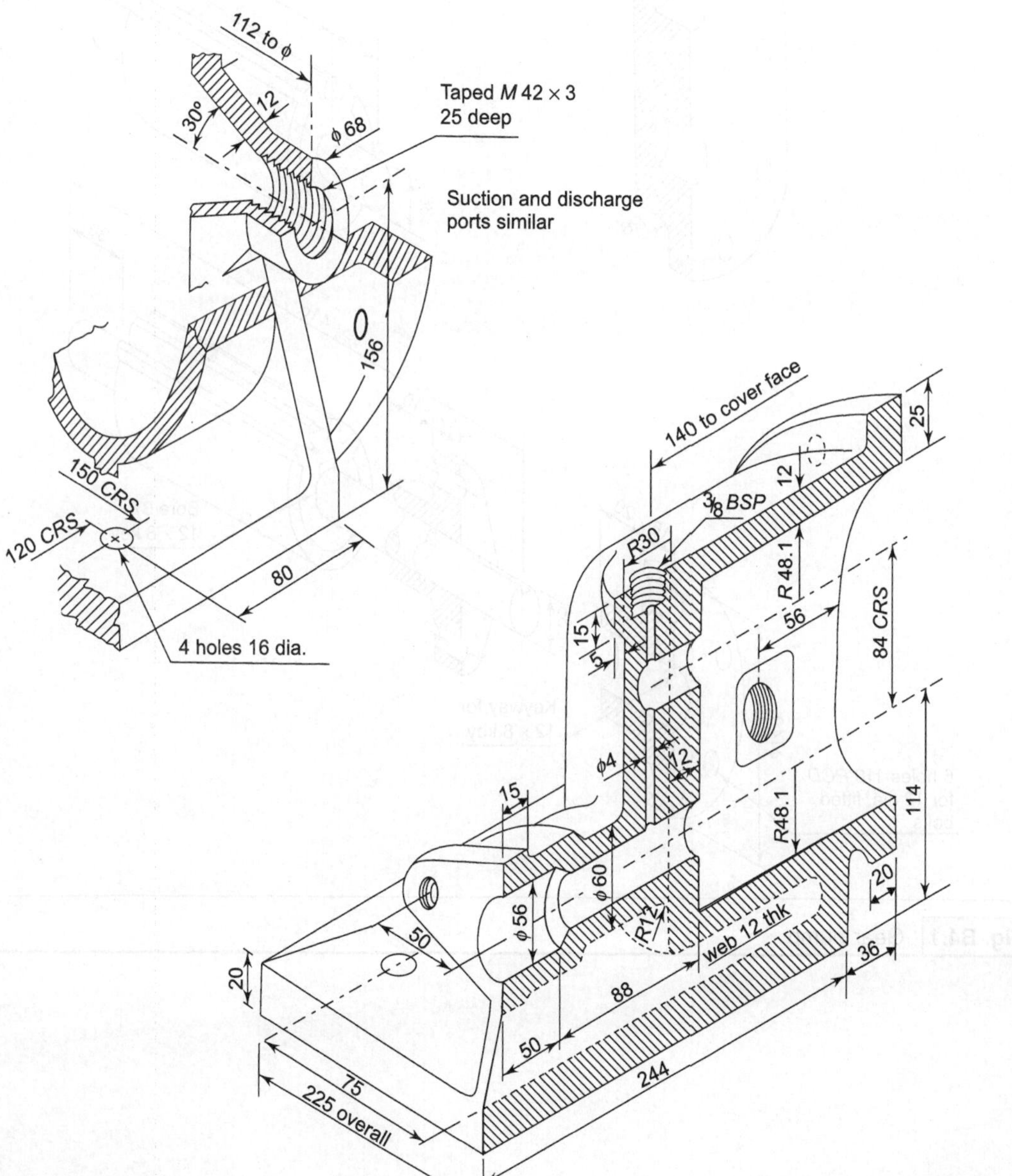

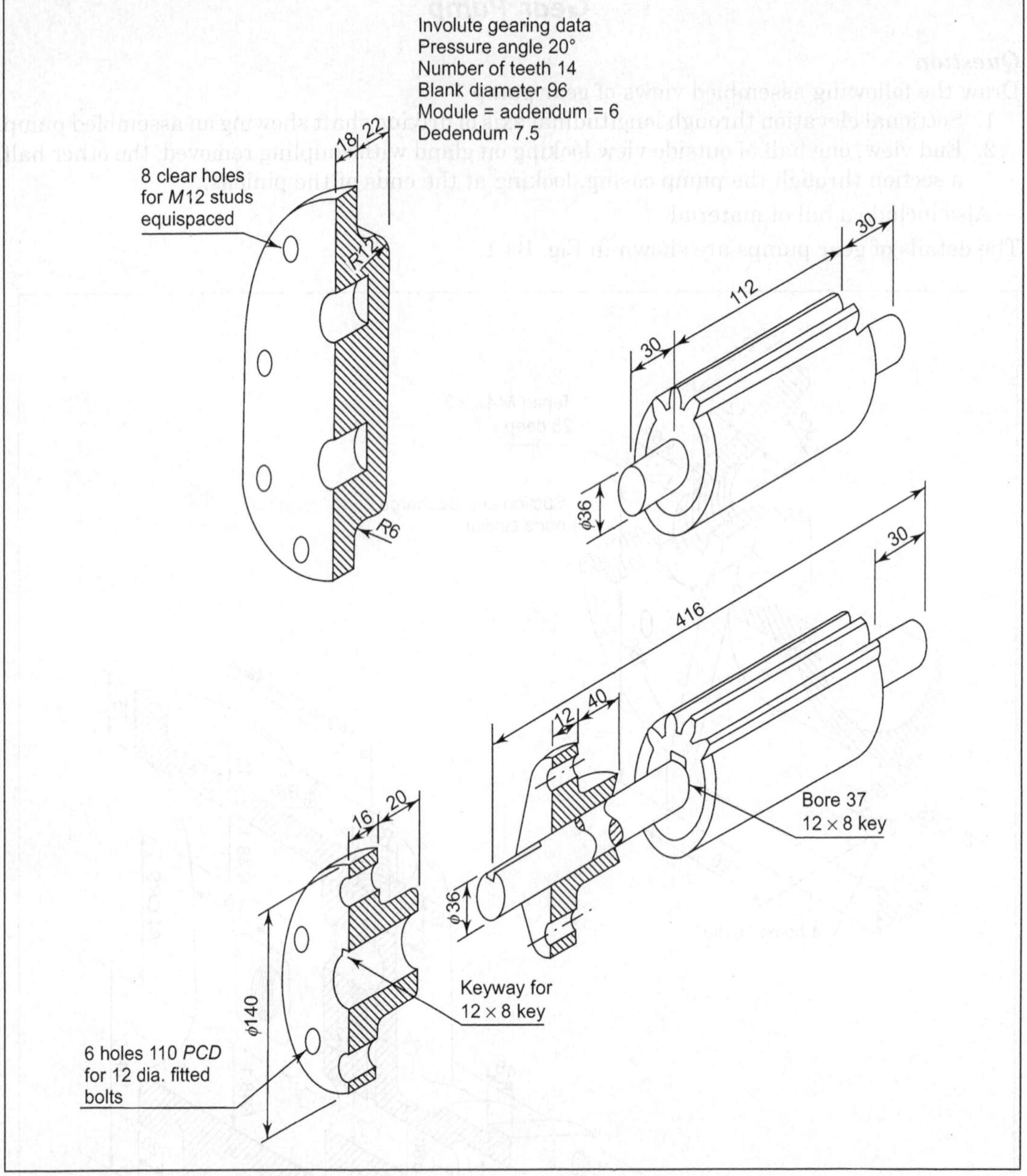

Fig. B4.1 Gear pump

Introduction

Gear pumps generally come under the category of constant displacement pumps, hence they do not require priming. The increase and decrease in the volume of suction and discharge of the pump is possible with the movement and action of gears in mesh. It is important that the direction of rotation of the gears is correct in relation to the suction and discharge ports. Gear pumps are best suited for viscous fluids and due to this reason they are commonly known as lubricating oil pumps.

Pump displacement is determined by:

1. Volume of the fluid between each pair of teeth
2. Speed of rotation

The pump merely delivers a fixed volume of fluid from the inlet port to the outlet port for each rotation. Outlet port pressure is determined solely by the design of the rest of the system.

The performance of any pump is limited by leakage and the ability of the pump to withstand the pressure difference between the inlet and outlet ports. The gear pump obviously requires

1. Closely meshing gears
2. Minimum clearance between the teeth land and housing and also between the gear face and side plates.

Lobe pump Gears when replaced by lobes give rise to lobe pump.

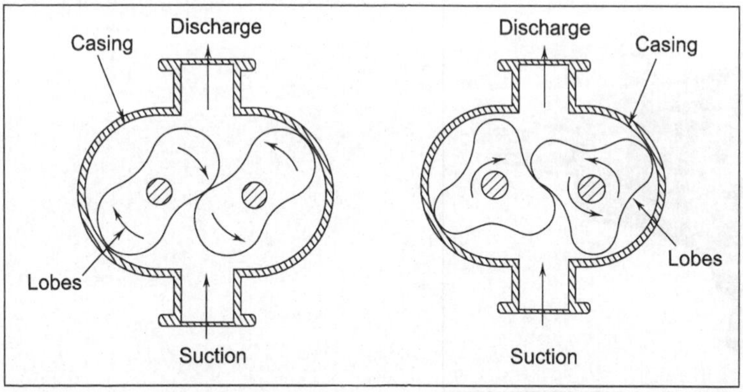

Fig. B4.2 Lobe pumps

Internal gear pump In this type of pump the driving gear, carried on the driving shaft, runs concentrically inside the pump casing. Between the two gears a crescent is formed. This space is occupied by a similar shaped crescent, which is cast integral. The crescent acts as a partition to prevent the liquid on the discharge side from entering the suction side.

Gerotor pump In these pumps the crescent-shaped moulding is dispensed with by using an internal gear with one tooth less than the outer gear wheel.

Internal gear pumps operate at lower capacities and pressures than other types of pumps.

Steps for Assembly Drawing

Commence the sectional elevation by drawing the horizontal centre line of the driving shaft and finally produce the end view. Note that this shaft has 37 mm diameter by way of the pinion, and 36 mm diameter for the remainder of its length. The driving pinion is located 30 mm from the right hand end of the shaft, and the centre line of the driven pinion is drawn in at 84 mm above the centre line of the driver. The pinions in this view and end view are illustrated as per the conventional representation and students are also advised to refer the conventional representation of machine parts with the terms used in involute gearing. Sufficient detail is given in the pictorial view, for the pump body and cover, and gland and coupling to be drawn while the details of the ports are given in a local section so that these can be shown in the end view. Note that there is a clearance of 1 mm between the outside diameter of the gears and pump casing, but this is of course too small to be shown on the drawing as per the dimensions given in the question.

Fig. B4.3 Gear pump

Answer

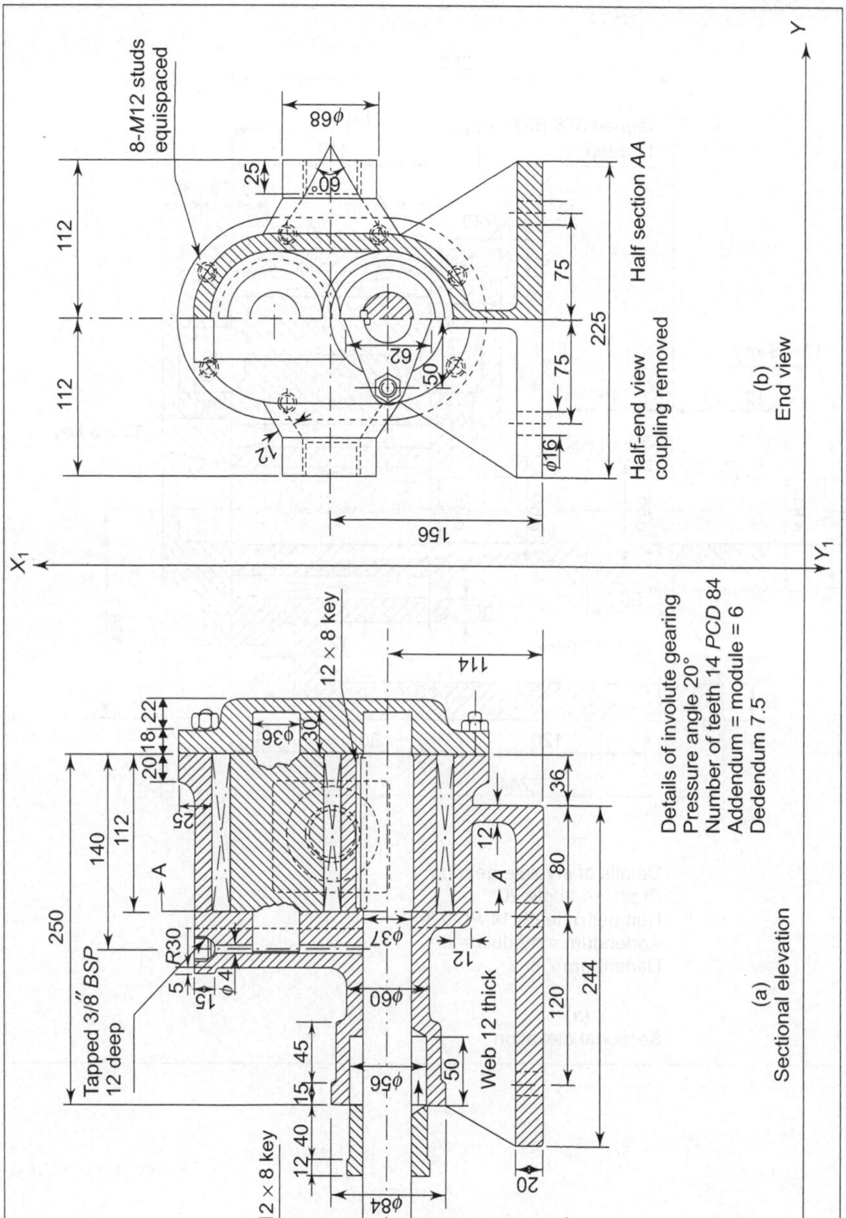

Fig.B4.4 Assembly of gear pump

The enlarged views of the Fig. B4.4 are shown as under:

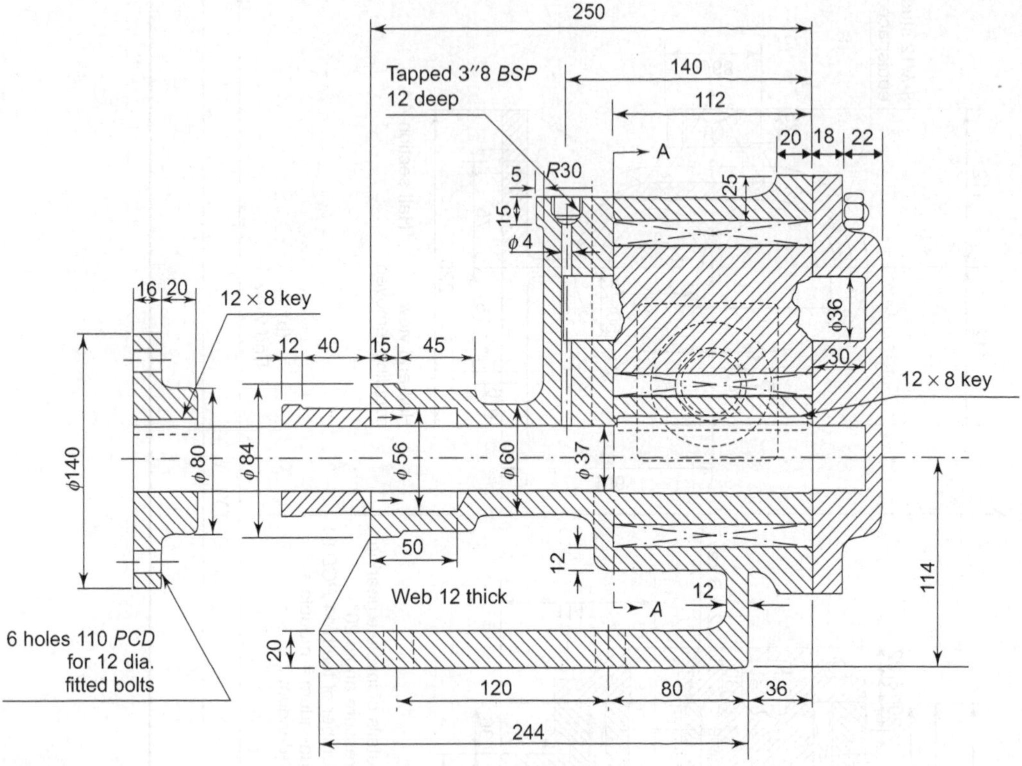

Details of involute gearing
Pressure angle 20°
Number of teeth 14 PCD 84
Addendum = module = 6
Dedendum 7.5

(a)
Sectional elevation

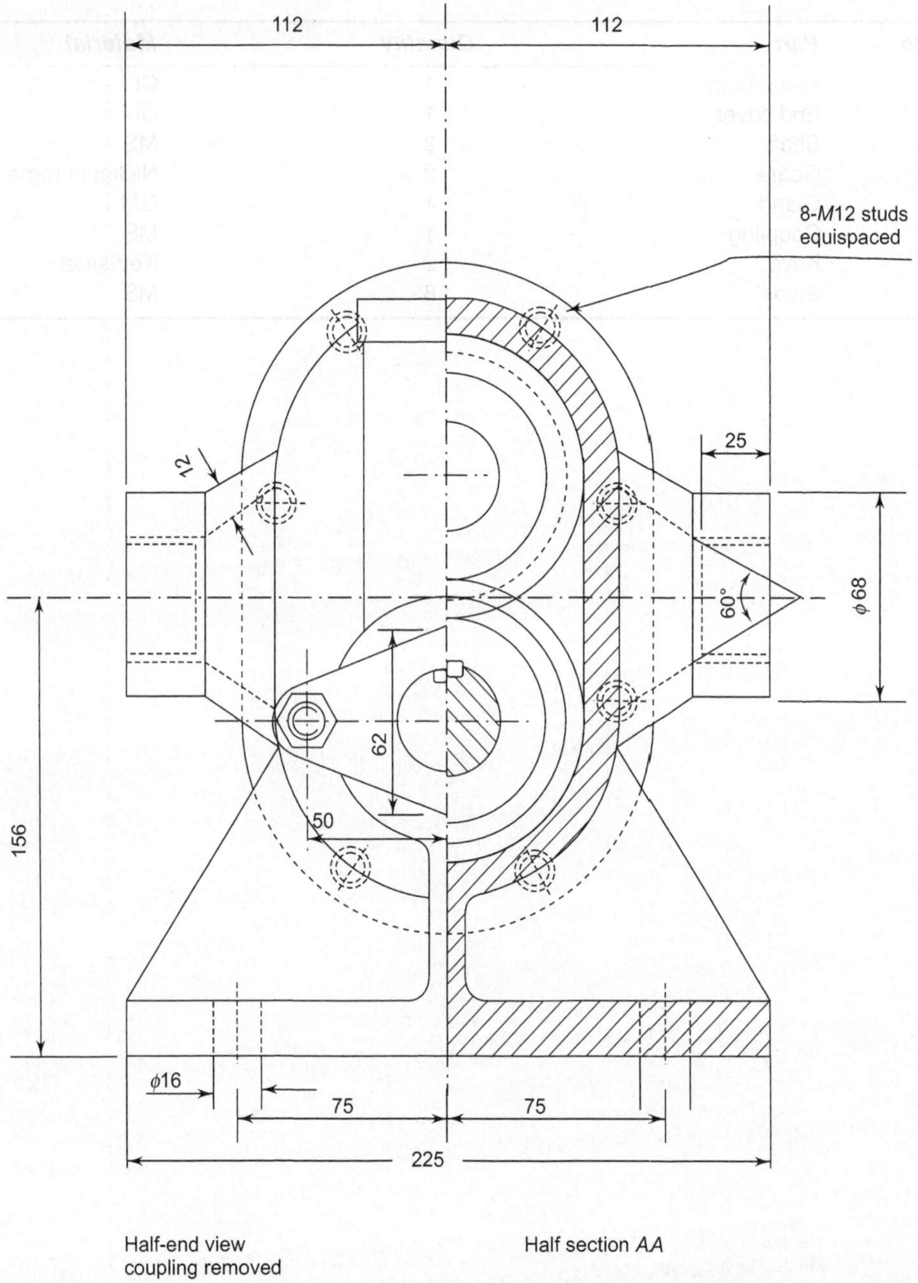
112
112
8-M12 studs
equispaced
25
12
60°
φ68
62
50
156
φ16
75
75
225
Half-end view
coupling removed
Half section AA

(b)
End view

Bill of Material

Sl. No.	*Part*	*Quantity*	*Material*
1	Pump body	1	CI
2	End cover	1	CI
3	Shaft	2	MS
4	Gears	2	Nickel chrome steel
5	Gland	1	GM
6	Coupling	1	MS
7	Keys	2	Key steel
8	Studs	8	MS

ASSEMBLY DRAWING 5

Valve

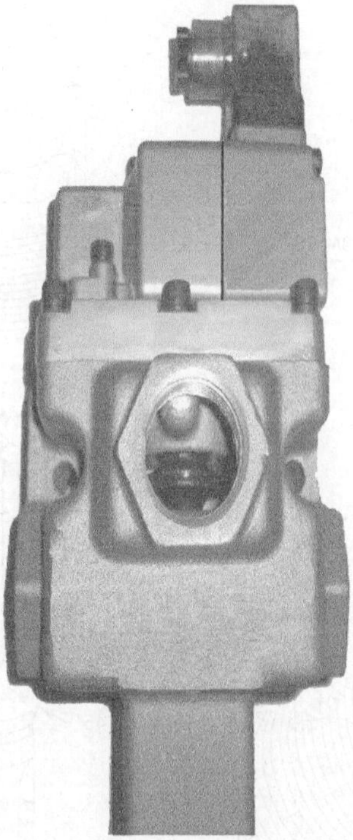

This section contains information on valves—starting air pilot valve and master air starting valve

After reading the section you will be able to

- understand the mode of operation and working
- understand the functioning of the master air starting valve and the starting air pilot valve
- draw assembly drawings

Valve

Question

Draw the following assembled views of the valve.

1. Elevation in section showing assembled valve
2. End view
3. Plan view, one half to be in section through atmospheric port and the other half, a section through port to automatic valve

Also include a bill of material.

The details of a starting air pilot valve are shown in Fig. B5.1.

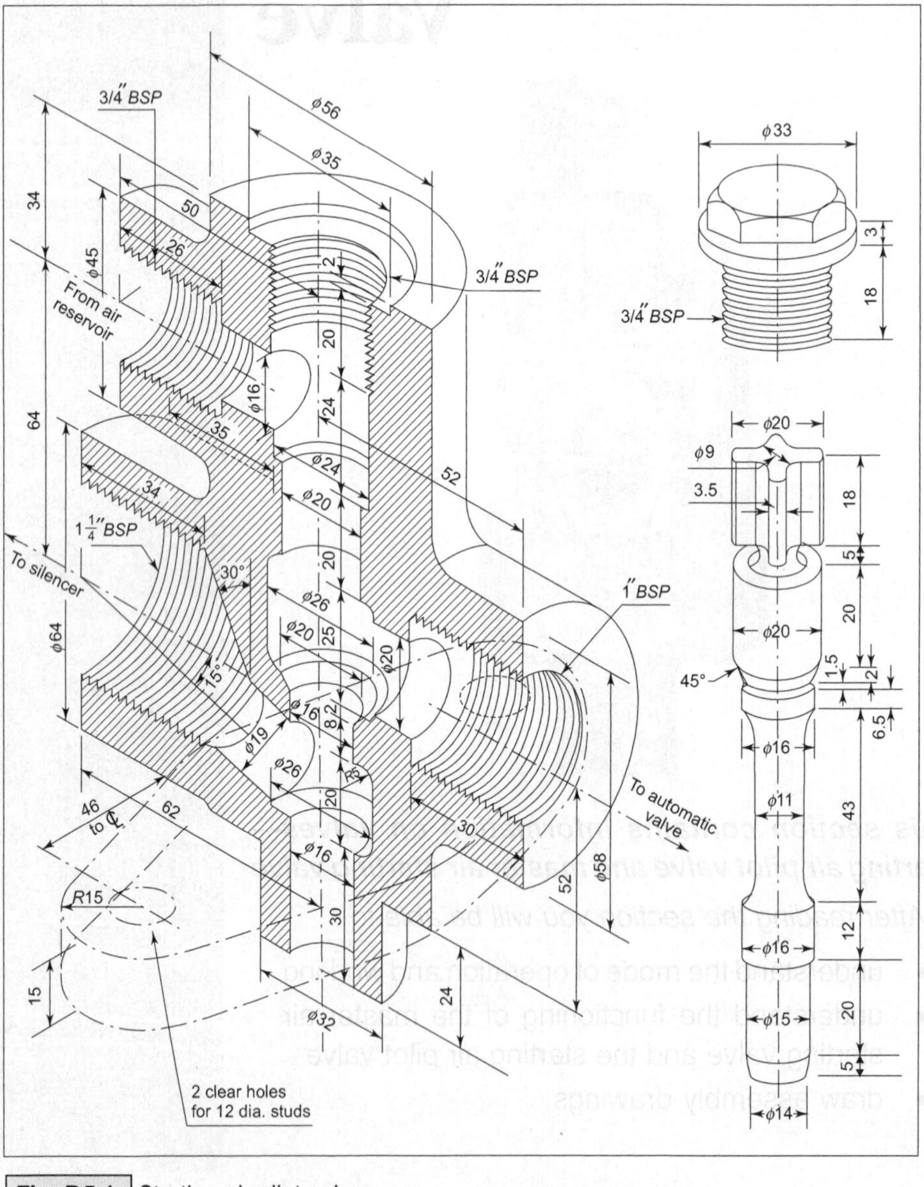

Fig. B5.1 Starting air pilot valve

Introduction

A valve is a mechanically, hydraulically, or pneumatically controlled device that is used to start, stop, and regulate the flow of fluid through pipelines or to machine parts. Thus, a valve is a very important and integral part of any machinery. As the valve plays such an important role, it has to be designed and selected according to its purpose.

Control Valve Selection

The selection of a control valve should be based on the availability of the best combination of valve, body style, material, and trim construction design for the required service. The system operating pressures and capacity requirements must also be considered in selecting a control valve. While consulting a manufacturer, the following information should be available.

- Temperature, viscosity, and specific gravity of the fluid
- Inlet and outlet pressure at the valve
- Pressure drop during normal flow and shut-off
- Inlet and outlet pipeline size and types of connections, degree of superheat or existence of flashing, if known, and any noise restriction that may exist

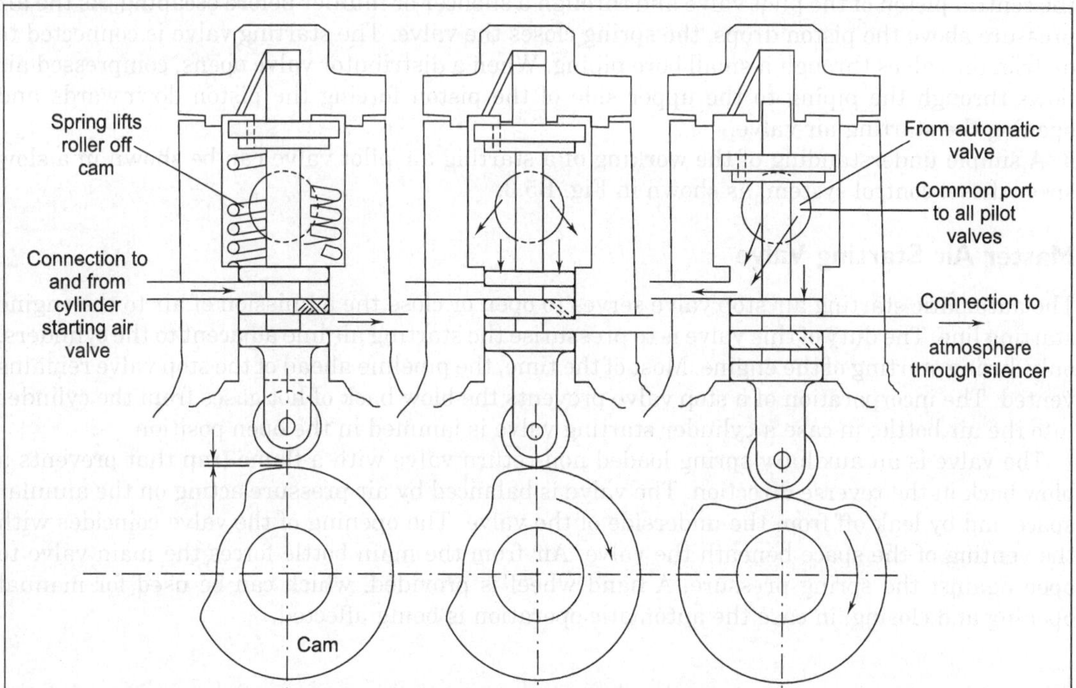

Fig. B5.2 Operation cycle of starting air valve

Starting Air Pilot Valve

This valve is used in the starting mechanism of heavy diesel engines. It is also referred as 'distribution valve'.

Operation and working A common form of a pilot air valve is shown in Fig. B5.3. Here three valves are shown, each being in a different operating position.

When an engine is in operation, stopped, or shutdown, the pilot valve takes up the position shown in Fig. B5.2. In this position the spring in the pilot valve lifts the roller of the follower and the valve remains inoperative.

When an engine is started, the automatic valve supplies compressed air to the pilot valve, which acts on the spring loaded piston and forces the roller on to the cam. Air cannot pass through the valve to operate the engine air starting valve as the roller is engaged on the idle sector of the cam, as shown in Fig. B5.2.

If the roller engages with the negative peak of the cam, as shown in the figure, compressed air passes through the engine air starting valve piston and causes the valve to open. The engine gets connected for the compressed air, supplied from the starting air manifold. As the engine gets connected, the roller comes on to the idle sector of the cam and the pilot valve reverts to the position, as shown in Fig. B5.2. Any compressed air trapped above the piston in the engine starting valve is quickly released to the atmosphere. The air passes through the holes drilled in the central piston of the pilot valve and through a silencer or muffler before escaping. As the air pressure above the piston drops, the spring closes the valve. The starting valve is connected to distributor valves through a small bore piping. When a distributor valve opens, compressed air flows through the piping to the upper side of the piston forcing the piston downwards and opening the starting air valve.

A simple understanding of the working of a starting air pilot valve can be shown in a slow speed diesel control system, as shown in Fig. B5.3.

Master Air Starting Valve

The automatic starting air stop valve serves to open or close the admission of air to the engine starting line. The duty of this valve is to pressurise the starting air line adjacent to the cylinders, only during starting of the engine. Most of the time, the pipeline ahead of the stop valve remains vented. The incorporation of a stop valve prevents the blow back of hot gases from the cylinder into the air bottle, in case a cylinder starting valve is jammed in the open position.

The valve is an auxiliary spring loaded non-return valve with a flame trap that prevents a blow back in the reverse direction. The valve is balanced by air pressure acting on the annular space and by leak off from the underside of the valve. The opening of the valve coincides with the venting of the space beneath the valve. Air from the main bottle forces the main valve to open against the spring pressure. A hand wheel is provided, which can be used for manual opening and closing, in case the automatic operation is being affected.

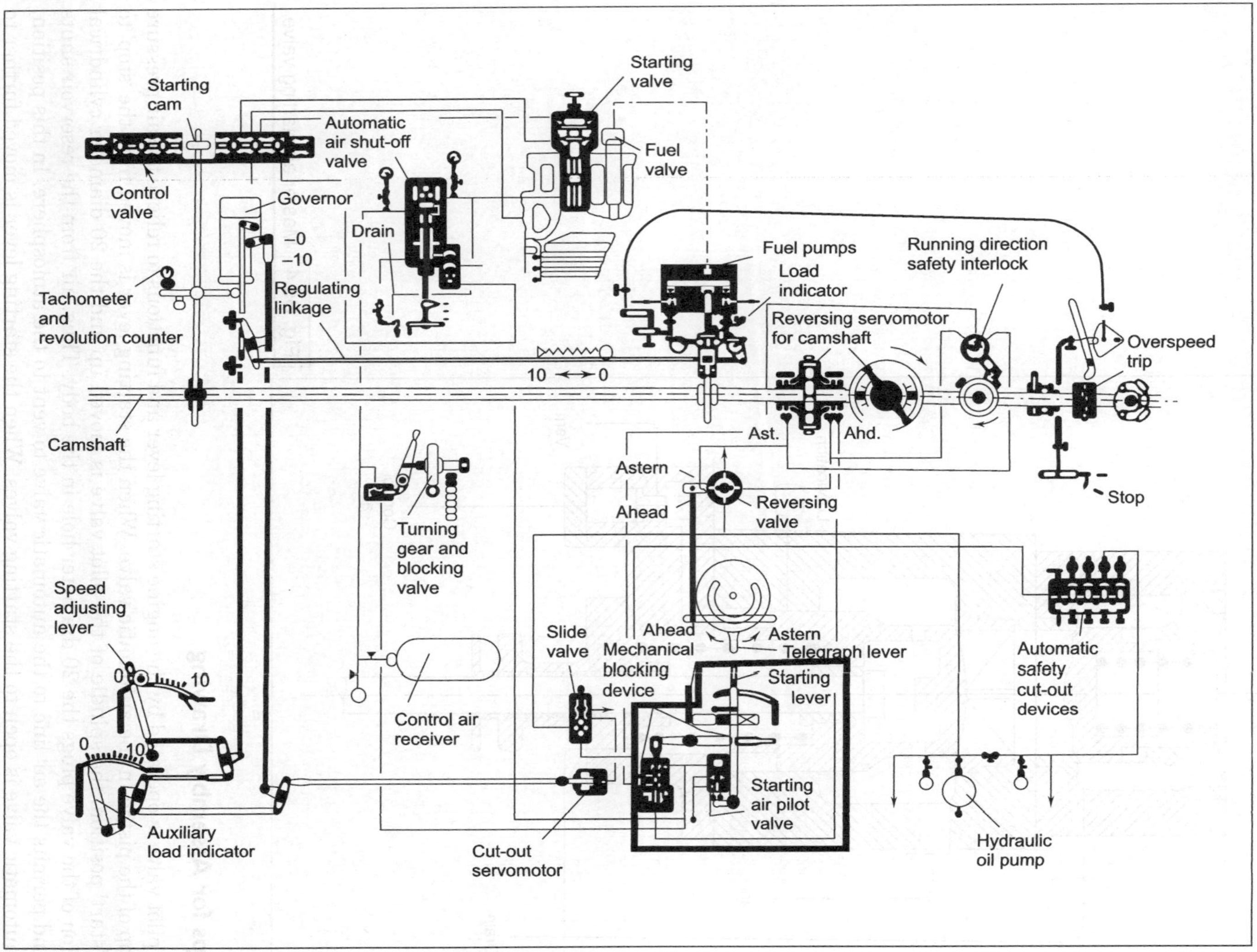

Fig. B5.3 Starting air pilot valve in a diesel engine

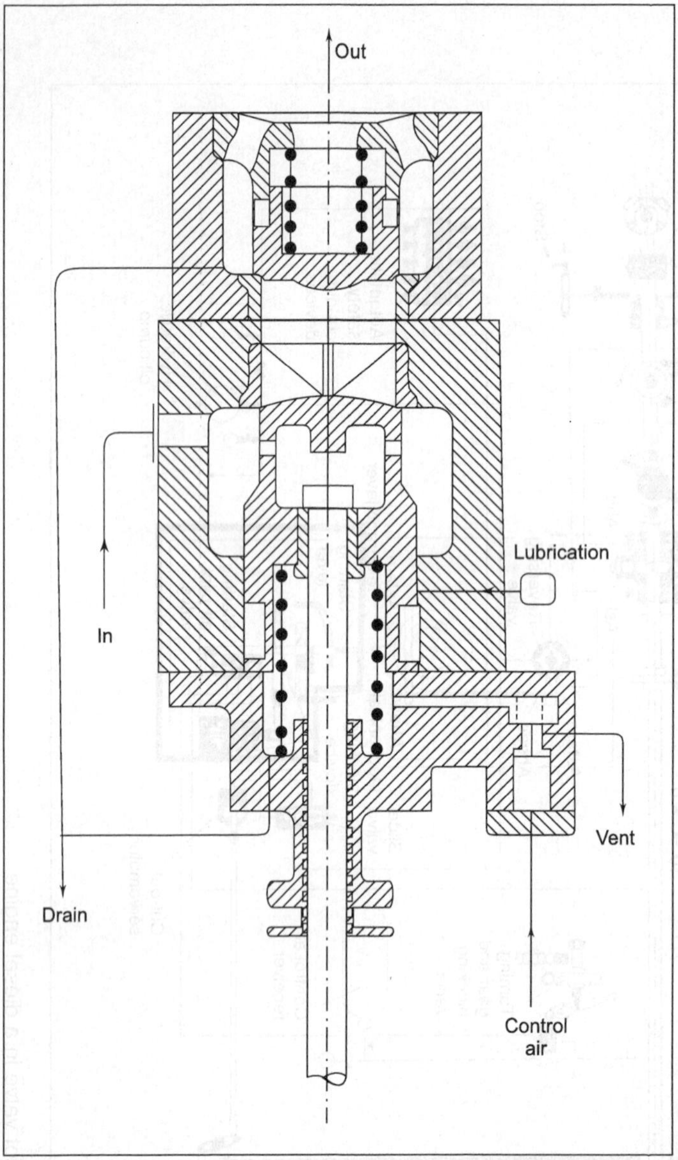

Fig. B5.4 Master air starting valve

Steps for Assembly Drawing

The pilot valve is operated by the engine starting lever and functions to relieve the air pressure on top of the piston in the automatic valve. When the starting lever is moved from the 'stop' to the 'start' position, the spindle of the pilot valve is moved up and the 20 diameter cylindrical portion of the valve plugs the 20 diameter hole in the body. Thus, air from the reservoir shuts off and permits the air line to the automatic valve to vent to the atmosphere. In this position, the automatic valve is open to the starting valves. When the starting lever is moved further to the 'fuel' position, the pilot valve spindle drops, thus shutting off the atmospheric port and following the line to the automatic valve to pressurise and, hence, shut the automatic valve.

The assembly of this valve is comparatively easy as there are only two components to locate. The 45° mitre on the valve locates on a similar mitre forming the valve seat machined in the valve body. The 3/4 *BSP* plug obviously fits in a similar hole tapped on top of the valve body.

Start with the sectional elevation by drawing the vertical centre line, build up the valve body around this, from the dimensions given above, and locate the valve spindle and plug. These two items will not be shown in section.

In the end view, the bosses for the inlet and atmospheric connections will appear as circles of 45 mm diameter and 64 mm diameter, respectively, with tapped holes in each boss indicated by the conventional method. In this view, the 19 mm diameter hole at the bottom of the atmospheric port will appear as an ellipse, the minor axis of which is found by projecting from the sectional elevation, whilst the major axis will be 19 mm and it should be noted that a part of the spindle will be seen through this hole.

The plan view will show on the left of the centre line, a section through the 11/4" *BSP* port and on the right, a section through the 1" *BSP* port. Also, in this view, the 19 mm diameter port at an angle of 15° to the horizontal will be seen as an ellipse. Finish off by showing the part of the fixing flange that is visible together with the 12 mm diameter clear holes for it bolting down.

Answer

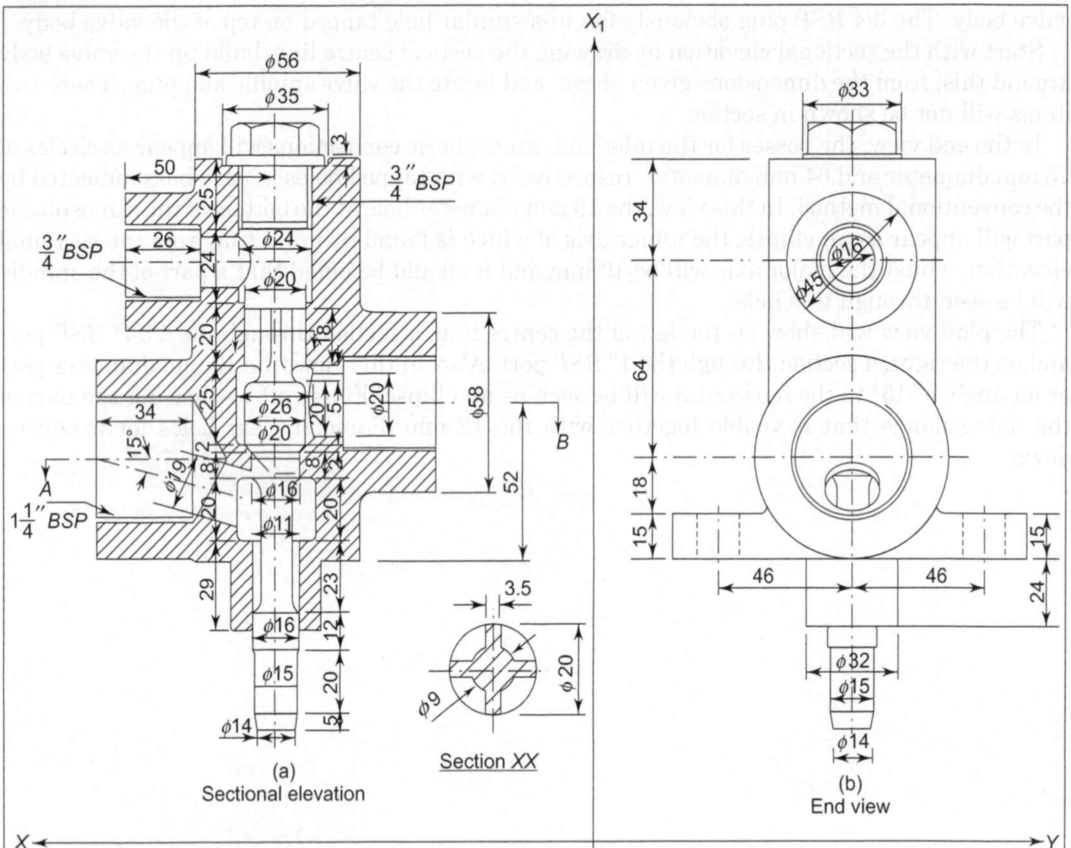

Item	Material
Valve body	GM
Valve	Stainless steel
Plug	Brass

Fig. B5.5 Assembly of starting air valve

The enlarged view of Fig. B5.5 is shown as under:

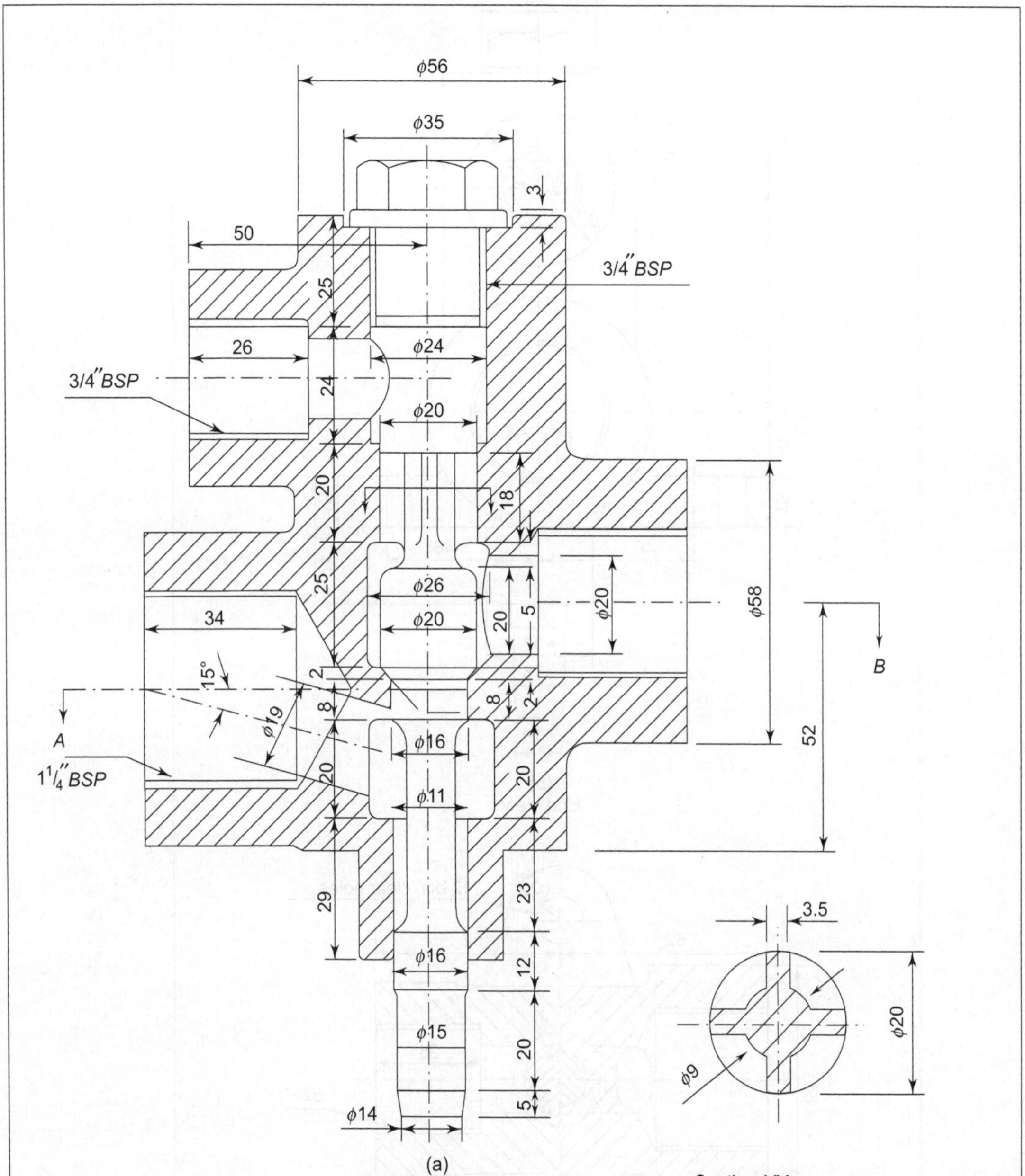

(a)
Sectional elevation

Section *XX*

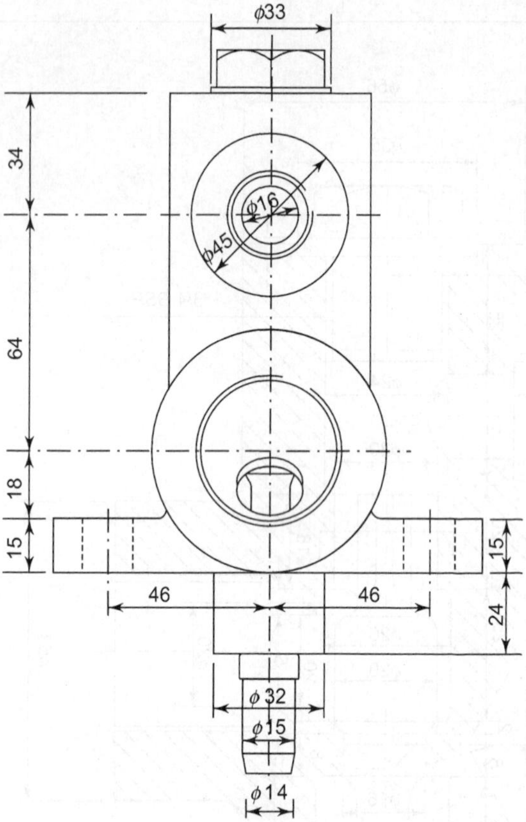

(b)
End view

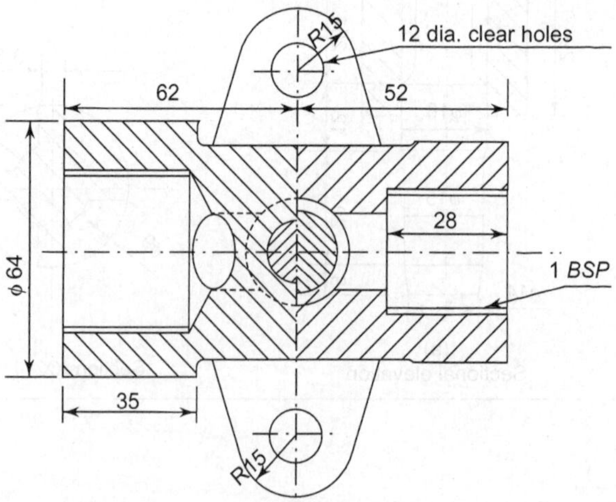

Section AB

(c)
Plan view

Bill of Material

Sl. No.	*Part*	*Quantity*	*Material*
1	Valve body	1	Gunmetal
2	Valve	1	Stainless steel
3	Plug	1	Brass

ASSEMBLY DRAWING 6

Fuel Pump

This section contains information on fuel pumps

After reading the section you will be able to

- understand the operation of a valve type plunger fuel pump
- make its assembly drawing

Fuel Pump

Question

Draw the following views of a fuel pump in 1:1 scale.

1. Section in full elevation
2. End view
3. Plan

Also include the bill of material.

The details of a fuel pump are shown in Fig. B6.1.

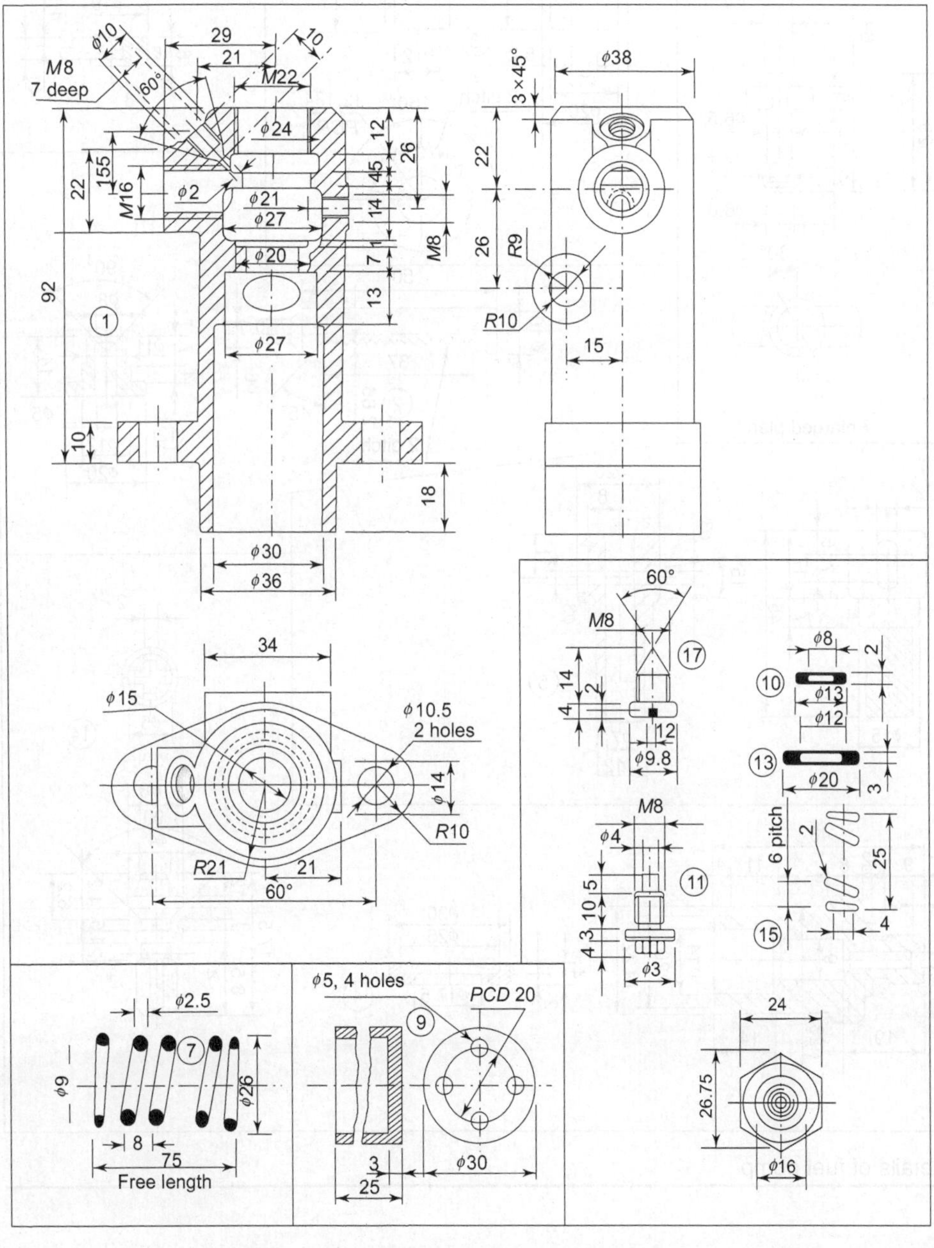

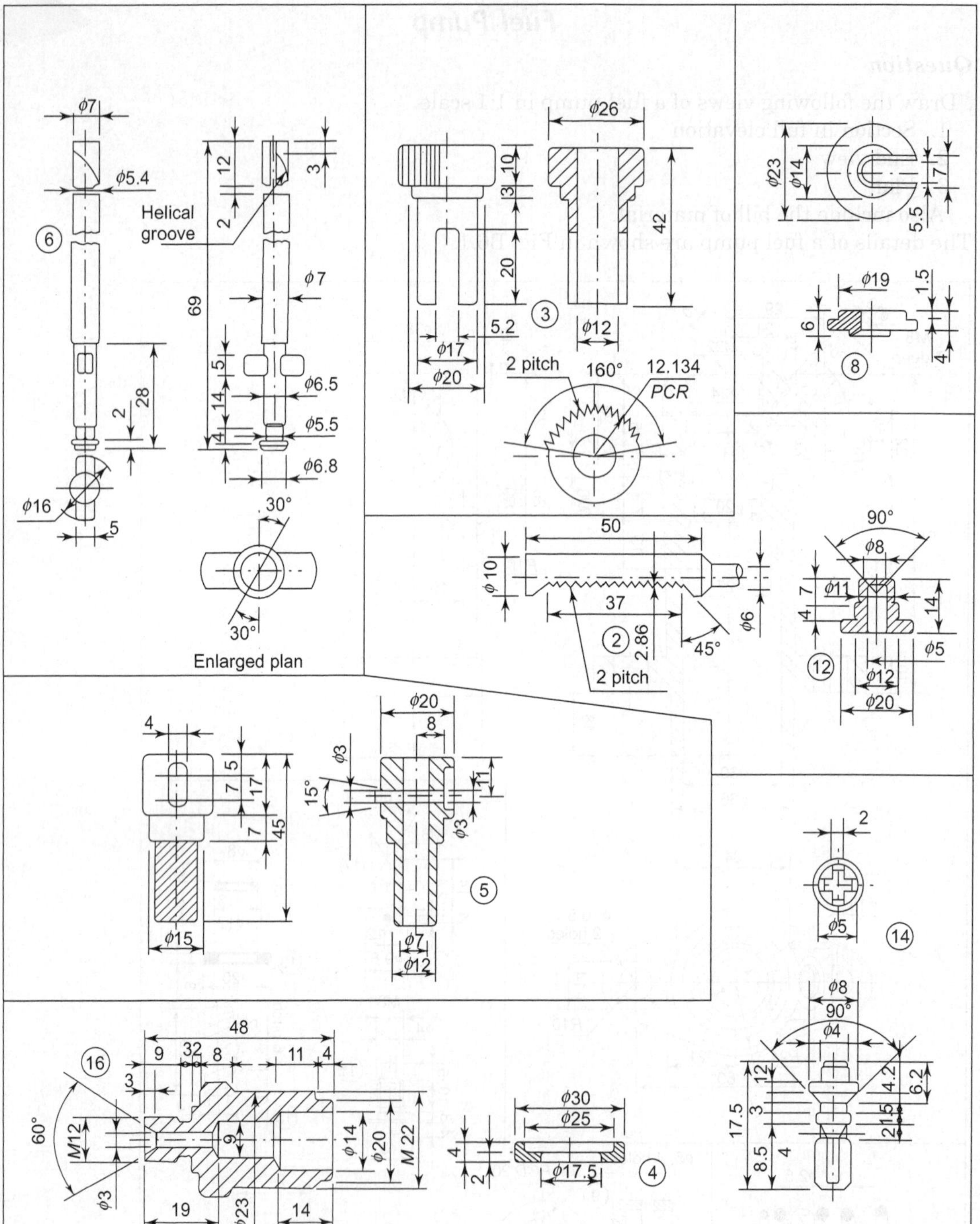

Fig. B6.1 Details of fuel pump

Introduction

In diesel engines, a plunger pump—called fuel pump driven by the main crankshaft of the engine is used to raise the pressure of the diesel in engine cylinder above the compression pressure in order to enable its injection through the injector in varying quantities to suit the changing loads on the engine.

The fuel system of a diesel engine includes nozzles, high pressure fuel pumps, stand by and service pumps, storage tanks, pipings, fittings, filters and separators, heaters, fuel feed pumps, control and measuring instruments, and so on. The reliable and economical operation of an engine depends upon the rational design of its fuel system and the good order thereof. Fuel pumps in airless injection engines are of the plunger type. The delivery stroke of the plunger is made under the action of a lobe of a cam secured on the camshaft and the suction stroke under the action of a spring. The amount of fuel delivered by the pump is regulated depending upon the load on the engine by means of a hand-drive from the control station or by means of a centrifugal governor.

Operation of a Valve Type Plunger Fuel-Pump

The fuel is supplied to the plunger through suction valve and it is directed to a nozzle through delivery valve, which is usually closed by spring. Supply is controlled by actuating the bypass valve by means of a hand-drive or centrifugal governor. The duration of supply of fuel and its amount depends on the moment when the bypass valve opens.

Hinged to the plunger is a horizontal lever, the right hand end of which can be moved by the eccentric. During the delivery stroke of the plunger, the left hand end of the lever rises and the fuel enters the nozzle until the striker is secured on the horizontal lever, which takes up the clearance and lifts the rod of the bypass valves. As a consequence, the pressure will rapidly drop in the pump, the delivery valve will close, and the fuel will enter the suction space through the bypass valve and along the duct, as shown in Fig. B6.2 by a dotted line. If by means of a lever 6, which is operated by a hand drive, is placed close to the governor, eccentric comes in a position at which the centre of the oscillation of the horizontal lever arm goes down and the clearance between the striker and the rod of the bypass valve will increase. When the eccentric raises the right hand end of the horizontal lever arm, the clearance will decrease, the bypass valve will open earlier, and the amount of fuel delivered will reduce. To adjust each pump of a multi-cylinder engine separately, (to obtain a uniform supply of fuel to all cylinders) the striker is screwed in or out and is secured in place with a nut. This will change the clearance and movement at which bypass valve open.

The pump plunger is relieved of the side forces generated by the cam, which is used to actuate the pump with the engine operating, ahead and astern.

The amount of fuel delivered depends on the position of the helical grove of the plunger. The plunger movement is controlled by rotary and reciprocating motion at the same time. This conforms quite well to the operating conditions of engines running at a constant speed (engines that drive generators). In this case the change in speed changes the advance of admission in time. For example, when the speed is reduced, advance of fuel admission in time (with the advance unchanged when measured in crank angle) increases, as a result the pressure in the cylinder during admission rises excessively.

To ensure a constant and timely fuel supply when an engine runs at a variable speed, it is a good practice to use fuel pumps with control fuel injection advance. Such pumps are extremely widespread in main marine diesel engines. Figure B6.2 refers to a fuel pump in which the quantity of fuel is regulated by changing the moment at which the injection begins. Here, the plunger is actuated by a cam. The pump is provided with suction valve 1, delivery valve 2, and bypass valve 3. When the plunger rises, the right hand end of the lever moves together with the rod and the bypass valve remains open until the clearance appears between the rod and the guide of valve 3. The spring will then close the bypass valve and fuel will be supplied to the nozzle through the delivery valve. The clearance between the rod and bypass valve guide is changed by means of an eccentric 6. This changes the moment when the injection begins and the amount of fuel that is supplied.

Slide valve fuel pumps have been introduced to a great extent in marine diesel engines. In this pump the amount of fuel injected is changed by changing the moment when bypassing of fuel begins but the bypass valve is absent. On turning off the plunger, which is provided with a curvilinear slot, the fuel supply is cut off.

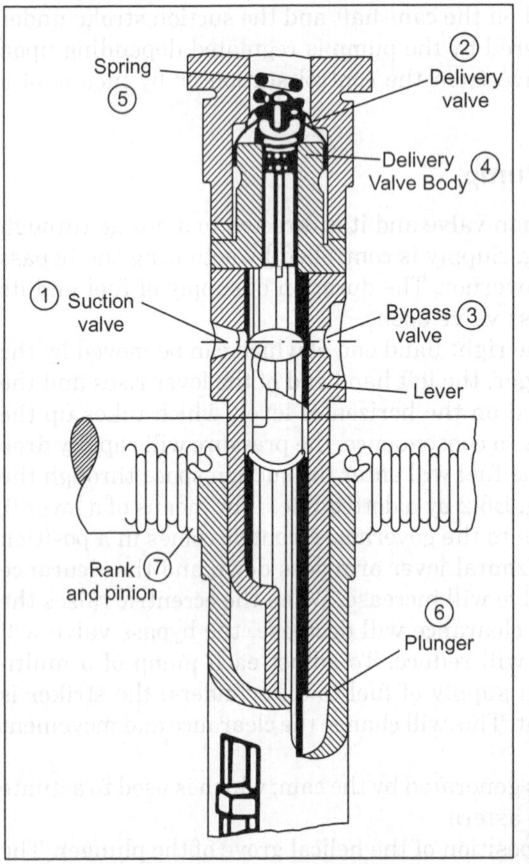

Fig. B6.2 Fuel pump used in marine diesel engines

Fuel pumps for larger engines incorporate variable injection timing for optimising fuel economy at part load. The start of fuel injection is controlled by altering the pump barrel position by means of a toothed rack and a servo unit. Individual adjustments can be made on each cylinder and collective adjustments in case of maximum pressure level of the engine, can be carried out to compensate for varying fuel qualities, wear, and so on. Both adjustments can be carried out while the engine is running.

The fuel oil pump is also furnished with a puncture valve, which prevents fuel injection during normal stopping and shutdown.

This variable timing fuel pump in assembled condition is shown in Fig. B6.3.

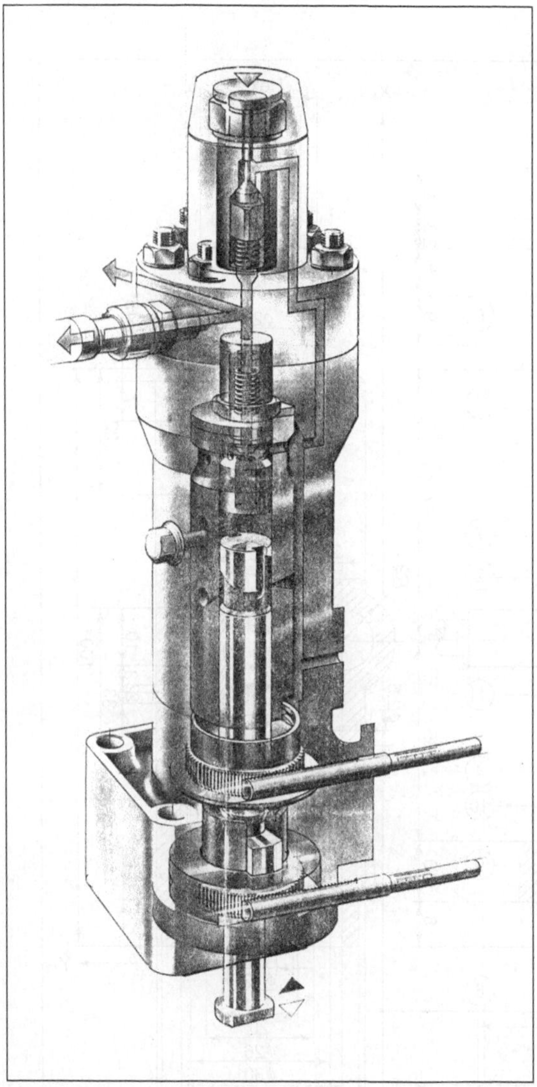

Fig. B6.3 Fuel pump

Steps for Assembly Drawing

First calculate the overall dimensions for deciding the space required to accommodate all the three views—elevation, plan, and end view. Start drawing with the elevation. Draw the vertical axis line given in the question. Try to identify the elevation of each and every component given in the question along the vertical axis. Start with the main body of the pump in question given. Except the delivery valve, almost all the components are assembled inside the valve body. For this, refer to the functional dimension of an individual component and try for the assembly with reference to male and female dimensions of the meeting components. Once the elevation is completed, project the end-elevation and plan view from the same. Fig. B6.4 shows all parts assembled in orthographic projection for ready reference.

Answer

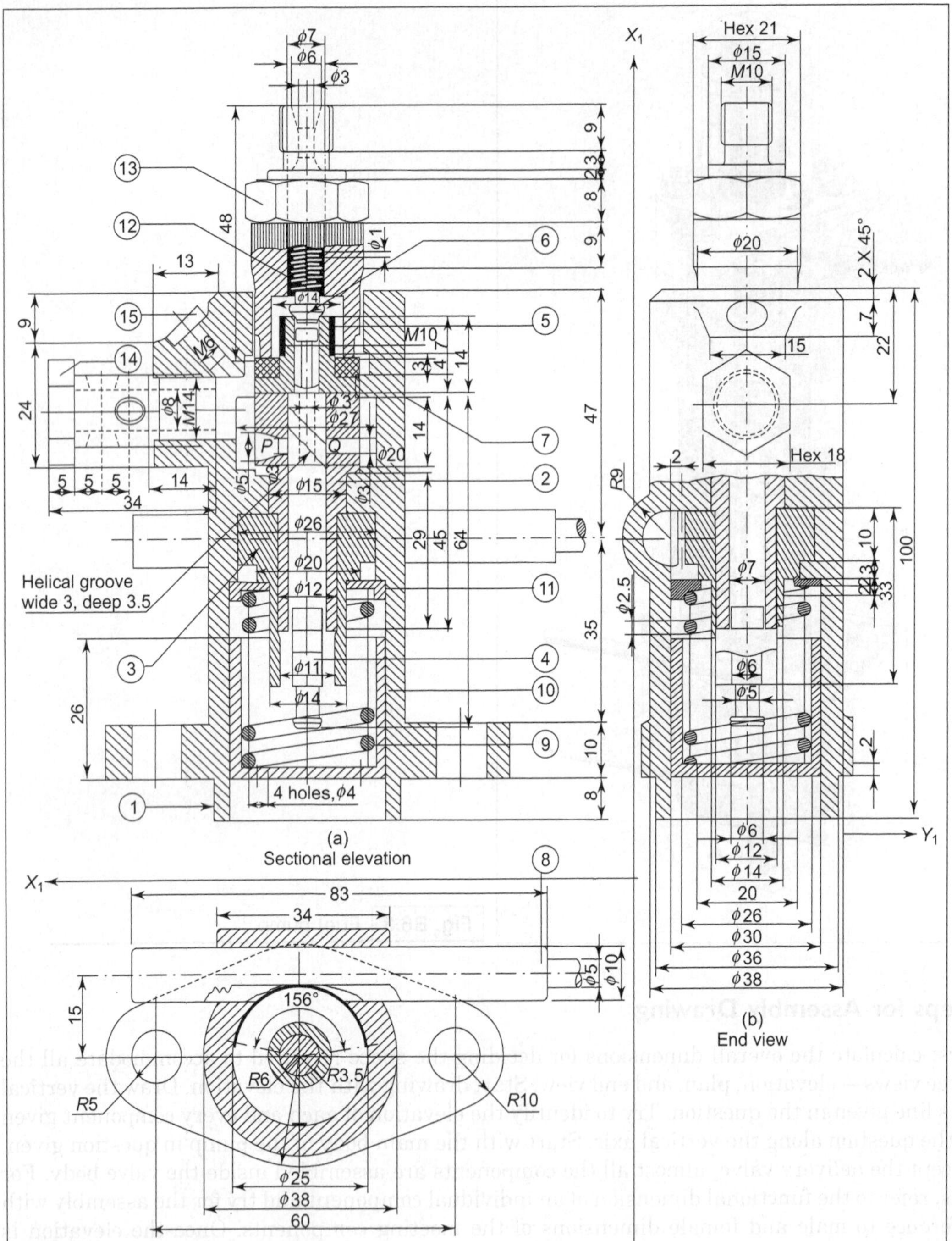

Fig. B6.4 Assembly of fuel pump

The enlarged views of Fig. B6.4 are shown as under:

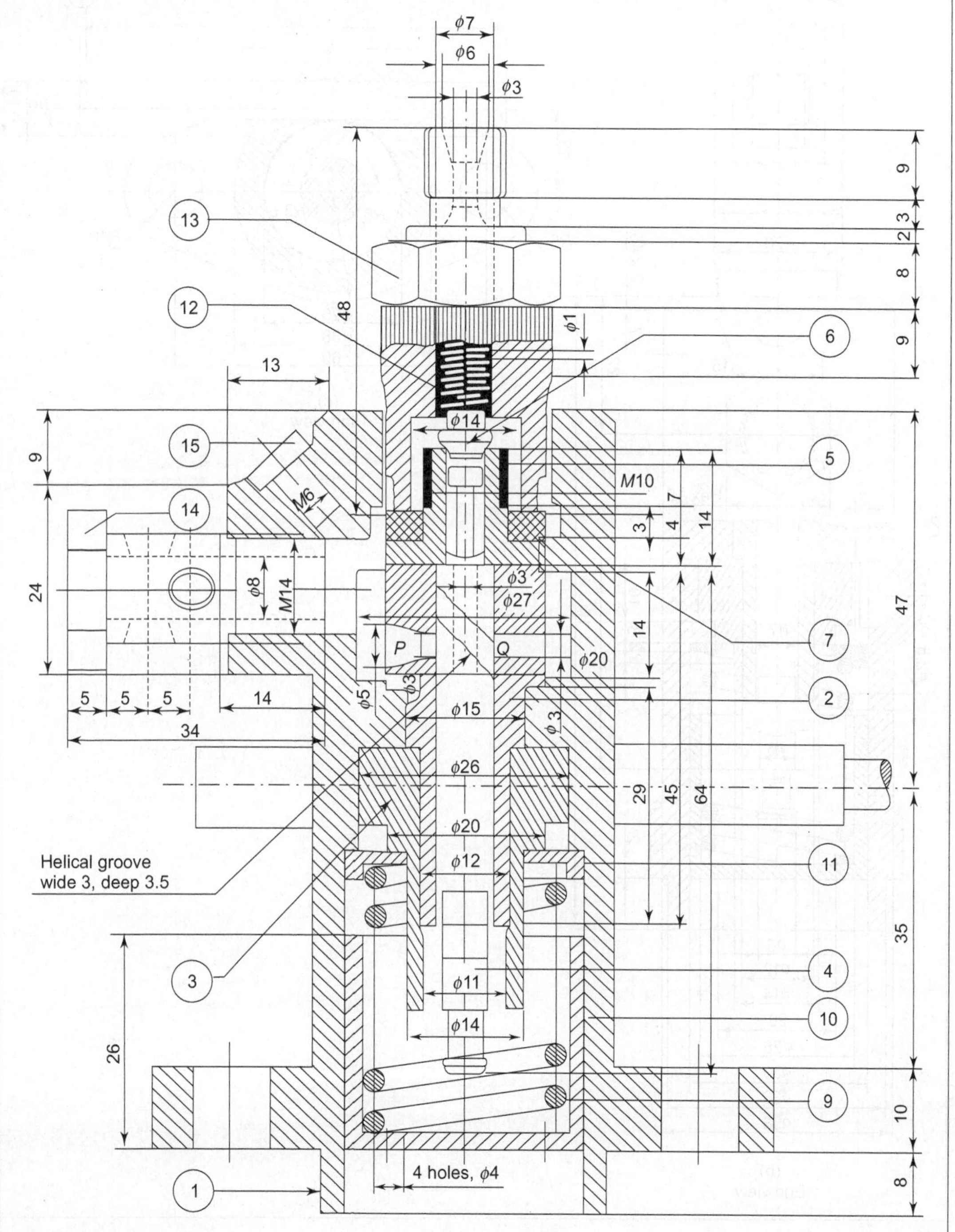

(a)
Sectional elevation

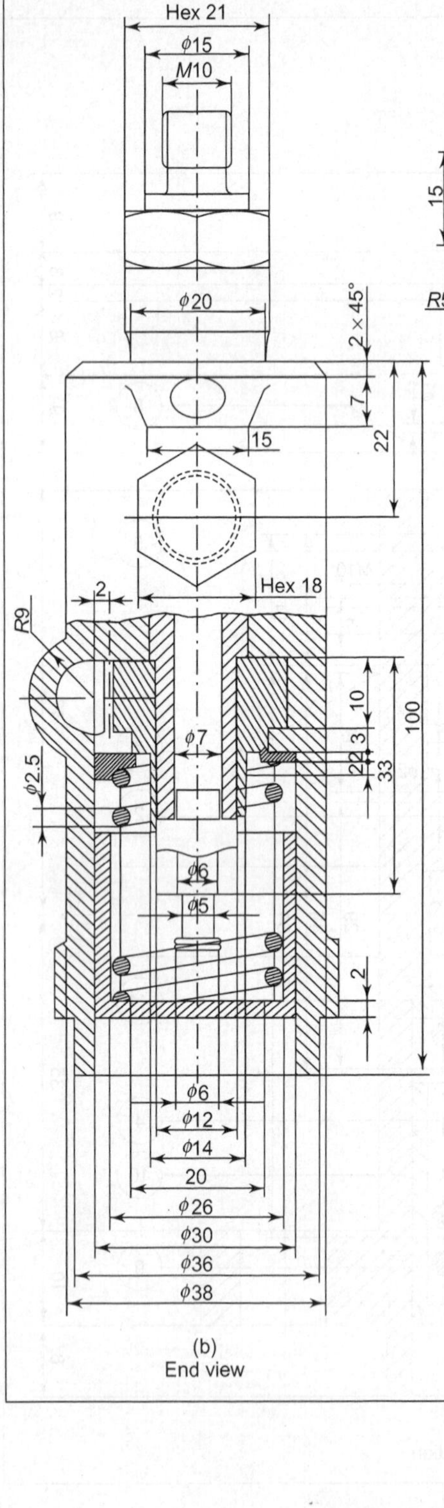

(b)
End view

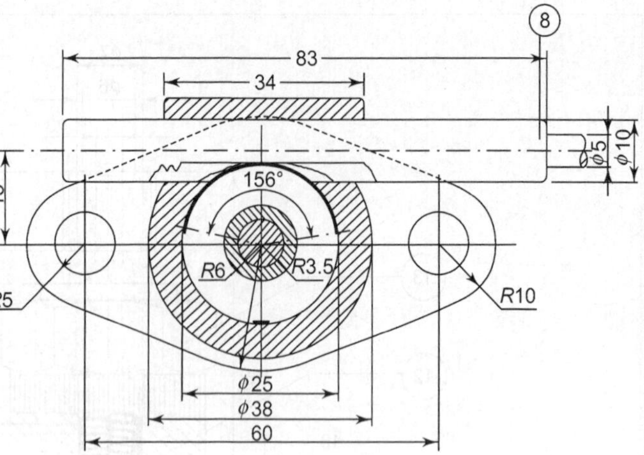

(c)
Plan view

Bill of Material

Sl. No.	*Part*	*Quantity*	*Material*
1	Pump body	1	Cast steel
2	Rack	1	Alloy steel
3	Pinion	1	Alloy steel
4	Spring seat top	1	Alloy steel
5	Barrel	1	Alloy steel
6	Plunger	1	Alloy hardened steel
7	Helical compression spring	1	Spring steel
8	Spring steel bottom	1	Alloy steel
9	Spring locator	1	Alloy steel
10	Plain washer	1	Alloy steel
11	Bypass pin	1	Alloy steel
12	Delivery valve	1	Alloy steel
13	Plain washer	1	Non-ferrous
14	Delivery valve	1	Alloy steel
15	Helical compression spring	1	Spring steel
16	Delivery valve body	1	Cast steel
17	Air release screw	1	Alloy steel

ASSEMBLY DRAWING 7

Starting Air Valve

This section contains information on starting air valve

After reading the section you will be able to

- understand the operation and maintenance of a starting air valve
- learn to make its assembly drawings

Starting Air Valve

Question

Draw the following views of a starting air valve.

1. Sectional elevation through inlet branch showing all the assembled parts
2. End elevation
3. Sectional plan through vent in valve body

Also include a bill of material.

The details of a starting valve are shown in Fig. B7.1.

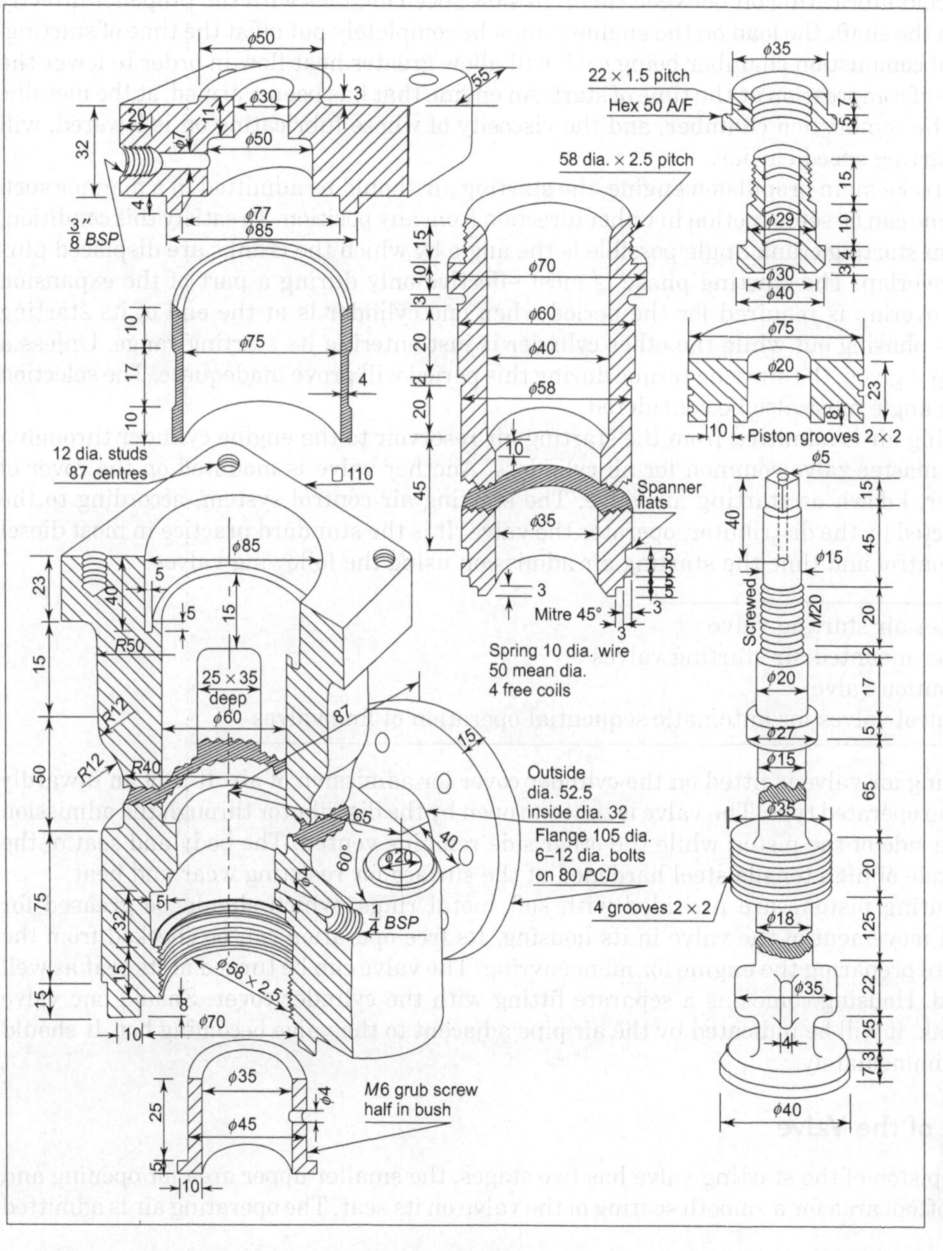

Fig. B7.1 Starting air valve

Starting and Reversing of the Engine

Almost all marine diesel engines are started by the admission of compressed air in the cylinder, which is in the starting position with respect to the direction of rotation. The compressed air acts on top of the piston and produces a starting torque at the crankshaft. The shaft is set to rotation upto a minimum firing speed when air is compressed in some other cylinder producing a temperature sufficiently high enough to ignite the fuel.

When a diesel engine is started after a period of idleness, a high initial torque at a low revolution is necessary to accelerate the engine's rotating and reciprocating masses. One of the principal factors is to overcome the forces of adhesion between the bearing surfaces due to the presence of cold lubricating oil between them. In slow speed engines with the propeller directly coupled with the shaft, the load on the engine cannot be completely put off at the time of starting. The valves of combustion chamber being cold, will allow greater heat flow in order to lower the temperature of compression at the time of start. An engine that has been warmed at the metallic surfaces of the combustion chamber, and the viscosity of whose lubricating oil is lowered, will reach the starting speed earlier.

In a two stroke main propulsion engine, the starting air should be admitted in a manner such that the engine can be set to motion in either direction from any position. To satisfy this condition, the minimum starting crank angle possible is the angle by which the cranks are displaced plus a period of overlap. The starting phase is most effective only during a part of the expansion stroke. The overlap is required for the period when one cylinder is at the end of its starting range and is phasing out while the other cylinder is just entering its starting range. Unless a large overlap is given, the starting torque during this period will prove inadequate. The selection of a starting angle must also be considered.

The starting air is admitted from the starting air reservoir to the engine cylinder through a starting air master valve common for all cylinders. Another valve is mounted on the cover of each cylinder, known as starting air valve. The starting air control system, according to the timings selected by the distributor, operates the valve. It is the standard practice in most diesel engines to control and time the starting air admission using the following valves.

1. A master air starting valve
2. Cylinder mounted air starting valves
3. Distribution valve
4. Air control valves for automatic sequential operation of line valves

The starting air valve is fitted on the cylinder cover for admission of air. It is of an inwardly opened piston operated type. The valve is timed to open by the distributor through the admission of air on one side of the piston while the other side remains vented. The body and seat of the valve are made of high tensile steel hardened at the surface for resisting wear and heat.

The operating pistons are provided with soft metal rings, which should be greased for unrestricted movement of the valve in its housing. Its free operation can be checked from the outside before preparing the engine for manoeuvring. The valve can be turned at its seat as well as depressed. Housing comes as a separate fitting with the cylinder cover. Should one valve develop a leak, it will be indicated by the air pipe adjacent to the valve becoming hot. It should be changed immediately.

Operation of the Valve

The central piston of the starting valve has two stages, the smaller upper area for opening and the larger bottom area for a smooth seating of the valve on its seat. The operating air is admitted to the piston through the distributor simultaneously with the opening of the master valve. The underside of the piston space is vented at the same time. The valve is forced to open due to the air pressure acting on the area and overcomes the spring and cylinder pressures. After the piston moves down a distance, the space on the top of the bottom piston is pressurised. As a result, the second phase of valve opening is more rapid. To close the valve, air is admitted on the other side of the bottom pistont; the top of the top piston is now vented through the distributor. The valve starts closing rapidly. Soon after the space is cut off, the central air acts on a small area. The damping effect of trapped air on the top of the larger piston, thus, controls the closing motion.

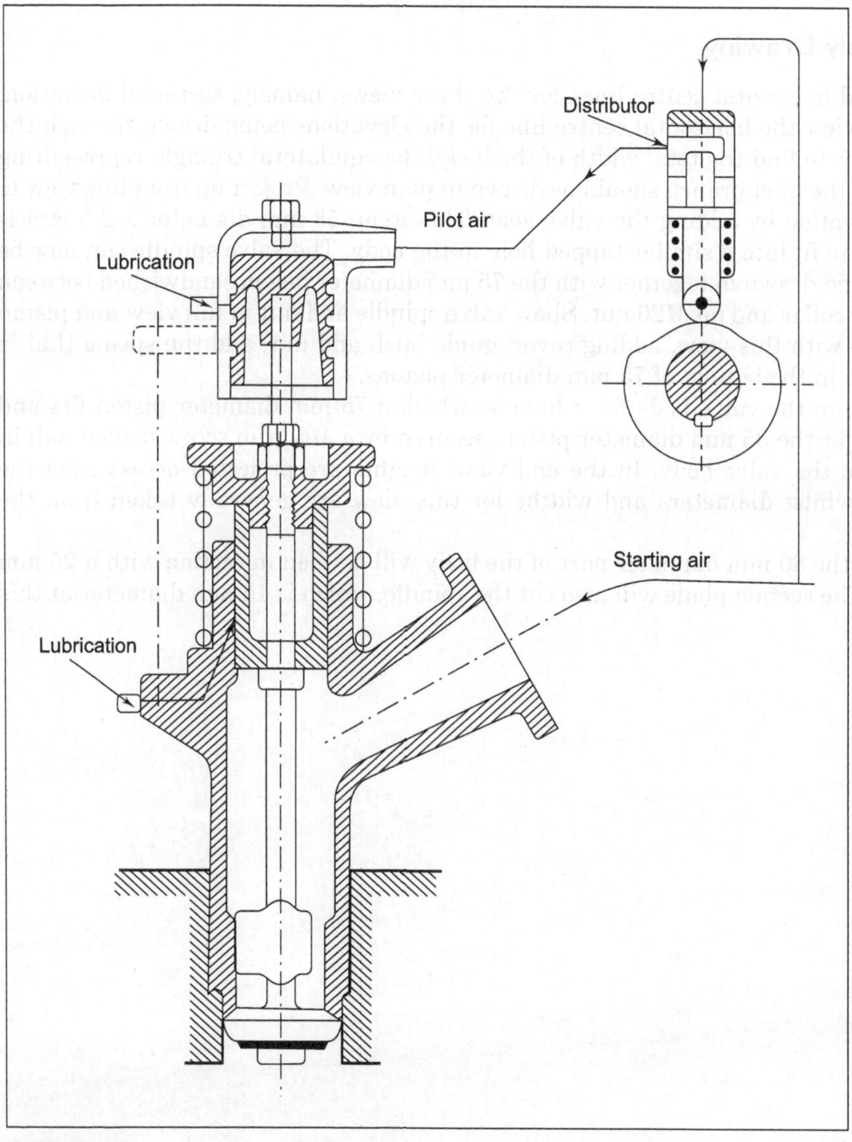

Fig. B7.2 Mechanism showing starting air supply

Maintenance

Starting air valves should be removed for overhauling at regular intervals. The control piston for the valve should be lubricated and checked for easy movability before preparing the engine for manoeuvring. Lubrication should be done by a heat resisting grease containing molybdenum disulphide.

The starting air pipe adjacent to the valve gets heated, indicating a leaky starting air valve. Combustion gases enter the starting pipeline. In such a case, the valve should be replaced at the earliest. Running the engine with a leaky valve will damage the valve seat due to the burning in the pipeline with consequent danger of explosion.

Steps for Assembly Drawing

Draw in vertical and horizontal centre lines for the three views, namely, sectional deviation, end view, and plan view the horizontal centre line for the elevations being drawn through the inlet branch. In order to find the total width of the body, the equilateral triangle representing the body flange, and the inlet branch should be drawn in plan view. Project up from this view to locate a body in elevation by adding the valve seat. This item, 58 mm diameter × 2.5 mm is screwed at one end, to fit into a similar tapped hole in the body. The valve spindle can now be located on the seat and drawn in together with the 75 mm diameter piston sandwiched between the 27 mm diameter collar and an *M*20 nut. Show valve spindle and nut in full view and piston in section. Continue with this view, adding cover, guide bush and nut, and the spring that is located in the groove in the bottom of 75 mm diameter pistons.

Provision is made in the valve body for a liner in which a 75 mm diameter piston fits and there is also a liner for the 35 mm diameter piston, secured by a *M*6 grub screw drilled half in the liner and half in the valve body. In the end view, heights are projected across from the sectional elevation whilst diameters and widths for this view are generally taken from the other two views.

In the plan view, the 80 mm diameter part of the body will be seen in section with a 25 mm wide port cut in it. The section plane will also cut the spindle, which is 15 mm diameter at this point.

Answer

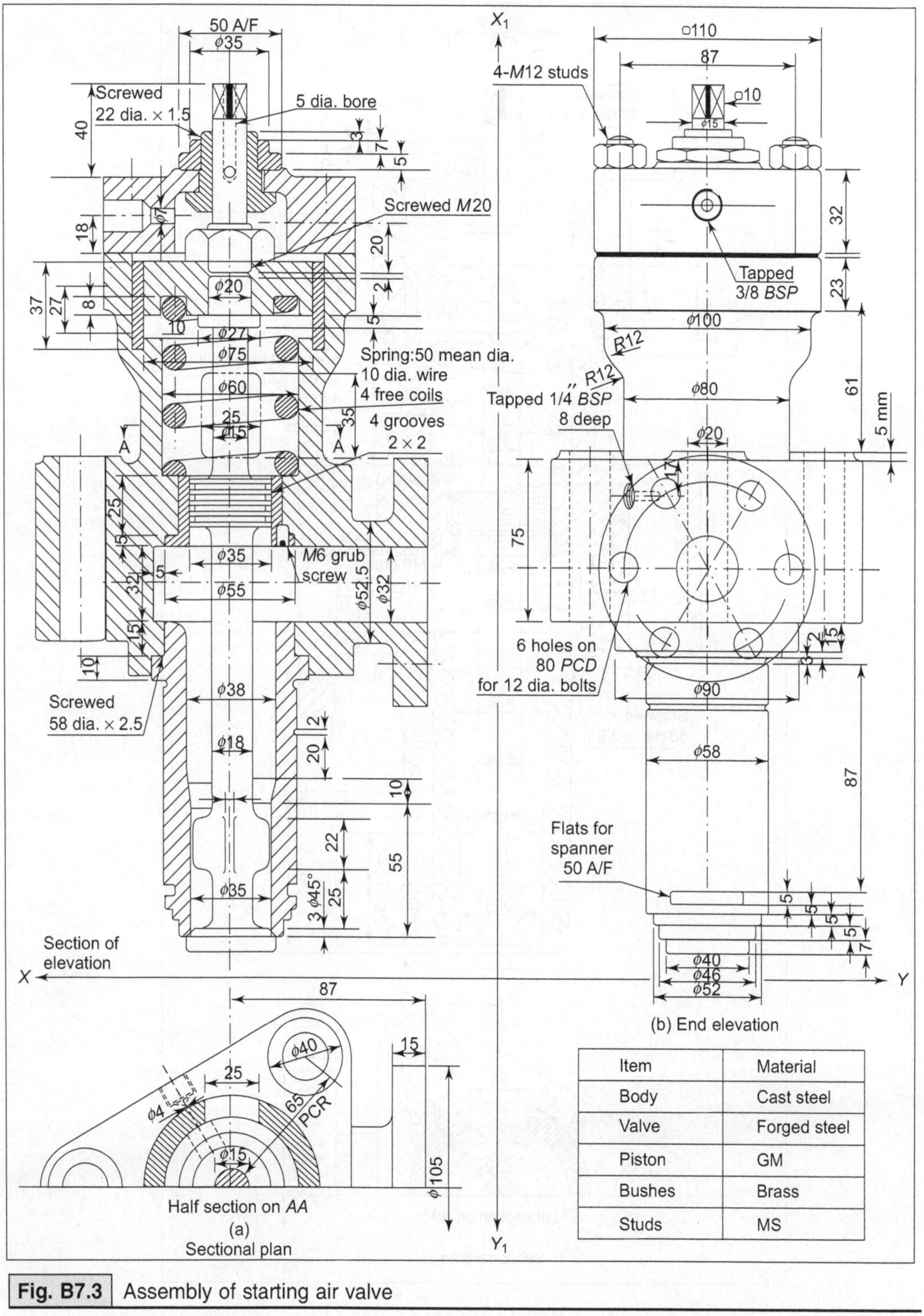

Item	Material
Body	Cast steel
Valve	Forged steel
Piston	GM
Bushes	Brass
Studs	MS

Fig. B7.3 Assembly of starting air valve

The enlarged views of Fig. B7.3 are shown as under:

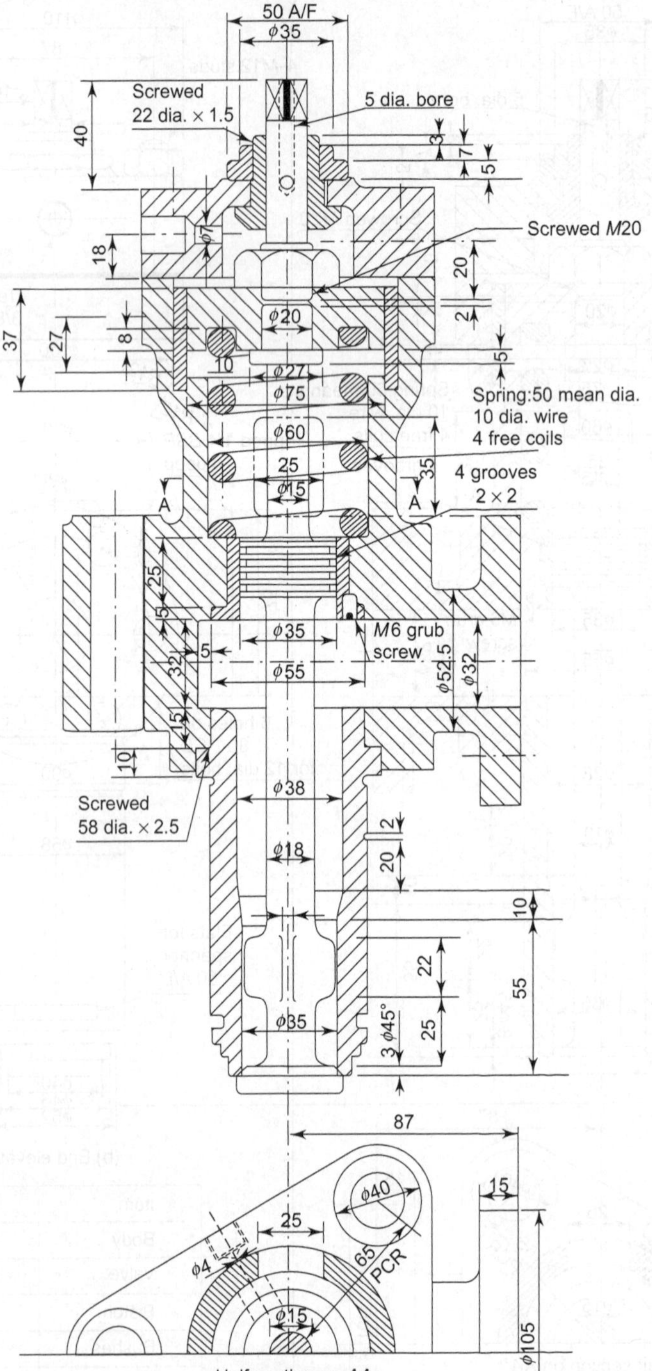

(a)
Sectional plan

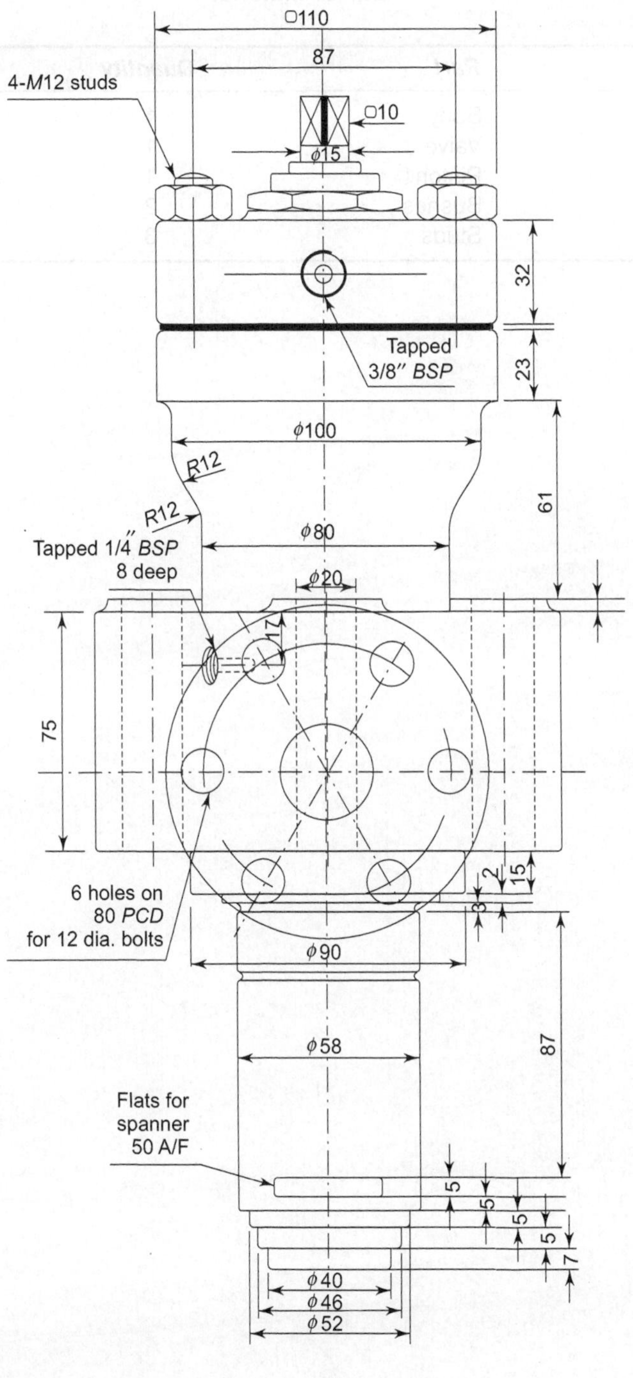

(b)
End elevation

Bill of Material

Sl. No.	Part	Quantity	Material
1	Body	1	Cast steel
2	Valve	1	Forged steel
3	Piston1	1	GM
4	Bushes	2	Brass
5	Studs	3	MS

ASSEMBLY DRAWING 8

Fuel Injector

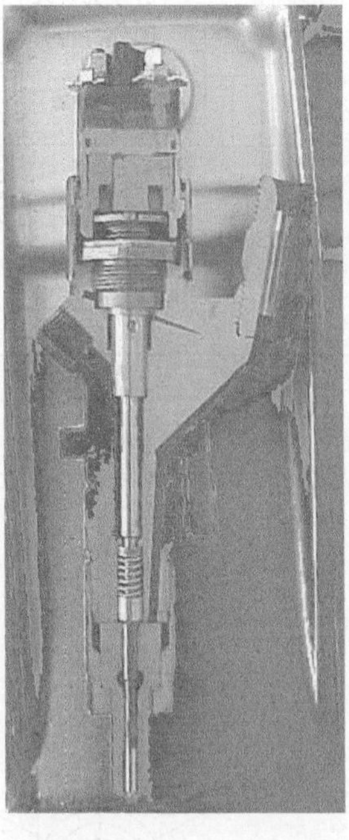

This section contains information on fuel injectors

After reading the section you will be able to

- understand the working of fuel injectors
- draw its assembled views

Fuel Injector

Question

Draw the following views of a fuel injector with all its parts assembled. Select suitable scale.

1. Half elevation in section
2. Plan view
3. End view

Also include a bill of material.

Fig. B8.1 shows the details of a fuel injector.

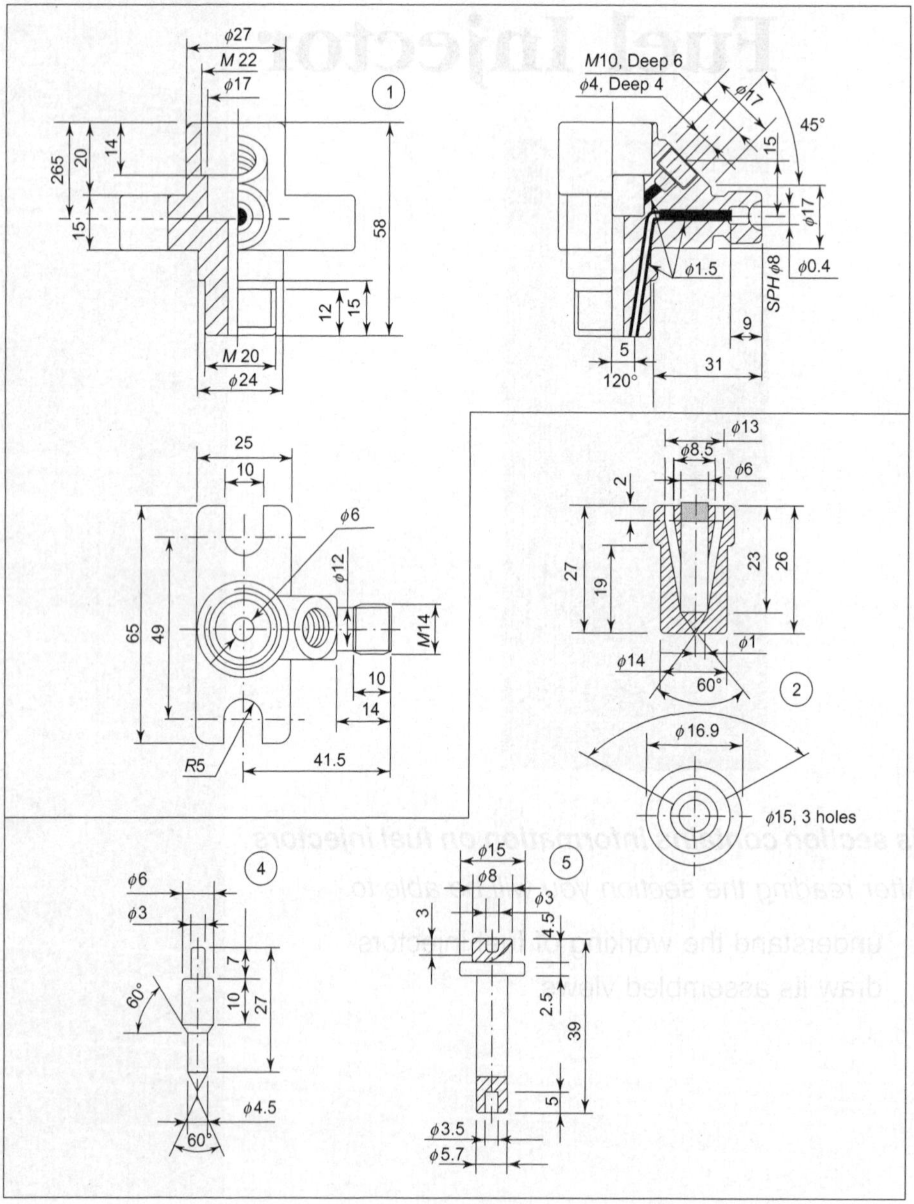

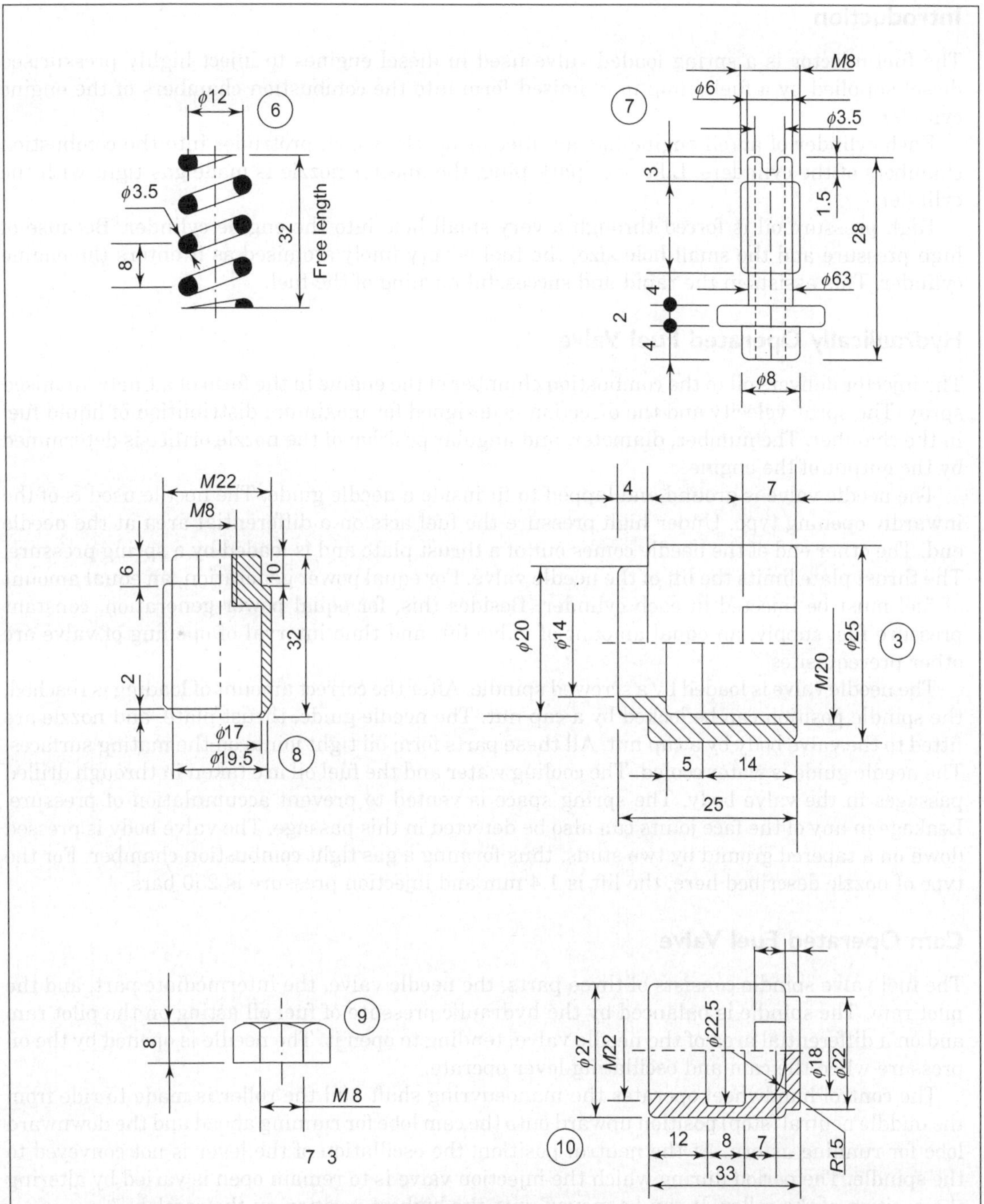

Fig. B8.1 Details of a fuel injector

Introduction

The fuel injector is a spring loaded valve used in diesel engines to inject highly pressurised diesel supplied by a fuel pump in atomised form into the combustion chambers of the engine cylinder.

Each cylinder of an oil engine has an injector nozzle, which protrudes into the combustion chambers of the cylinders. Like the spark plug, the injector nozzle is made gas tight with the cylinder.

High pressure oil is forced through a very small hole into the engine cylinder. Because of high pressure and the small hole size, the fuel is very finely atomised as it enters the engine cylinder. This assists in the rapid and successful burning of the fuel.

Hydraulically Operated Fuel Valve

The injector delivers oil to the combustion chamber of the engine in the form of a finely atomised spray. The spray velocity and the direction is designed for maximum distribution of liquid fuel in the chamber. The number, diameter, and angular position of the nozzle orifice is determined by the output of the engine.

The needle valve is ground and lapped to fit inside a needle guide. The needle used is of the inwardly opening type. Under high pressure the fuel acts on a differential area at the needle end. The other end of the needle comes out of a thrust plate and is loaded by a spring pressure. The thrust plate limits the lift of the needle valve. For equal power generation, an equal amount of fuel must be injected in each cylinder. Besides this, for equal power generation, constant pressure fuel supply, an equal amount of valve lift, and time interval of opening of valve are other prerequisites.

The needle valve is loaded by a screwed spindle. After the correct amount of loading is reached, the spindle position can be locked by a cap nut. The needle guide, thrust plate, and nozzle are fitted to the valve body by a cap nut. All these parts form oil tight joints on the mating surfaces. The needle guide is water cooled. The cooling water and the fuel oil are taken in through drilled passages in the valve body. The spring space is vented to prevent accumulation of pressure. Leakage in any of the face joints can also be detected in this passage. The valve body is pressed down on a tapered ground by two studs, thus forming a gas tight combustion chamber. For the type of nozzle described here, the lift is 1.4 mm and injection pressure is 250 bars.

Cam Operated Fuel Valve

The fuel valve spindle consists of three parts, the needle valve, the intermediate part, and the pilot ram. The spindle is balanced by the hydraulic pressure of fuel oil acting on the pilot ram and on a differential area of the needle valve, tending to open it. The needle is opened by the oil pressure when the cam and oscillating lever operate.

The control handwheel operates the manoeuvring shaft and the roller is made to ride from the middle neutral (stop) position upward onto the cam lobe for running ahead and the downward lobe for running astern. At the neutral position, the oscillation of the lever is not conveyed to the spindle. The period during which the injection valve is to remain open is varied by altering the position of the roller. It can be moved upto the highest position on the camlobe.

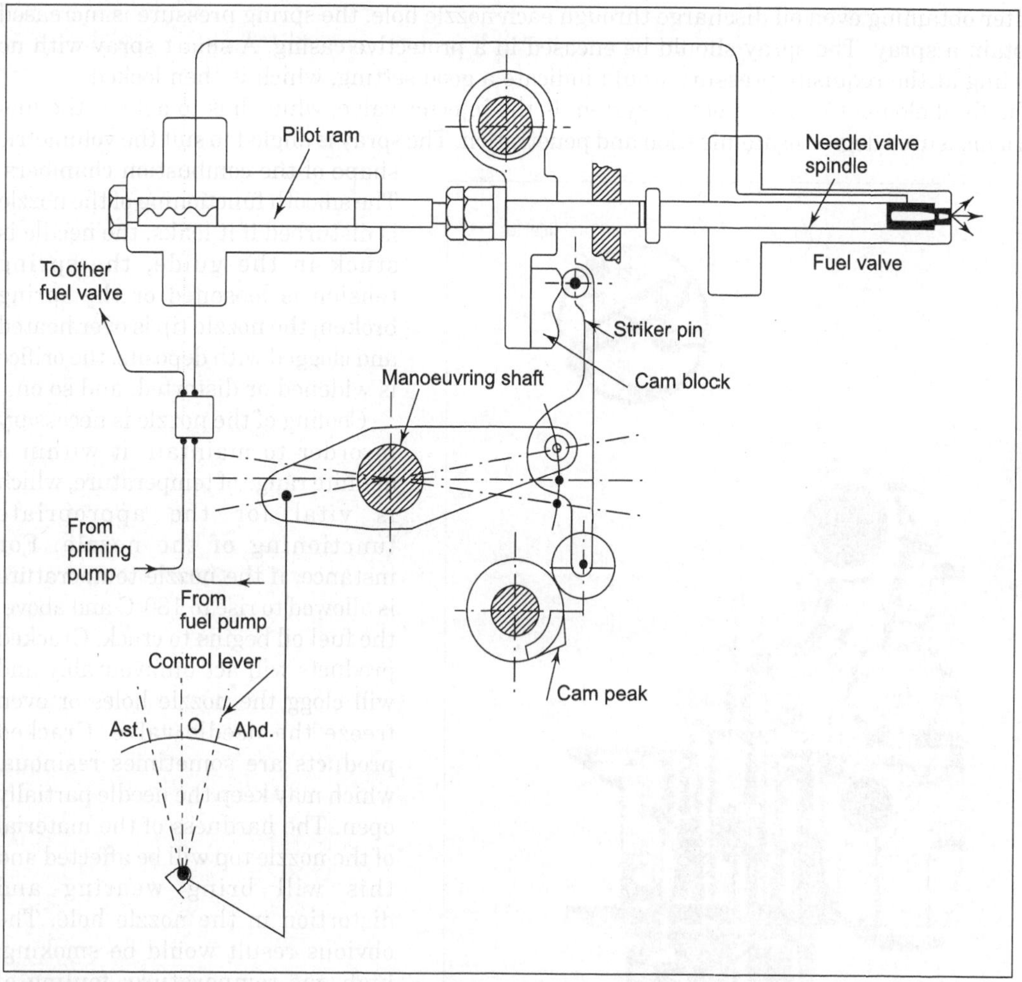

Fig. B8.2 Machanism layout showing fuel valve

Injector setting The injector should be withdrawn periodically, cleaned, and reset to the correct lifting pressure. Carbonisation on the exterior portion of the nozzle tip should be removed. All orifices should be made clear, lapping surfaces be made leak free, the cooling water space cleaned, and the needle valve free from dribbling.

The nozzle nut is removed after slackening the spring loading. The various parts are cleaned and reassembled. No lubricant needs to be applied on the lapping surfaces. All parts are reassembled by tightening the cap nut. (The maker's caution against applying any excessive torque, to be as it would distort the needle guide causing bending or sticking of the valve. It is recommended that a torque spanner be used.)

The opening pressure is adjusted by setting the spring tension. The injector is clamped on the testing stand and connected to the oil pump fitted with a pressure gauge. The passages are primed by pumping oil freely when the spring tension is low.

After obtaining even oil discharge through each nozzle hole, the spring pressure is increased to obtain a spray. The spray should be encased in a protective casing. A smart spray with no dribbling at the requisite pressure would indicate a good setting, which is then locked.

The final element in the injection system is the injector valve, which has to deliver the fuel with a measured degree of atomisation and penetration. The spray is angled to suit the volumetric shape of the combustion chambers. The smooth functioning of the nozzle is disturbed if it leaks, the needle is stuck in the guide, the spring tension is loosened or the spring broken, the nozzle tip is over heated and clogged with deposits, the orifice is widened or distorted, and so on.

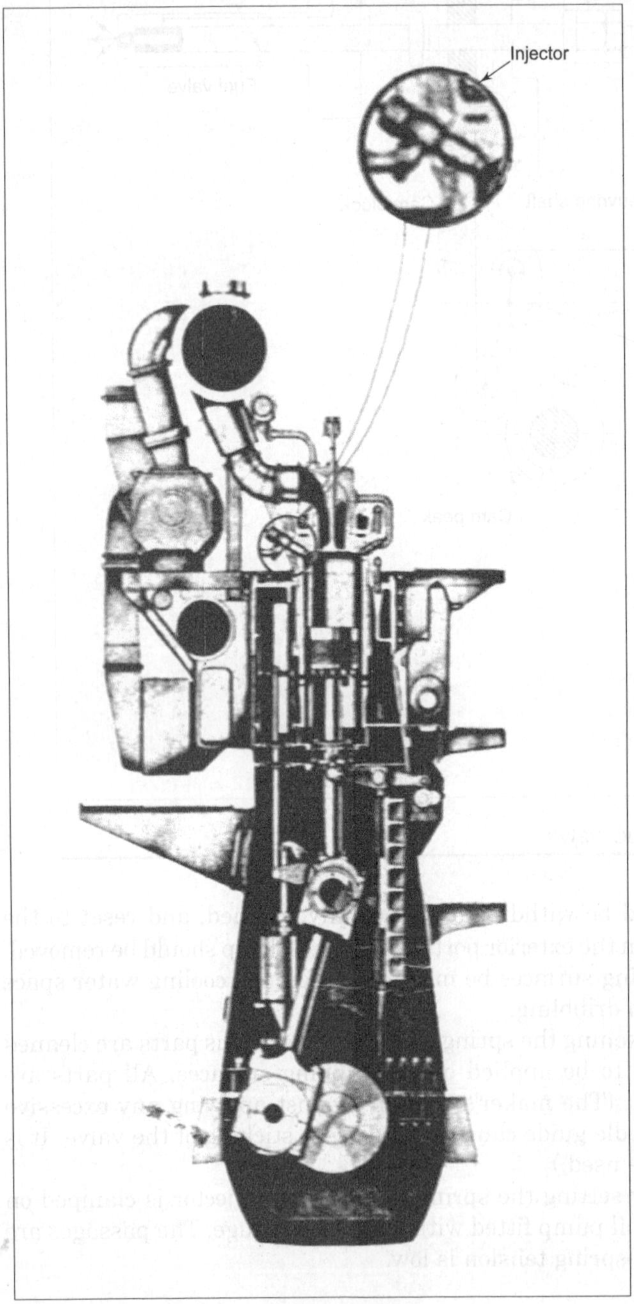

Fig. B8.3 Diesel engine showing application of injector

Cooling of the nozzle is necessary in order to maintain it within a definite range of temperature, which is vital for the appropriate functioning of the nozzle. For instance, if the nozzle temperature is allowed to rise to 180°C and above, the fuel oil begins to crack. Cracked products will act unfavourably and will clogg the nozzle holes or even freeze the needle valve. Cracked products are sometimes resinous, which may keep the needle partially open. The hardness of the material of the nozzle top will be affected and this will bring wearing and distortion in the nozzle hole. The obvious result would be smoking; high gas temperature; fouling of exhaust passage, turbine blades, and so on.

If the nozzle temperature goes below 110°C, formation of water is possible. Here, sulphur is always present and it may attack the metal in contact aggressively. Therefore, the deposits in the nozzle tips will indicate faulty cooling (temperature that is too high or too low); insufficient flow quantity, degree of atomisation too high temp. (local burning and over heating), and the deposits may contain cracked and resinous products, metallic salts, corrosion products, oil ash, and so forth.

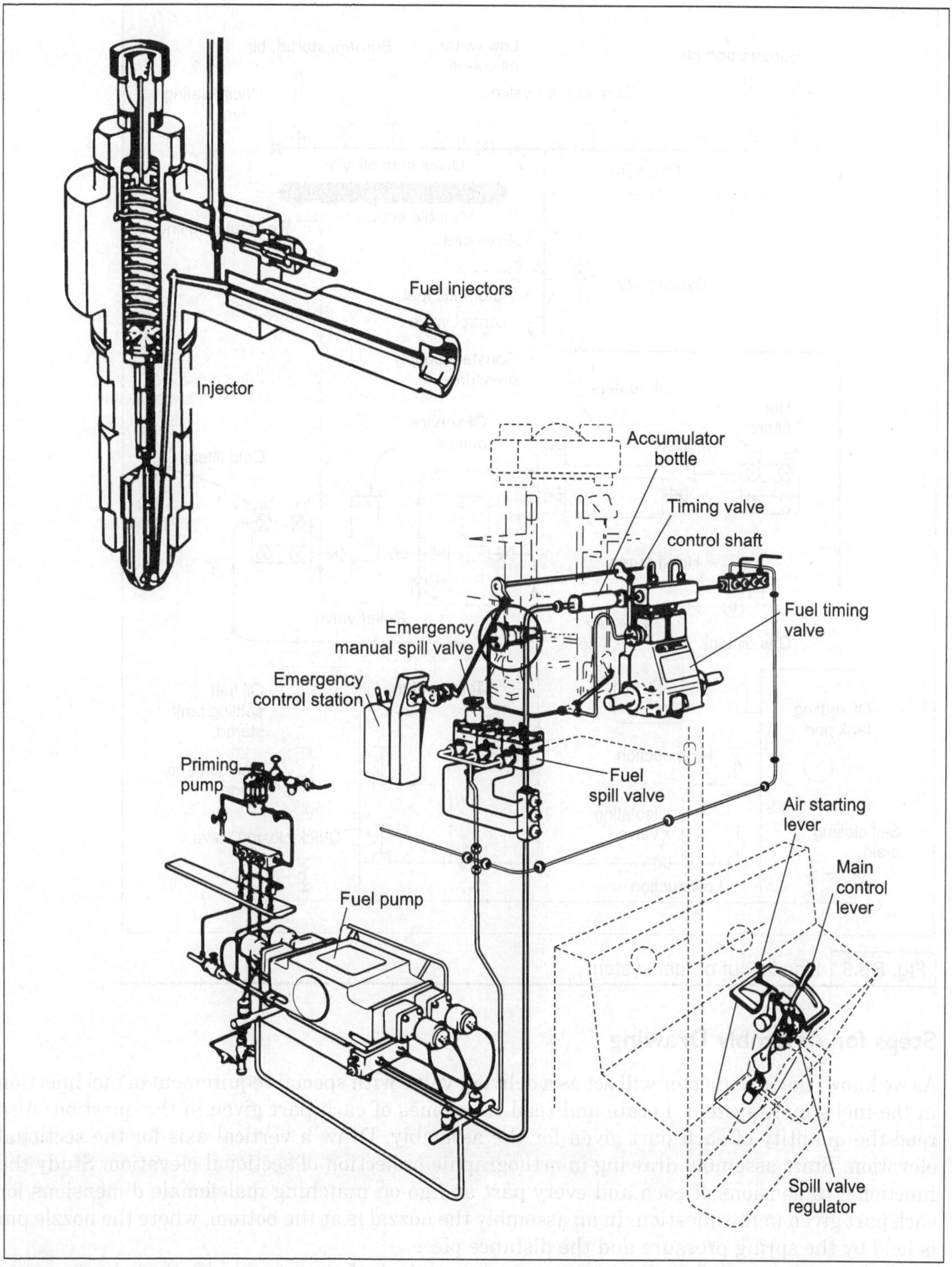

Fig. B8.4 Fuel system

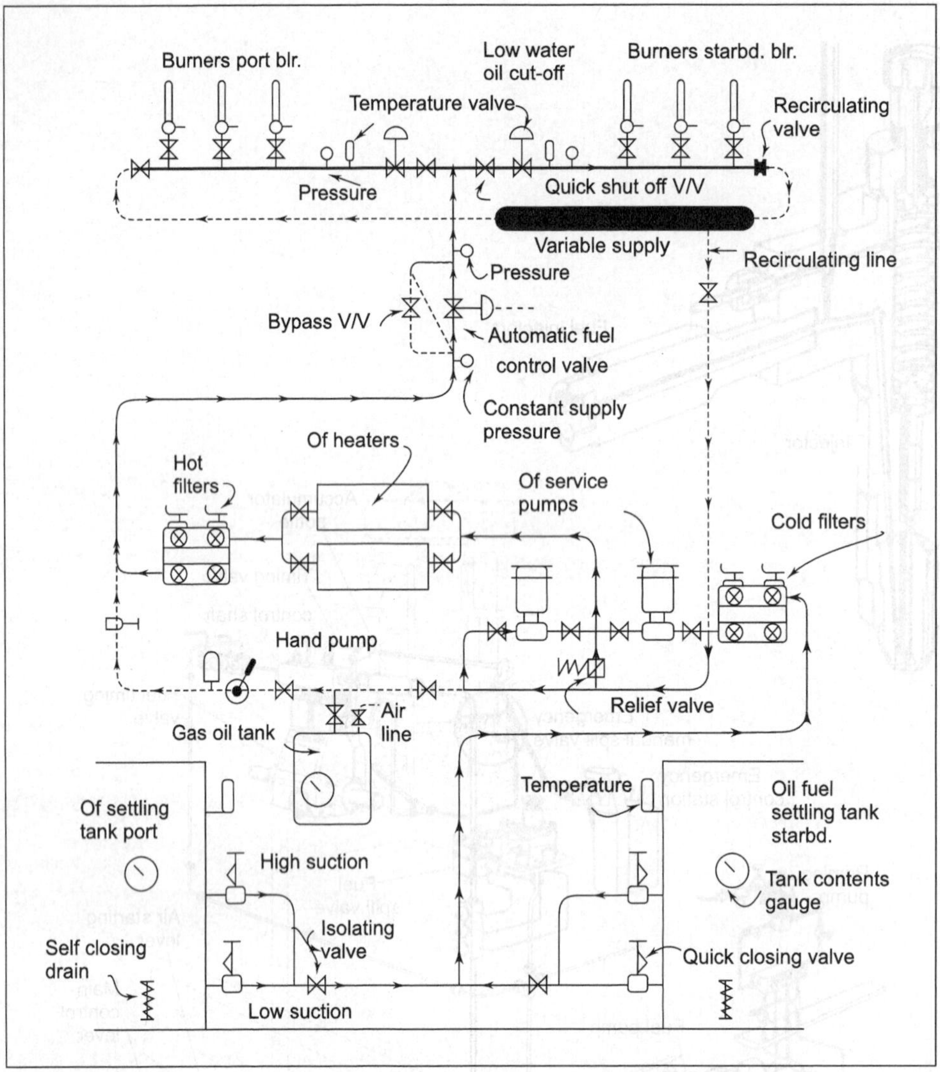

Fig. B8.5 Line layout of fuel system

Steps for Assembly Drawing

As we know, the fuel injector will act as a delivery valve with special requirement of fuel injection in the fuel supply system. Locate and read the names of each part given in the question. Also read the quantity of each part given for the assembly. Draw a vertical axis for the sectional elevation. Start assembly drawing in orthographic projection of sectional elevation. Study the functional dimensions of each and every part and go on matching malefemale dimensions for each part given in the question. In an assembly the nozzal is at the bottom, where the nozzle pin is held by the spring pressure and the distance piece.

All these parts are held inside the main body and the body is covered with the help of a cap at the top.

Answer

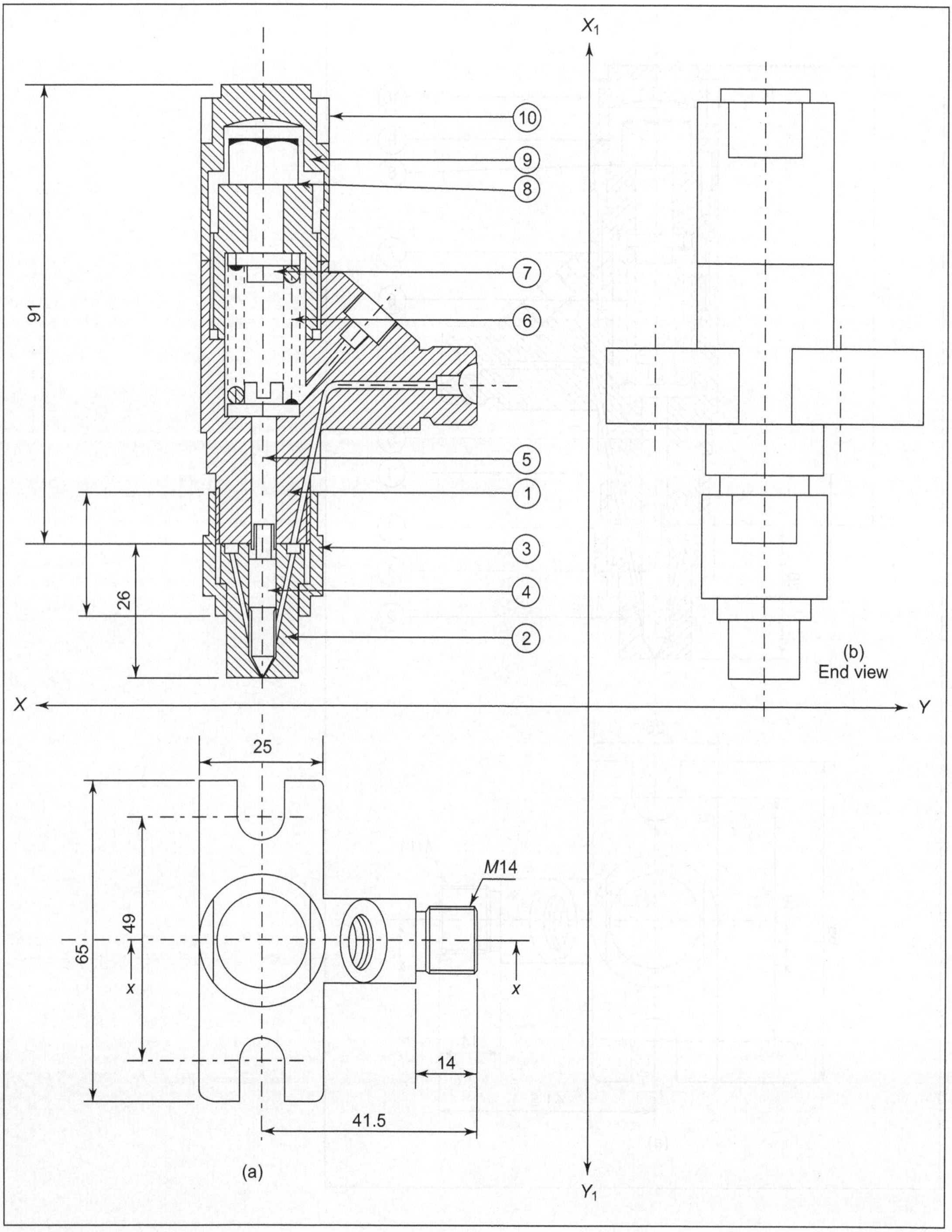

Fig. B8.6 Fuel injector details

The enlarged views of Fig. B8.6 are shown as under:

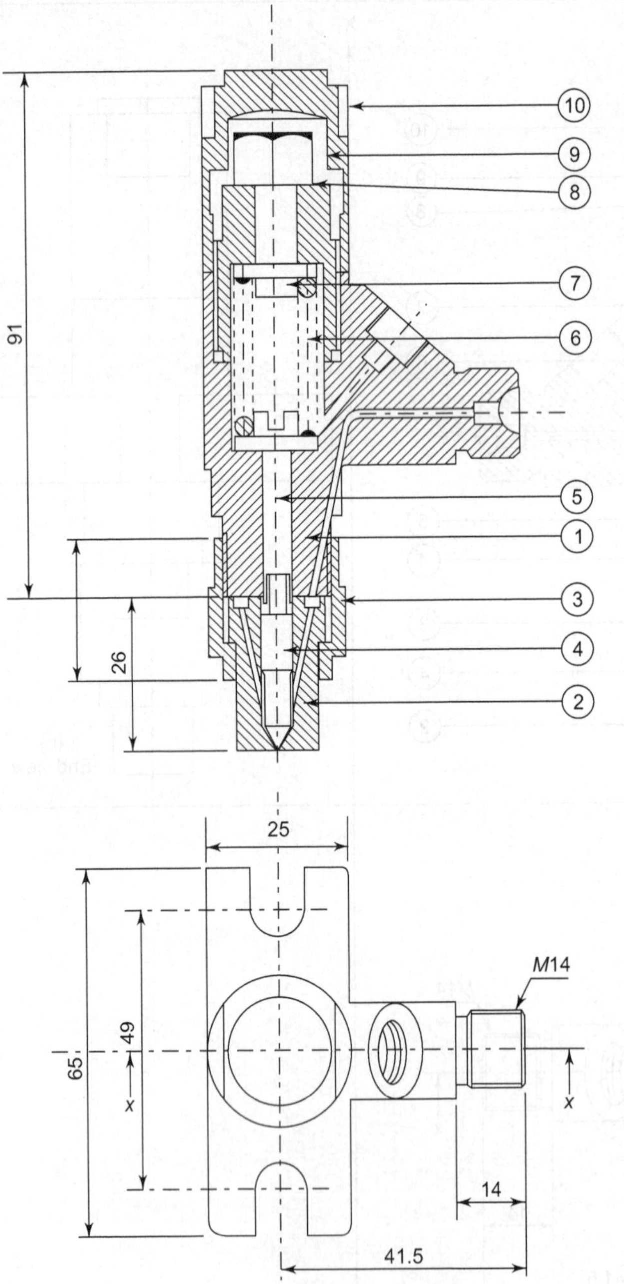

(a)

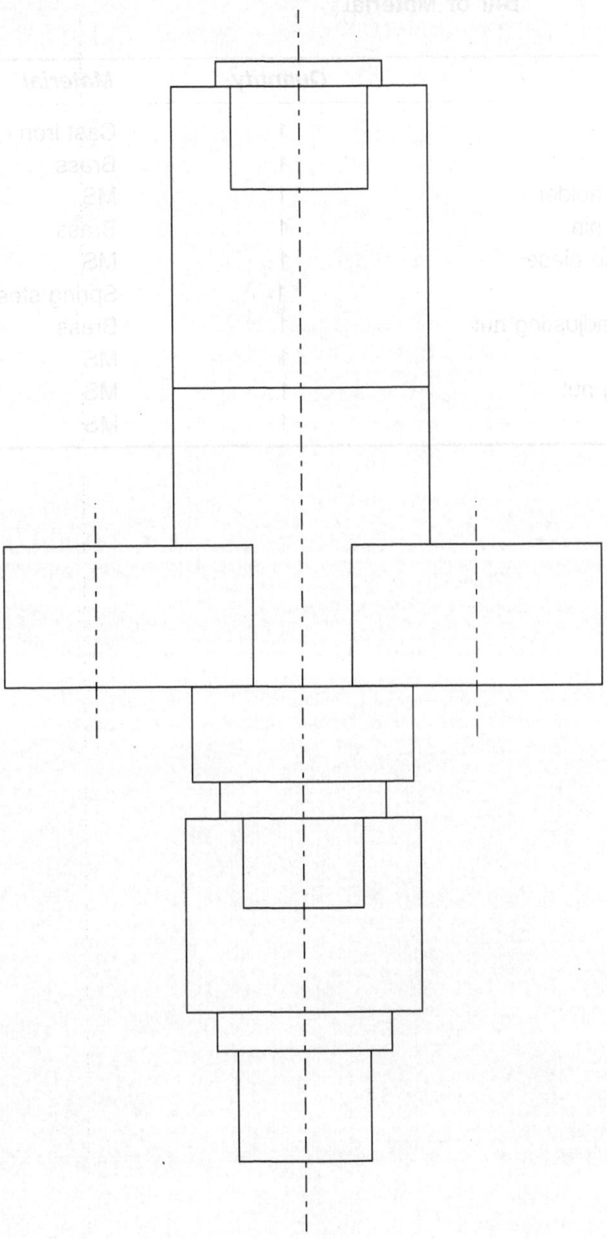

(b)
End view

Bill of Material

Sl. No.	Part	Quantity	Material
1	Body	1	Cast iron
2	Nozzle	1	Brass
3	Nozzle holder	1	MS
4	Nozzle pin	1	Brass
5	Distance piece	1	MS
6	Spring	1	Spring steel
7	Screw adjusting nut	1	Brass
8	Sleeve	1	MS
9	Locking nut	1	MS
10	Cap	1	MS

ASSEMBLY DRAWING 9

Cylinder Relief Valve

This section contains information on the cylinder relief valve

After reading the section you will be able to

- understand the working, applications and types of relief valves
- draw assembled views of the relief valve

Cylinder Relief Valve

Question

Draw the assembled views of the following:

1. Elevation in section showing assembled valve
2. End elevation (outside view)
3. Plan view (outside view)

Also include a bill of material.

The details of a cylinder relief valve are shown in Fig. B9.1.

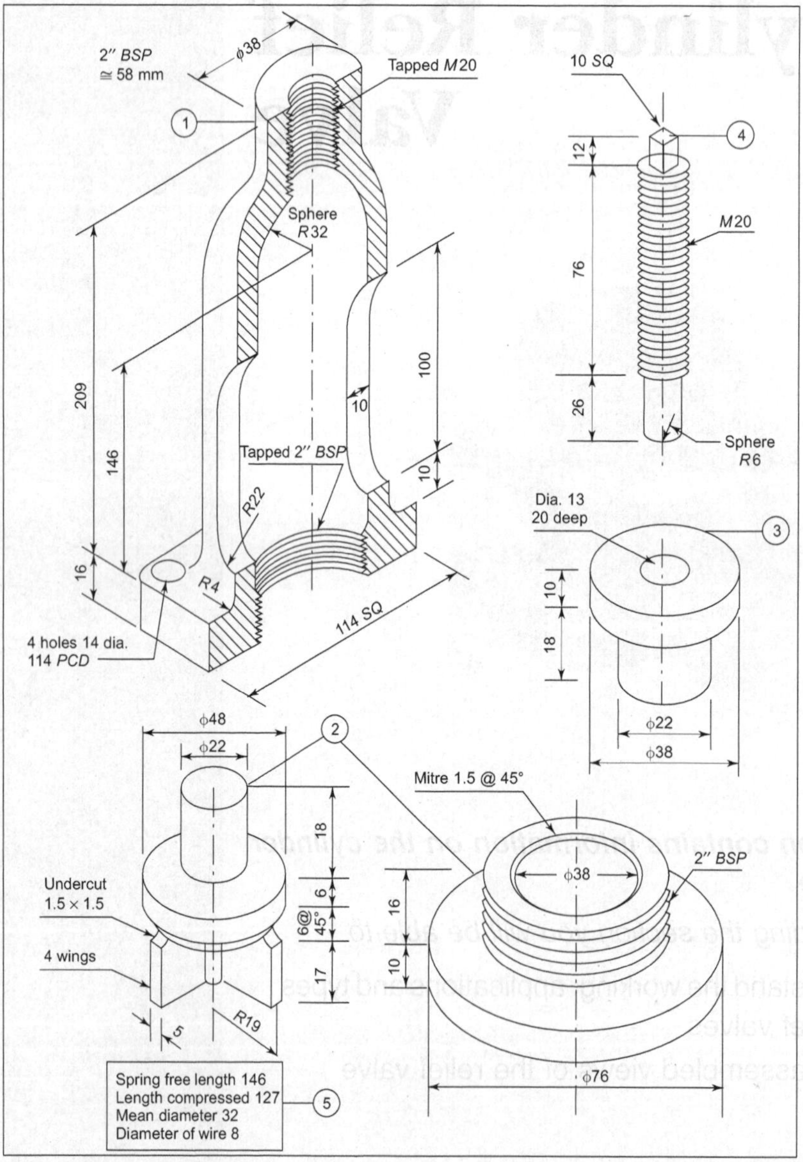

Fig. B9.1 Cylinder relief valve

Introduction

Relief valves are generally used at both the ends of steam engine cylinders. They are meant to release the excessive pressure inside the cylinder at the end of the position stroke due to the presence of water formed by the condensation of steam.

Relief valves are generally used with high pressure vessels. They are meant to release the excessive pressure inside vessels. Pressure in a vessel can be increased by

1. Overheating
2. Compression

This will maintain the downstream pressure within the defined limits over a range of flow, despite any changes in supply pressure. Compressibility of the spring (spring constant k) is calculated keeping in mind the maximum pressure that can be sustained in the vessel.

The main body is a casting of gun metal, which is strong enough to withstand the maximum pressure experienced in the vessel. It has two openings in the sides for the escape of excessive pressure causing fluid. At the bottom of the casing we have the valve seat that facilitates the introduction of a valve. The valve is tapered and under closed condition, it has a perfect fit on the valve seat. The spring is loaded on the top of the valve and its upper end is attached to the spring collar. A compression screw is fitted on the spring collar and it has a locknut arrangement.

Working

Closed condition The valve sits securely on the valve seat. The spring is under no compression and the locknut is screwed tightly on the compression screw. There is no relief action under this condition.

Open condition When the pressure in the vessel exceeds the maximum tolerable value of spring pressure by the locknut, there is a danger of the vessel getting damaged. Under this condition, the locknut is loosened, allowing the spring to get compressed. As the bottom part of the spring attached to the valve moves up, the valve is raised from the valve seat. Since the valve is tapered, a gap occurs between the valve and the valve seat.

The pressurised fluid moves up through the gap of the valve and valve seat and goes out of the main body through the openings provided. Due to release of fluid, pressure in the vessel falls down and the spring expands to its original length. With this action the valve gets pressed due to spring pressure and once again rests on the valve seat. To maintain the same gap consistently, the locknut is tightened on the screw from the top.

The relief valve can also be of 'the automatic type', wherein instead of a locknut arrangement, relay valve or poppet valve can be used. When the pressure in the vessel exceeds sufficiently, it overcomes the force of the spring attached to the poppet valve. As the poppet valve spring compresses, the piston is raised. Excessive fluid then passes through the bottom of the piston to the tank. When pressure in the vessel falls down to a tolerable value, the poppet valve resets, thereby pushing the piston back to its initial position.

Applications

The relief valve is found in, virtually, every hydraulic system. It is normally a closed valve connected between the pressure line and the reservoir. Its purpose is to limit the pressure in the system, to preset the maximum by diverting some or all of the pump's output to the tank, when the pressure setting is released.

Many rotary pumps require a relief valve on the discharge side to protect the casing and piping from excessive pressure that may build up when the pump operates against a closed valve or some obstruction develops in the piping. Some pumps are fitted with a built-in relief valve that circulates the liquid from the discharge side back to the suction. When there is a possibility of frequent shut off of the pump discharge valve, or frequent reduction of flow, the built-in relief valve should not be relied upon to perform all the pressure relieving duties, instead a separate relief valve should be used at the discharge end or outlet end of the fluid supply tank. This is more safe as compared to a built-in relief valve, especially when overheating of the liquid takes place due to lengthy and continous operation of the pump.

Types of Relief Valve

1. Direct acting relief valve
2. Pilot operated relief valve
3. Electrically modulated relief valve
4. Multiple preset pressure relief valve

Steps for Assembly Drawing

Figure B9.1 represents the components that comprise a cylinder relief valve. Though this type of valve does not illustrate modern mechanical engineering practice, from a preliminary drawing point of view, it is worth including.

Having decided upon a suitable scale for the size of paper, draw two vertical centre lines for the sectional elevation and plan view, and also for the end elevation. These are used as datum

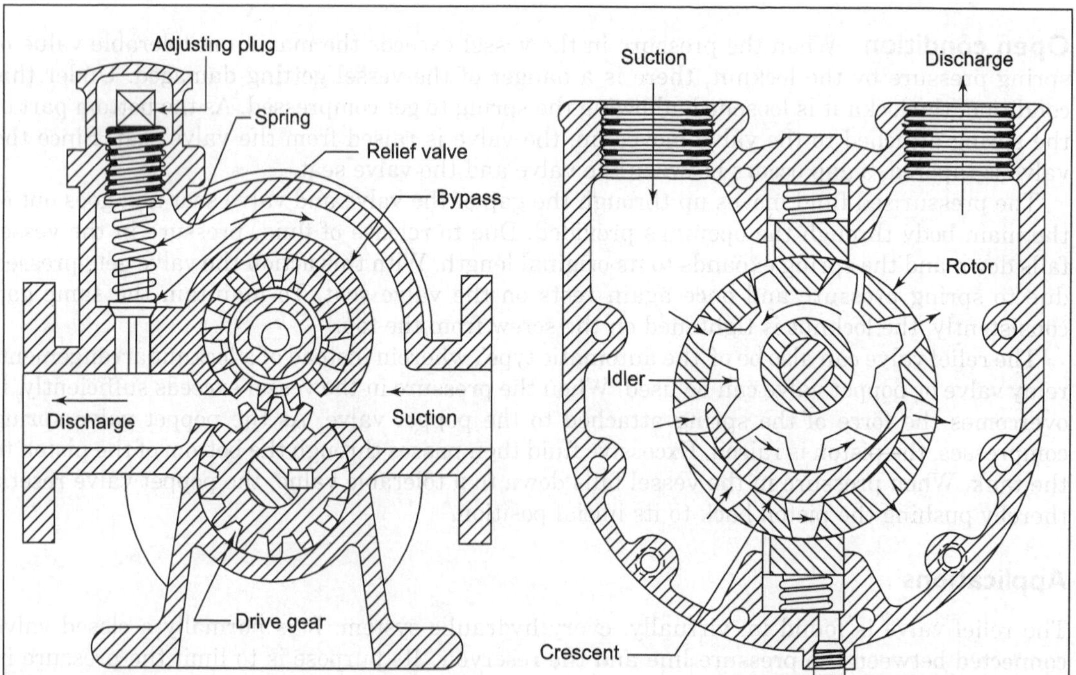

Fig. B9.2 Application of cylinder relief valve

lines for dimensions in the horizontal direction. The base of the valve cannot be drawn now. Taking into account the overall height, this line is used as the datum line for vertical dimensions. At 162 mm up from the base line in both elevations, draw a spherical radius of 42 mm. On the sectional view where the thickness of the material will be seen, draw another radius of 32 mm to represent the inside of the casing. The radii are now joined to the base by vertical lines and radii at the bottom. Note that it is easier to blend a straight line with a radius, rather than a radius with a line. The base flange 114 mm square and 16 mm thick is now drawn and a 2″ *BSP* tapped hole is shown in the sectional view. Note that a 2″ *BSP* thread measures about 57 mm diameter and was earlier used in pipe threads where 1/4 inch was added to the tap size to give the actual diameter. The seat can now be added to the sectional elevation and the 76 mm diameter flange that projects below the base is shown in the end elevation. The valve, which is not sectioned is now added together with two coils of the spring in section, one at each end of its 146 mm free length. Finish off the top of the body to 38 mm diameter and a suitable radius, the height from the base being 225 mm. The spring cap in section and the adjusting screw and locknut shown in full could now be drawn and the elevations completed by showing the escape ports.

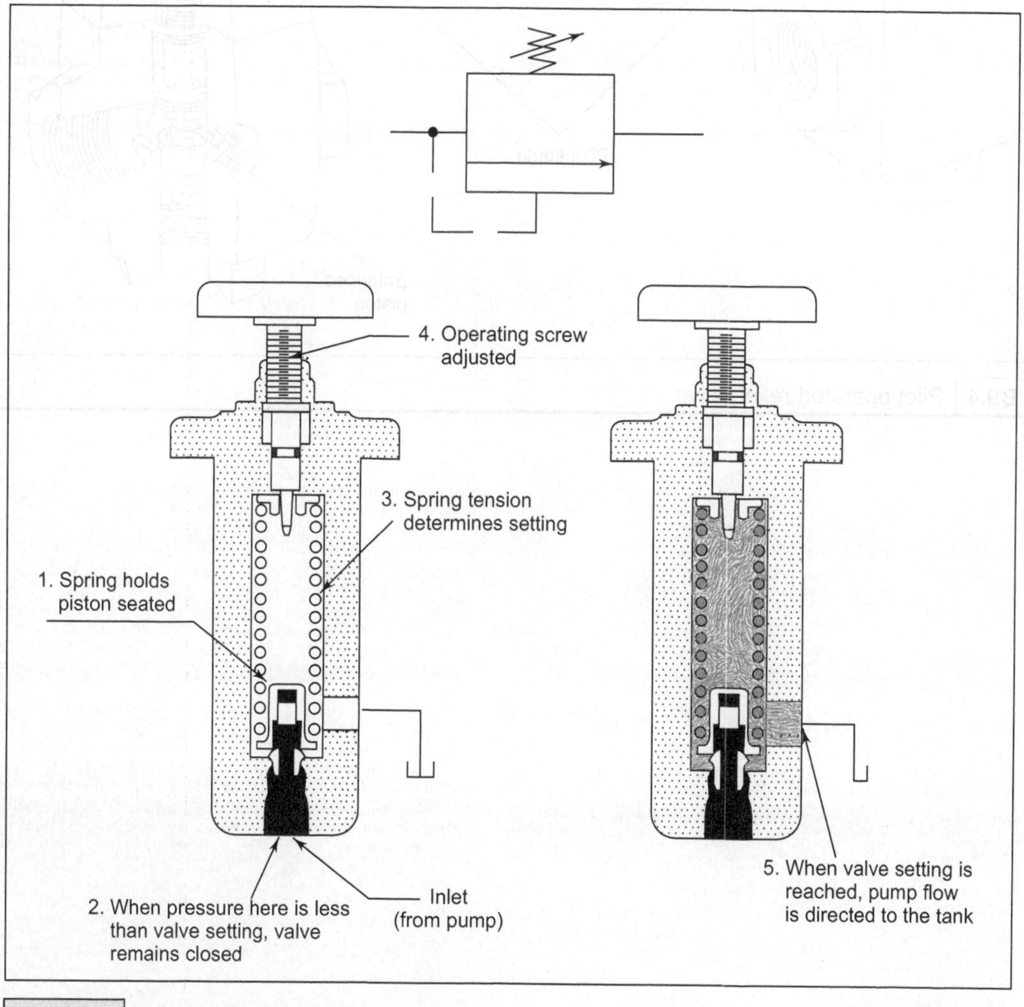

Fig. B9.3 Direct acting relief valve

The plan view consists of a 114 mm square to represent the bottom flange with 4-16 mm diameter holes on a 114 mm *PCD*. No radius is given for the corners on the square. But as this is not important, it is left to the discretion of the student. A circle 84 mm diameter represents the outside of the body and the dotted circle 64 mm diameter, the inside of the body. The two dotted lines 44 mm apart represent the ports. Finish off the plan view with a hexagon 39 *A/F* to represent the locknut, concentric circles (the inner circle broken), and a square of 10 mm to illustrate the end of the 20 mm diameter adjusting screw.

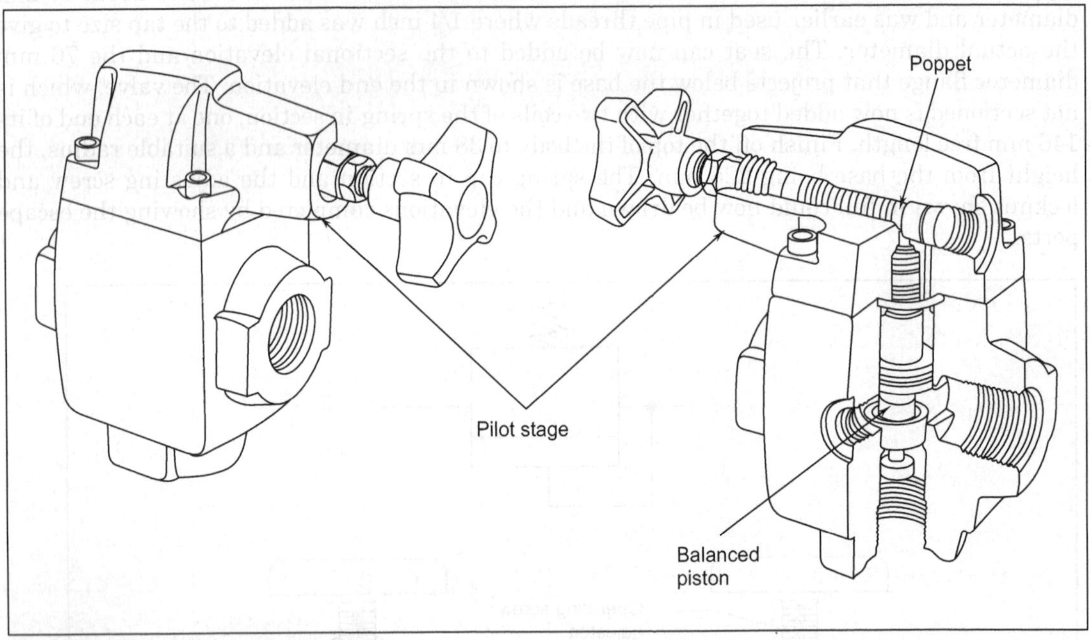

Fig. B9.4 Pilot operated relief valve

Answer

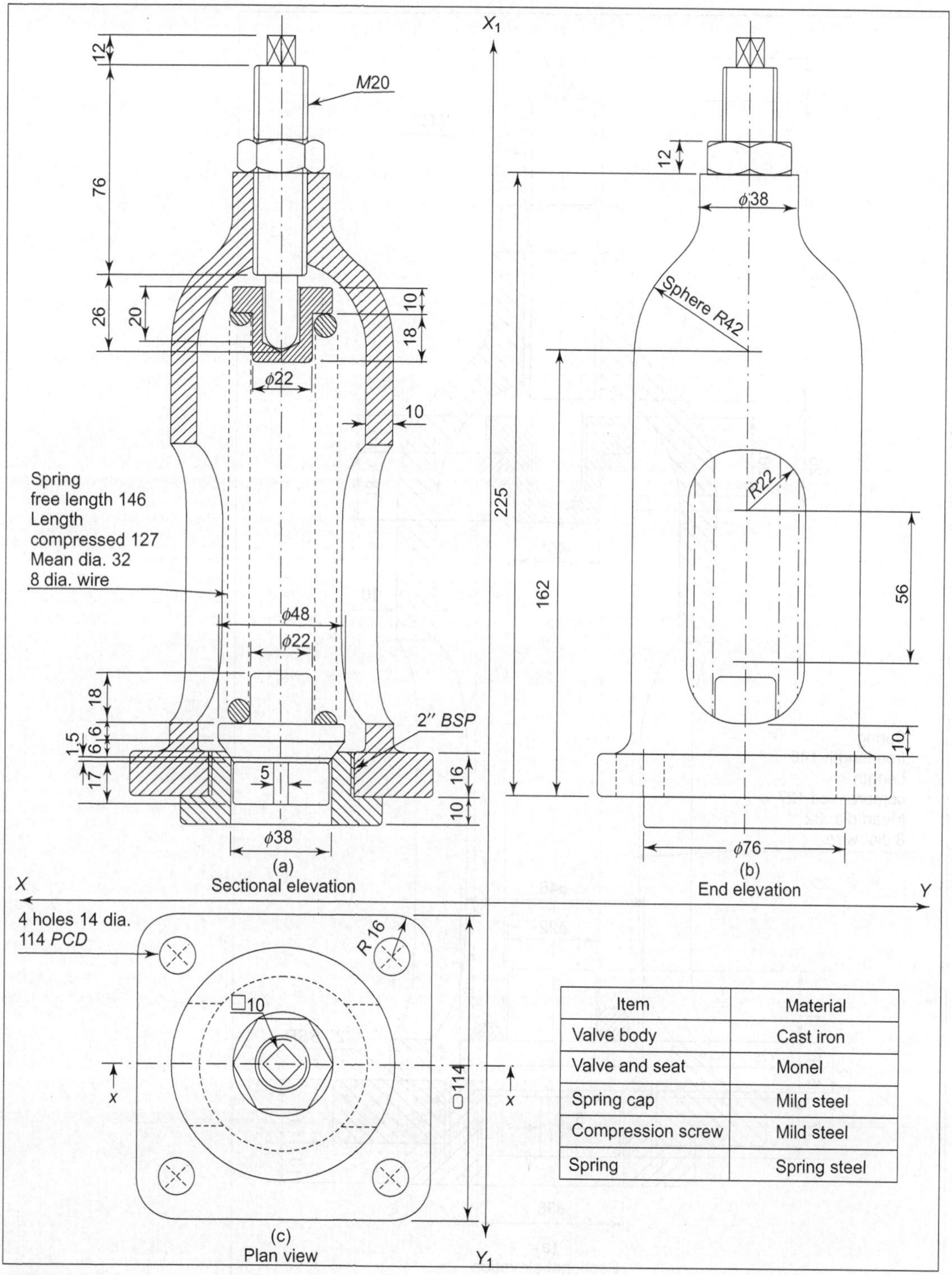

Item	Material
Valve body	Cast iron
Valve and seat	Monel
Spring cap	Mild steel
Compression screw	Mild steel
Spring	Spring steel

Fig. B9.5 Assembly of cylinder relief valve

The enlarged view of the Fig. B9.5 is shown as under:

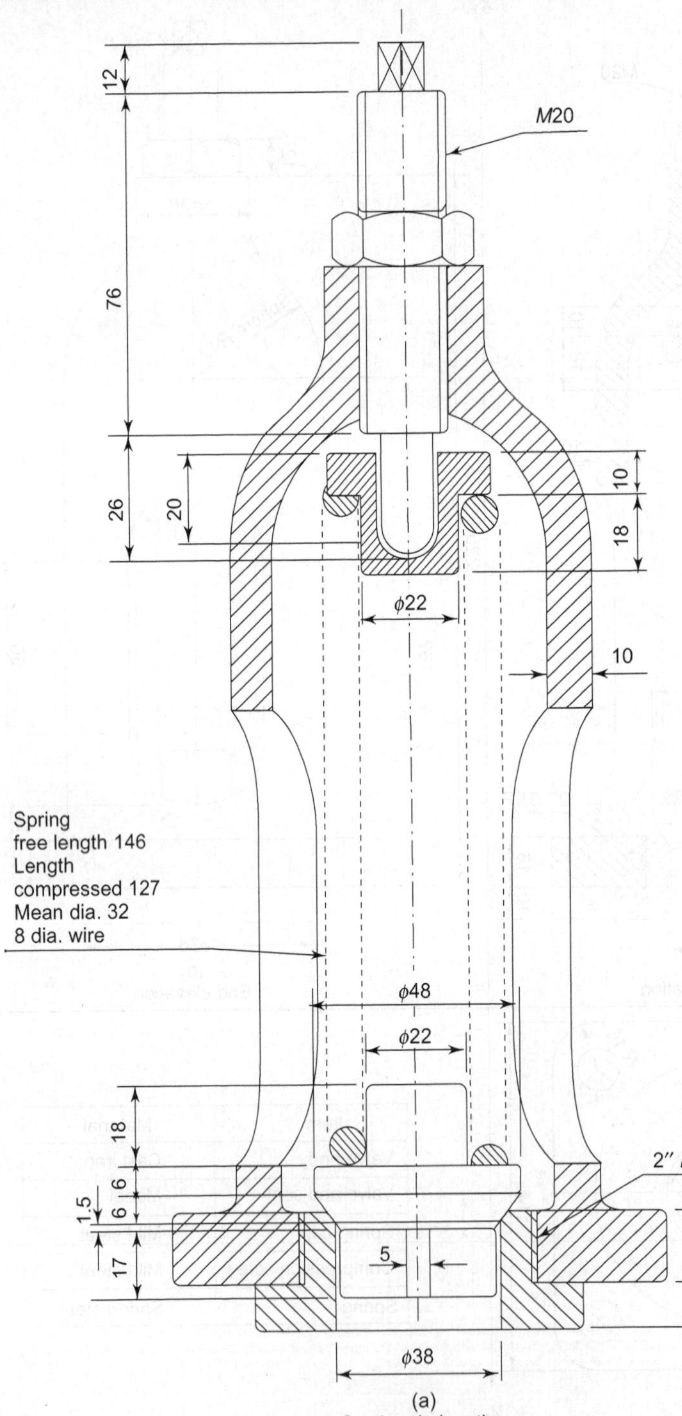

(a)
Sectional elevation

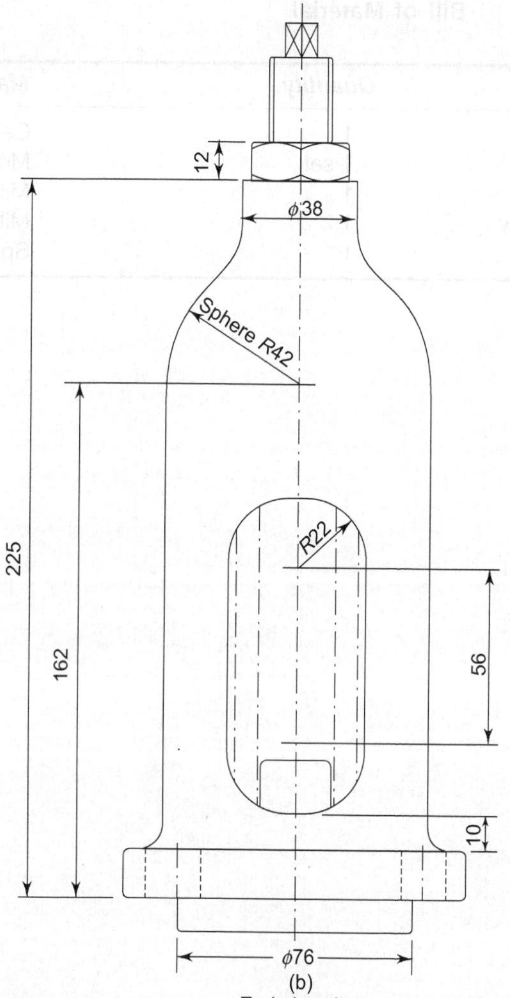

(b)
End elevation

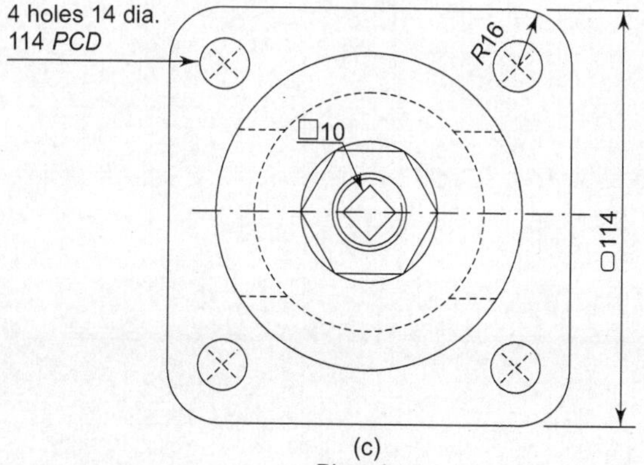

(c)
Plan view

Bill of Material

Sl. No.	Part	Quantity	Material
1	Valve body	1	Cast iron
2	Valve and seat	1 set	Monel metal
3	Spring cap	1	Mild steel
4	Compression screw	1	Mild steel
5	Spring	1	Spring steel

ASSEMBLY DRAWING 10

Feed Check Valve

This section contains information on feed check valve

After reading the section you will be able to

- understand the functions of safety fittings and control fittings
- functions of the various components of feed check valve
- draw its assembled views

Feed Check Valve

Question

Draw the following assembled views of feed check valve.

1. Sectional elevation of the assembled valve with inlet branch on the right, the section plane being taken through the centre of the valve
2. End elevation (outside view)
3. Plan (outside view)

Also include a bill of material.

The details of a feed check valve are shown in Fig. B10.1.

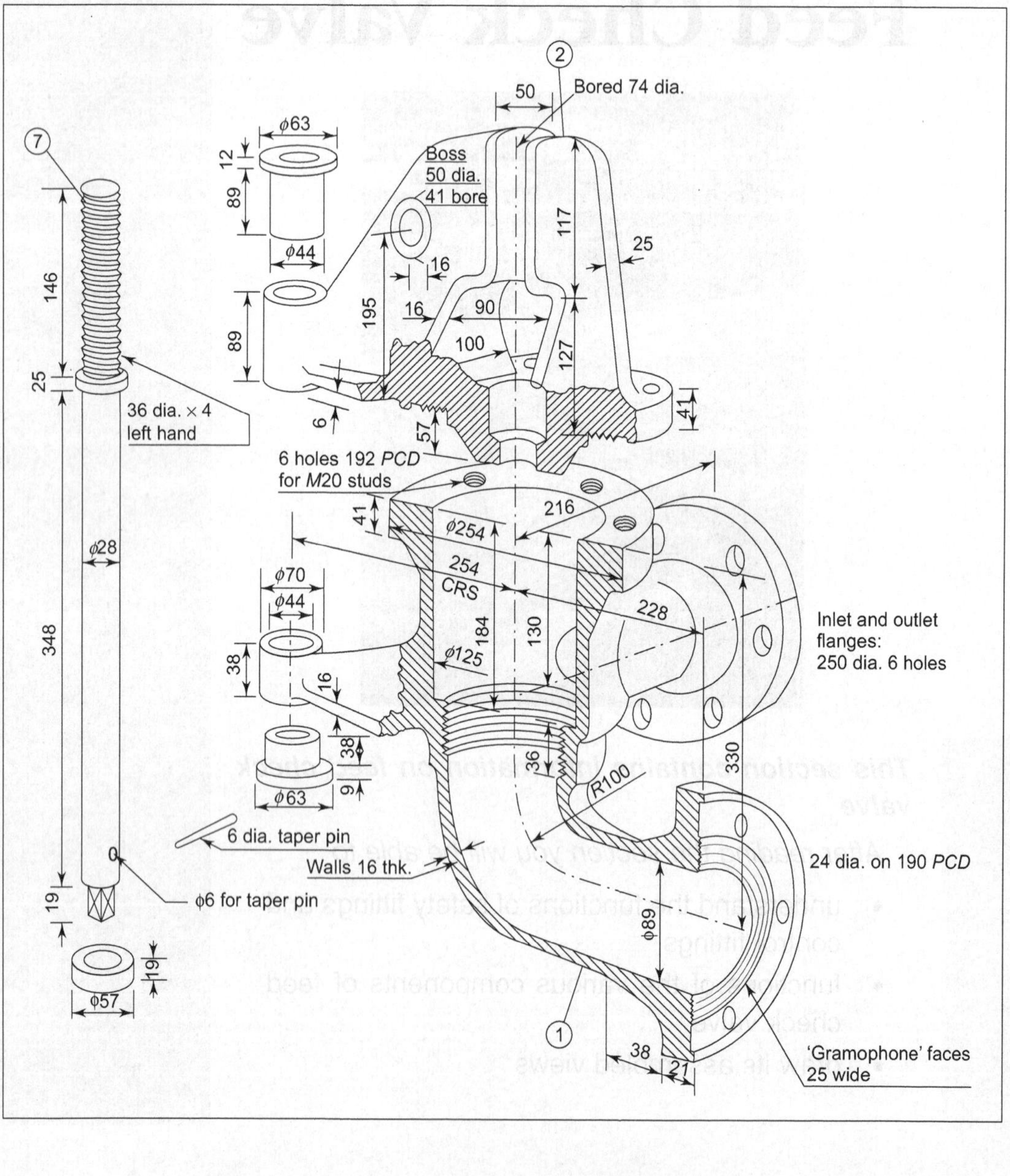

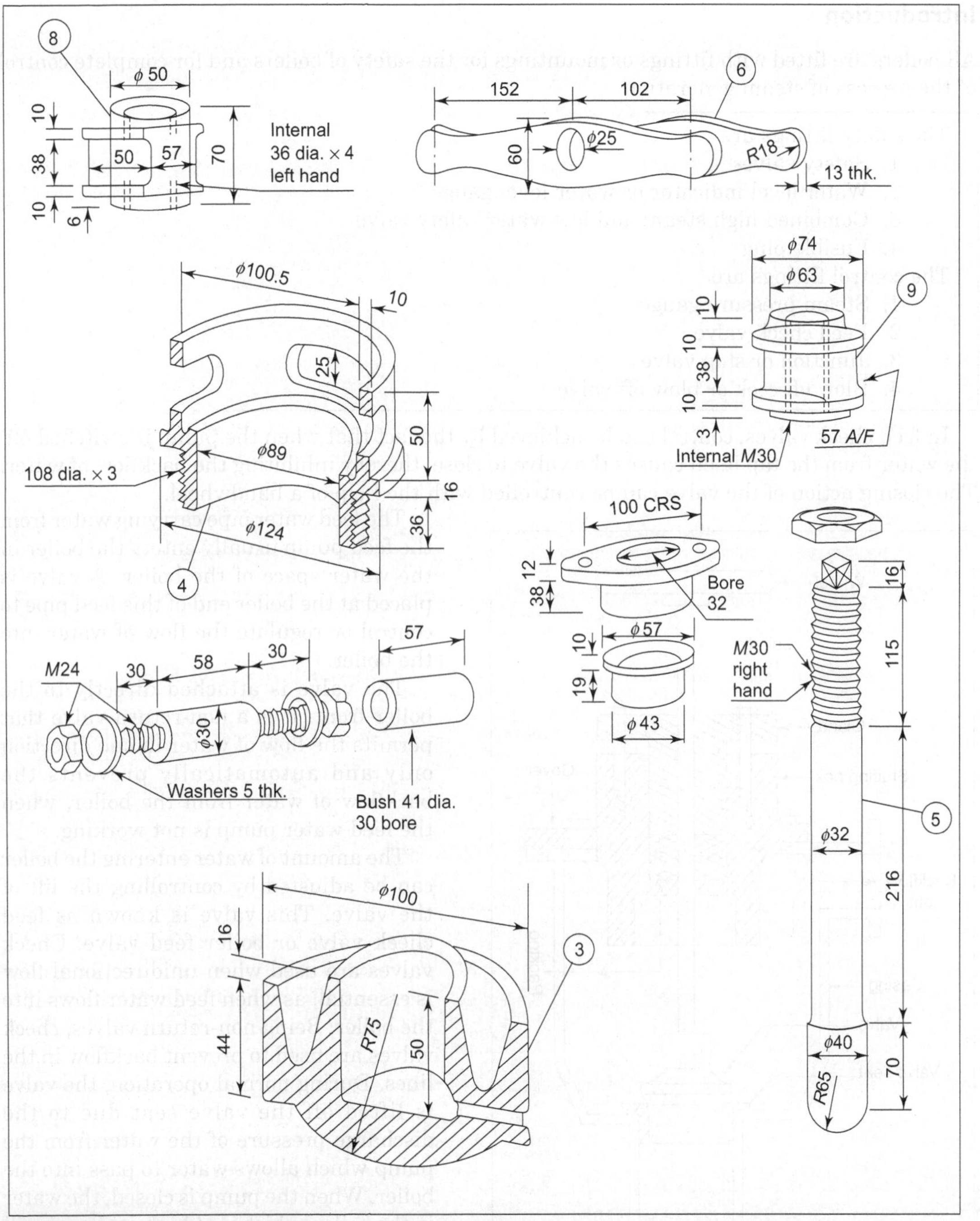

Fig. B10.1 Feed check valve

Introduction

All boilers are fitted with fittings or mountings for the safety of boilers and for complete control of the process of steam generation.

The safety fittings are
1. Safety valves
2. Water level indicator or water level gauge
3. Combined high steam and low water safety valve
4. Fusible plug

The control fittings are
1. Steam pressure gauge
2. Feed check valve
3. Junction or stop valve
4. Blow off cock or blow off valve

In feed check valves, control can be achieved by the fact that when the pump is switched off, the water from the tap itself causes the valve to close, thereby inhibiting the backflow of water. The closing action of the valve can be controlled with the help of a handwheel.

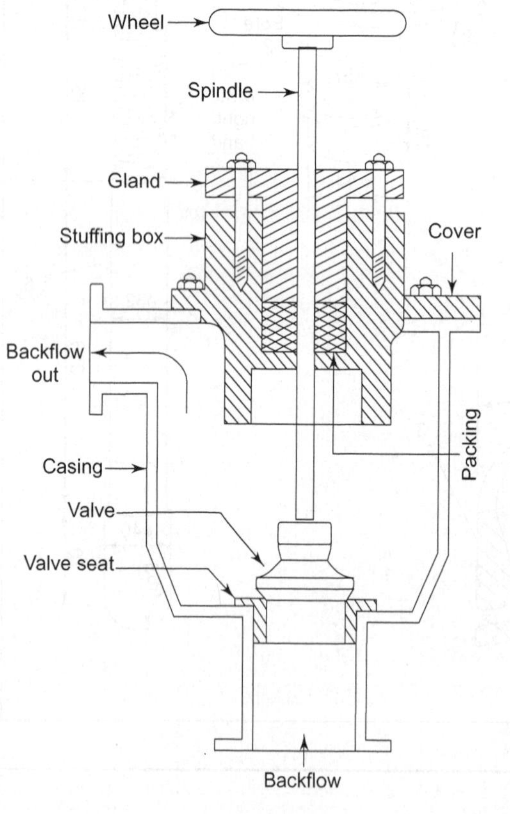

Fig. B10.2 Feed check valve

The feed water pipe carrying water from the feed pump usually enters the boiler in the water space of the boiler. A valve is placed at the boiler end of this feed pipe to control or regulate the flow of water into the boiler.

The valve is attached directly to the boiler front. It is a non-return valve that permits the flow of water in one direction only and automatically prevents the backflow of water from the boiler, when the feed water pump is not working.

The amount of water entering the boiler can be adjusted by controlling the lift of the valve. This valve is known as feed check valve or boiler feed valve. Check valves are used when unidirectional flow is essential, as when feed water flows into the boiler. Being non-return valves, check valves are used to prevent backflow in the lines. During normal operation, the valve is lifted off the valve seat due to the discharge pressure of the water from the pump which allows water to pass into the boiler. When the pump is closed, the water in the boiler, which is at high pressure, will rush back and press the valve tightly upon its seat. The flow can also be controlled with the aid of a handwheel that can regulate the lift of the valve by means of a spindle connected to it.

Function of Components

Valve body: It is the main structure comprising of all parts, holding each component in position.

Cover: It is used to cover the valve body.

Valve: It is part of the valve body that restricts the flow amount of fluid that is not required.

Valve seat: The seat on which the valve rests.

Spindle: The main component carrying the valve on which a force is applied manually or automatically, in order to make the valve work.

Lever: It connects to the spindle directly or by means of a gear meshing. It is a link which provides connectivity between the main spindle and operating stud.

Operating studs: It is used to operate the spindle.

Operating nuts: It is screwed onto the external threaded stud and external threading of the spindle.

Steps for Assembly Drawing

In the feed check valve given in the question, the piston formation of the valve eliminates hammering and pulsating, common to ordinary valves. The valve allows only a sufficient amount to pass through. The rocking lever gear for remote operation eliminates the low mechanical efficiency of the spur gear arrangement.

In the sectional elevation, draw the vertical centre line of the valve and the vertical centre line of the operating spindle 254 mm apart. Horizontal centre lines for the inlet and outlet branches are now drawn 200 mm apart and the valve body is built around these from the dimensions given. Locate the valve seat in the screwed part of the body and then the valve in the seat. Before drawing the valve spindle, the cover and the operating spindle should be shown. Locate the operating nut, which is tapped 36 mm diameter × 4 *LH*. This could be drawn at the bottom of the screwed portion of the operating spindle nut and the other end on a collar of 74 mm diameter cylindrical nut screwed on the valve spindle. This nut is capable of being adjusted on the valve spindle and when in the correct position, it is locked by a *M*30 locknut. It can be seen from the pictorial view that the cover of the valve has guides cast on it and these are bored to 74 mm diameter to suit the valve spindle nut, which slides up and down in these guides. The valve spindle can now be drawn, corresponding with the valve in the open position. It will be seen that there is a neck bush in the bottom of the stuffing box so that when excessive wear occurs, this bush, rather than the whole cover, is renewed. The plane of the required section in part (1) of the question cuts through a 25 mm thick web on the cover and a similar web on the body supporting the boss for the operating spindle, but being webs these are not shown in section.

The various heights are obtained in the end view by projecting from the sectional elevation and in this view rocker levers will be seen end on and 57 mm apart, with the operating spindle nut sandwiched between them.

Answer

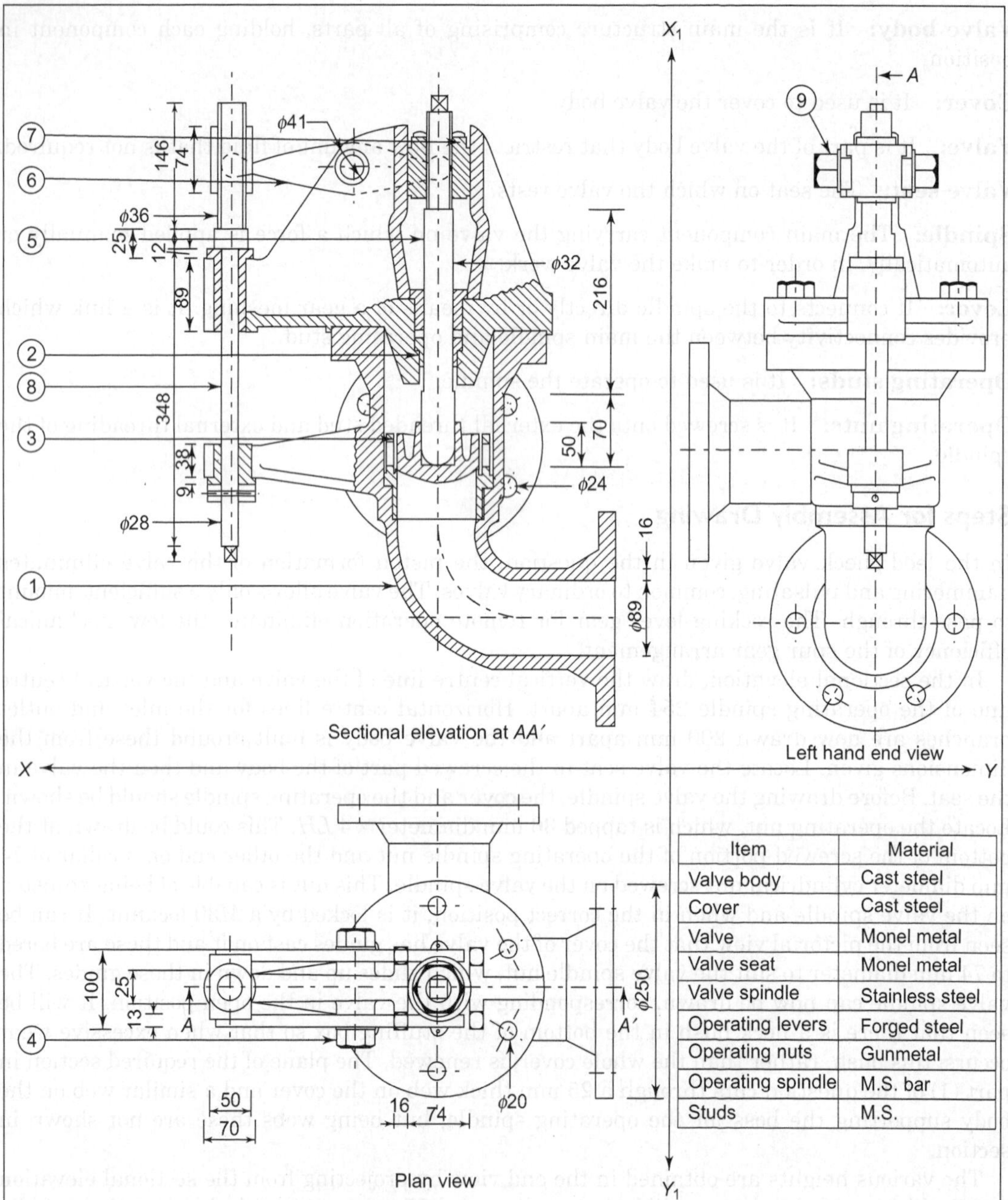

Item	Material
Valve body	Cast steel
Cover	Cast steel
Valve	Monel metal
Valve seat	Monel metal
Valve spindle	Stainless steel
Operating levers	Forged steel
Operating nuts	Gunmetal
Operating spindle	M.S. bar
Studs	M.S.

Fig. B10.3 Assembly of feed check valve

The enlarged view of Fig. B10.3 are shown as under:

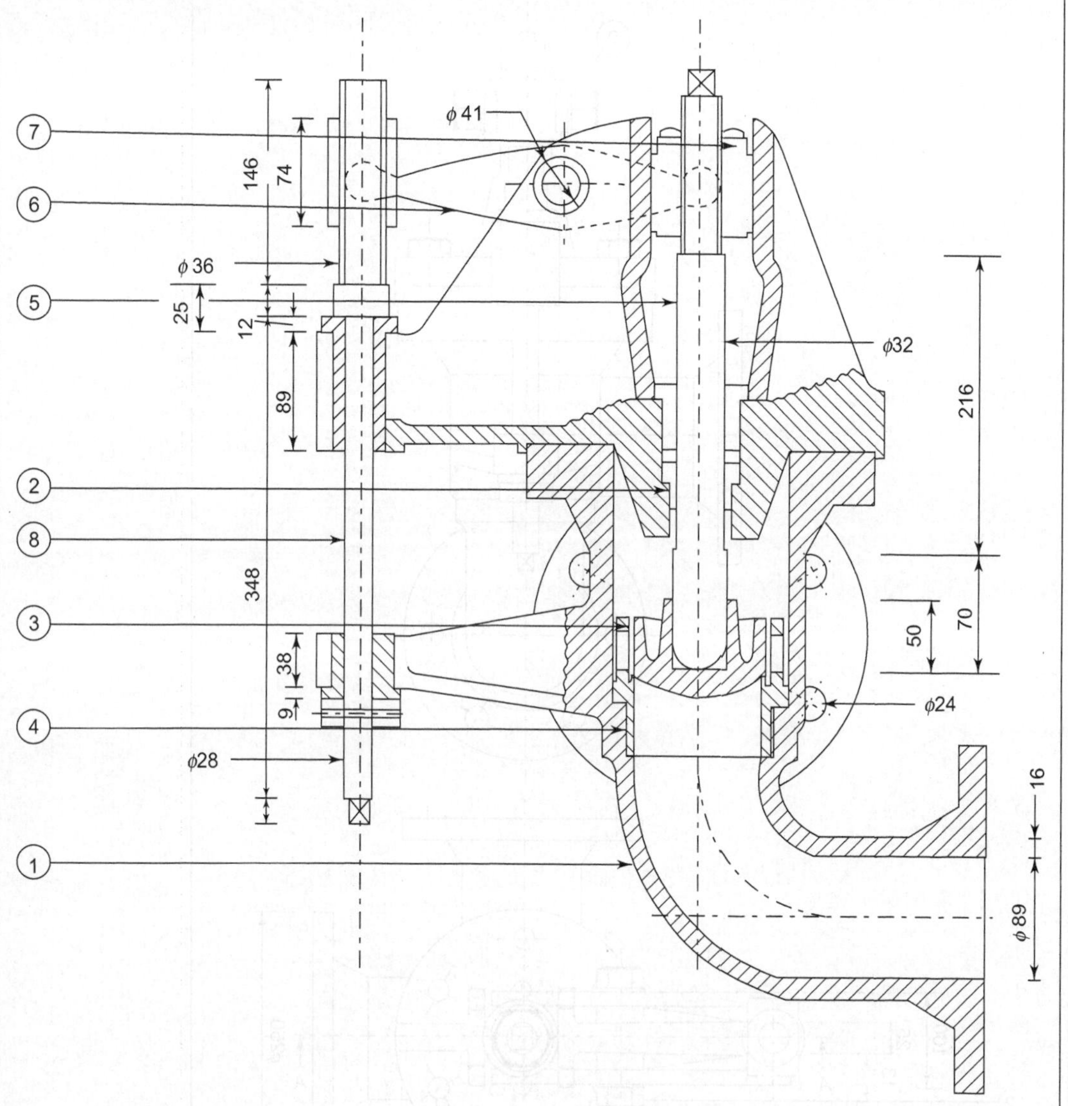

Sectional elevation at *AA′*

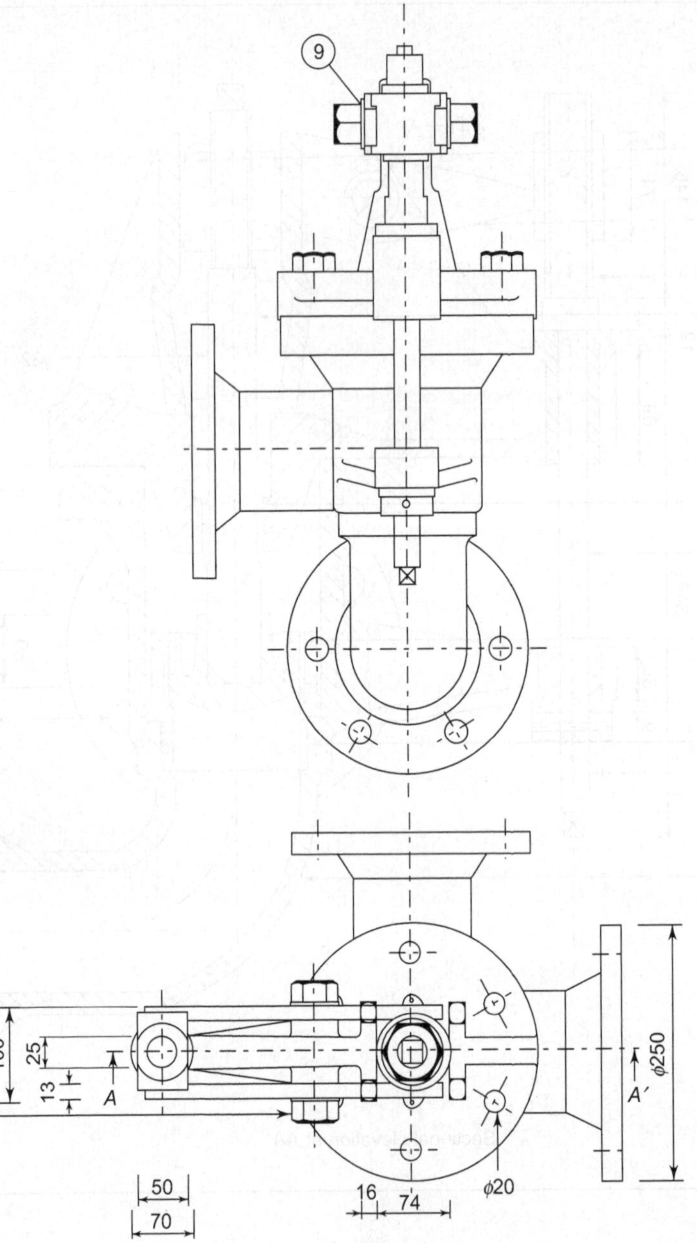

Plan view

Bill of Material

Sl. No.	*Item*	*Quantity*	*Material*
1	Valve body	1	Cast steel
2	Cover	1	Cast steel
3	Valve	1	Monel metal
4	Valve seat	1	Monel metal
5	Valve spindle	1	Stainless steel
6	Operating levers	1set	Forged steel
7	Operating nuts	2	Gunmetal
8	Operating spindle	2	MS bar
9	Studs	6	MS

ASSEMBLY DRAWING 11

Control Valve

This section contains information on control valves

After reading the section you will be able to

- understand the different elements that constitute a control valve
- understand its setup and applications
- make assembly drawings of control valves

Control Valve

Question

Draw the following assemled views of a control valve.

1. Longitudinal section through assembled valve with fulcrum pin vertically above right hand branch
2. End elevation (outside view)
3. Plan view (outside view)

Also include a bill of material.

The details of a control valve are shown in Fig. B11.1.

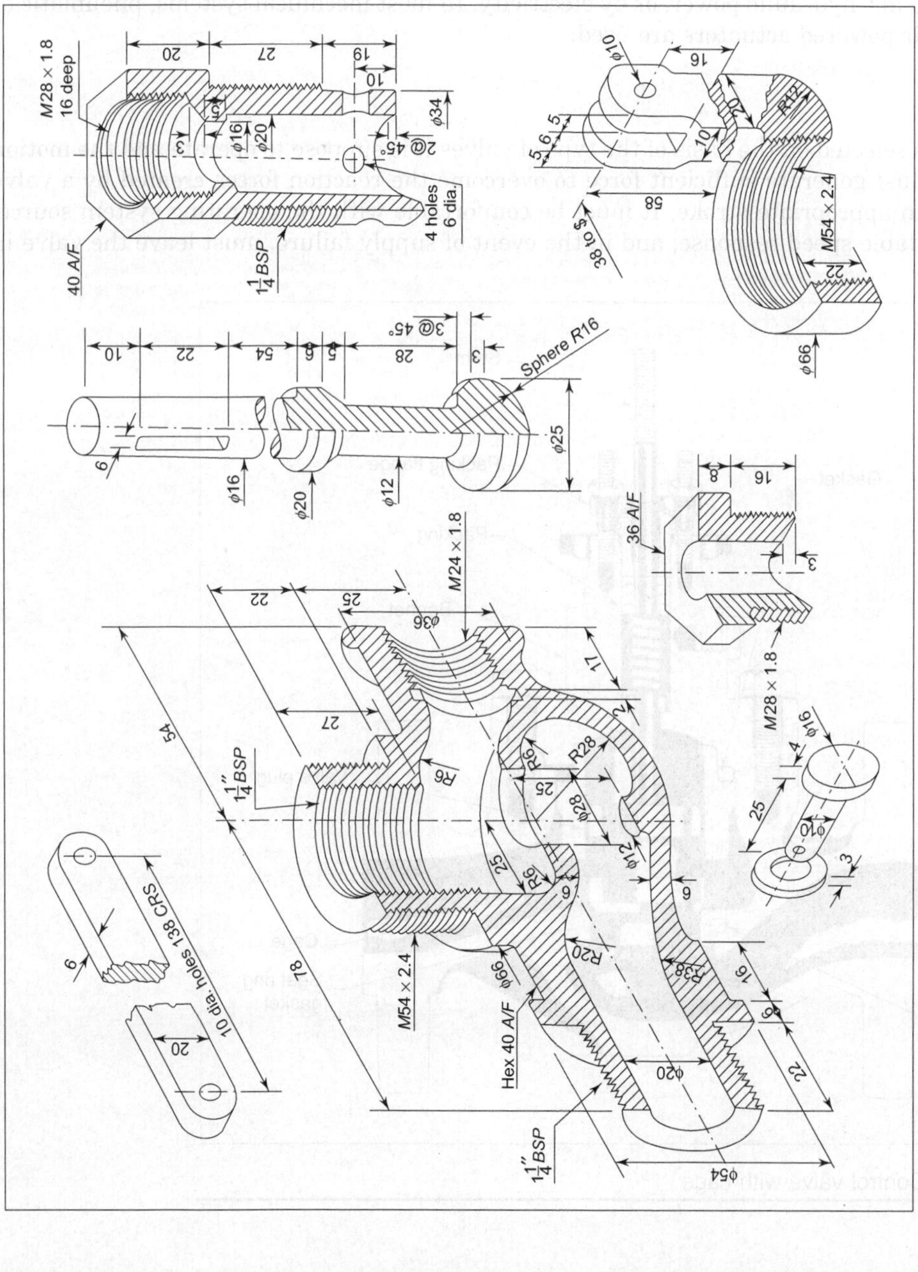

Fig. B11.1 Control valve

Introduction

In all the control loops, a correcting unit is employed, which alters the output signal of the controller in order to return the system to its desired value. This correcting unit can be a valve, a motor or a damper. Most mechanical systems employ a control valve for this purpose.

A control valve is basically made up of two distinct units

- Actuator or Motor Element—It provides the required operating force
- Valve Body—It is used to regulate the flow of liquid

Movement of the valve or other units require considerable amount of force. This can be provided by compressed air, hydraulic power, or by electricity. In most mechnical systems, pneumatic or compressed air powered actuators are used.

Actuators

An actuator is selected on the basis of the type of valves it is suppose to operate and the motion required. It must generate sufficient force to overcome the reaction forces created by a valve and provide an appropriate stroke. It must be comfortable with the operating system source; act with a suitable speed response; and in the event of supply failure, must leave the valve in

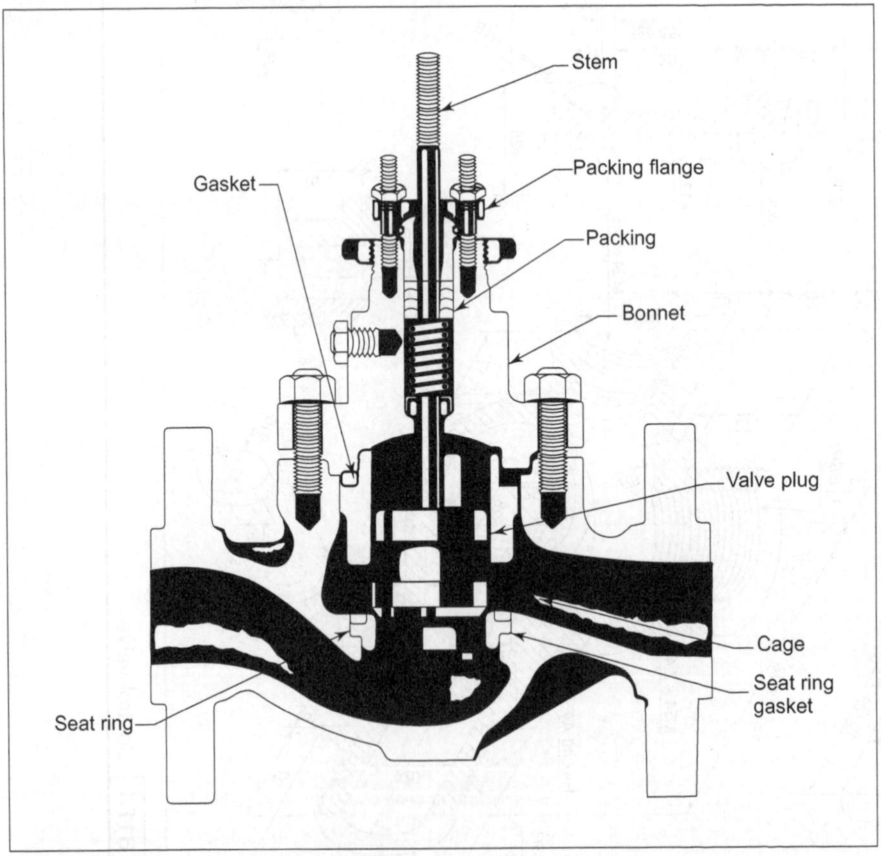

Fig. B11.2 Control valve with cage

the most desirable condition. There are three major types of actuators, depending on the nature of the operating medium.

- Pneumatic—They operate on compressed air.
- Hydraulic—They operate on hydraulic oil.
- Electric—They operate on electricity.

Valve body The actuator yoke is fastened on to the valve body and the valve spindle fastens to the actuator stem. Depending on the shape of the valve body, valves are of two types:

- Globe valves
- Butterfly valves

The selection of the valve body material is based upon the pressure, temperature, and corrosive and erosive properties of the flowing liquid. Usually cast iron and cast carbon steel are used and in some cases non-femous alloy can be used depending on the applications.

Construction Various components of a control valve are as follows.

Valve disk or plug: It is a movable part, which provides a variable restriction to the fluid flow. Various plug or disk designs are available, which produce different flow characteristics, enable different methods of guiding or alignment with the seat ring, or have a certain shut off capability. Initially, the valve disc is designed for either of the two position type or throttled position type controlled. If it is of the two-position type, then the actuator will locate the valve to the particular points of its travel, usually open or closed. In throttled position type, the valve plug may be positioned at any point within the spindle of the valve, as required.

Gland and gland sealing: The gland filled around the valve spindle must effectively seal against the leakage of the controlled fluid. The packing used must not, however, create high friction load on the spindle or a load, which varies as the spindle moves. The type of packing may be detected by upper or lower temperature limits; it may be adjustable or spring loaded when the screw is fully down. Mostly Polyterafluroethyne (PTFE), known by the trade name of teflon, which is the plastic material with the inherent property to minimise friction, is moulded into *V*-shaped rings and is used as packing material within the stuffing box.

Valve body bonnet: It forms a pressure tight cover at the top of the valve body. The actuator yoke normally fastens onto the bonnet. It is made up of material similar to that of the body and has a bolted flange connected to it. The stuffing box and packing arrangement are housed with the bonnet and the lubricating opening is usually provided in the side. Extension bonnets are used for either high temperature or low temperature situations to protect the valve stem packing from extreme temperatures. They are mostly made up of stainless steel as it has low coefficient of thermal conductivity. There are three types of valve bonnets.

- Plain bonnet
- Extension bonnet
- Bellows seal bonnet

A bellow seal bonnet is used where absolutely no leakage is allowed along the valve stem.

Operation In a control valve the valve disc can be made to increase or decrease the flow as the controller pressure signal increases. A direct acting valve is arranged to decrease the flow as the pressure on the diaphragm (pneumatic actuator) is increased. Another term used to describe this is air to close (*ATC*) and such an arrangement means that the valve will open in the event of an air failure. A reverse acting valve will open as the pressure on the diaphragm is increased, that is, air flow increases and the valve closes. The particular application of the control valve in a system will decide the choice of action and fail say condition required. Some valve bodies are arranged with a removable bottom flange below the valve assembly. The actuator yoke assembly can be attached here and the valve, in effect, operated upside down. The direct acting valve now becomes reverse acting.

Valve positioners A valve positioner is needed when:

- there is a high pressure drop across the valve
- the valve is far away from the controller
- the medium being controlled is viscous
- high fluid pressures are required

Accessories The number of items related to control valves come under this general heading. They are handwheels for manual operation, hydraulic snubbers for damper valve disc movement, and cock up and fail-safe arrangement in the event of supply failure.

Facility action of manual operation of control valves is essential. This is usually provided by a handwheel, which may be mounted on the top or side of the valve. Pneumatic piston actuate valves have side mounted handwheels since the valve positioner is mounted on the top.

A hydraulic snubber is usually provided on the control valve, which operates under severe conditions. A piston attached to the valve stem moves in the oil filled cylinder and provides a damping action on the sum movement. A snubber may be fitted on a pneumatic diaphragm or on piston actuated control valves.

Applications

1. Pressure regulator
2. Temperature regulator
3. Two way control valves are used on pipelines between cargo space and the 'Pysene' smoke detecting cabinet in the wheel house of a merchant ship. When fire is located in a protected space, the two way valve is turned, thus, isolating the space from the smoke cabinet and opening it manifold to the CO_2 cylinder, for the passage of CO_2 gas that can extinguish the fire. By this means the same system of piping is used for detecting and extinguishing the fire.
4. The control valve illustrated in the question could be used for a variety of purposes where simple automatic control, are required. The operating lever would be connected at its free end to actuators, recieving pneumatic signals and operating the valve accordingly. This valve could be used in the steam inlet of heaters in air conditioning units, the air pressure to the actuator being controlled by a thermostat in the room. It should be noted that as the valve opens downwards, loss of operating air pressure will allow the valve to close, giving it a fail-safe-shut characteristic.

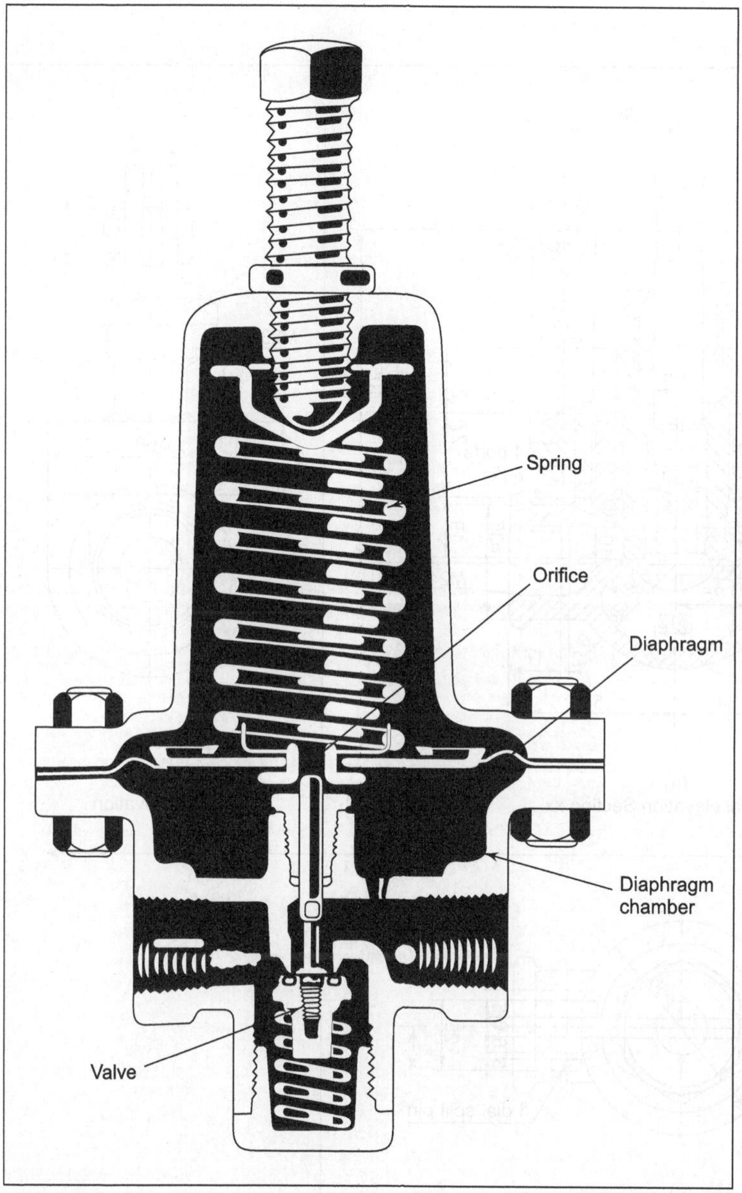

Fig. B11.3 Pressure regulator

Maintenance Maintenance of control valves is a must to ensure regular attention to all major valve components and gaskets. The various tasks to be performed during the maintenance of the valve are:

- Actuator diaphragm replacement
- Stem packing replacement
- Threaded seat ring replacement
- Seat grinding
- Valve packing lubrication
- Stem connection and adjustment

Steps for Assembly Drawing

For the required sectional view, draw the horizontal and vertical centre lines and produce the end view and plan view, respectively. From the dimensions given, draw the valve body in section, omitting the horizontal division separating the inlet and outlet branches, until the valve seat is drawn in. As the valve seat locates on this division, it can now be drawn 6 mm thick below the seat. The valve spindle could now be drawn with the valve mitre, locating it on the seat mitre. Finish off the sectional elevation by drawing the gland, fulcrum nut, and operating lever.

If the hexagons on the valve seat and gland nut are shown across the flats in the front elevation, they will, ofcourse, be seen across the corners in the end view, and the width across the corners can be found when the hexagons are drawn in plan view.

Answer

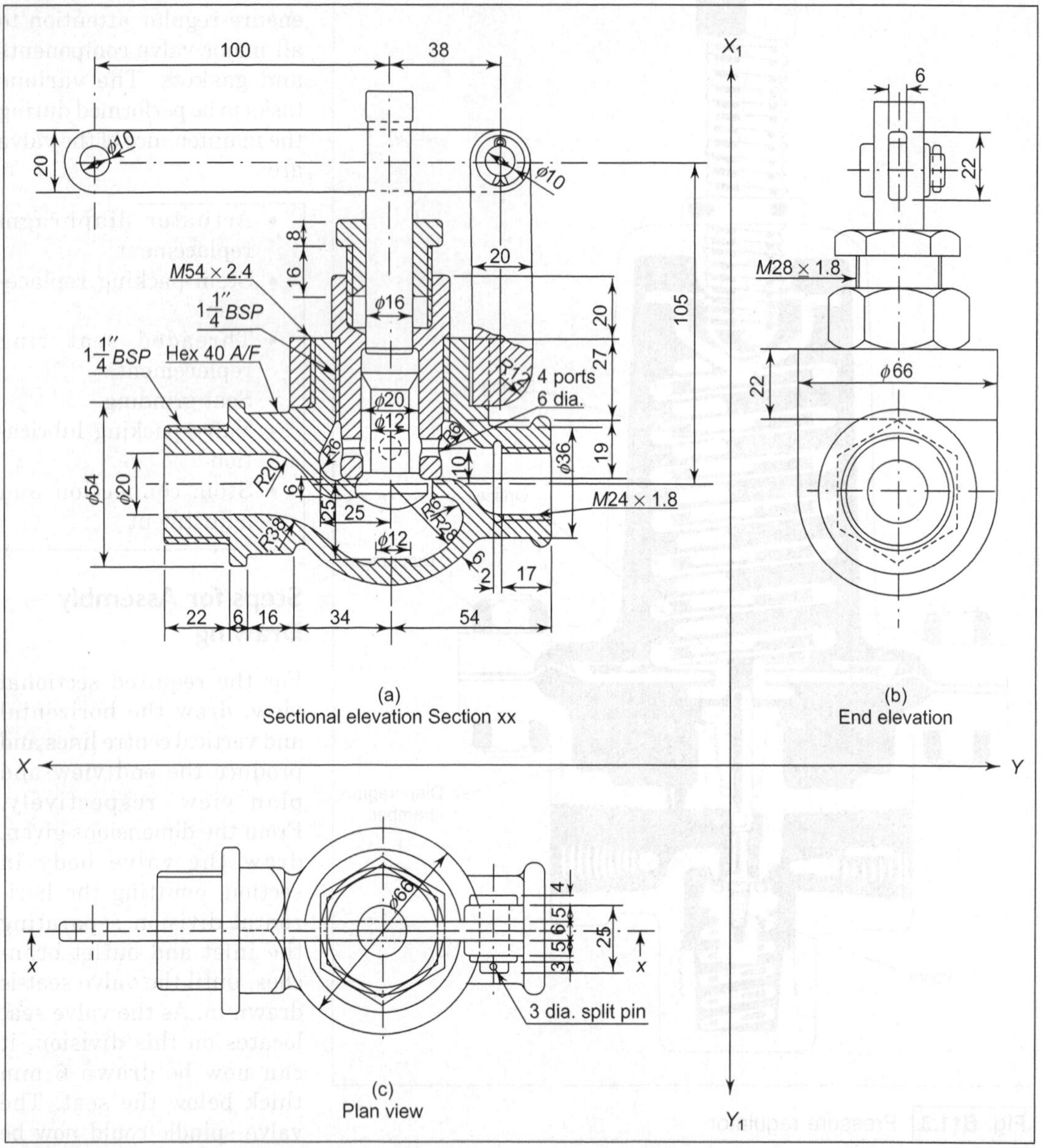

Fig. B11.4 Control valve

The enlarged views of Fig. B11.4 are shown as under:

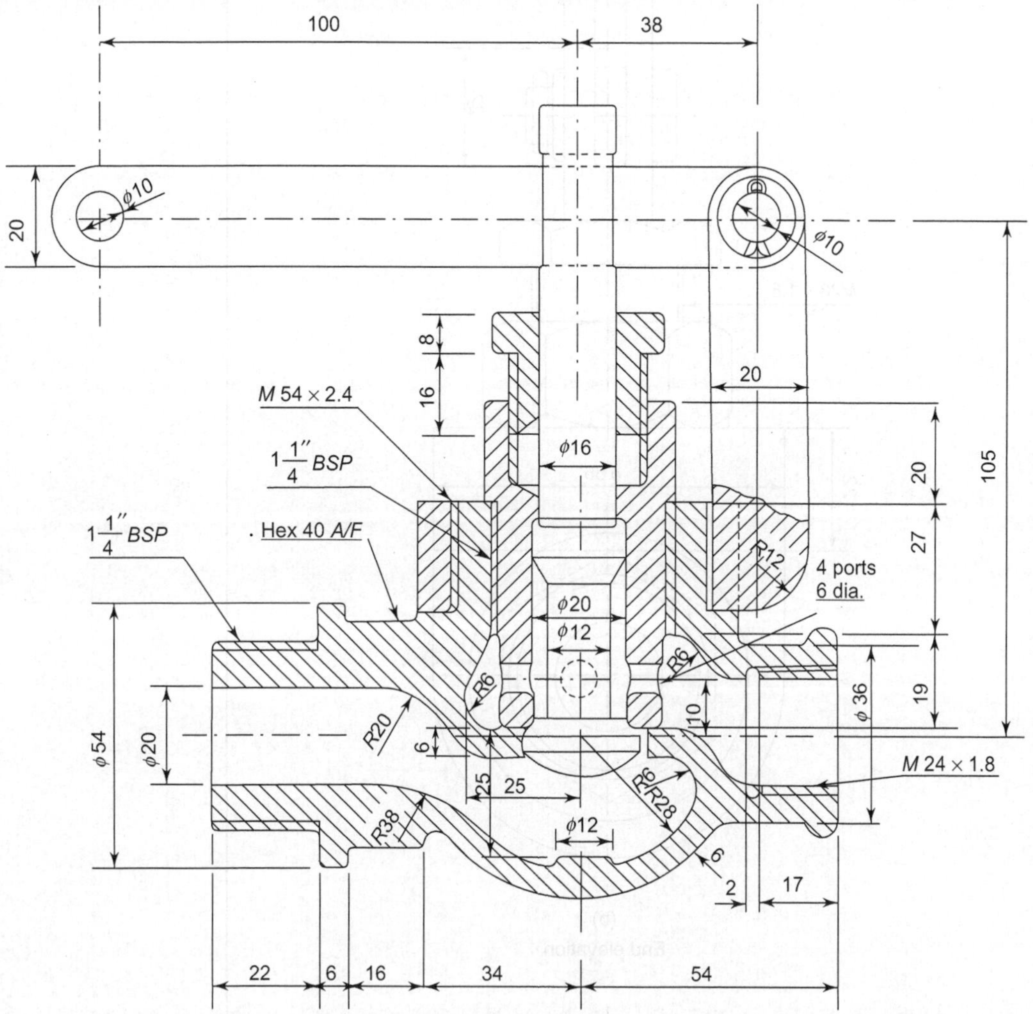

(a)
Sectional elevation

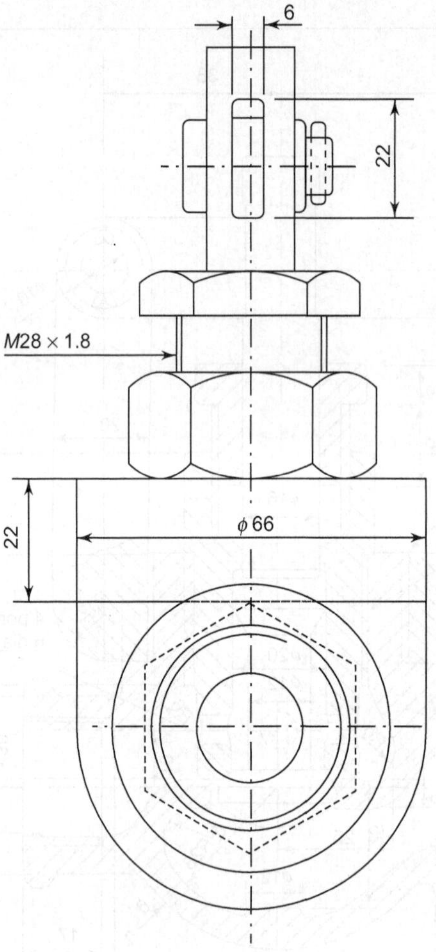

(b)
End elevation

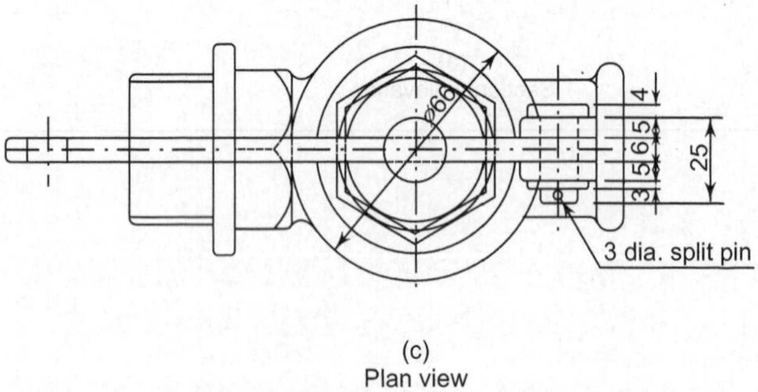

(c)
Plan view

Bill of Material

Sl. No.	*Part*	*Quantity*	*Material*
1	Valve body	1	GM
2	Valve spindle	1	Stainless steel
3	Valve seat	1	Monel metal
4	Gland	1	GM
5	Fulcrum nut	1	GM
6	Operating lever	1	MS
7	Fulcrum pin	1	MS

ASSEMBLY DRAWING 12

Bilge Suction Strainer

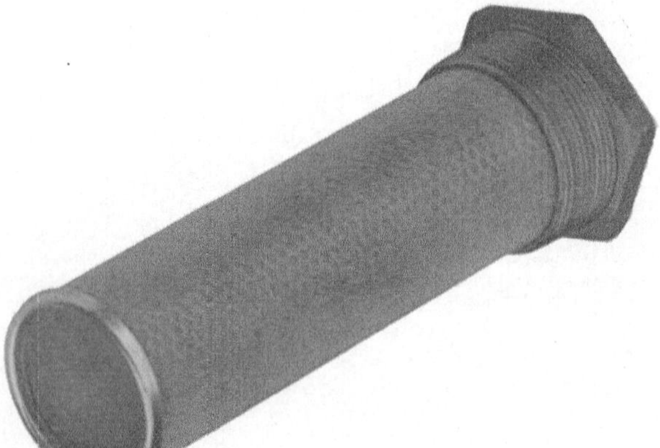

This section contains information on bilge suction strainers

After reading the section you will be able to

- understand the working and operation of bilge suction strainer
- differentiate between various types of strainers
- draw its assembled views

Bilge Suction Strainer

Question

Draw the following assembled views of bilge suction strainer.

1. Sectional elevation through pillars showing all assembled parts
2. End elevation
3. Plan view

Also include a bill of material.

The details of a bilge suction strainer are shown in Fig. B12.1.

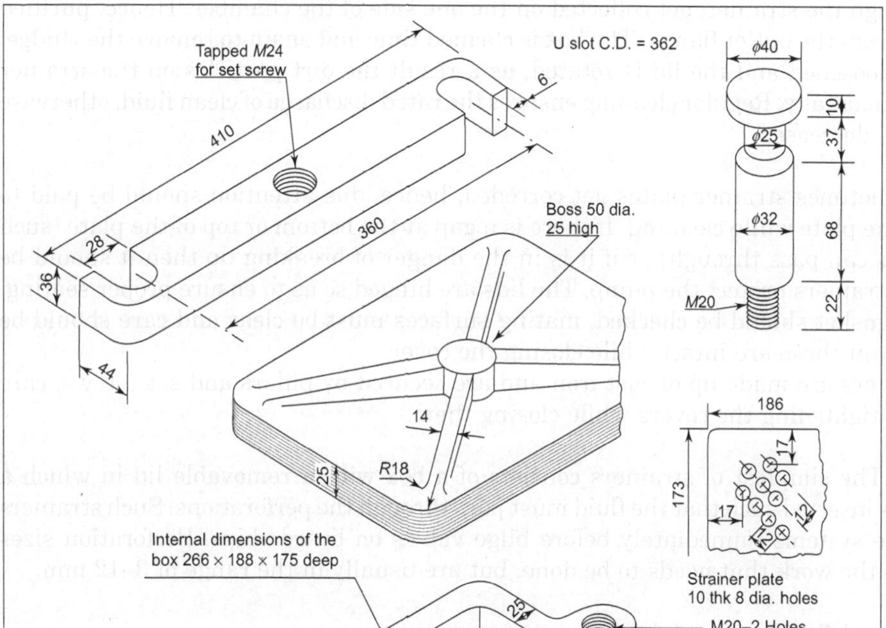

Fig. B12.1 Bilge suction strainer

Introduction

The term strainer is used specifically for a simple device made up of a single layer of coarse gauze, a coarse wire mesh or a drilled or perforated plate. Thus, the strainer is installed to hold large foreign objects that could damage or lead to blockage of pumps.

The term filter describes a device designed to prevent the passage of unwanted solids into or further along a system, but the implication is that the particles are very small and the filter cartridge or element is more complex.

Working The strainer works on the simple principle that when the fluid (oil, water, and the like) is forced to pass through a thin mesh of wires, all dirt particles that are present, and which cannot pass through the strainer, get collected on the one side of the chamber. Hence, purified fluid is obtained from the outlet flange. The box is cleaned time and again to remove the sludge. The set screw is loosened and the lid is rotated, as a result the dirt particles on the strainer plate are cleaned manually. Regular cleaning ensures the rated discharge of clean fluid, otherwise the discharge will decrease.

Operation Sometimes strainer plates get corroded, hence, due attention should be paid to the condition of the plate while cleaning. If there is a gap at the bottom or top of the plate (such that dirt particles can pass through) or if it is in the danger of breaking up then it should be removed. These strainers protect the pump. The lids are hinged so as to ensure proper seating. The state of the gasket should be checked, mating surfaces must be clear and care should be taken to ensure that these are intact while closing the cover.

Since these covers are made up of cast iron and are secured by pillars and set screws, care must be taken in tightening the covers while closing them.

Construction The simplest of strainers consists of a box with a removable lid in which a perforated plate is inserted such that the fluid must pass through the perforations. Such strainers are found in bilge systems immediately before bilge valves on board ships. Perforation sizes vary according to the work that needs to be done, but are usually in the range of 3–12 mm.

Different Types of Strainers

Simple basket strainer: For high pressure water, steam, or oil services simple basket strainers are used. These consist of a cylindrical container in which a perforated metal or wire basket is suspended. Flow through these units is from the top, into the basket, and out from the outer side of the basket. They may be installed as duplex units with three way cocks at the inlet and outlet so that one or both baskets can be used, but one of them can be shut down for cleaning. Similar basket strainers are installed close to sea water suction valves.

Knife edge strainer: Lubricating oil systems are fitted with a wide variety of strainers, some of which can be cleaned. The knife edged strainer has a series of discs hanged to a shaft. The solid particles are trapped on the edges of, and between, the discs. By rotating the disc shaft the particles are cleared by the fingers and fall to a sump, which is drained periodically. It is essential to operate these strainers regularly to prevent clogging when rotation may be difficult. Some strainers have been fitted with electric motors for continuous operation.

Cartridge filters: These are usually of the duplex type and have various types of elements. Some cartridges can be removed for cleaning; others have renewable elements. Cartridges are generally made of gauge layers with different mesh sizes, while others are made up of layers of mesh and felt.

Magnetic filters: Magnetic filters provide extra protection for engines and gearboxes where iron and steel wear particles are likely to be present.

Steps for Assembly Drawing

All marine engineers examine the bilge suction strainer, illustrated in the given question early in their career. A high vacuum on the bilge pump suction (little or no water through the pump) usually indicates that the strainer in the strum box is choked and it falls to the engineer to clear the rags, waste, apple cores, and so on that have found their way into the bilges, and eventually into the strainer.

Choose a suitable scale, commence the sectional elevation by drawing a vertical line to represent the face of the outlet range. The left hand wall of the box is then drawn in 100 mm from this face. The overall length of the box is 266 mm + 2 mm × 14 mm = 294 mm, so another line could be drawn at this distance to represent the right hand wall. Before drawing in the horizontal line to represent the bottom of the box, work out the overall height from the bottom to the top of the *M*24 mm the set screw as follows.

14 mm + 175 mm + 68 mm + 37 mm = 294 mm to top of the strong back. Allow another 60 mm for the head of the set screw and a few threads for tightening purposes. The thickness of the wall is given as 14 mm, so the inside of the casting are now to be drawn. The top flange, 25 mm deep and 25 mm from the walls, is now drawn and the lugs to take *M*20 mm holes for the pillars. The overall length of strong back = 410 mm – 50 mm = 360 mm. The pillars (shown full), cover (in section), strong back (in section) and *M*24 mm set screw (shown full) are now drawn. Thereafter, the inlet branch, the centre line, which is 200 mm from the left hand wall of the box and 55 mm down from bottom are drawn. The outlet branch in this view can be drawn chain dotted, as the section plane is behind the centre line of this branch.

The plan view will show the strong back and *M*20 mm set screw cover with its stiffening webs and also the outlet branch, the centre line of which is 33 mm below the horizontal centre line of the cover.

Answer

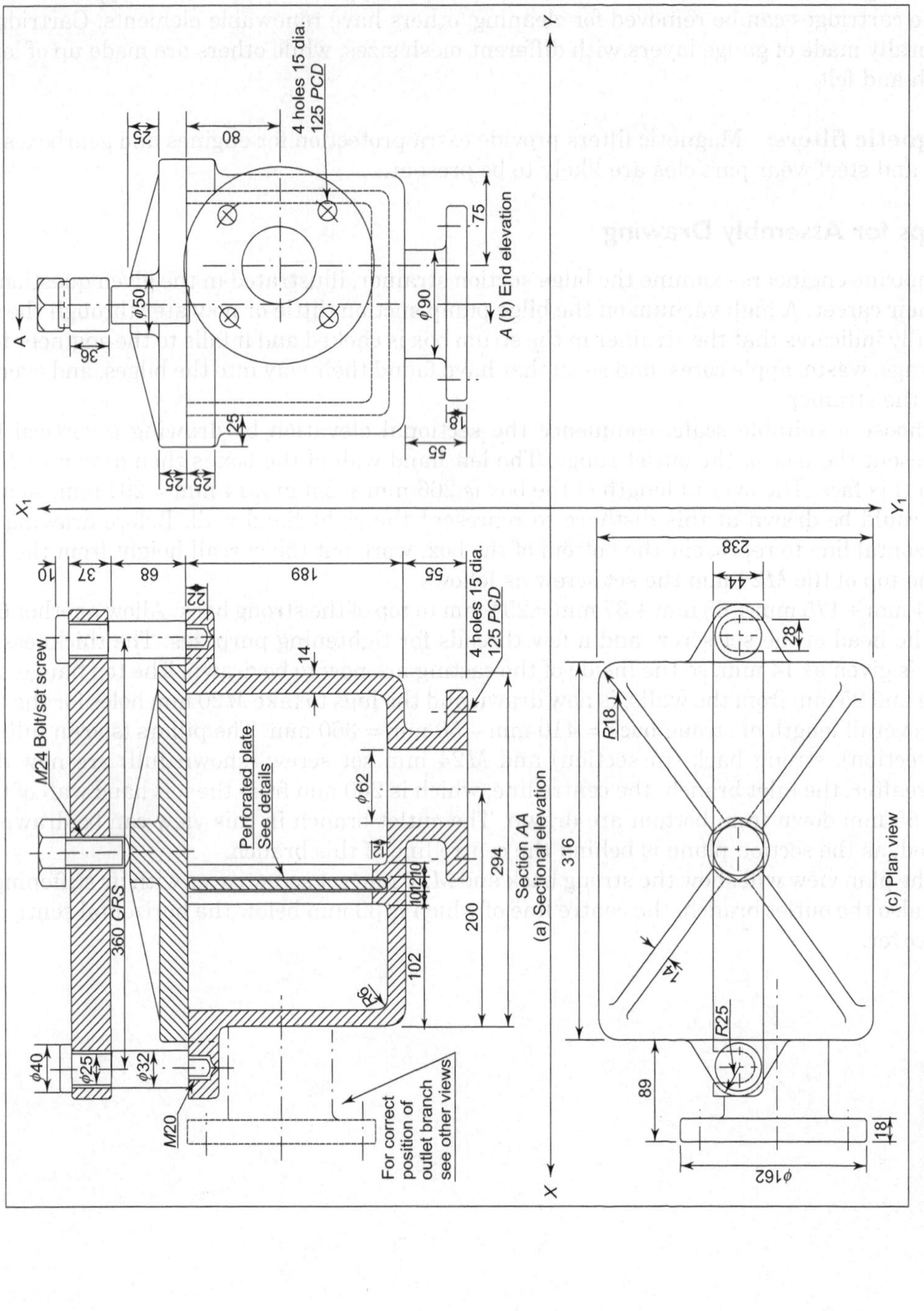

Fig. B12.2 Bilge suction strainer

The views of Fig. B12.2 are shown as under:

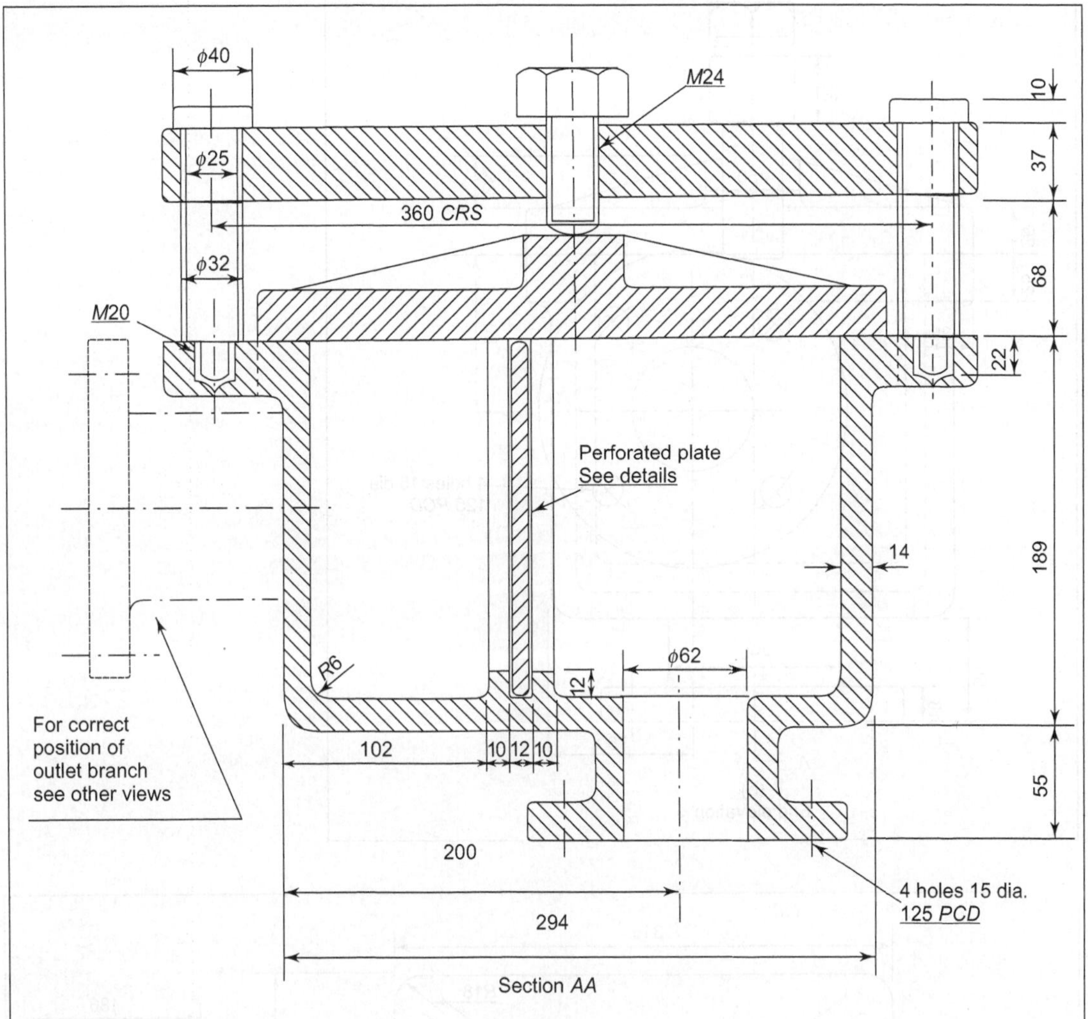

(a)
Sectional elevation

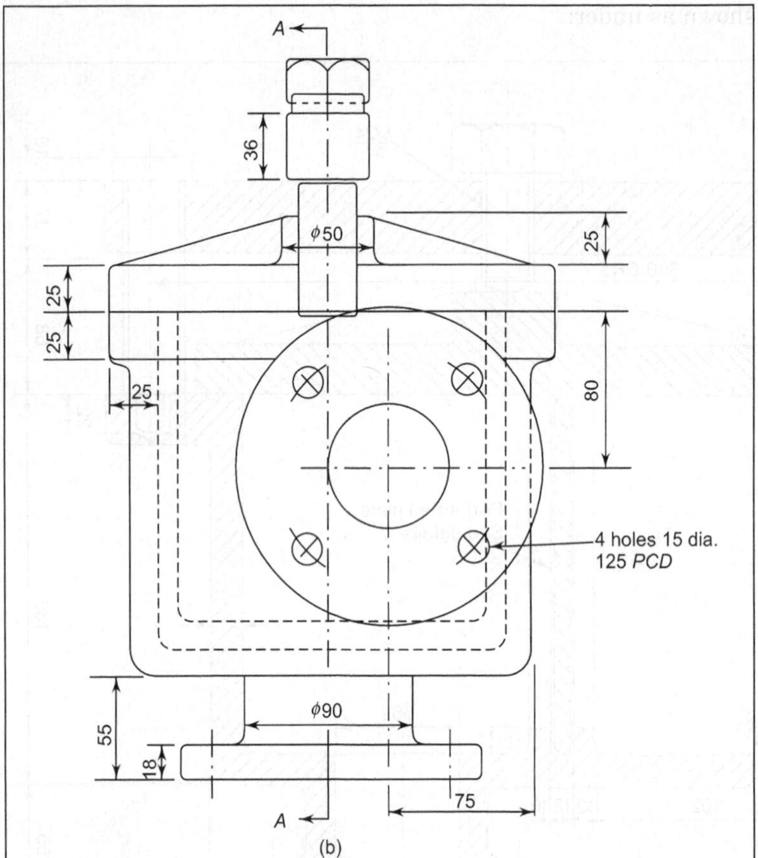

(b)
End elevation

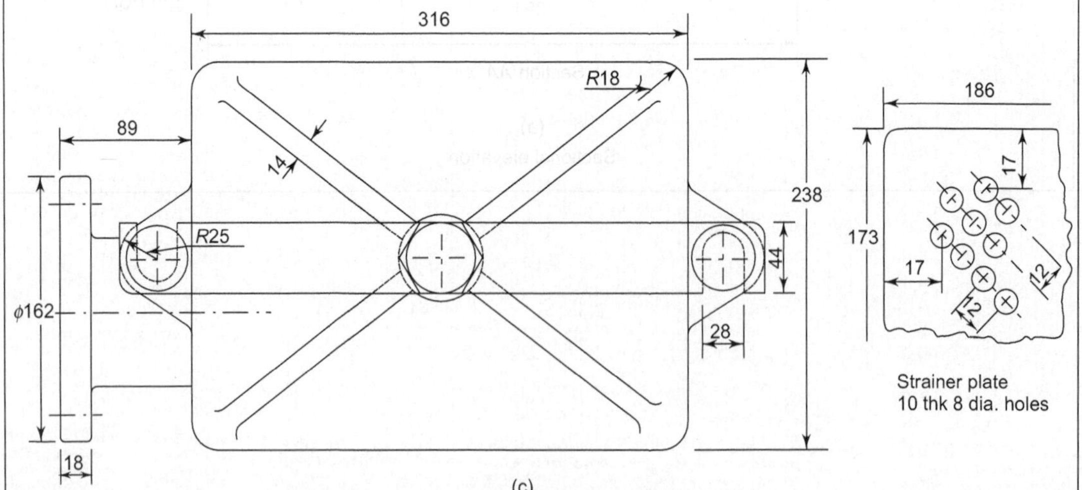

(c)
Plan view

Bill of Material

Sl. No.	*Part*	*Quantity*	*Material*
1	Box	1	CI
2	Lid	1	CI
3	Strong back	1	MS
4	Pillars	2	MS
5	Set screw	1	MS
6	Strainer plate	1	MS

ASSEMBLY DRAWING 13

Quick Closing Sluice Valve

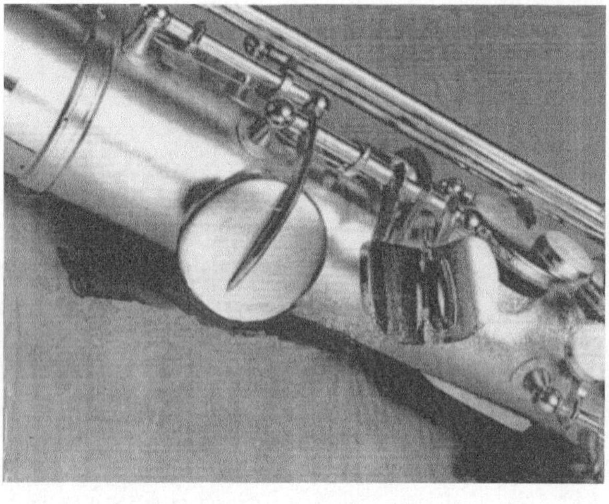

This section contains information on the quick closing sluice valve

After reading the section you will be able to

- understand the operational aspects of the valve
- draw its various assembled views

Quick Closing Sluice Valve

Question

Draw the following assembled view of a quick closing sluice valve.

1. Elevation in sections showing the assembled valve and link gear upright.
2. End view in section
   (The link gear in both the above views need not be shown in section)
3. Plan view with handwheel removed

Also include a bill of material.

The details of a quick closing sluice valve are shown in Fig. B13.1.

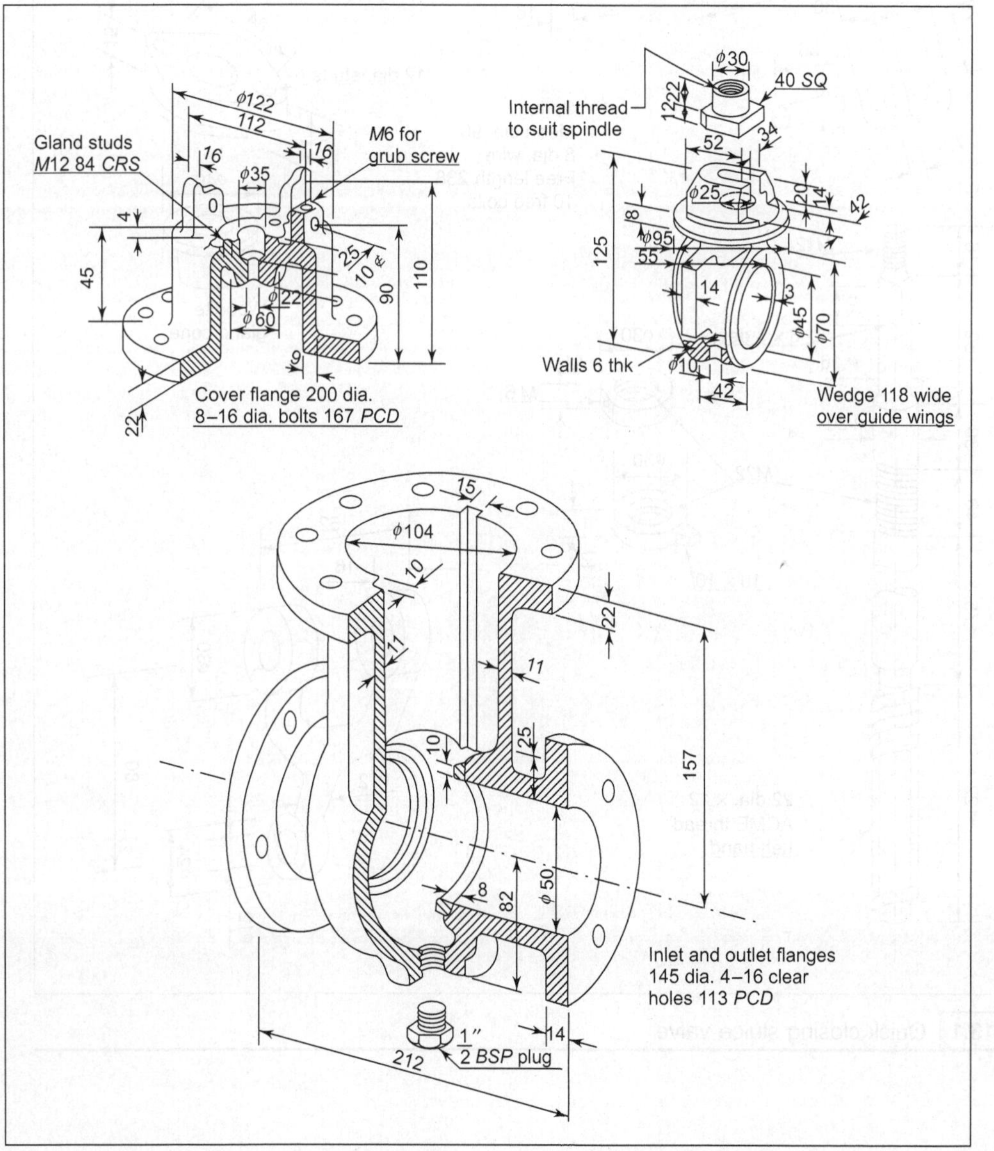

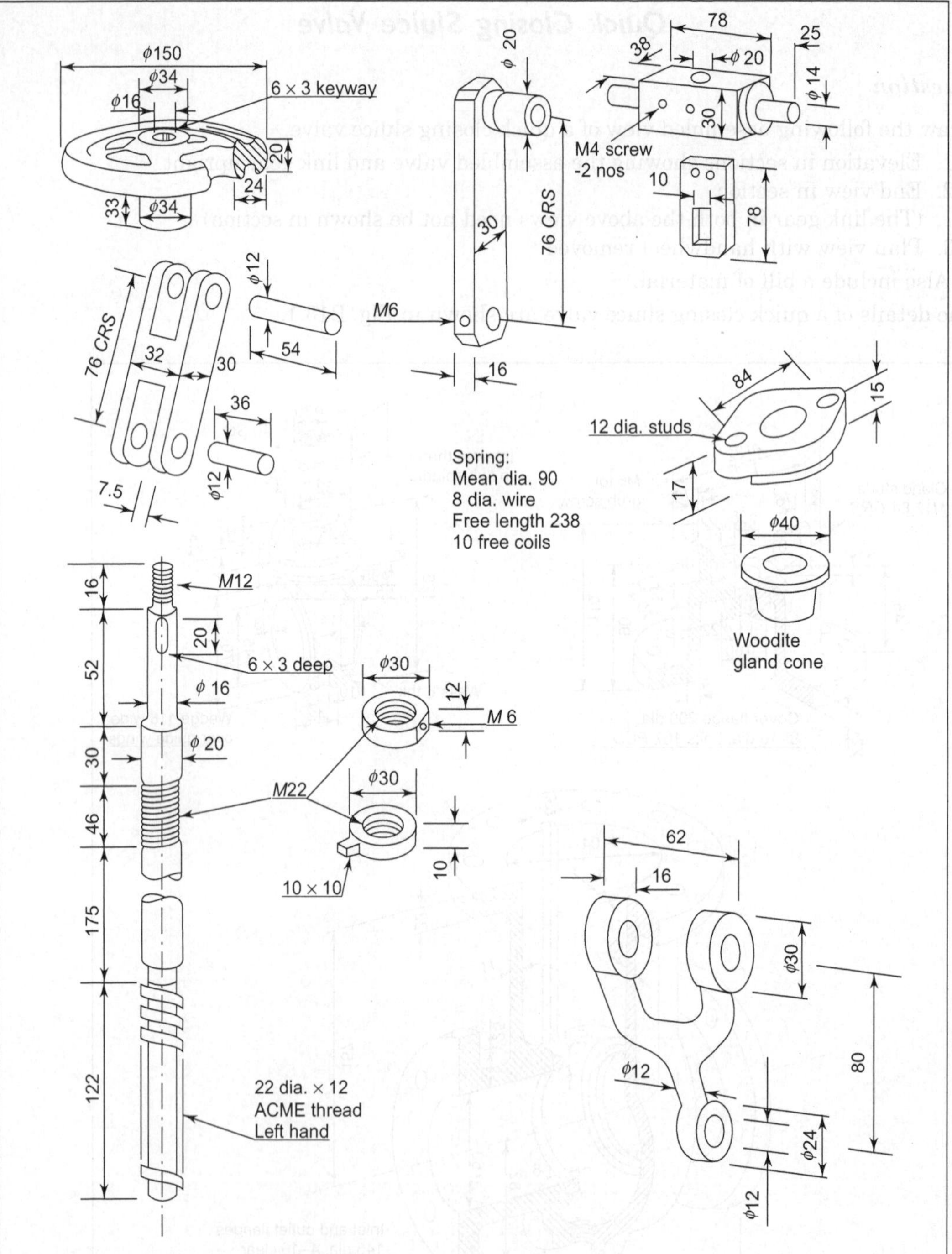

Fig. B13.1 Quick closing sluice valve

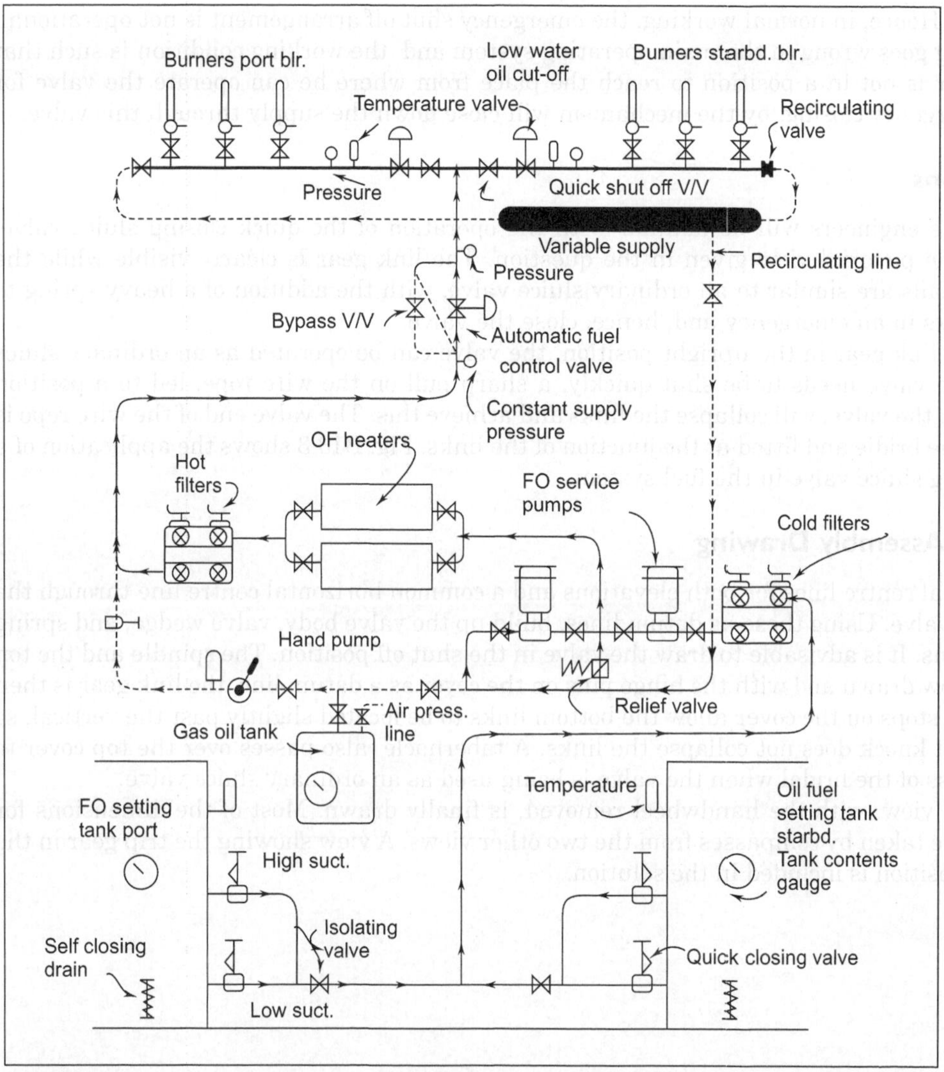

Fig. B13.2 System layout showing application of quick closing sluice valve

Introduction

This is a gate type valve with a special mechanism to enable quick closing. The valve cover has a split link bridge arrangement. It is opened against a spring fitted within the cover, with cover links acting as a rigid bridge. A fork ended pull rod attached to the centre of the range is connected to the wire, which leads to a slope area, normally out of machinery spaces. When the wire is pulled, the bridge links collapse and the compressed spring forces the valve and the spindle into the closed position. It is used as an emergency device, usually in fuel supply lines in the engine rooms where fuel oil tanks are placed to shut off the fuel from a remote position, in the event of fire in the engine room. This valve is basically a normal spring operated screw down valve. Initially, the lift of the valve can be set by manual operation, by revolving the handwheel. Hence, in normal working, the emergency shut off arrangement is not operational. If something goes wrong in the main operating system and the working condition is such that the operator is not in a position to reach the place from where he can operate the valve for closing,emergency closing, by the mechanism will close down the supply through this valve.

Applications

Most marine engineers will be familiar with the operation of the quick closing sluice valve, shown in the pictorial view given in the question. The link gear is clearly visible while the internal details are similar to an ordinary sluice valve, with the addition of a heavy spring to collapse links in an emergency and, hence, close the valve.

With the link gear in the upright position, the valve can be operated as an ordinary sluice valve. If the valve needs to be shut quickly, a sharp pull on the wire rope, led to a position remote from the valve, will collapse the links and achieve this. The valve end of the wire rope is spliced to the bridle and fitted at the junction of the links. Fig. B13.3 shows the application of a quick closing sluice valve in the fuel system.

Steps for Assembly Drawing

Draw vertical centre lines for both elevations and a common horizontal centre line through the bore of the valve. Using these as datum lines, build up the valve body, valve wedge, and spring all in sections. It is advisable to draw the valve in the shut off position. The spindle and the top cover are now drawn and with the hinge pins on the cover as a datum line, the link gear is then drawn. The stops on the cover allow the bottom links to be located slightly past the vertical, so that a slight knock does not collapse the links. A tabernacle also passes over the top cover to take the boss of the bridal when the valve is being used as an ordinary sluice valve.

The plan view, with the handwheel removed, is finally drawn. Most of the dimensions for this view are taken by compasses from the two other views. A view showing the trip gear in the collapsed position is included in the solution.

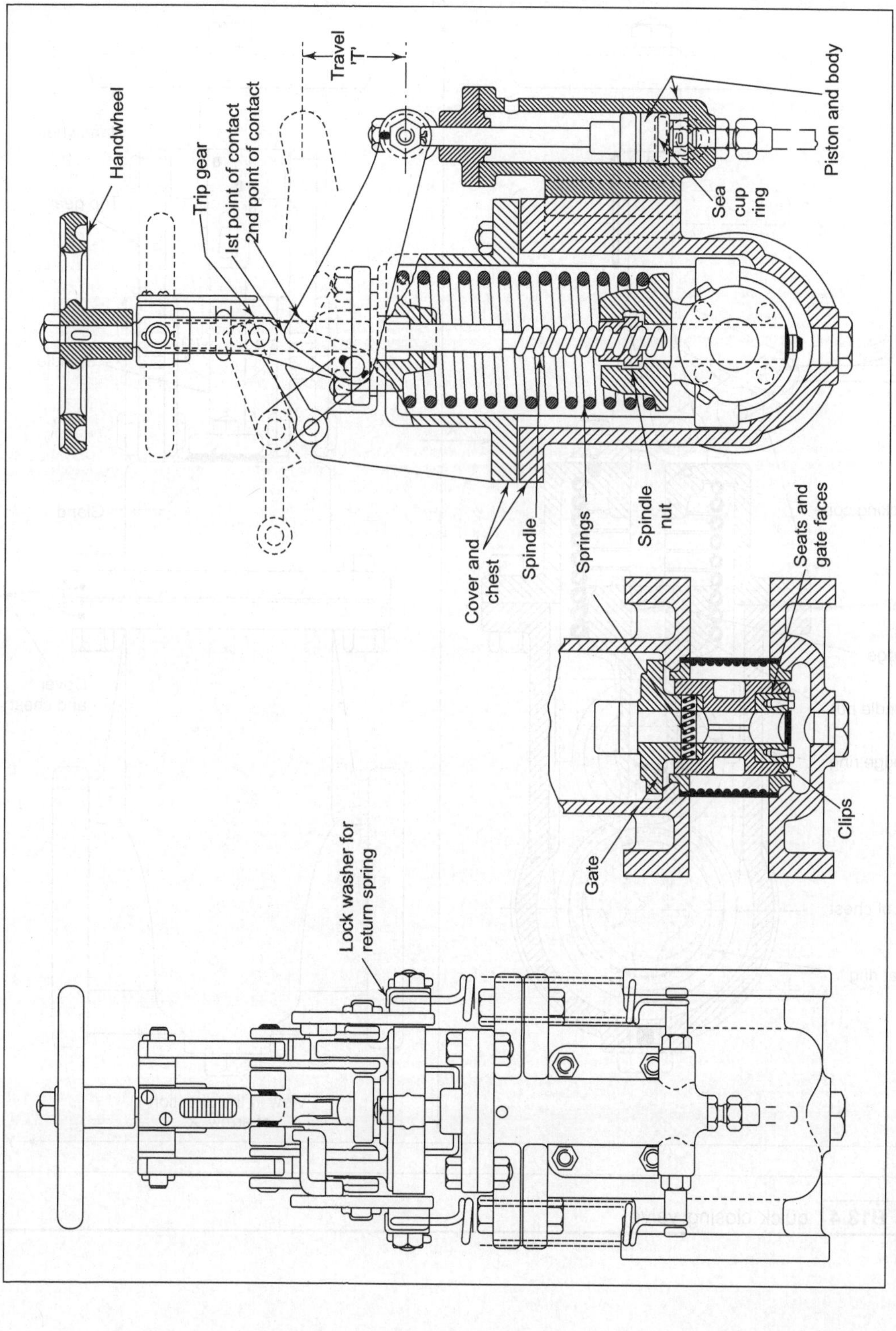

Fig. 13.3 Arrangement of quick closing parallel side valve with oil operating cylinder

Answer

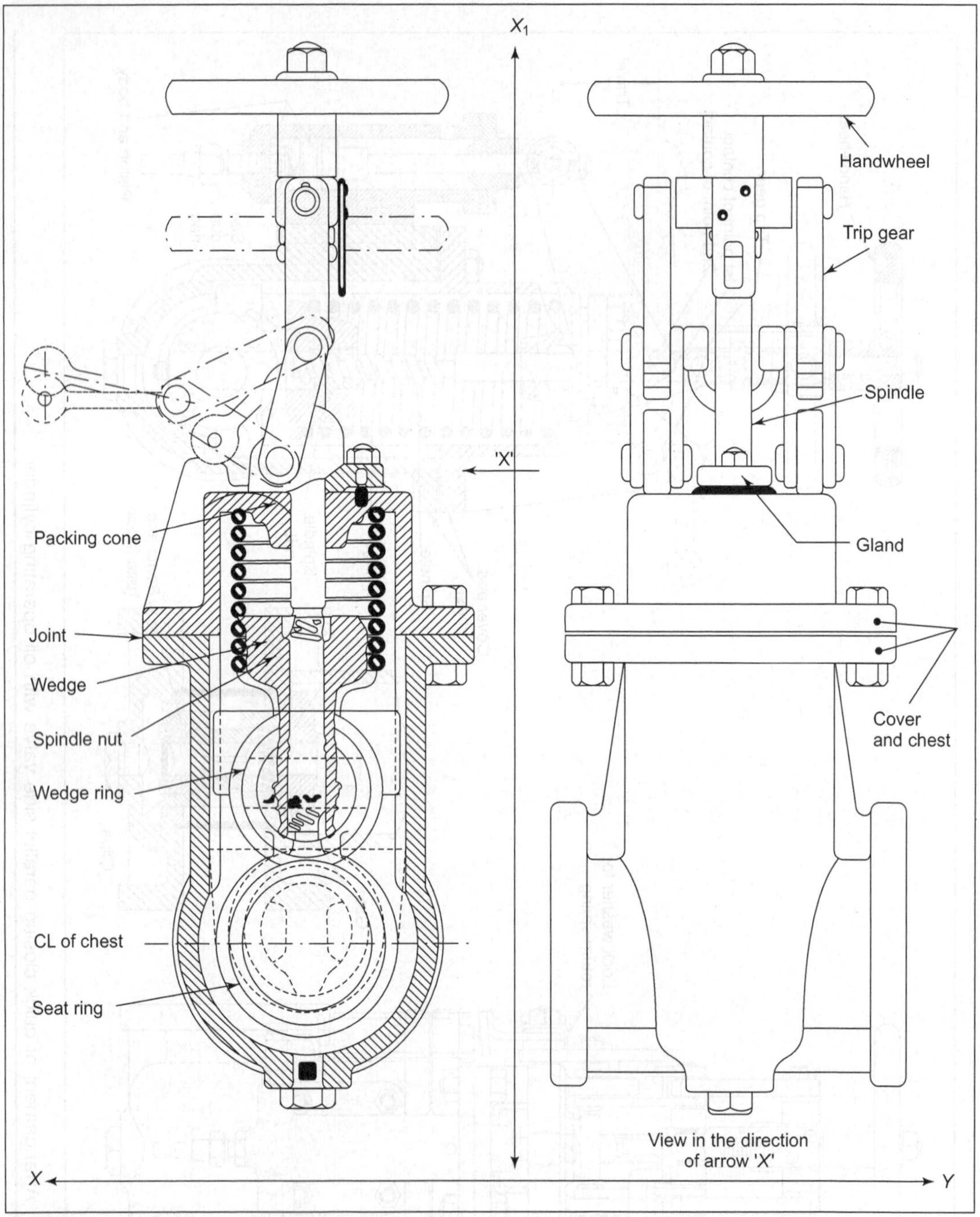

Fig. B13.4 quick closing valve

The enlarged views of Fig. B13.4 are shown as under:

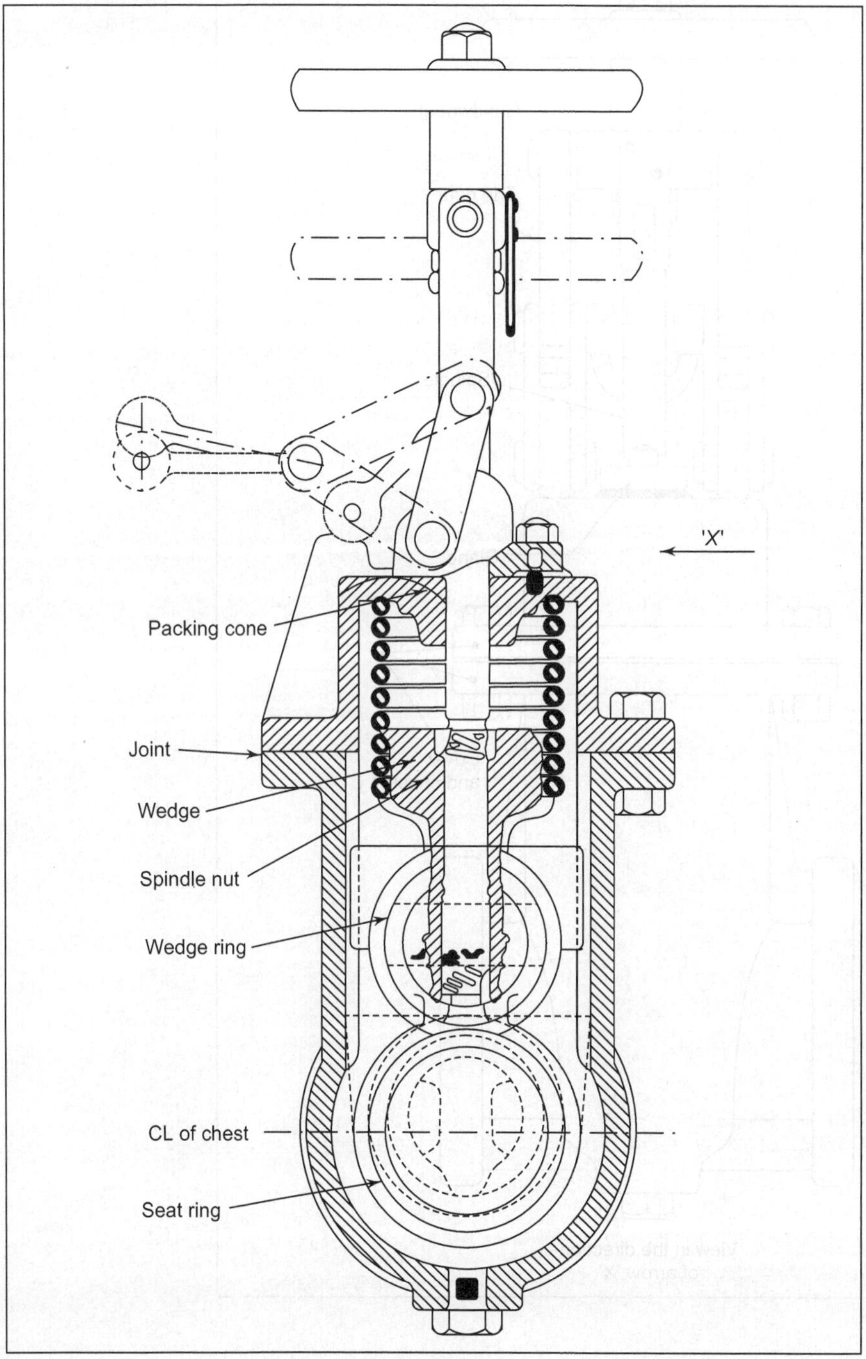

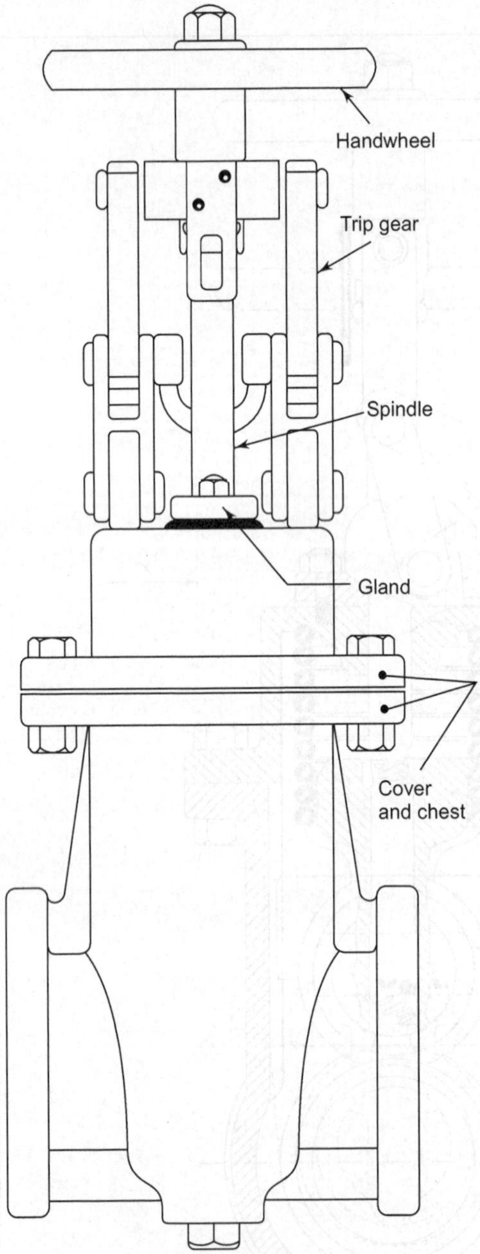

View in the direction of arrow 'X'

Bill of Material

Sl. No.	Part	Quantity	Material
1	Chest and cover	1 set	Cast steel
2	Spildle	1	Alloy steel
3	Wedge and seat ring	1 set	Gun metal
4	Cross head and links	1 set	MS
5	Pins	1 set	MS
6	Thrust washer and bushes	1 set	Alloy steel
7	Indicator nut	1	Gun metal
8	Spindle nut	1	Gun metal
9	Packing cone	1	Wood
10	Gland	1	MS
11	Hand wheel	1	CI

Assembly Drawing 14

Sealed Ball Joint

This section contains information on sealed ball joint

After reading the section you will be able to

- understand the uses of a sealed ball joint
- learn the various steps in making assembly drawing

Sealed Ball Joint

Question

Draw the following assembled views of a sealed ball joint.

1. Sectional elevation through the assembly
2. End elevation
3. Plan view

Also include a bill of material.

The details of a sealed ball joint are shown in Fig. B14.1.

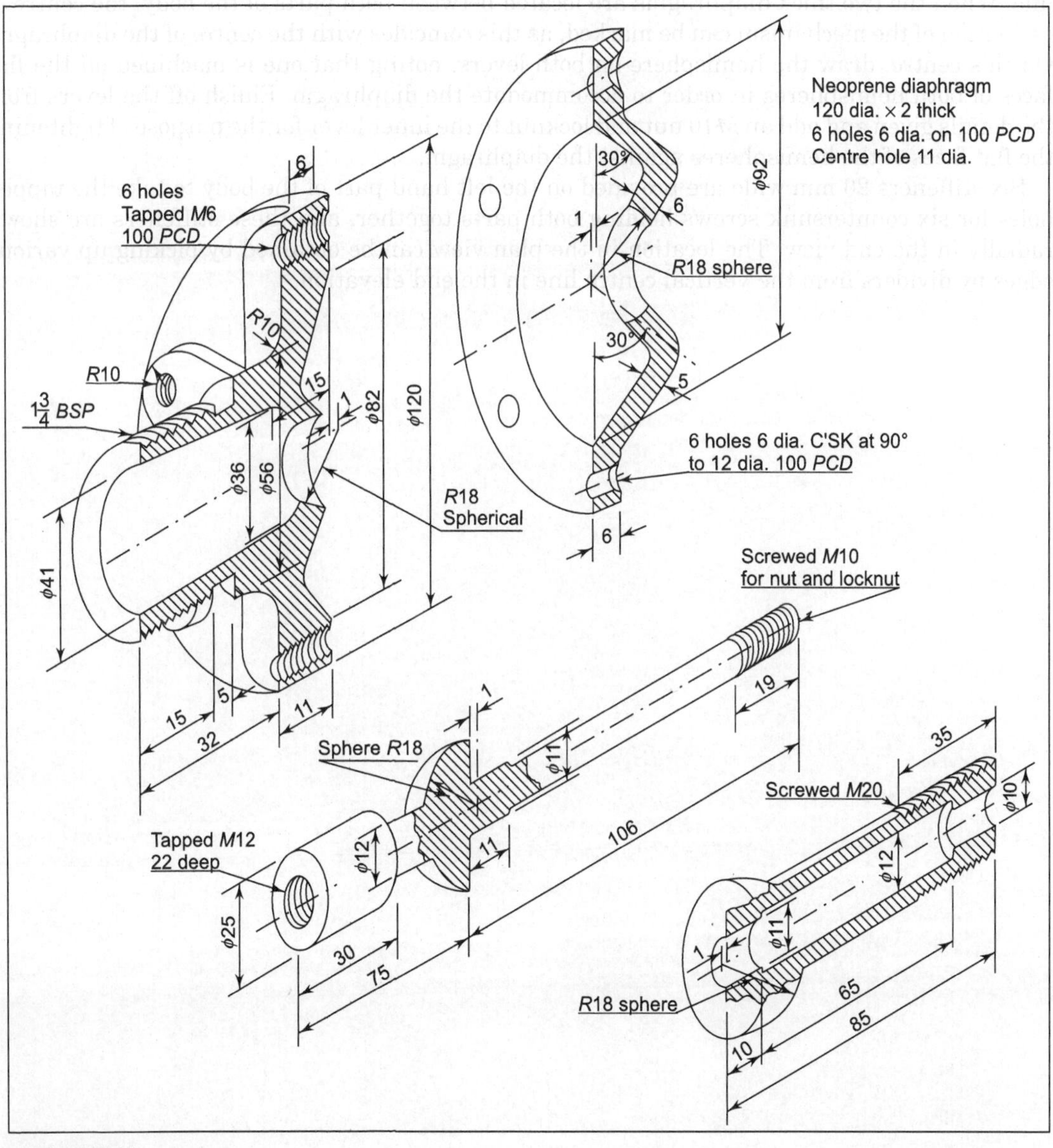

Fig. B14.1 Sealed ball joint

Introduction

The sealed ball joint could be used in conjunction with a float, as gas tight mechanism for tank. The body of the fitting is screwed 1 3/4″ *BSP* to suit a tapped boss on the tank and a gas tight joint on the 56 mm diameter spigot, whilst that part of lever in the tank is sealed from the atmosphere by a neoprene diaphragm sandwiched between the hemispheres on, both, the male and female parts of the lever.

Steps for Assembly Drawing

Commence the sectional elevation by building the body of the fitting around its horizontal centre line. When the two thick diaphragms are located between both parts of the body, the centre of oscillation of the mechanism can be marked, as this coincides with the centre of the diaphragm. On this centre, draw the hemisphere on both levers, noting that one is machined off the flat faces of both hemispheres in order to accommodate the diaphragm. Finish off the levers from the details given and add an *M*10 nut and locknut to the inner lever for the purpose of tightening the flat faces of the hemispheres against the diaphragm.

Six stiffeners 20 mm wide are provided on the left hand part of the body to take the tapped holes for six countersunk screws holding both parts together, and these stiffeners are shown radially in the end view. The location in the plan view can be obtained by picking up various edges by dividers from the vertical centre line in the end elevation.

Answer

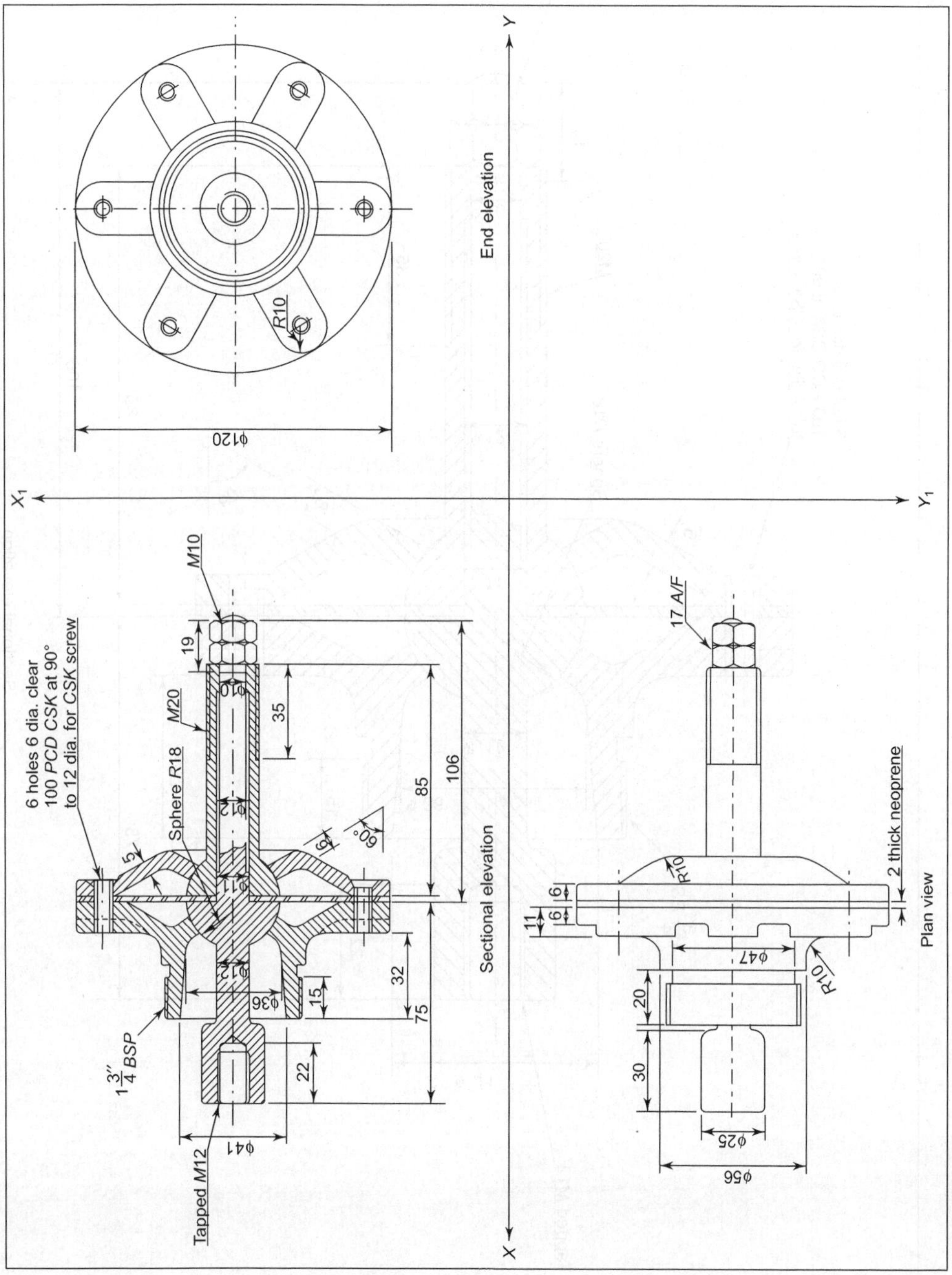

Fig. B14.2 Sealed ball joint

The enlarged view of Fig. B14.2 is shown as under:

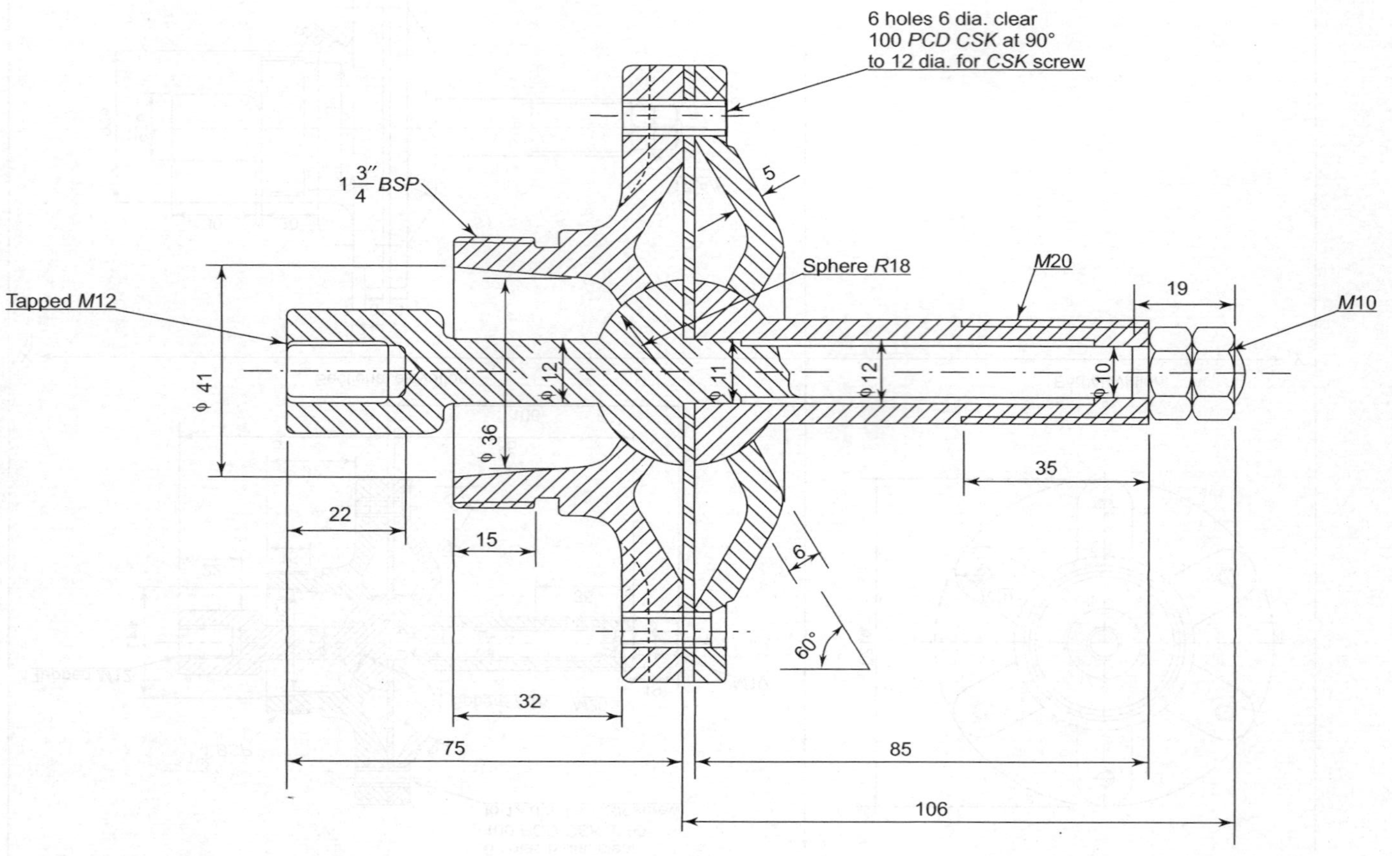

Sectional elevation

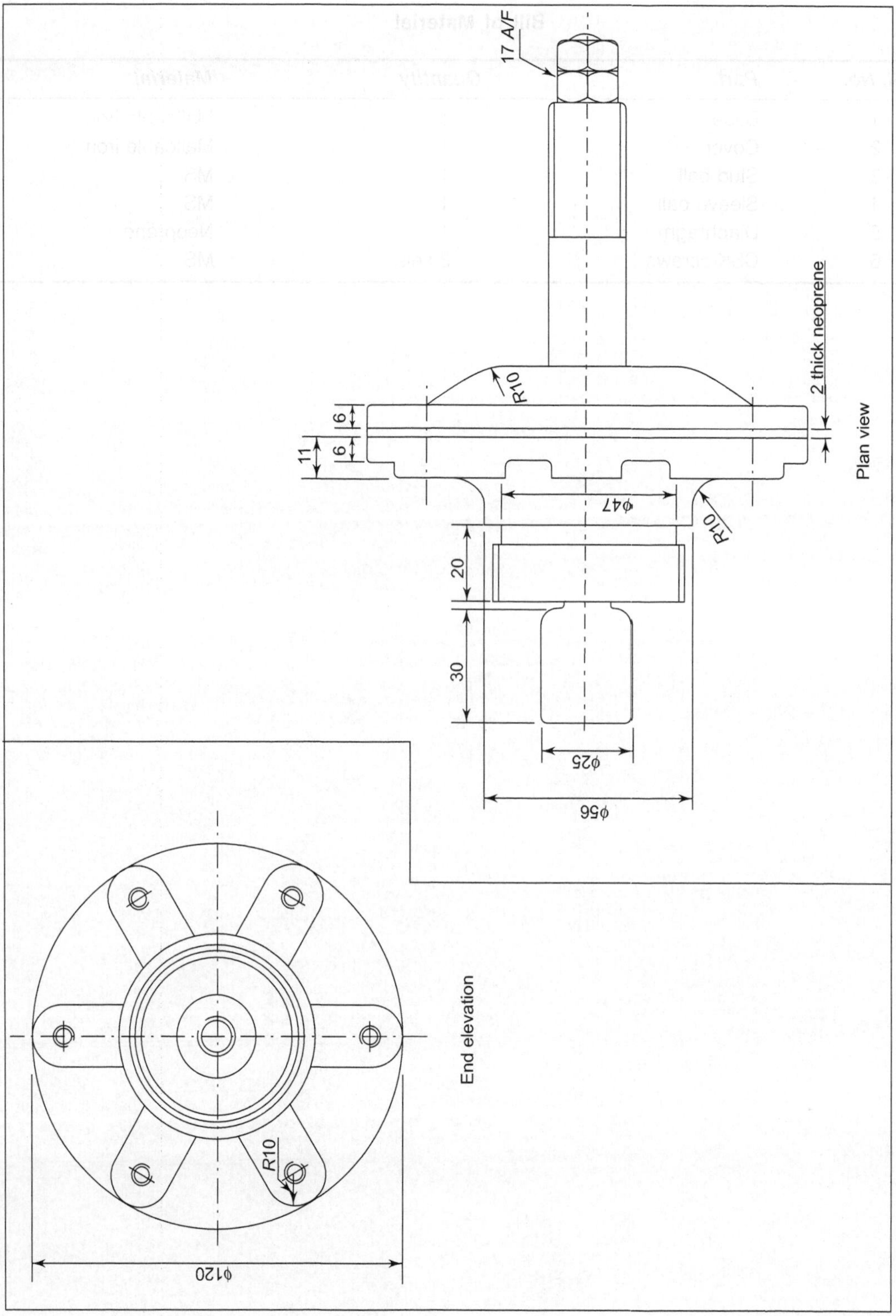
17 A/F
R10
2 thick neoprene
Plan view
6
6
11
φ47
R10
20
30
φ25
φ56
End elevation
R10
φ120

Bill of Material

Sl. No.	Part	Quantity	Material
1	Base	1	Malleable iron
2	Cover	1	Malleable iron
3	Stud ball	1	MS
4	Sleeve ball	1	MS
5	Diaphragm	1	Neoprene
6	CSK screws	3 sets	MS

PART C

Questions on Assembly Drawings

1 *Gear Detacher*
2 *Drill Machine Vice*
3 *Crane Truck*
4 *Pipe Holding Device*
5 *Screw Operated Lifting Machine*
6 *Lifting Machine Hook*
7 *Tool Carrier*
8 *Shaping Machine Tool Holder*
9 *Elastic Film Regulator*
10 *Connecting Rod for Small Engine*
11 *Connecting Rod for Big Engine*
12 *Spark Plug*
13 *Variable Supply Fuel Pump*
14 *Fuel Injector*
15 *Job Holding Device*
16 *Force Pump*
17 *Feed Check Valve*
18 *Flow Check Valve*
19 *Blow Off Cock*
20 *Stop Vlave*
21 *125 mm Stop Valve*
22 *Screw Down Stop Valve*
23 *Plunger Type Stop Valve*
24 *Automatic Valve*
25 *Safety Valve*
26 *Air Injection Valve*
27 *Full Bore Safety Valve*
28 *Flow Regulator*
29 *Strainer*
30 *Ship Rudder Bearing*
31 *Universal Coupling (Type I)*
32 *Jig*
33 *Universal Coupling (Type II)*
34 *Clamping Fixture*
35 *Non-Return Valve*
36 *Diaphragm Regulator*
37 *Gate Valve*
38 *3-Jaw Chuck*
39 *Water Circulating Pump*
40 *Manual Drilling Machine*

1 Gear Detacher

The details of a gear detacher are shown in Fig. C1.1. Assemble all the parts and draw the following views of the assembly.

1. Elevation
2. End view
3. Plan

Enlist a bill of material.

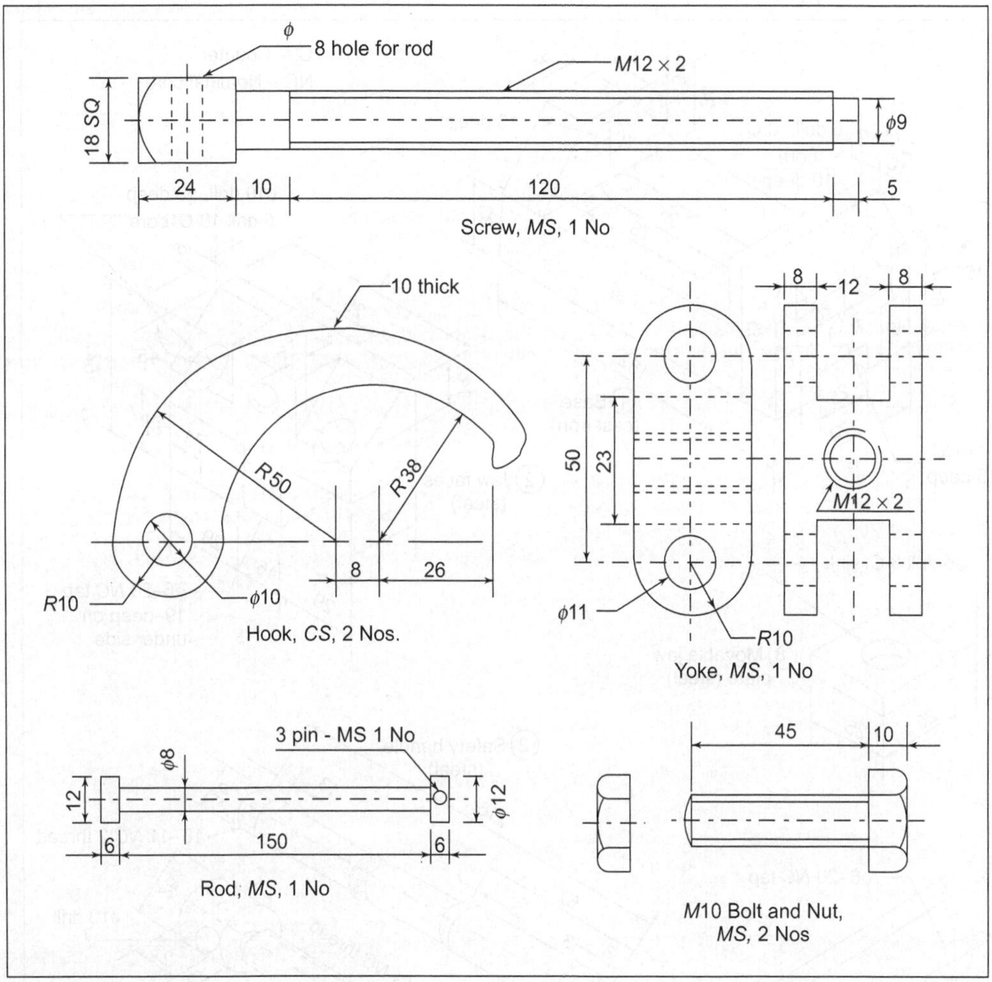

Fig. C1.1 Details of gear detacher

Sl. No.	Part	Quantity	Material
1	Screw	1	MS
2	Hook	2	CS
3	Yoke	1	MS
4	Rod	1	MS
5	Nut and Bolt	2	MS

2 Drill Machine Vice

The details of a drill machine vice are shown in Fig. C2.1. Assemble all the parts and draw the following views of the assembly.

1. Elevation
2. End view
3. Plan

   Enlist a bill of material.

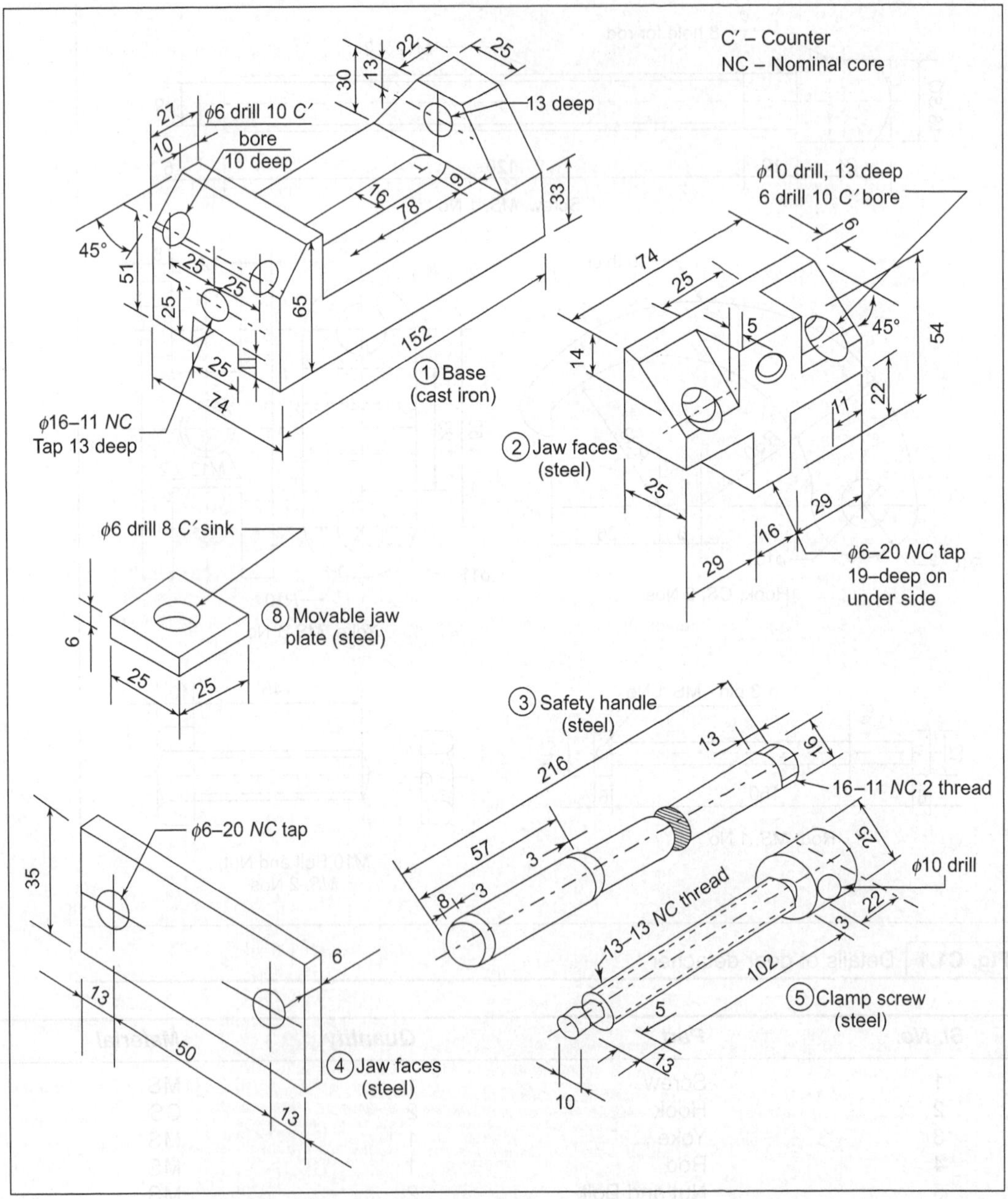

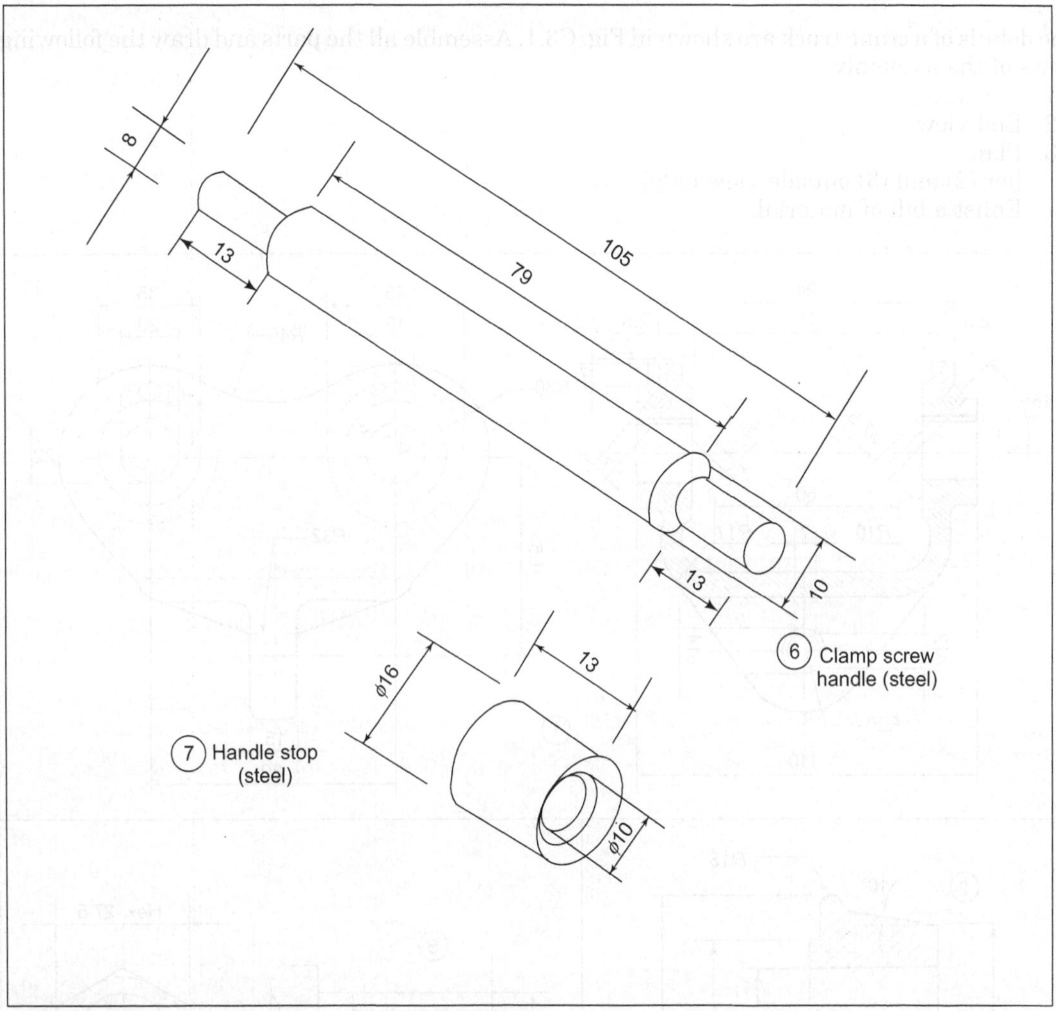

Fig. C2.1 Drill machine vice

Sl. No.	*Part*	*Quantity*	*Material*
1	Base	1	CI
2	Jaw faces	1	MS
3	Safety handle	1	MS
4	Jaw faces	2	MS
5	Clamp screw	1	MS
6	Clamp screw handle	1	MS
7	Handle stop	2	MS
8	Movable jaw plate	1	MS

3 Crane Truck

The details of a crane truck are shown in Fig. C3.1. Assemble all the parts and draw the following views of the assembly.

1. Elevation
2. End view
3. Plan

   [for (2) and (3) outside view only]

   Enlist a bill of material.

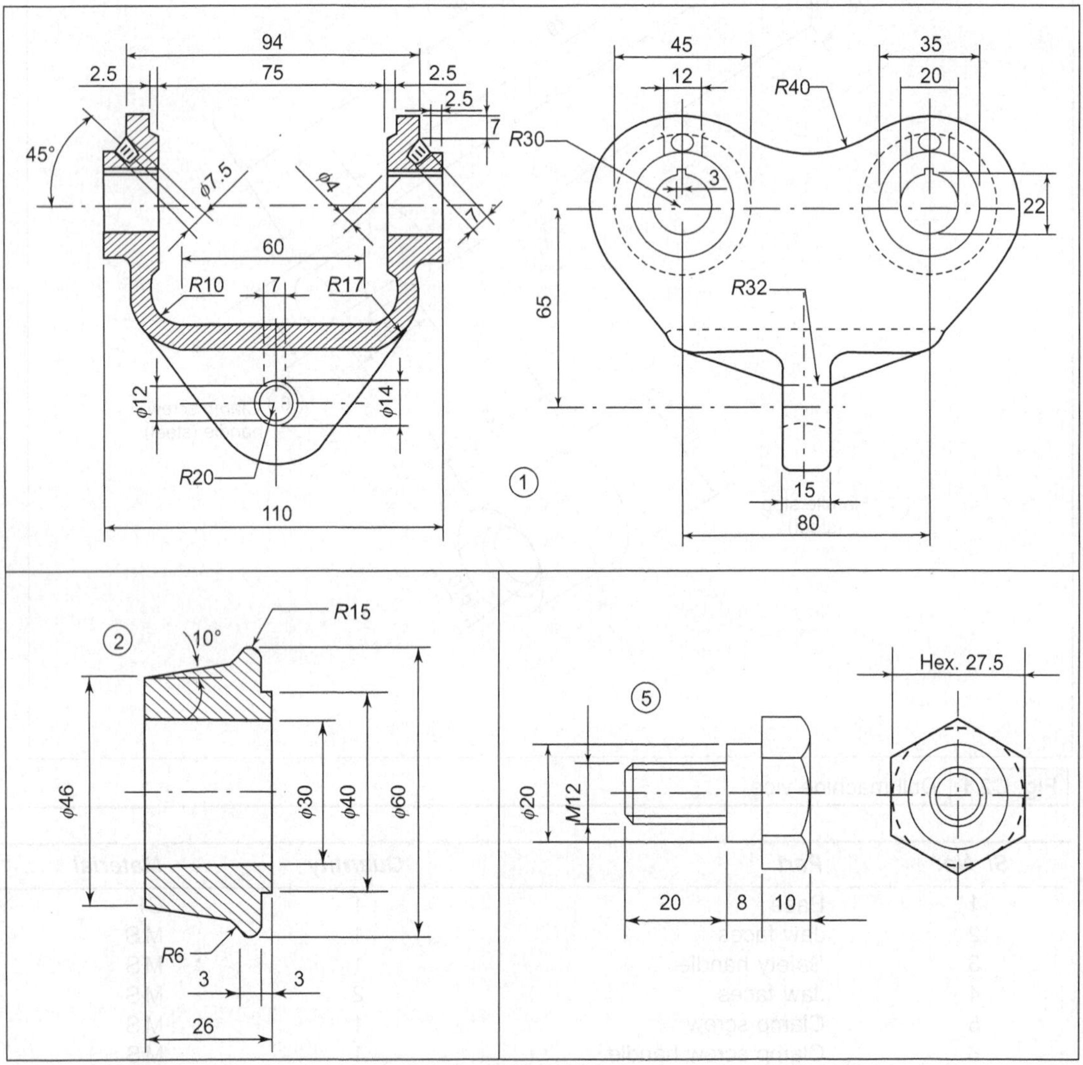

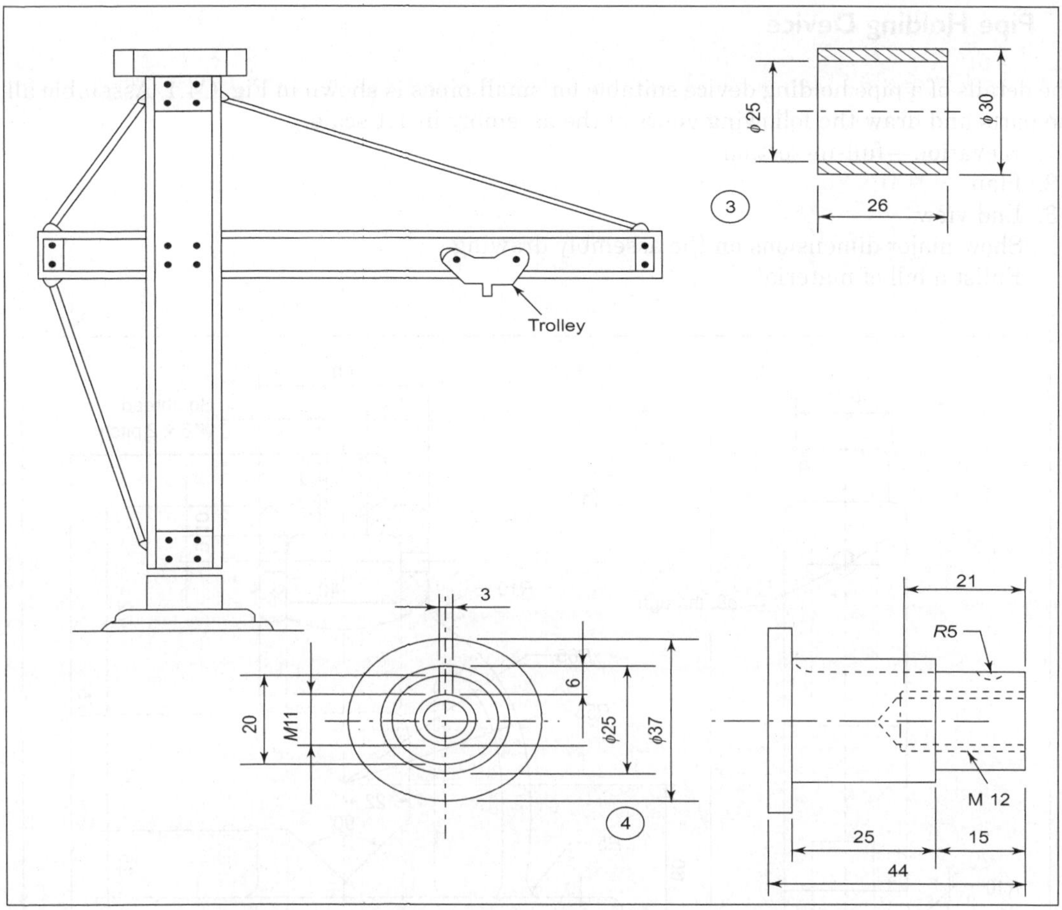

Fig. C3.1 Crane truck

Sl. No.	*Part*	*Quantity*	*Material*
1	Trolley frame	1	CI
2	Trolley wheel	4	CI
3	Trolley wheel bushing	4	Brass
4	Trolley wheel stud	4	Cast steel
5	Trolley wheel screw	4	Cast steel

4 Pipe Holding Device

The details of a pipe holding device suitable for small pipes is shown in Fig. C4.1. Assemble all the parts and draw the following views of the assembly in 1:1 scale.

1. Elevation—full-in-section
2. Plan
3. End view

   Show major dimensions on the assembly drawing.

   Enlist a bill of material.

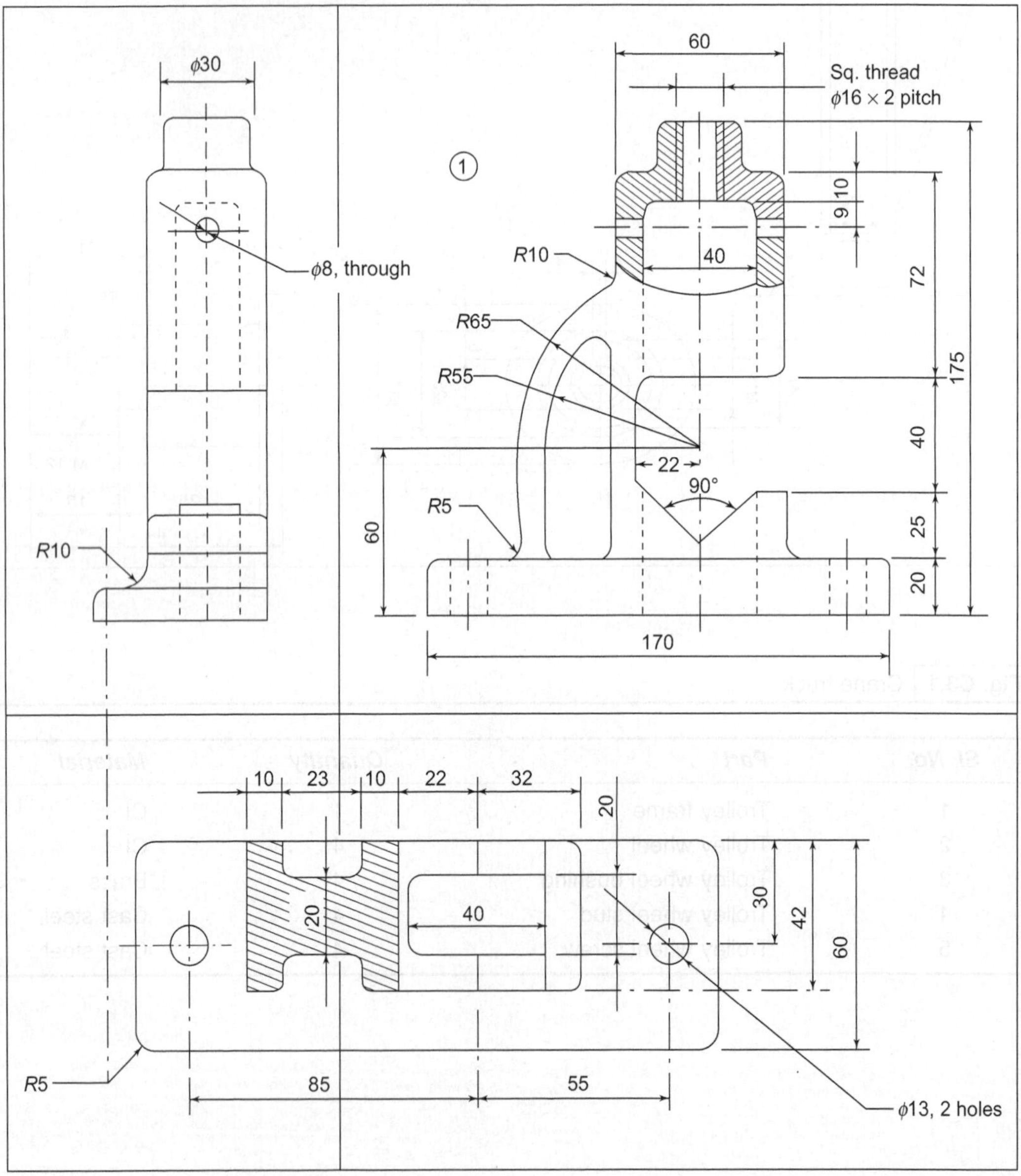

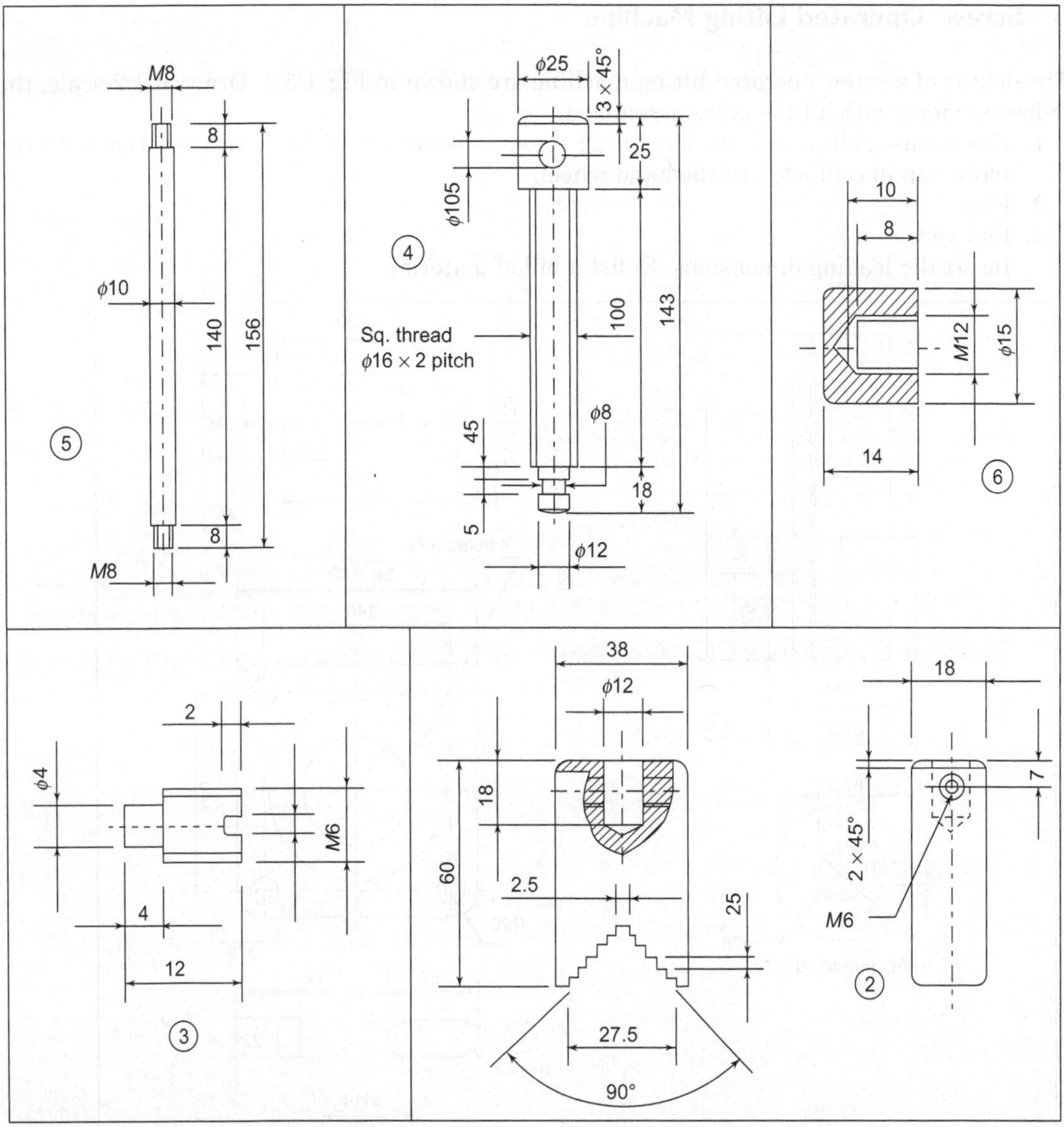

Fig. C4.1 Details of pipe holding device

Sl. No.	*Part*	*Quantity*	*Material*
1	Frame	1	CI
2	'V' jaw	1	CI
3	Screw *M* 6	1	MS
4	Main screw	1	MS
5	Handle rod	1	MS
6	Handle stop	1	MS

5 Screw Operated Lifting Machine

The details of a screw operated lifting machine are shown in Fig. C5.1. Draw in 1:2 scale, the following views with all the parts assembled.

1. Elevation—half-in-section. The lifting screw to be drawn in the lowest position with the screw cap in contact with the hand wheel.
2. Plan
3. End view

Insert the leading dimensions. Enlist a bill of material.

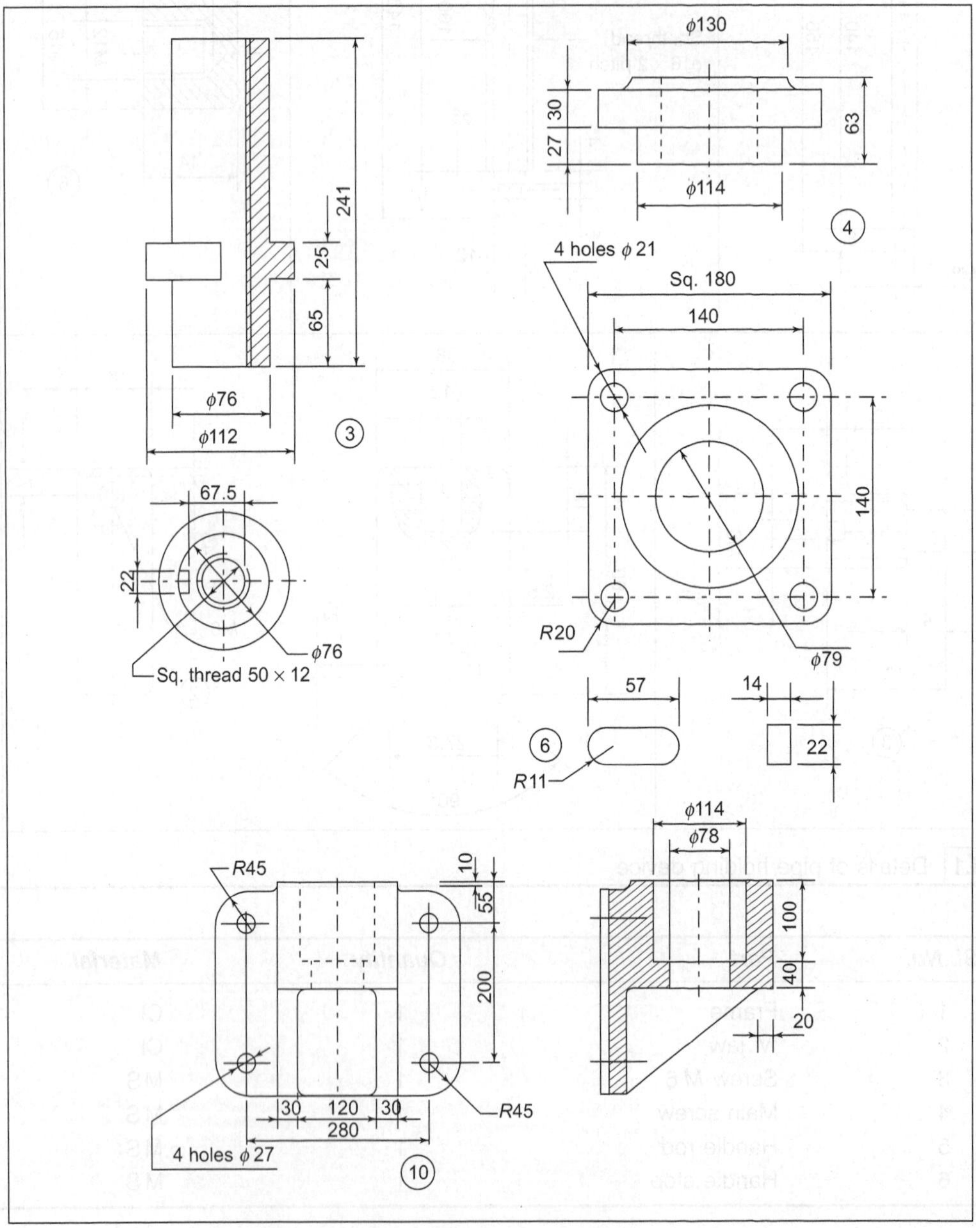

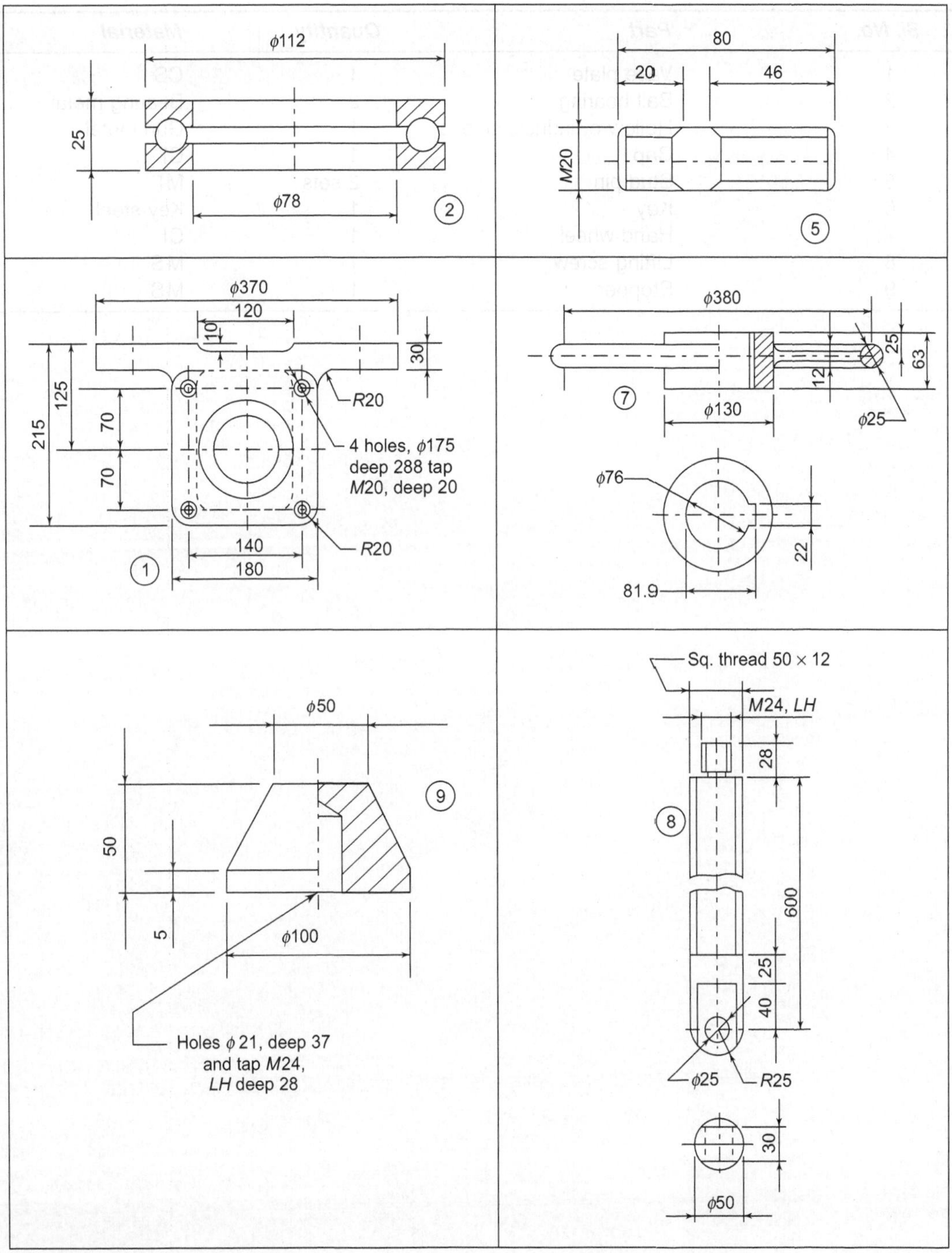

Fig. C5.1 Details of screw operated lifting machine

Sl. No.	Part	Quantity	Material
1	Vase plate	1	CS
2	Ball bearing	2	Bearing metal
3	Hollow cylindrical sleeve	1	Gun metal
4	Cap	1	CI
5	Stud pin	2 sets	MI
6	Key	1	Key steel
7	Hand wheel	1	CI
8	Lifting screw	1	MS
9	Stopper	1	MS

6 Lifting Machine Hook

The details of a lifting machine hook with pulley block are shown in Fig. C6.1. Draw the following views of the assembled block in 1:2 scale.

1. Sectional view of the assembly with the section plane containing the axis of the pulley pin.
2. End view
3. Plan—half-in-section, with the section plane passing through the axis of the pulley pin. Show important dimensions on the assembly drawing. Enlist a bill of material.

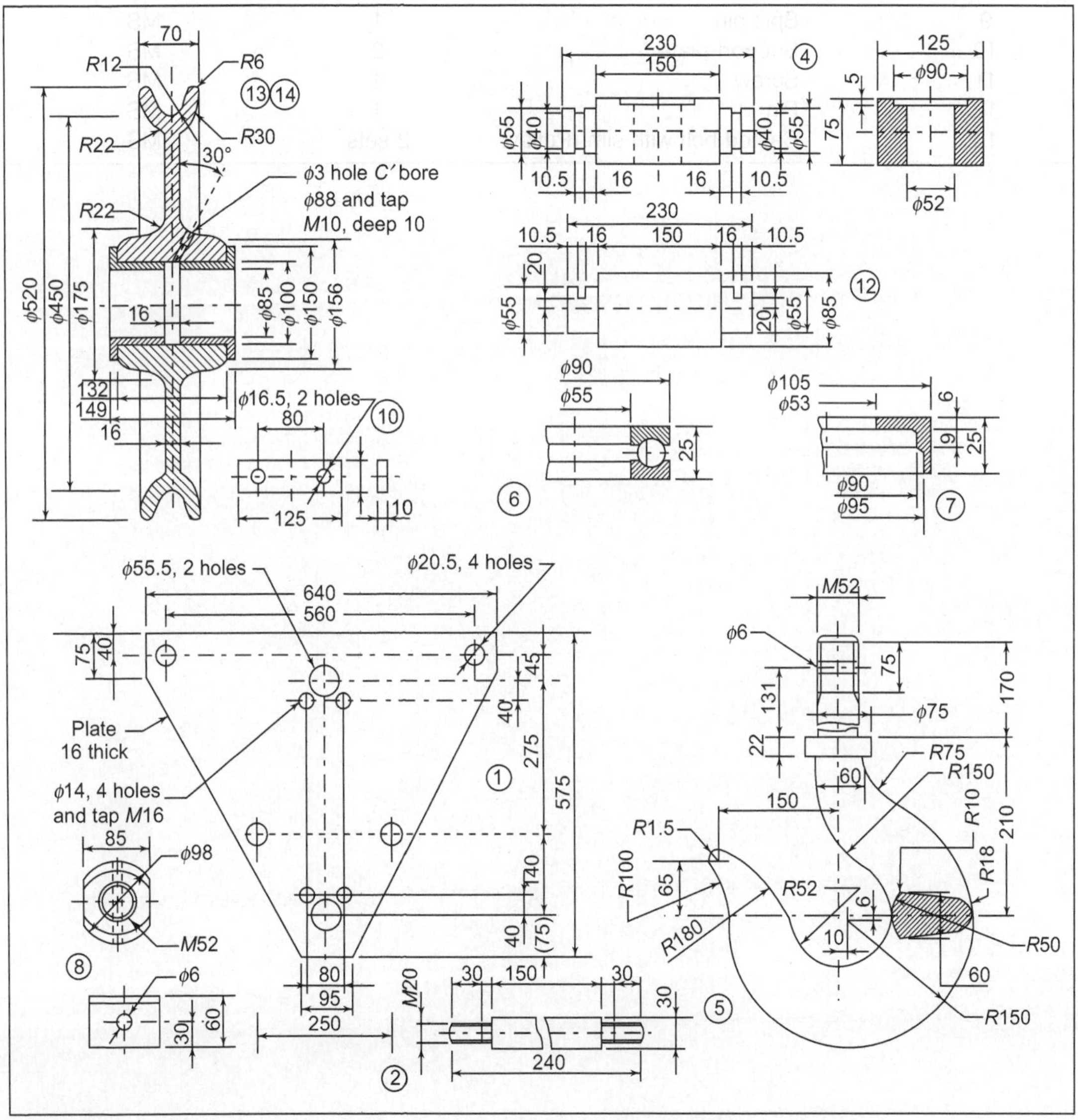

Fig. C6.1 Details of lifting machine hook with pulley block for hoisting tackle

Sl. No.	Part	Quantity	Material
1	Side plates	1 set	MS
2	Pin with nut	4 + 4	MS
3	Hexagonal nut	4 + 4	MS
4	Bush	1	MS
5	Hook	1	CS
6	Ball bearing	1	Bearing metal
7	Cover plate	1	MS
8	Nut	1	MS
9	Split pin	1	MS
10	Support plate	2	MS
11	Screw	4	MS
12	Pin	1	MS
13	Casted bolt with sim (Pulley)	2 sets	MS

7 Tool Carrier

The details of a tool carrier are shown in Fig. C7.1. Assemble all the parts and draw the following views of the assembly.

1. Elevation
2. Plan
3. End elevation

Enlist a bill of material and mark important dimensions in assembly drawing.

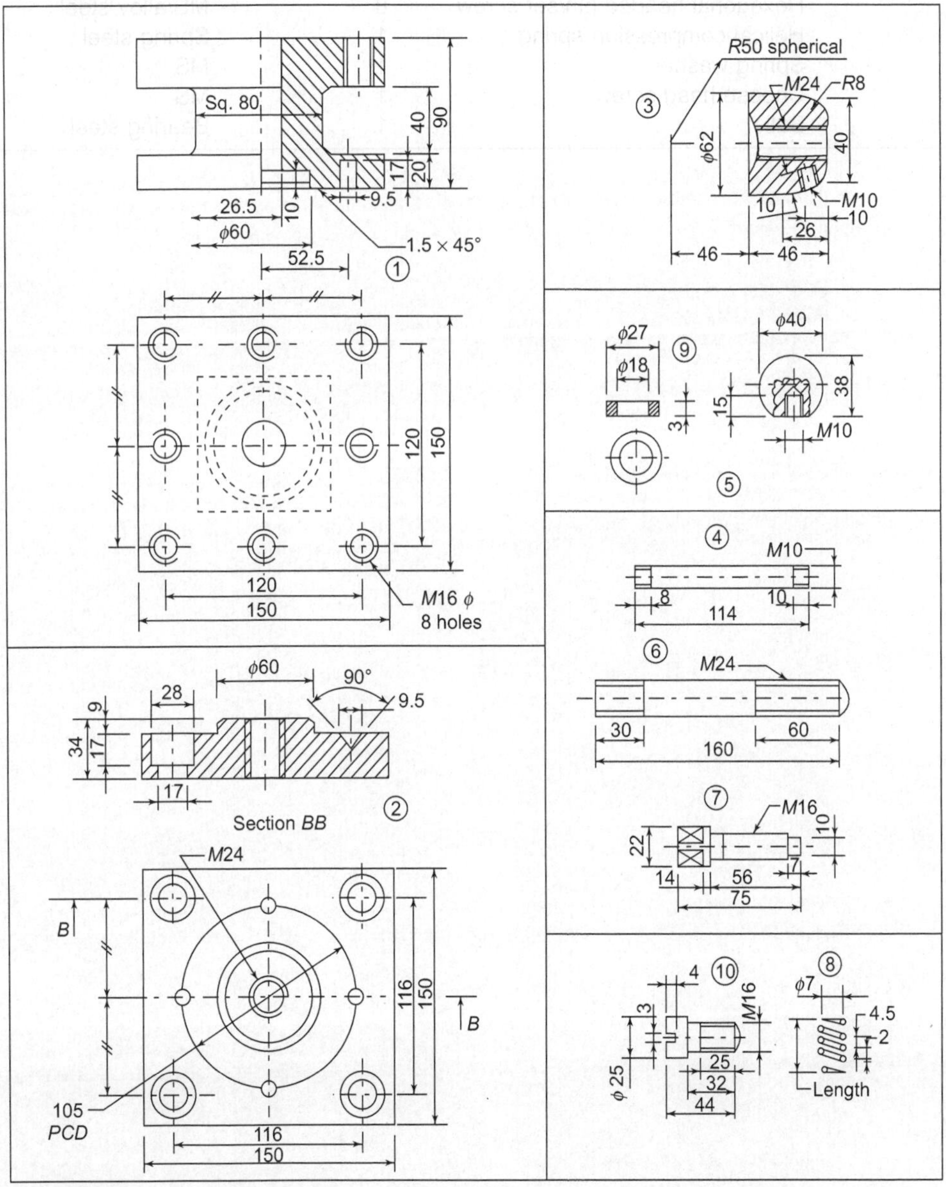

Fig. C7.1 Tool carrier

Sl. No.	Part	Quantity	Material
1	Tool holder body	1	MS/alloy steel
2	Base plate	1	MS/alloy steel
3	Mover	1	MS/alloy steel
4	Handle	1	MS/alloy steel
5	Knot	1	MS/ebonite
6	Stud pin	1	MS/alloy steel
7	Hexagonal headed pin/set screw	8	MS/alloy steel
8	Helical compression spring	1	Spring steel
9	Spring washer	1	MS
10	Cheese head screw	1	MS
11	Ball	1	Bearing steel

8 Shaping Machine Tool Holder

The details of a tool holder of a shaping machine are shown in Fig. C8.1. Draw the following views of the assembly.

1. Elevation
2. End view

   Show important dimensions in assembly drawing and enlist a bill of material.

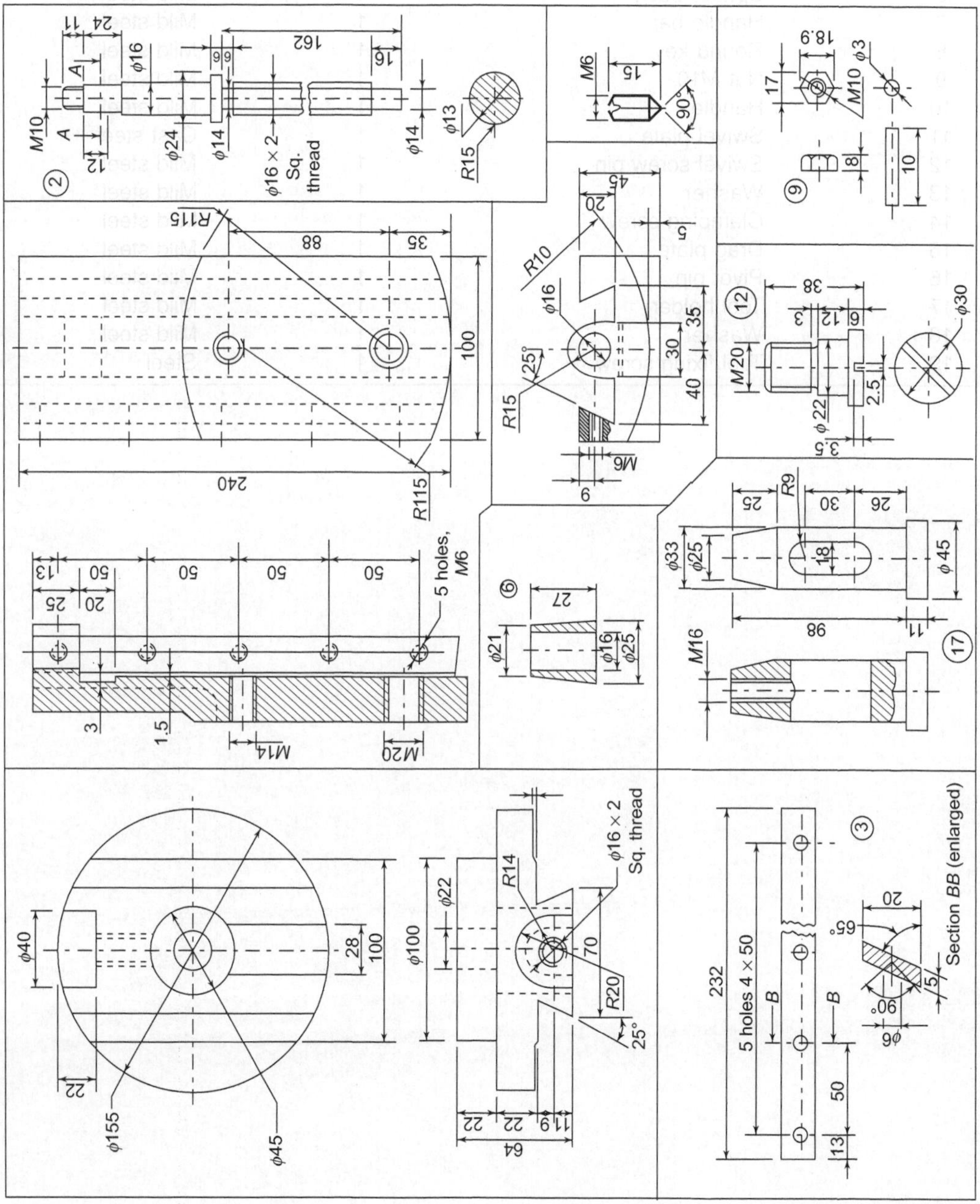

Fig. C8.1 Shaping machine tool holder

Sl. No.	Part	Quantity	Material
1	Black plate	1	Cast steel
2	Vertical slide	1	Cast steel
3	Adjustable strip	1	Mild steel
4	Grub screw	5	Mild steel
5	Screw rod	1	Mild steel
6	Spacer bush	1	Mild steel
7	Handle bar	1	Mild steel
8	Round key	1	Mild steel
9	Nut *M*10	1	Mild steel
10	Handle	1	Mild steel
11	Swivel plate	1	Cast steel
12	Swivel screw pin	1	Mild steel
13	Washer	1	Mild steel
14	Clamping screw	1	Mild steel
15	Drag plate	1	Mild steel
16	Pivot pin	1	Mild steel
17	Tool holder	1	Mild steel
18	Washer	1	Mild steel
19	Tool fixing screw	1	Steel

9 Elastic Film Regulator

Fig. C9.1 shows details of a elastic film regulator. Assemble all the parts and draw the following views of the assembly in 1:1 scale.

1. Elevation—full-in-section
2. End view–outside

   Enlist a bill of material.

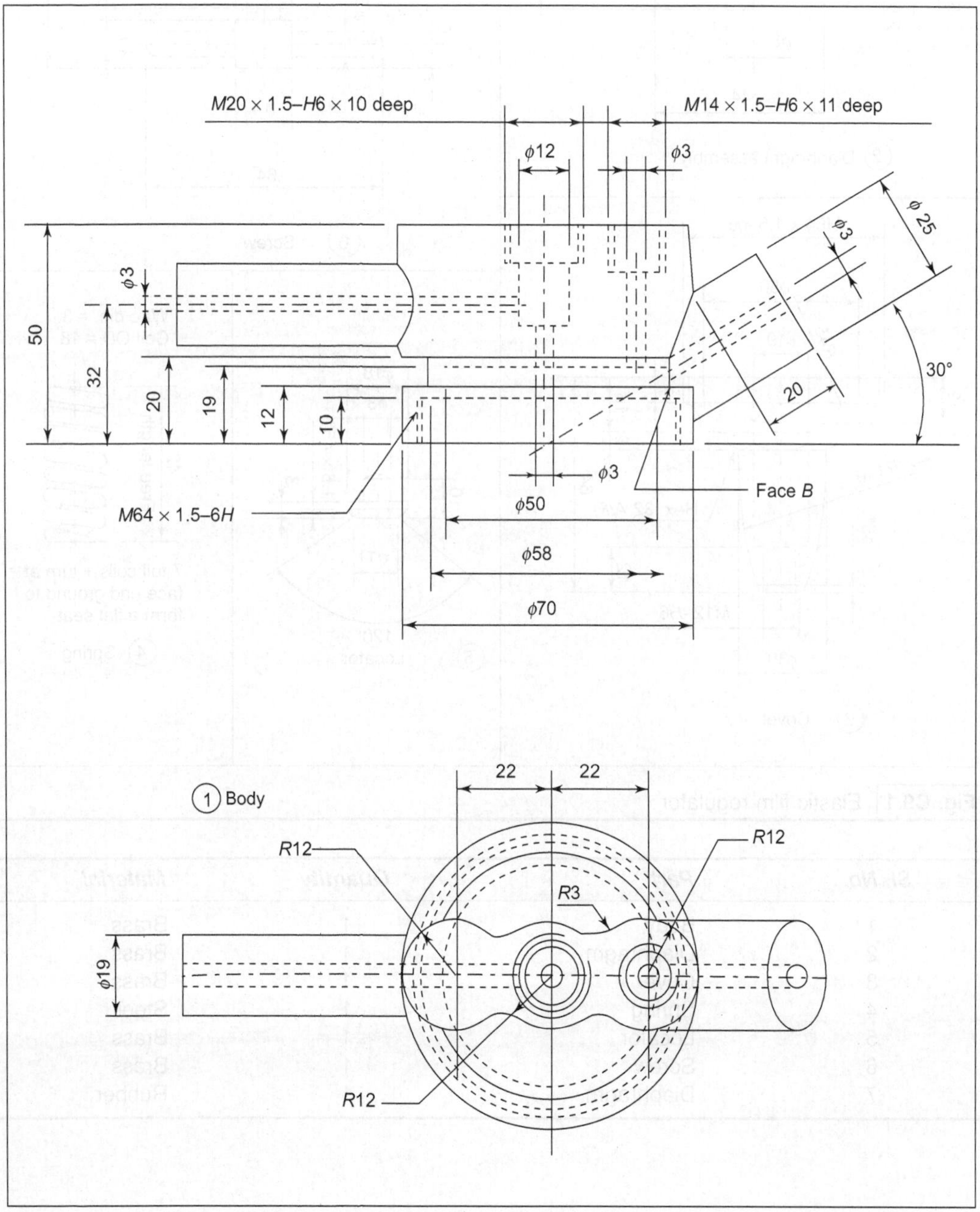

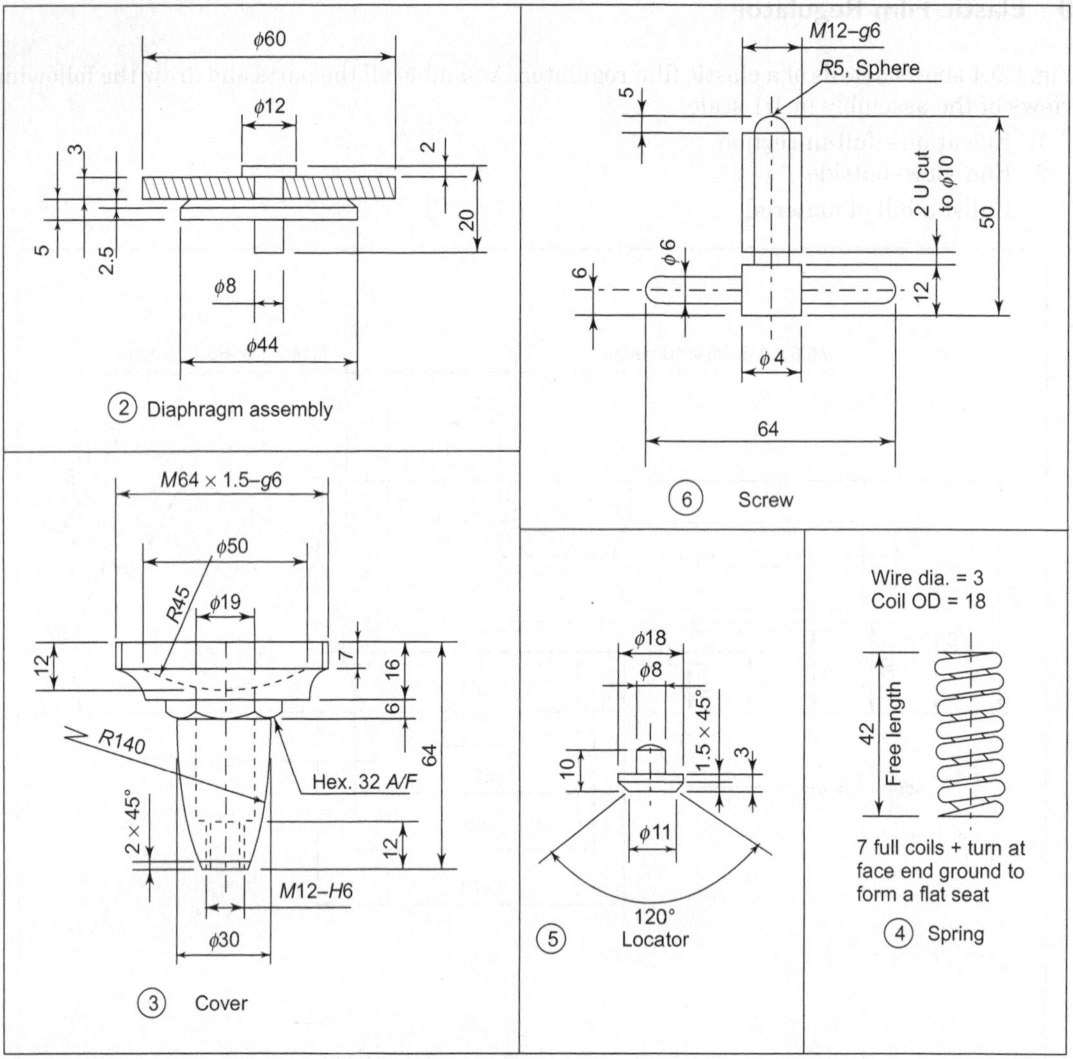

Fig. C9.1 Elastic film regulator

Sl. No.	*Part*	*Quantity*	*Material*
1	Body	1	Brass
2	Diaphragm	1	Brass
3	Cover	1	Brass
4	Spring	1	Steel
5	Locator	1	Brass
6	Screw	1	Brass
7	Diaphragm	1	Rubber

10 Connecting Rod for Small Engines

Fig. C10.1 shows details of a connecting rod for small engine. Assemble all the parts and draw the following views of the connecting rod.

1. Elevation
2. Plan—full-in-section
   Scale—double full size
   Enlist a bill of material.

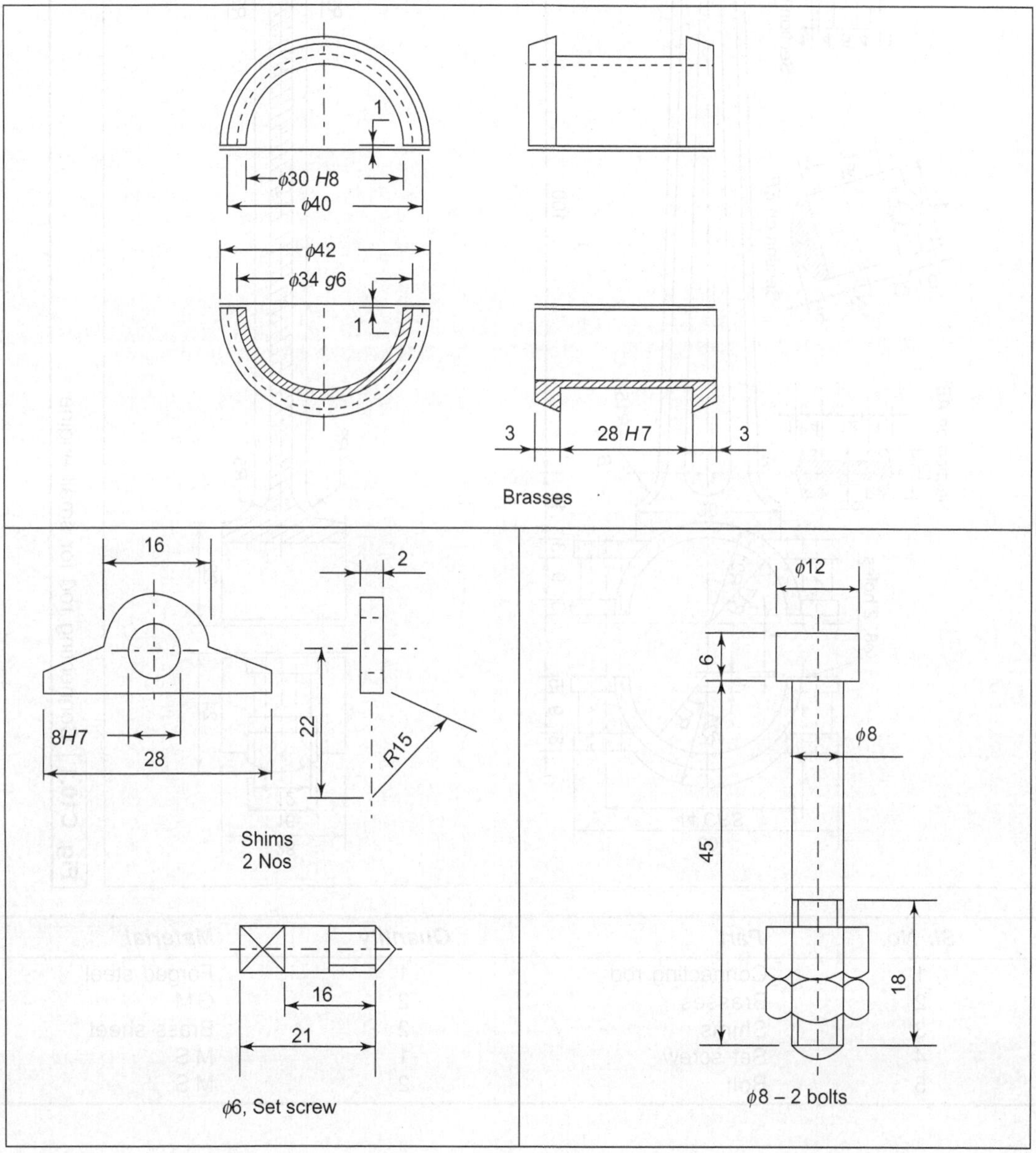

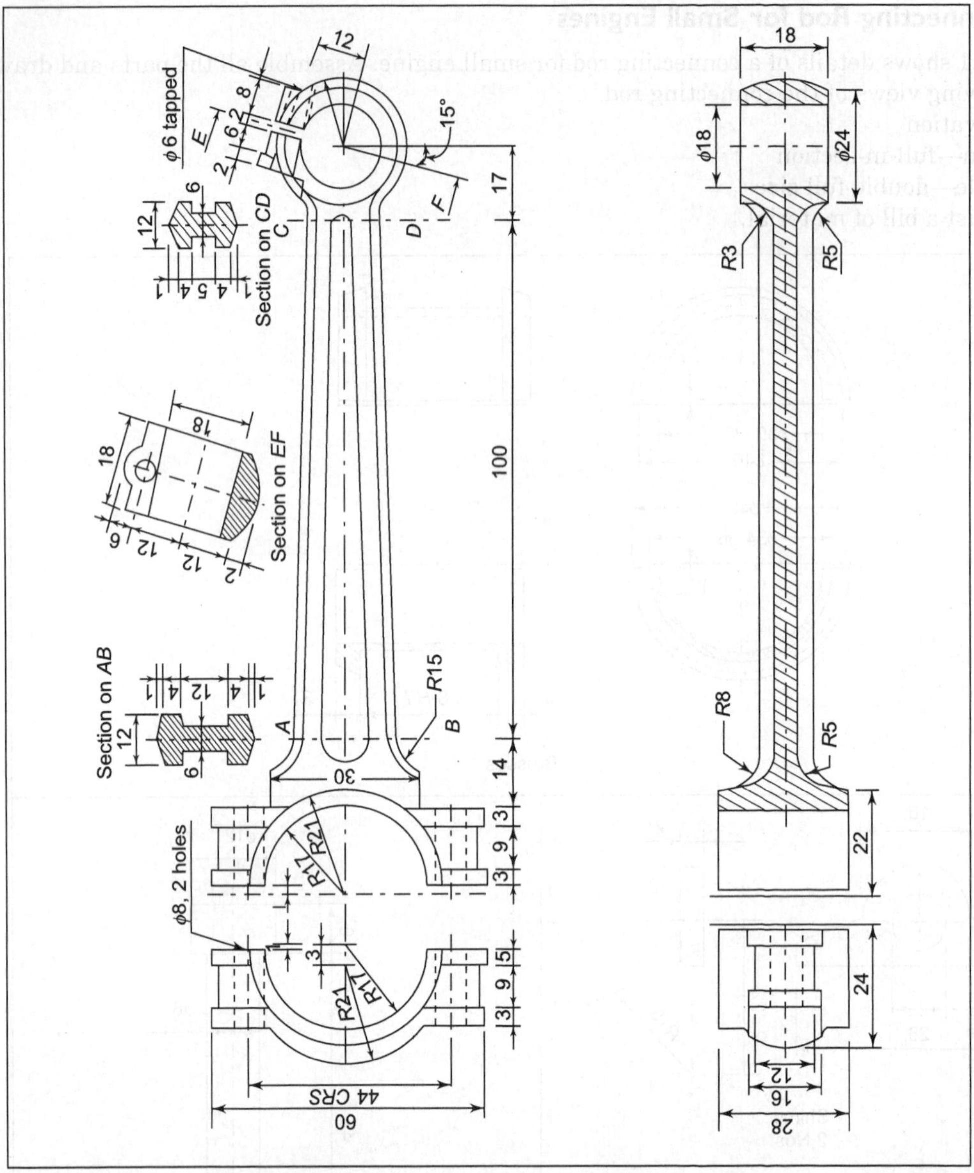

Fig. C10.1 Connecting rod for small engine

Sl. No.	*Part*	*Quantity*	*Material*
1	Connecting rod	1	Forged steel
2	Brasses	2	G M
3	Shims	2	Brass sheet
4	Set screw	1	M S
5	Bolt	2	M S

11 Connecting Rod for Big Engine

Fig. C11.1 shows details of connecting rod for big engine. Assemble all the parts and draw the following views of the assembly.

1. Elevation—full-in-section
2. Plan

Give major dimensions and provide a bill of material.

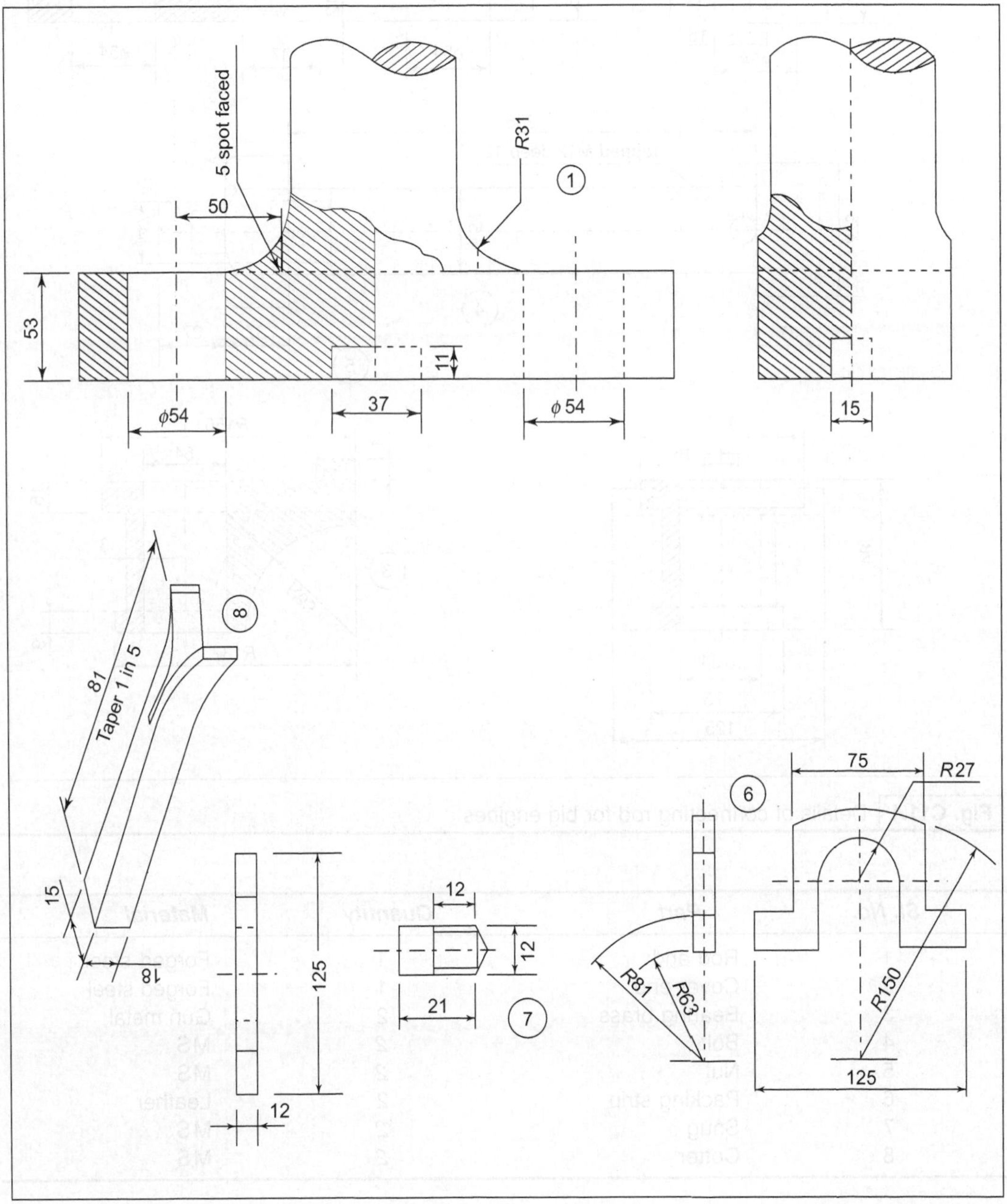

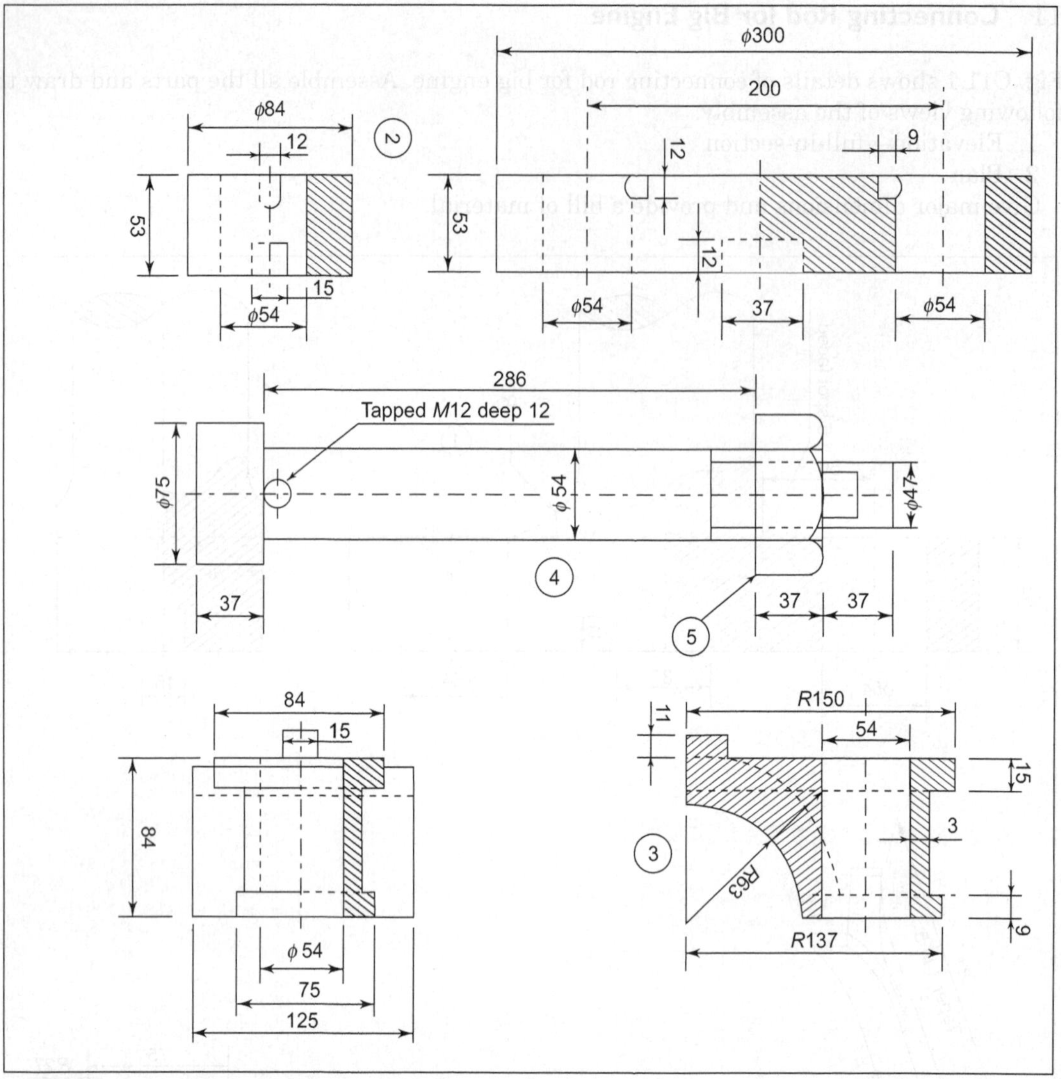

Fig. C11.1 Details of connecting rod for big engines

Sl. No.	Part	Quantity	Material
1	Rod end	1	Forged steel
2	Cover end	1	Forged steel
3	Bearing brass	2	Gun metal
4	Bolt	2	MS
5	Nut	2	MS
6	Packing strip	2	Leather
7	Snug	2	MS
8	Cotter	2	MS

12 Spark Plug

The details of a spark plug are given in Fig. C12.1. Draw to a scale of 2.5:1, the following assembled views.

1. Front view—right half-in-section
2. Top view

Add all the necessary dimensions and enlist a bill of material.

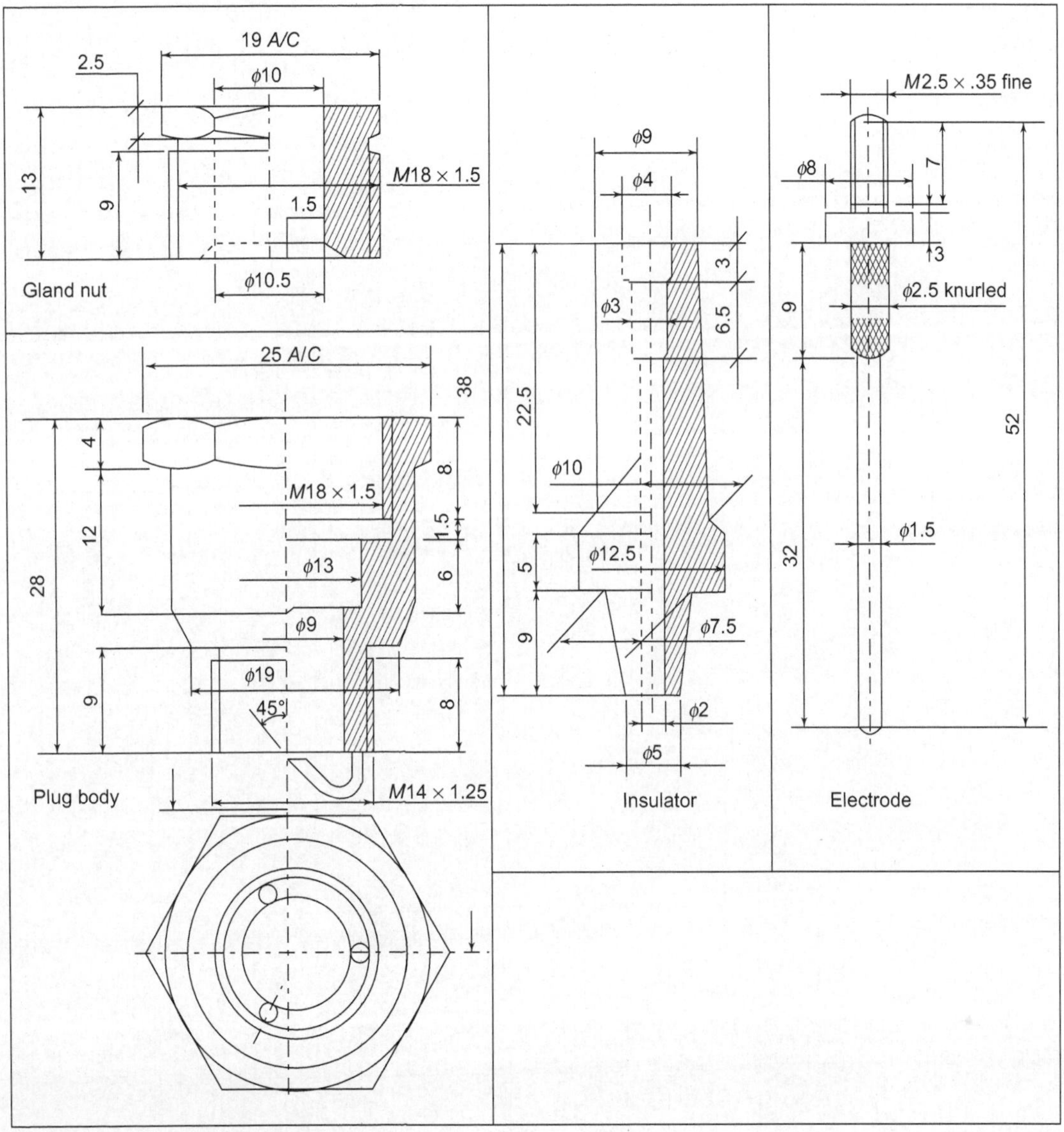

Fig. C12.1 Spark plug

Sl. No.	*Part*	*Quantity*	*Material*
1	Plug body	1	Alloy steel
2	Gland nut	1	Alloy steel
3	Insulator	1	Ceramic material
4	Electrode	1	Electric current conductor (Non-ferrous)

13 Variable Supply Fuel Pump

The details of a variable supply fuel pump are shown in Fig. C13.1. Draw the following views of the pump in 1:1 scale.

1. Elevation—full-in-section
2. End view
3. Plan

Enlist a bill of material.

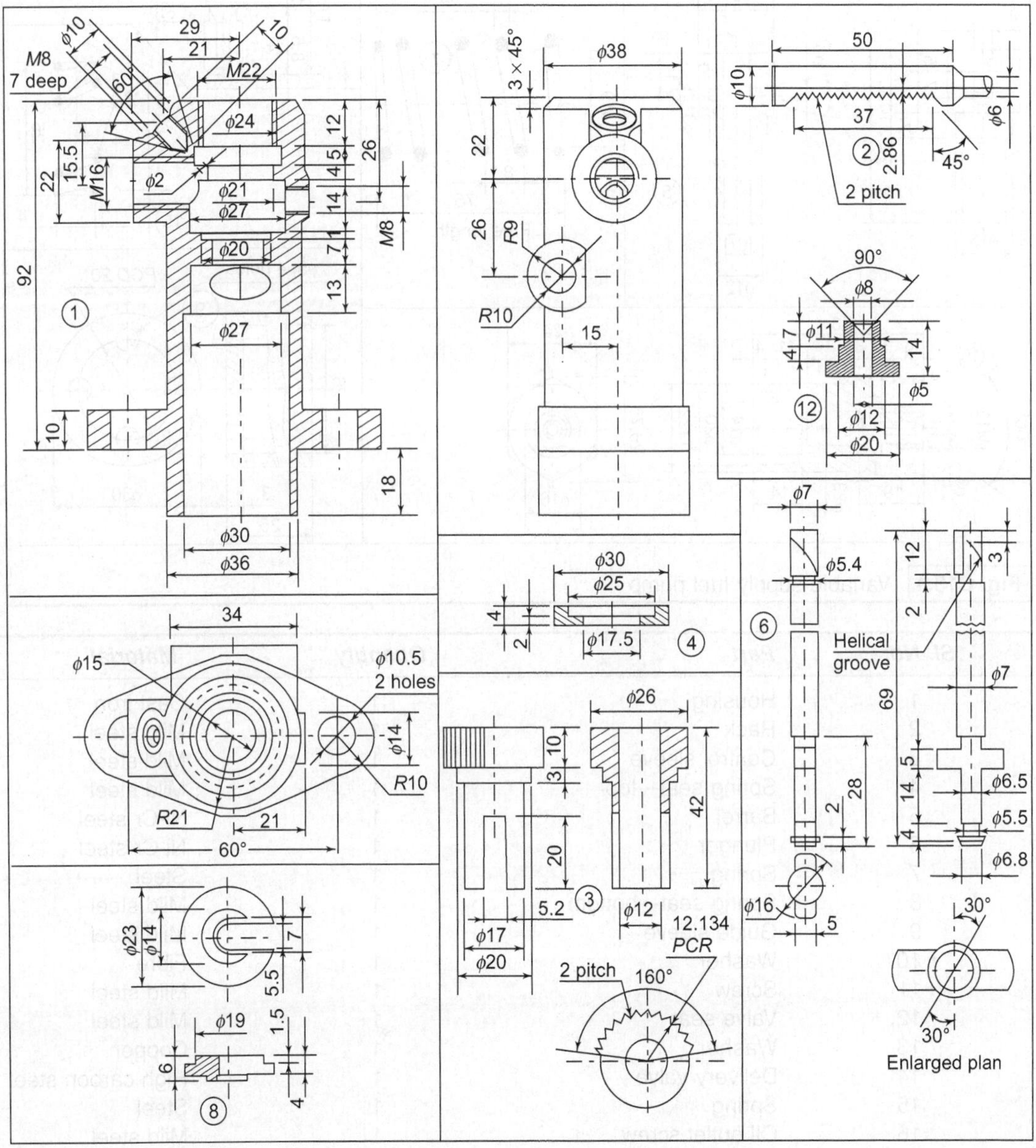

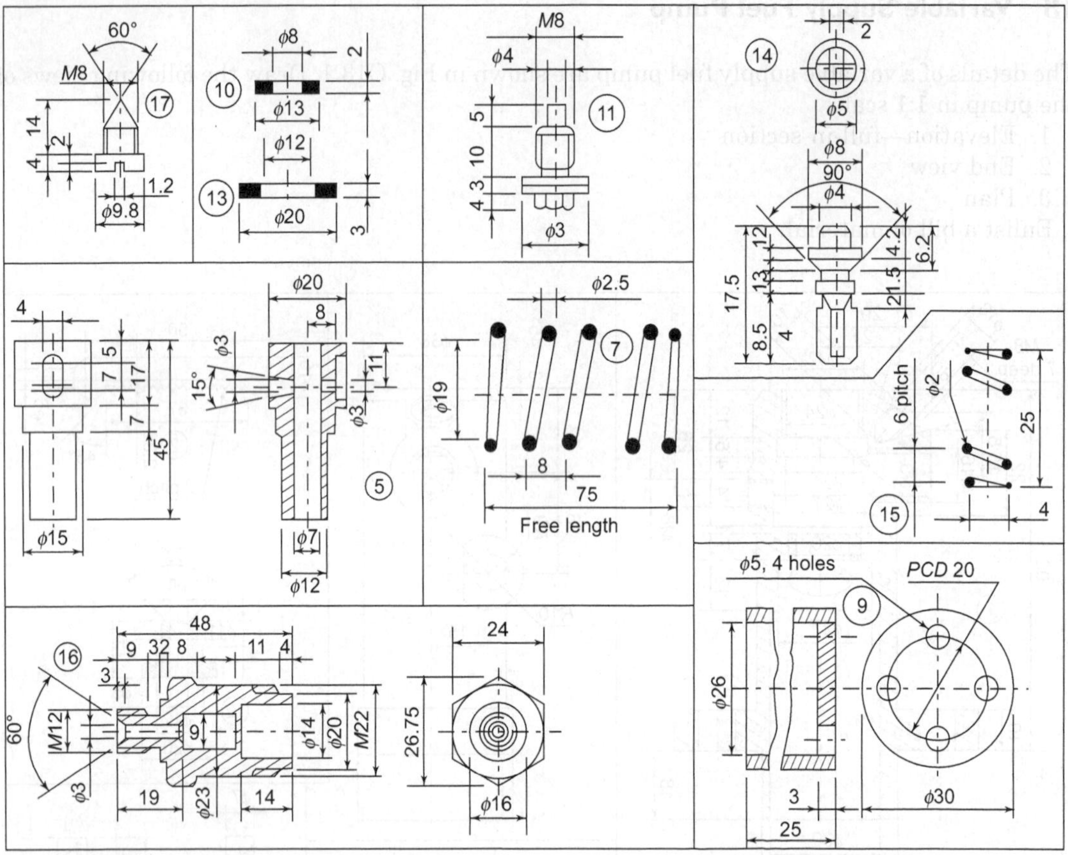

Fig. C13.1 Variable supply fuel pump

Sl. No.	Part	Quantity	Material
1	Housing	1	Cast iron
2	Rack	1	Mild steel
3	Control sleeve	1	Mild steel
4	Spring seat—top	1	Mild steel
5	Barrel	1	Ni-Cr steel
6	Plunger	1	Ni-Cr steel
7	Spring	1	Steel
8	Spring seat—bottom	1	Mild steel
9	Guide sleeve	1	Mild steel
10	Washer	1	Fibre
11	Screw	1	Mild steel
12	Valve seat	1	Mild steel
13	Washer	1	Copper
14	Delivery valve	1	High carbon steel
15	Spring	1	Steel
16	Oil outlet screw	1	Mild steel
17	Air release screw	1	Brass

14 Fuel Injector

Fig. C14.1 shows the details of a fuel injector. Draw the following views of the injector with all the parts assembled. Select 2:1 scale.

1. Elevation—half-in-section
2. Plan
3. End view

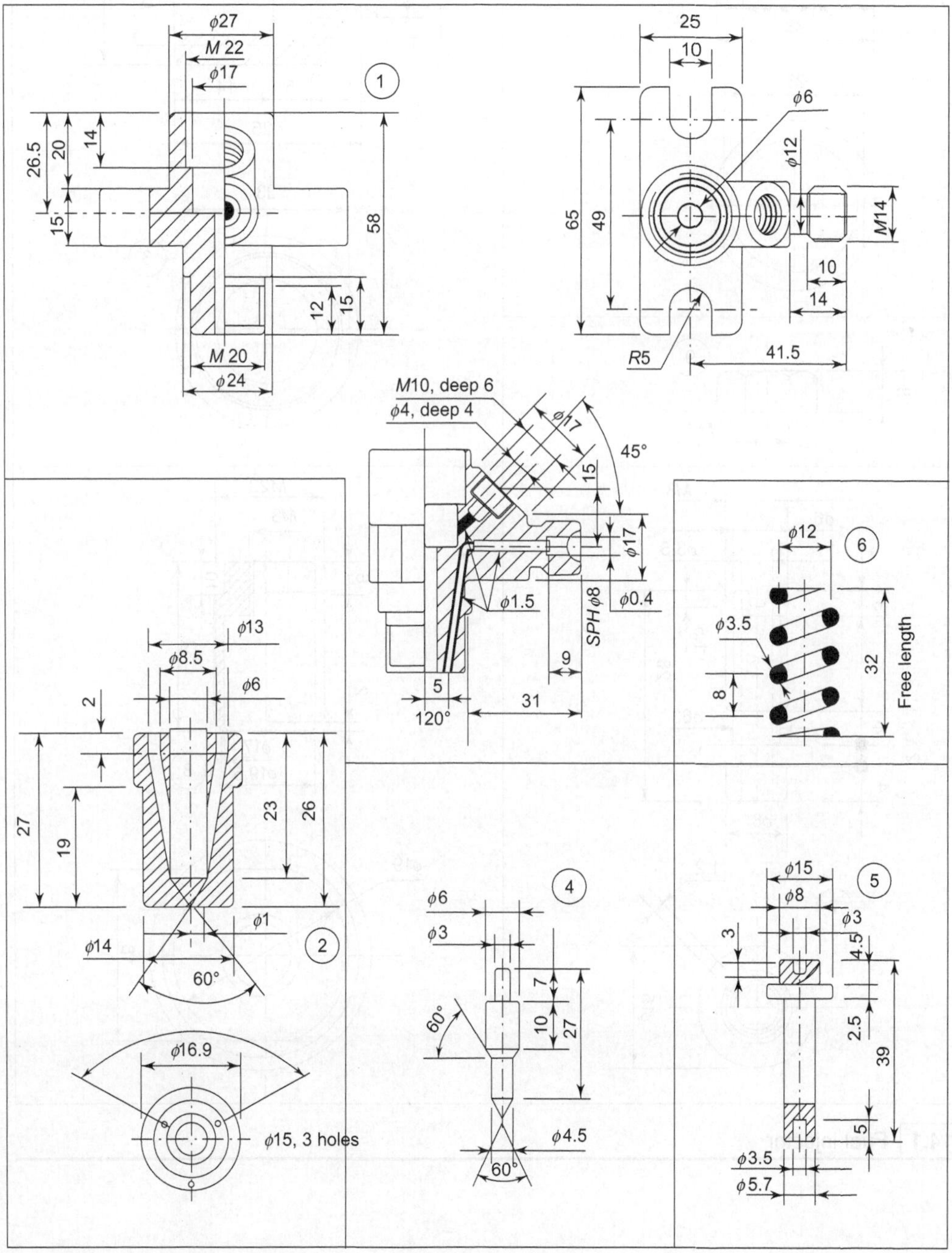

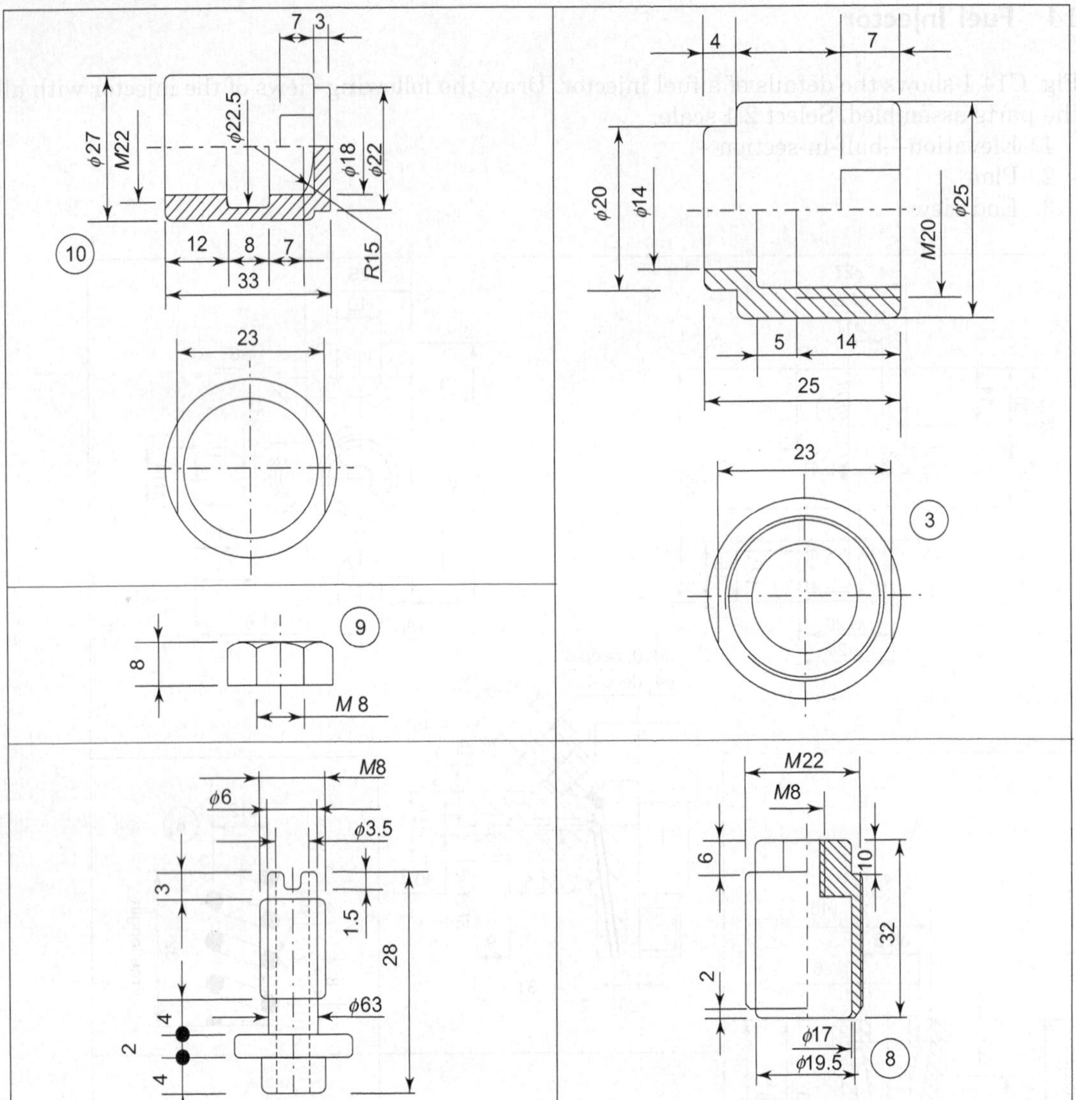

Fig. C14.1 Fuel injector

Sl. No.	*Part*	*Quantity*	*Material*
1	Body	1	CI
2	Nozzle	1	Brass
3	Nozzle holder	1	MS
4	Nozzle pin	1	Brass
5	Distance piece	1	MS
6	Spring	1	Spring steel
7	Screw adjusting piece	1	Brass
8	Sleeve	1	MS
9	Locknut	1	MS
10	Cap	1	MS

15 Job Holding Device

Fig. C15.1 shows the details of a small job holding device and a key to its assembly. Draw full size and orthographic projection of the following views of the assembled vice, the sliding jaw being approximately 25 mm from the fixed jaw.

1. Sectional elevation on a vertical plane passing through the axis of the square threaded screw, in the direction indicated by *XX* in the key.
2. Plan projected from the above.

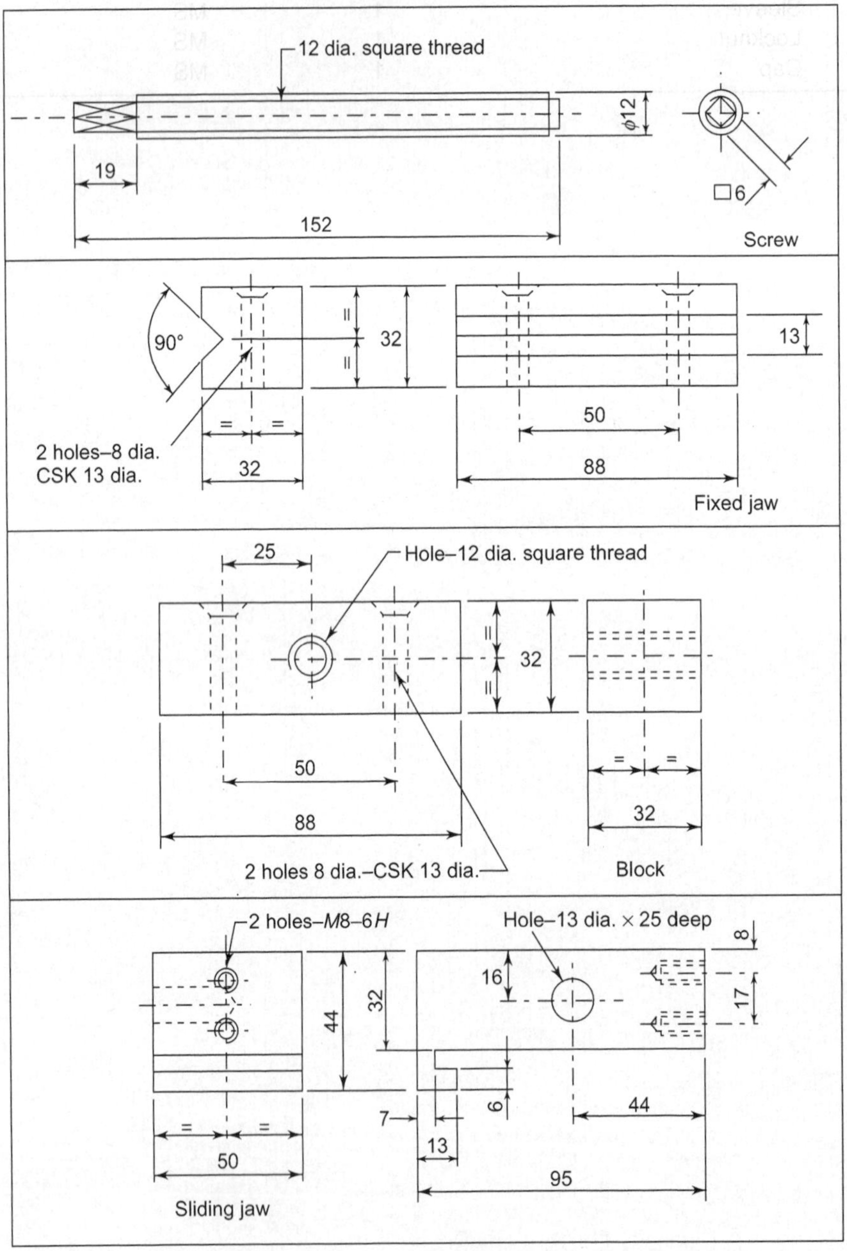

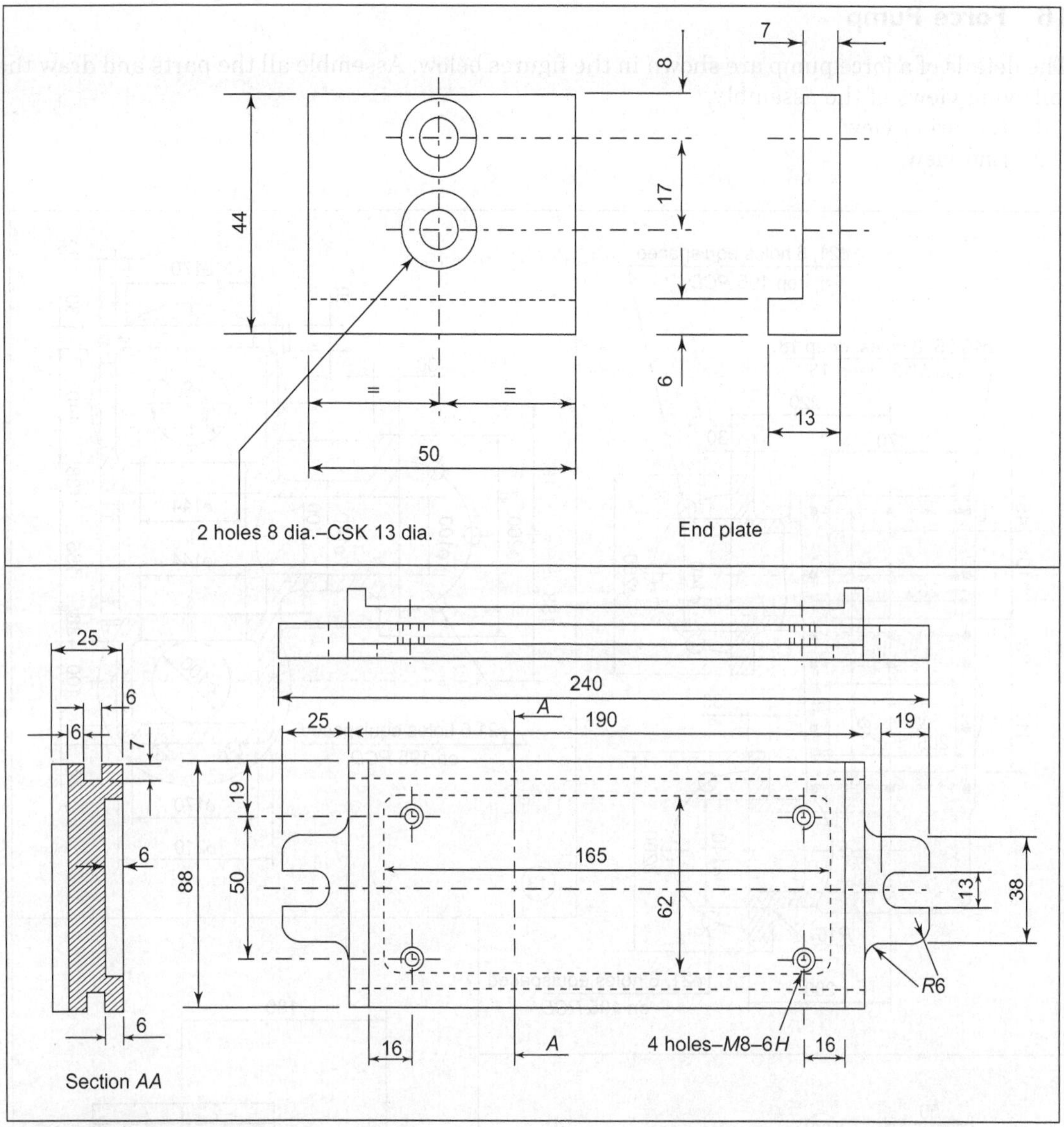

Fig. C15.1 Details of job holding device

Sl. No.	*Part*	*Quantity*	*Material*
1	Base	1	CI
2	Sliding jaw	1	CI
3	Fixed jaw	1	CI
4	End plate	1	MS
5	Block	1	MS
6	Screw	1	MS

16 Force Pump

The details of a force pump are shown in the figures below. Assemble all the parts and draw the following views of the assembly.

1. Elevation view
2. End view

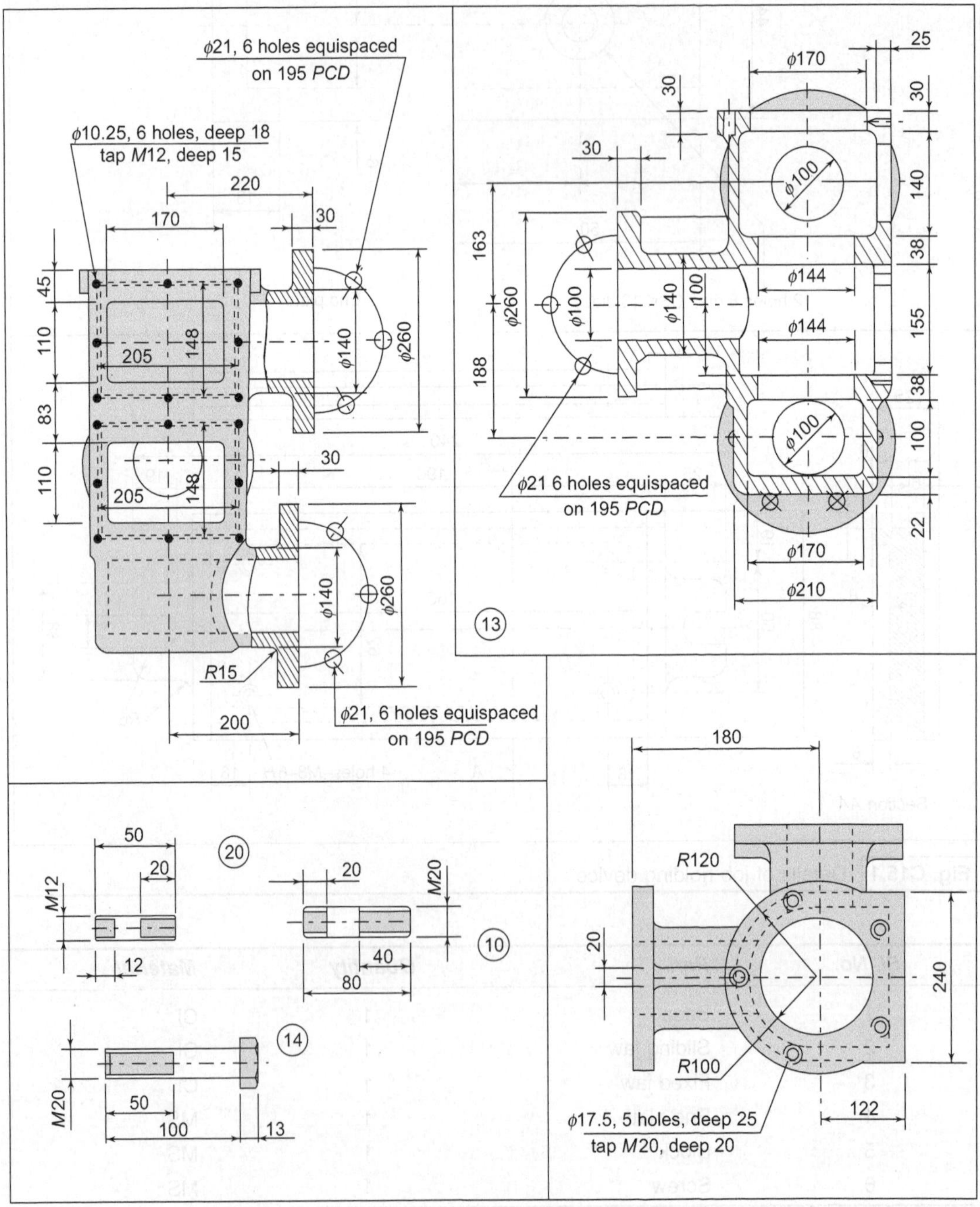

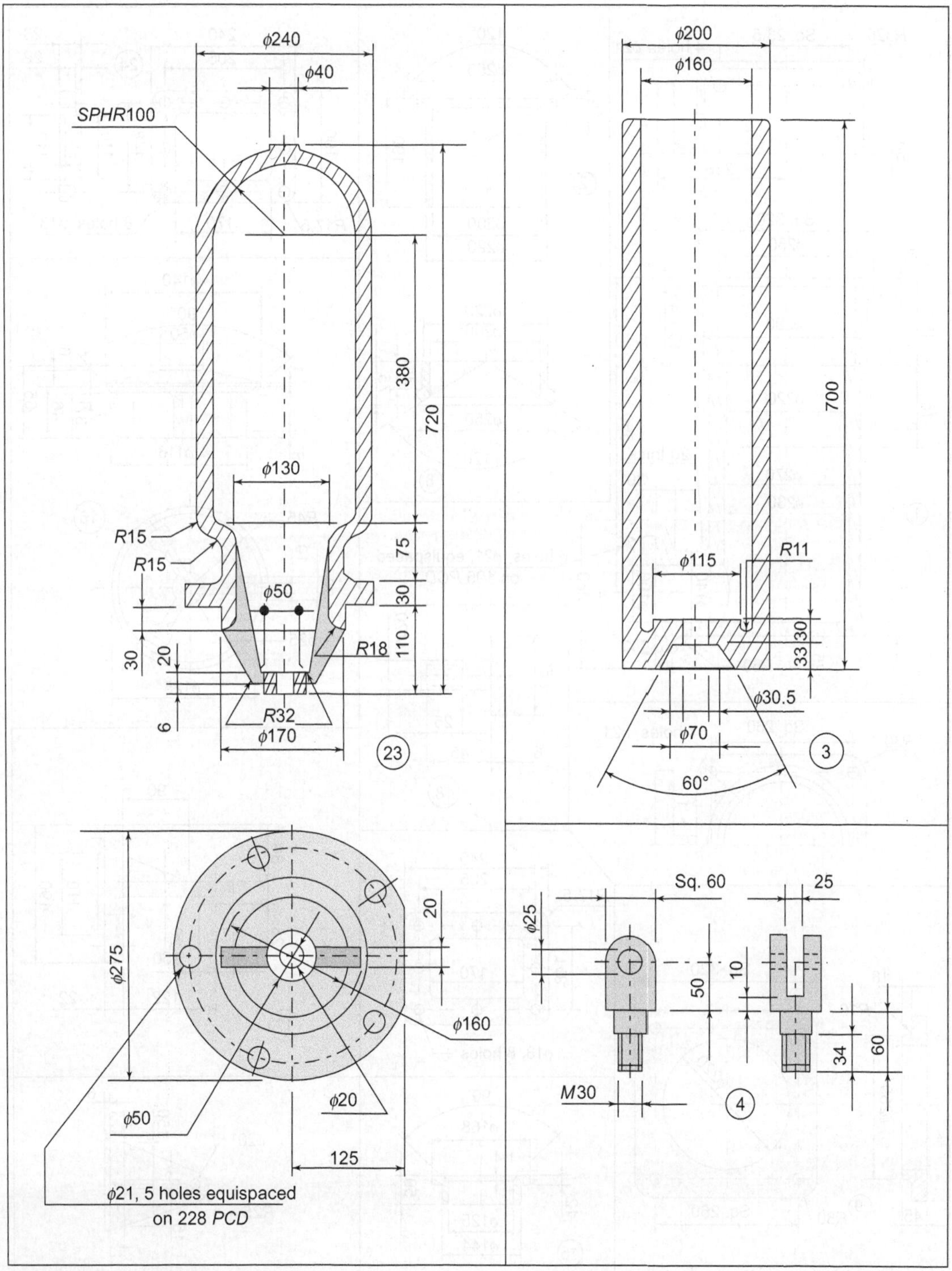

Fig. C16.1 Details of force pump

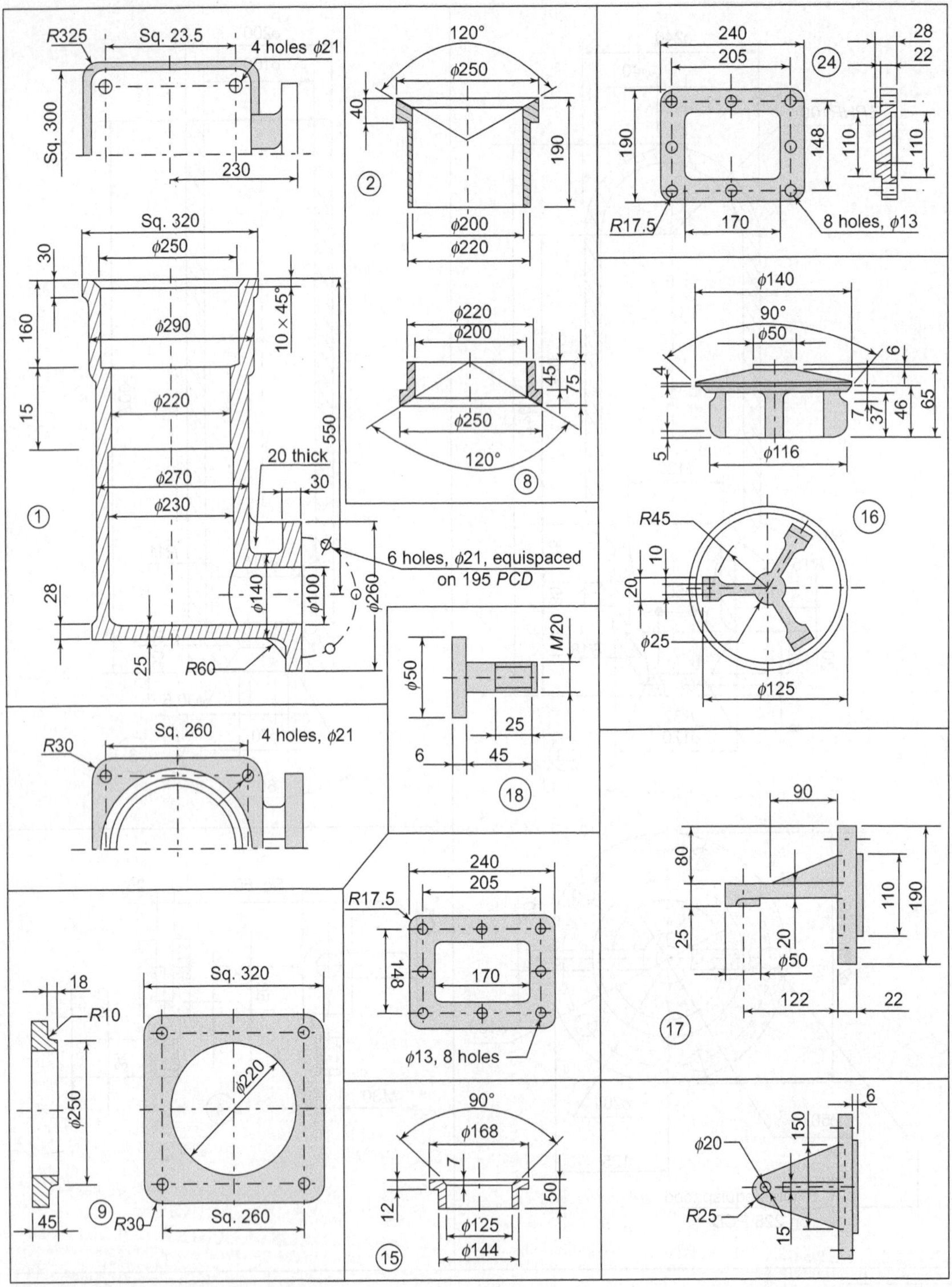

Fig. C16.2 Details of force pump

Sl. No.	Part
1	Pump barrel
2	Neck bush
3	Plunger
4	Split pin
5	Clamp bolt
6	Castle nut
7	Fibre packing
8	Gland bush
9	Fibre packing
10	Studs, nuts, lock nuts
11	Bolts
12	Nuts
13	Valve chest
14	Lock nuts
15	Valve seat
16	Valve
17	Valve shopper
18	Valve adjuster
19	Nut
20	Studs
21	Nuts
22	Lock nuts
23	Air vessel
24	Cover plate

***Note*:** The Quantity and material depends upon the application of force pump.

17 Feed Check Valve

The details of a feed check valve are shown in Fig. C17.1. Assemble all the parts and draw the following views of the assembly in 1:2 scale.

1. Elevation view—right half-in-section
2. Plan view—removing the handle

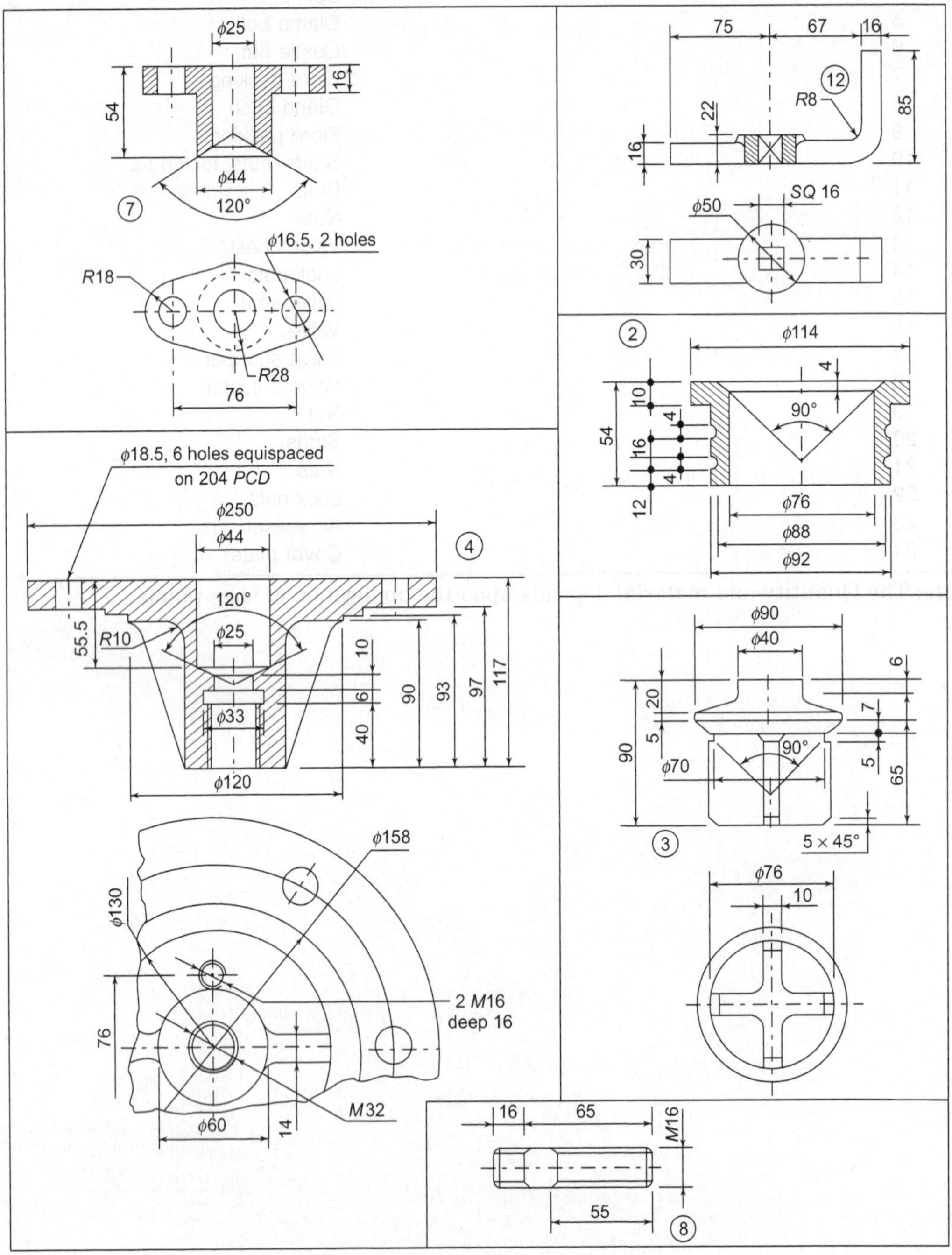

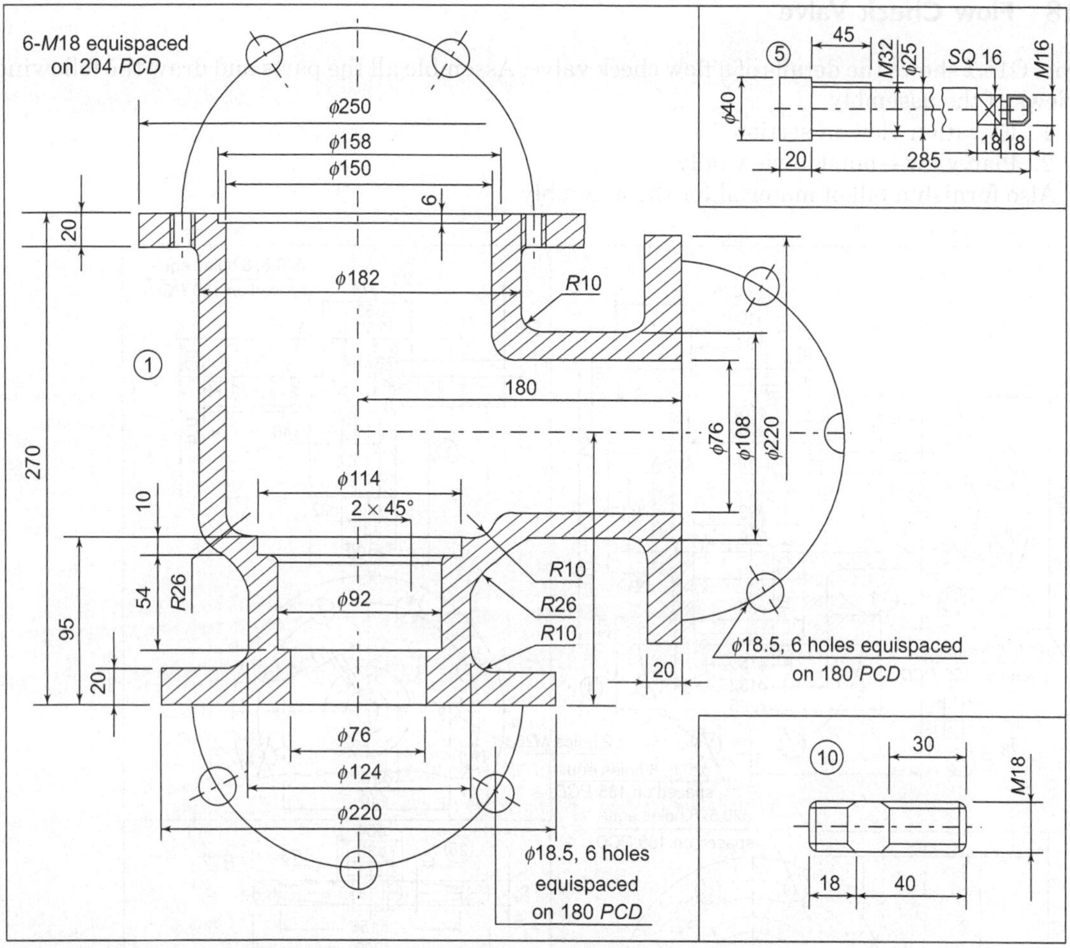

Fig. C17.1 Feed check valve

Sl. No.	Part	Quantity	Material
1	Valve body	1	CI
2	Valve seat	1	GM
3	Valve	1	GM
4	Cover	1	CI
5	Spindle	1	MS
6	Packing	As required	Fibre
7	Gland	1	GM
8	Stud *M*16	2	MS
9	Nut *M*16	3	MS
10	Stud *M*18	6	MS
11	Nut *M*18	6	MS
12	Handle	1	MS

18 Flow Check Valve

Fig. C18.1 shows the details of a flow check valve. Assemble all the parts and draw the following views of the assembly.

1. Elevation view in section
2. Plan view—outside view only.

Also furnish a bill of material for the assembly.

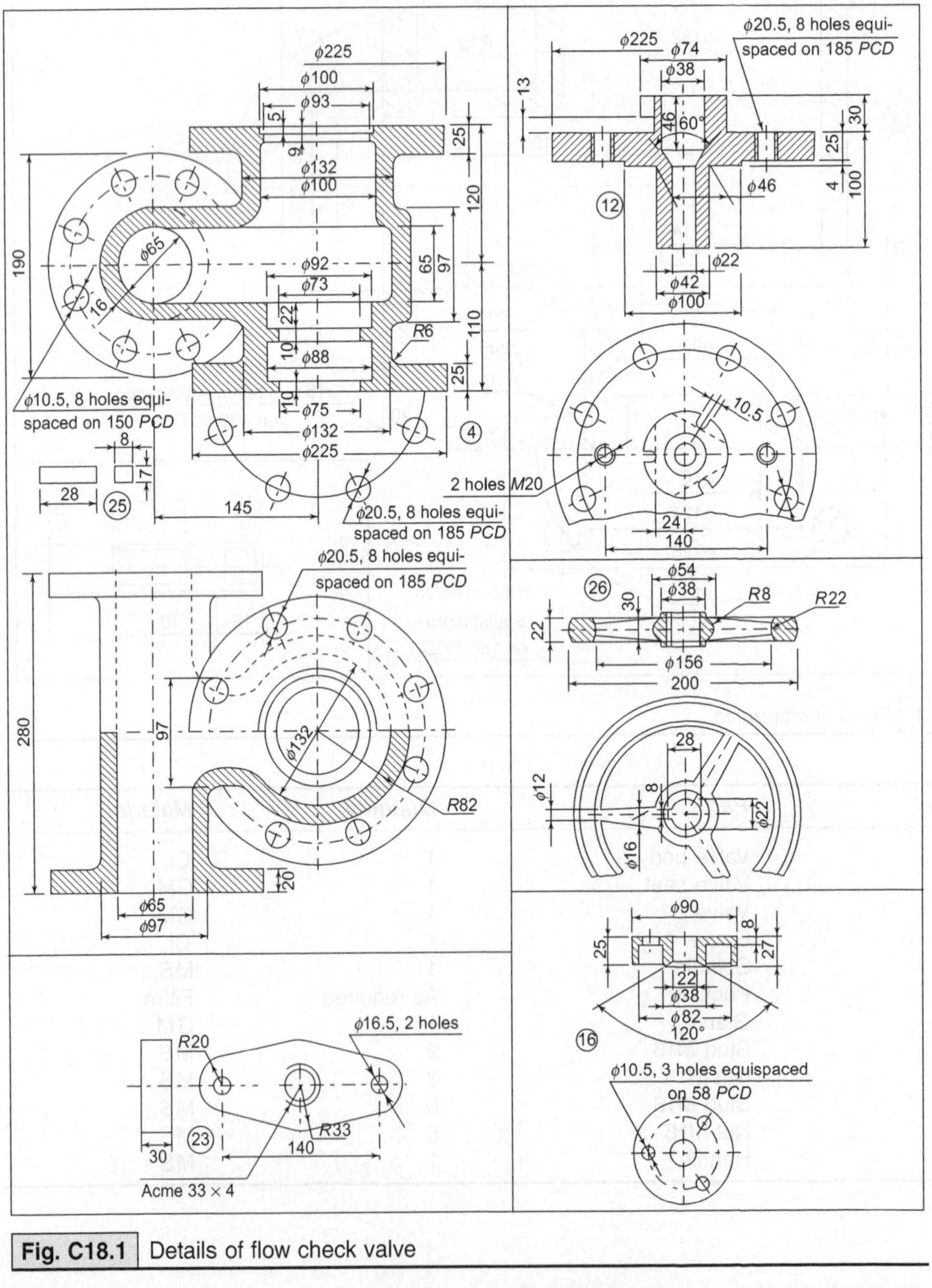

Fig. C18.1 Details of flow check valve

19 Blow Off Cock

Fig. C19.1 shows details of a blow off cock. Assemble all the parts and draw the following views of the assembly.

1. Elevation view
2. Plan view

Give important dimensions and enlist a bill of material.

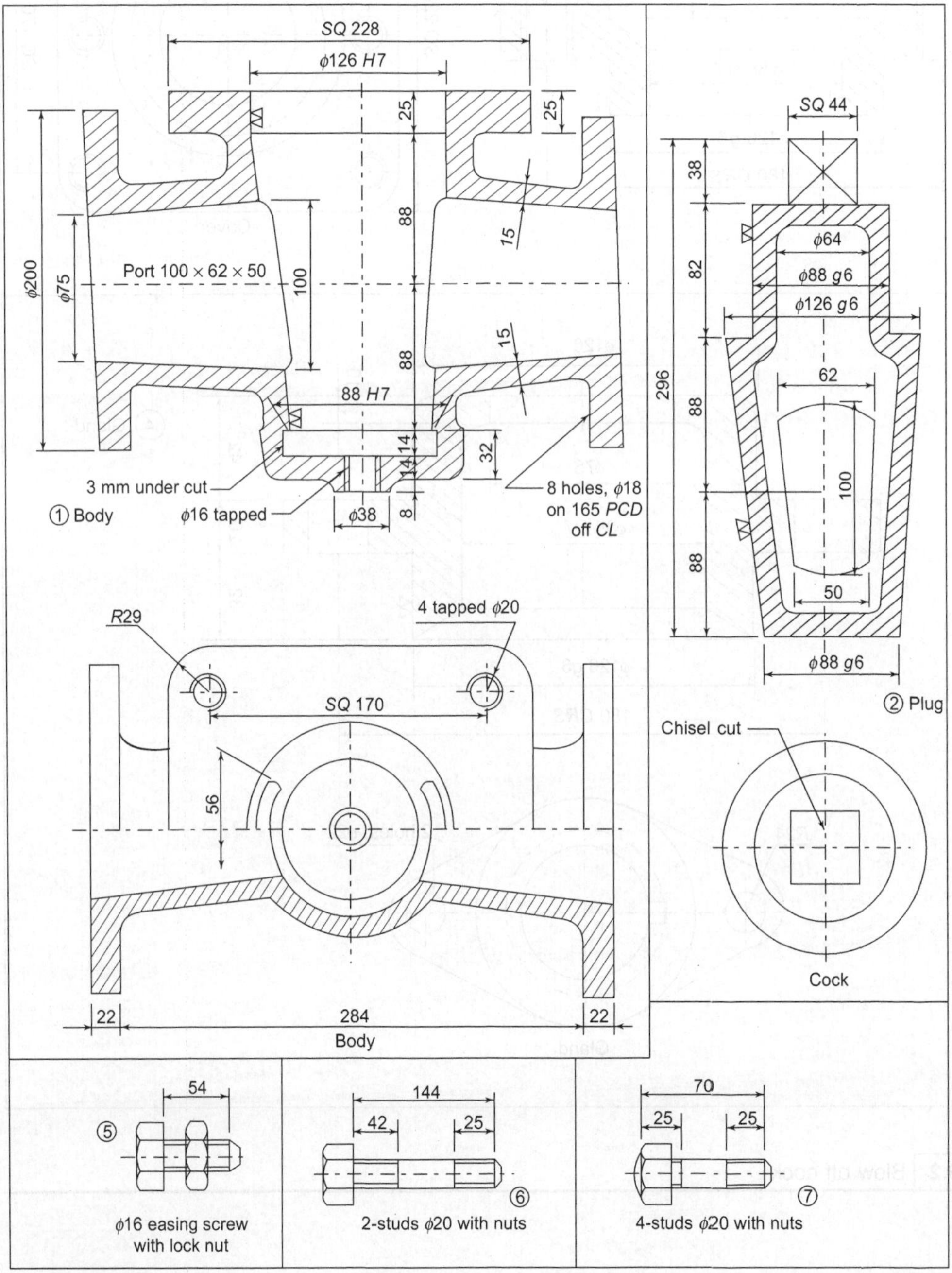

Fig. C19.1 Blow off cock

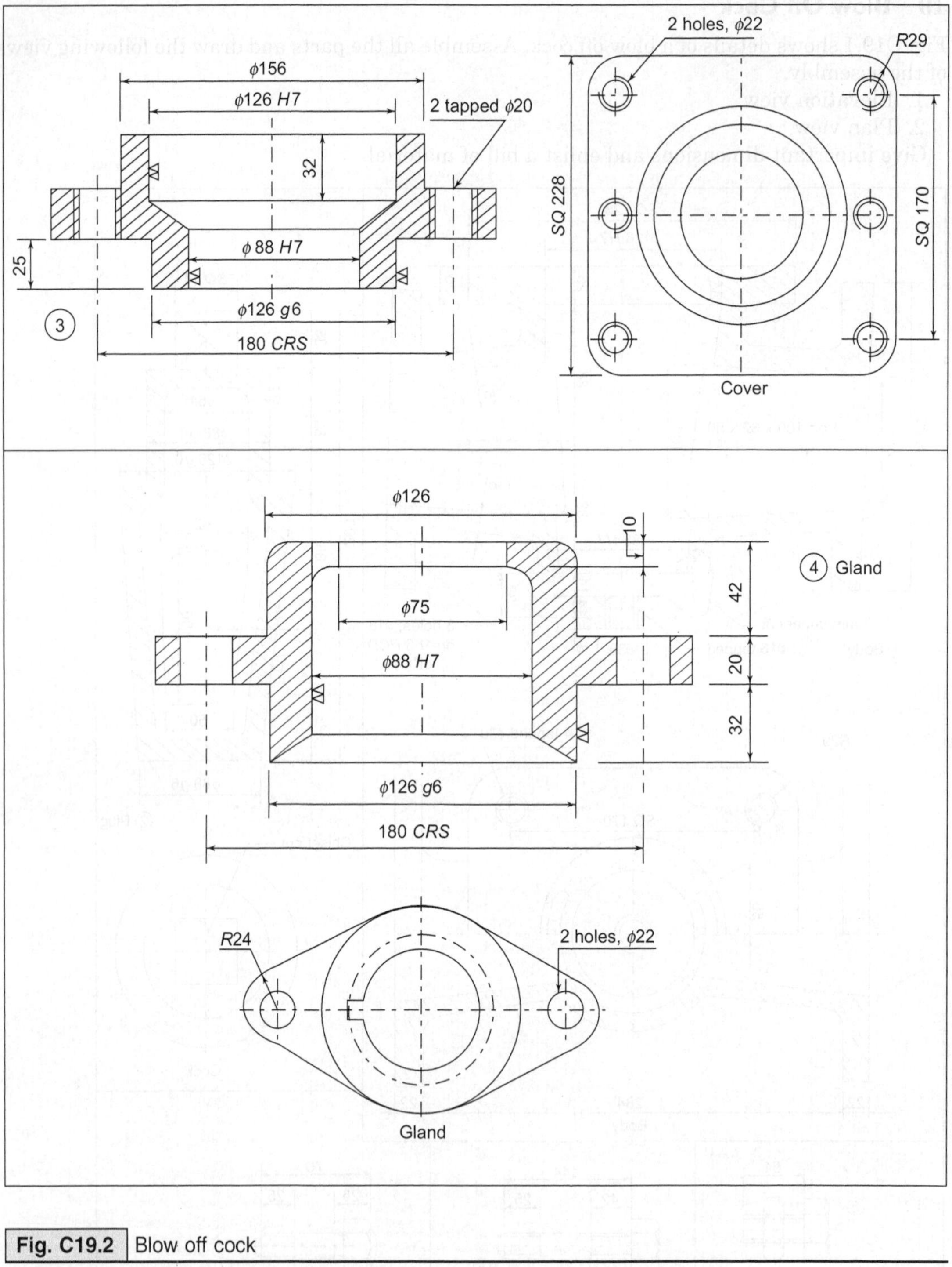

Fig. C19.2 Blow off cock

Sl. No.	Part	Quantity	Material
1	Valve body	1	GM
2	Valve plug	1	GM
3	Valve cover	1	GM
4	Gland	1	GM
5	Easing screw	4 sets	MS
6	Stud no. 1	1 sets	MS
7	Stud no. 2	1 sets	MS

20 Supply Stop Valve

Fig. C20.1 shows details of a stop valve. Assemble all the parts and draw the following assembly views.

1. Elevation view
2. Plan view

Mention important dimensions and enlist a bill of material.

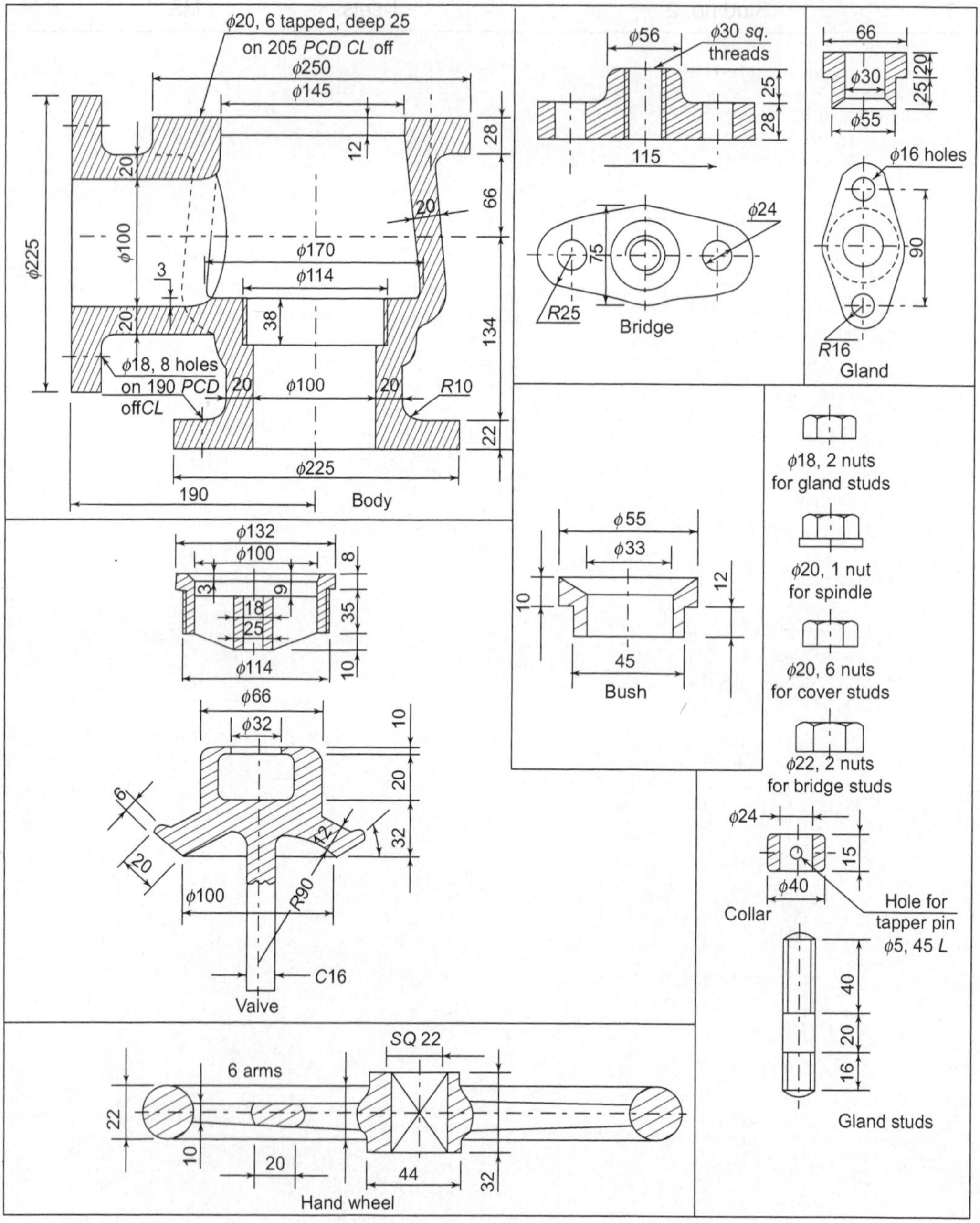

Fig. C20.1

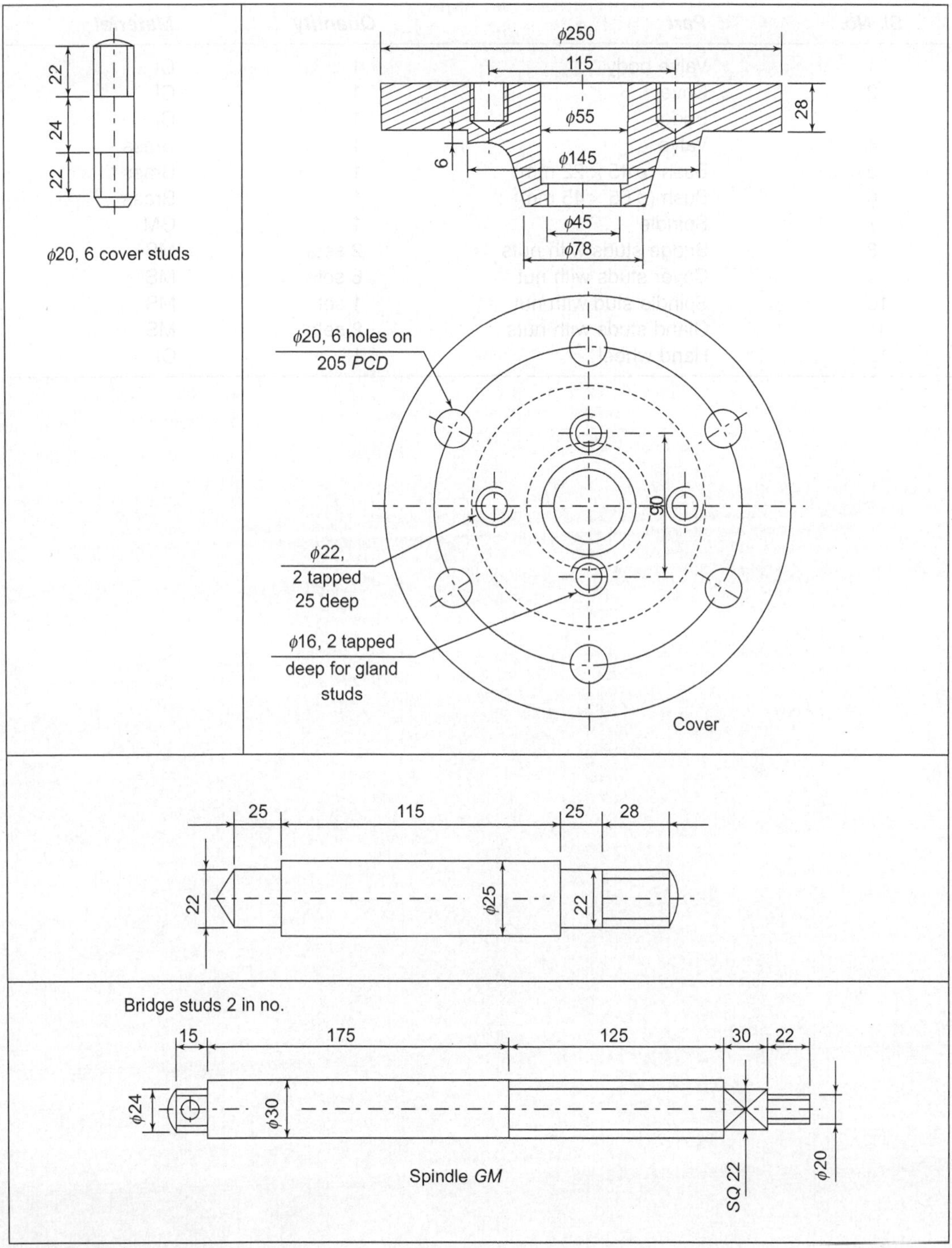

Fig. C20.1 Details of stop valve

Sl. No.	Part	Quantity	Material
1	Valve body	1	CI
2	Cover	1	CI
3	Bridge	1	CI
4	Valve	1	Brass
5	Bush (ϕ 45 × 22 mm)	1	Brass
6	Bush (ϕ 55 × 45 mm)	1	Brass
7	Spindle	1	GM
8	Bridge studs with nuts	2 sets	MS
9	Cover studs with nut	6 sets	MS
10	Spindle stud with nut	1 set	MS
11	Gland studs with nuts	2 sets	MS
12	Hand wheel	1	CI

21 125 mm Stop Valve

Fig. C21.1 shows details of a 125 mm stop valve. Assemble all the parts and draw the following views.

1. Elevation view
2. Plan view

Mention important dimensions on the drawing and enlist a bill of material.

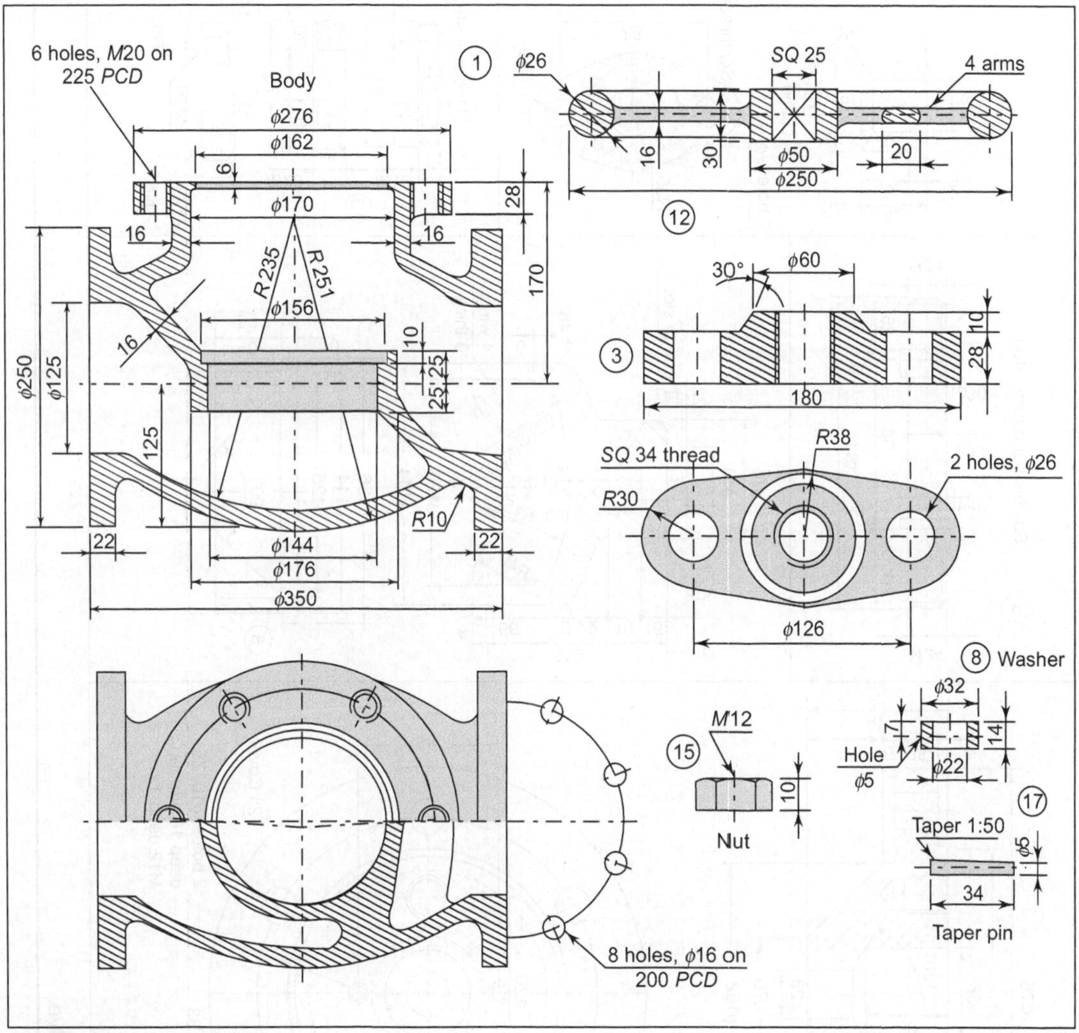

Fig. C21.1 125 mm stop valve

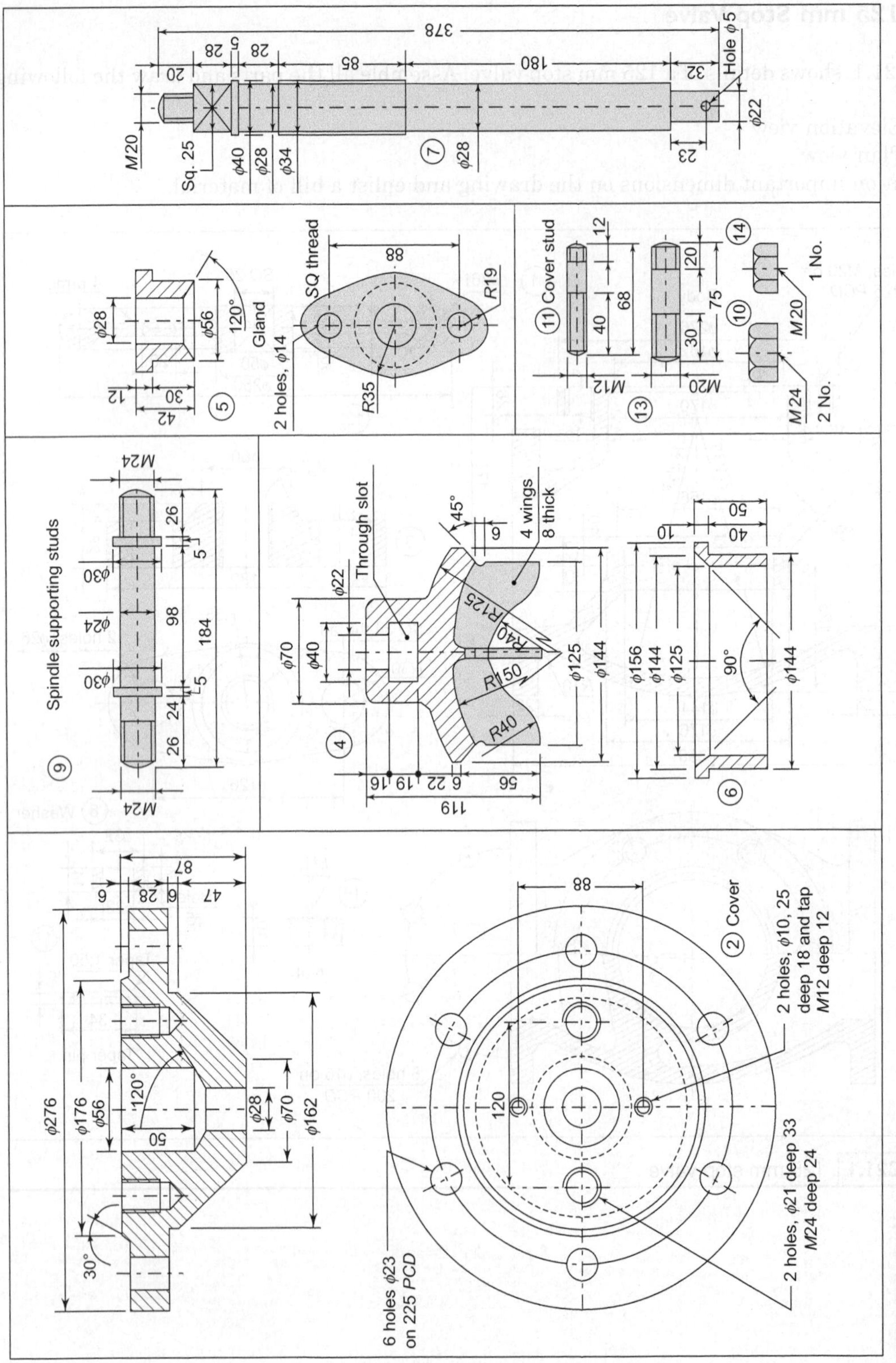

Fig. C21.2 Stop valve

Sl. No.	*Part*	*Quantity*	*Material*
1	Valve body	1	CI
2	Cover	1	CI
3	Support flange	1	CI
4	Valve	1	Brass
5	Gland bush	1	Brass
6	Bush	1	Brass
7	Operating spindle	1	Alloy steel
8	Washer	1	MS
9	Studs	2	M S
10	Stud pins	1 set	M S
11	Stud pins	1 set	MS
12	Nuts for pin 9	2 sets	MS
13	Nuts for pin 10	2 sets	M S
14	Nuts for pin 11	2 sets	MS
15	Nuts for pin 9	2 sets	MS
16	Packing	1	Fibre
17	Taper pin	1	MS

22 Screw Down Stop Valve

Fig. C22.1 shows details of a screw down stop valve. Draw the following assembly views.

1. Elevation-in-section through branches, showing assembled valve.
2. End elevation

Enlist a bill of materials

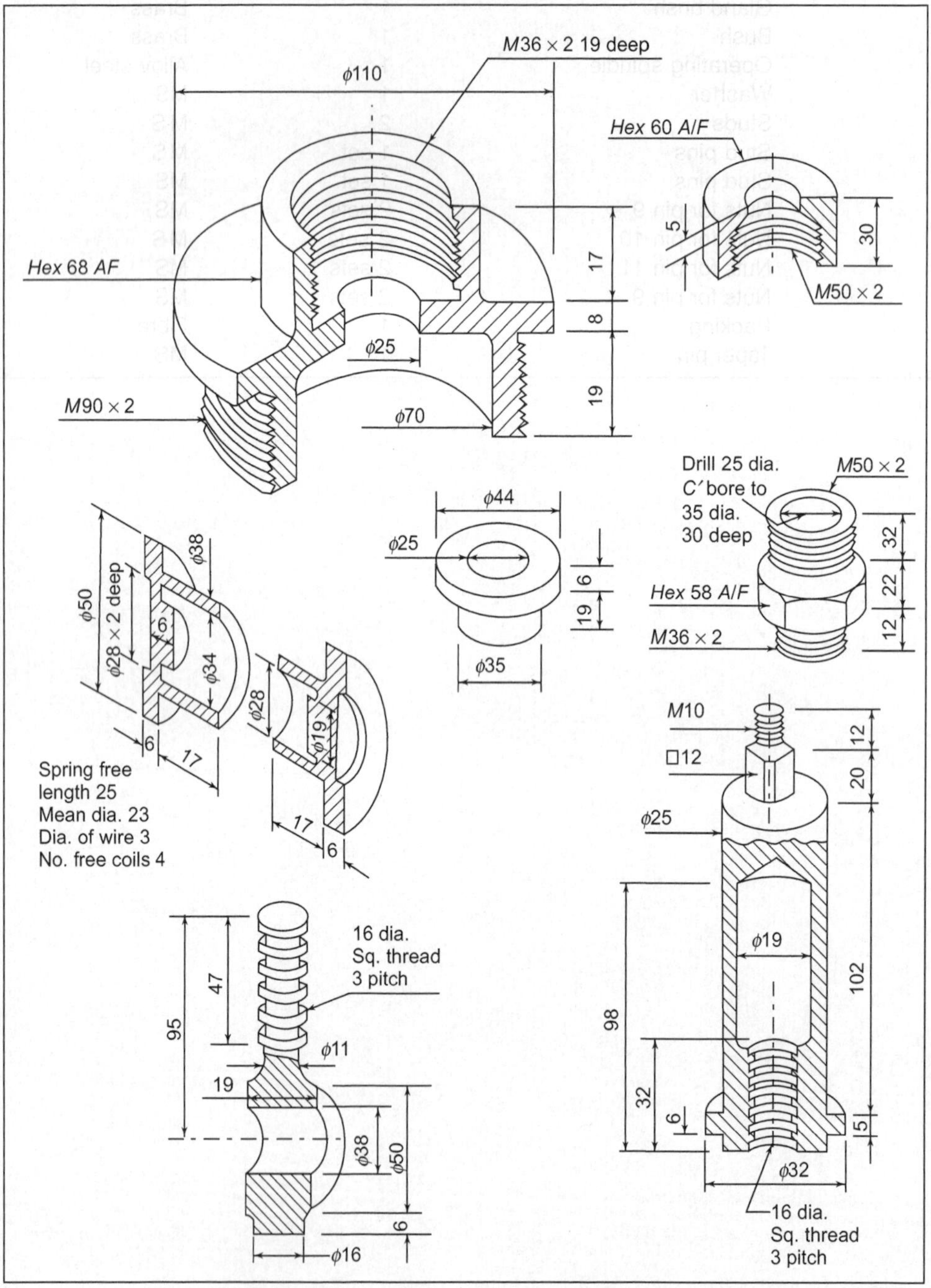

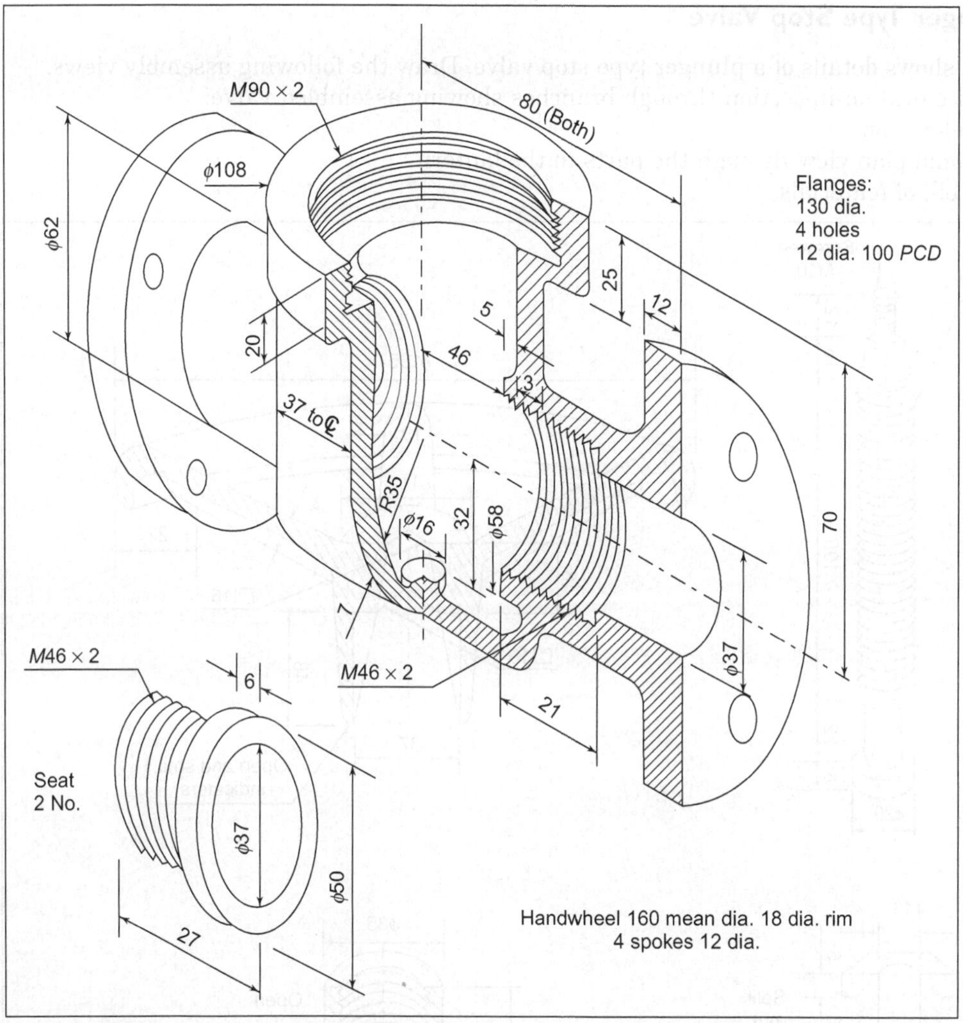

Fig. C22.1 Screw down stop valve

Sl. No.	*Part*	*Quantity*	*Material*
1	Valve chest	1	Cast steel
2	Valves and seats	2 sets	Monel metal
3	Valve spindle	1	Forged steel
4	Sleeve nut	1	Monel metal
5	Cap for chest	1	Mild steel
6	Stuffing box	1	Bronze
7	Gland	1	Bronze
8	Gland nut	1	Bronze
9	Hand wheel	1	Mild steel
10	Spring	1	Spring steel

23 Plunger Type Stop Valve

Fig. C23.1 shows details of a plunger type stop valve. Draw the following assembly views.

1. Front elevation-in-section through branches showing assembled valve.
2. End elevation
3. Sectional plan view through the parts in the lantern.

Enlist a bill of materials

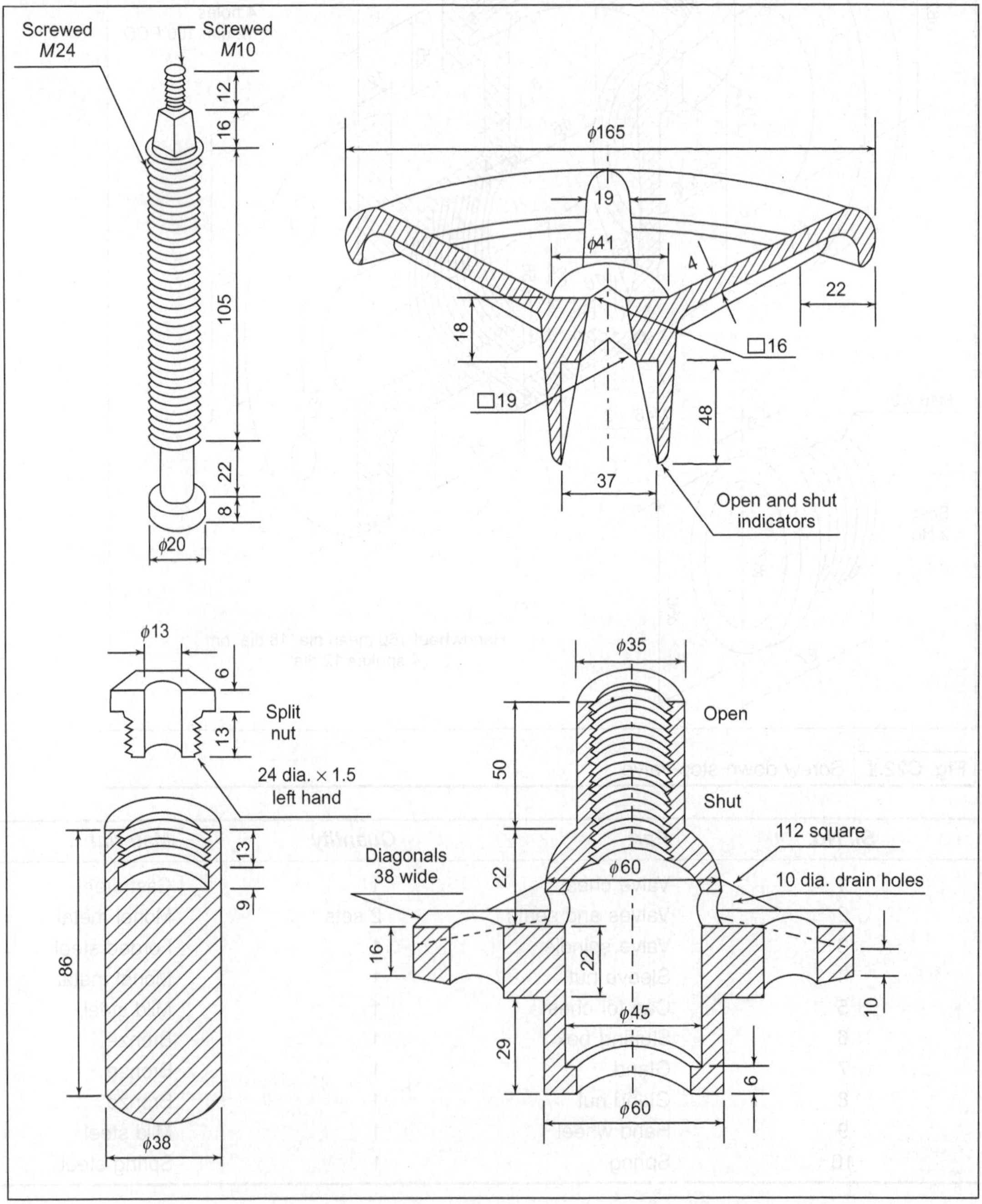

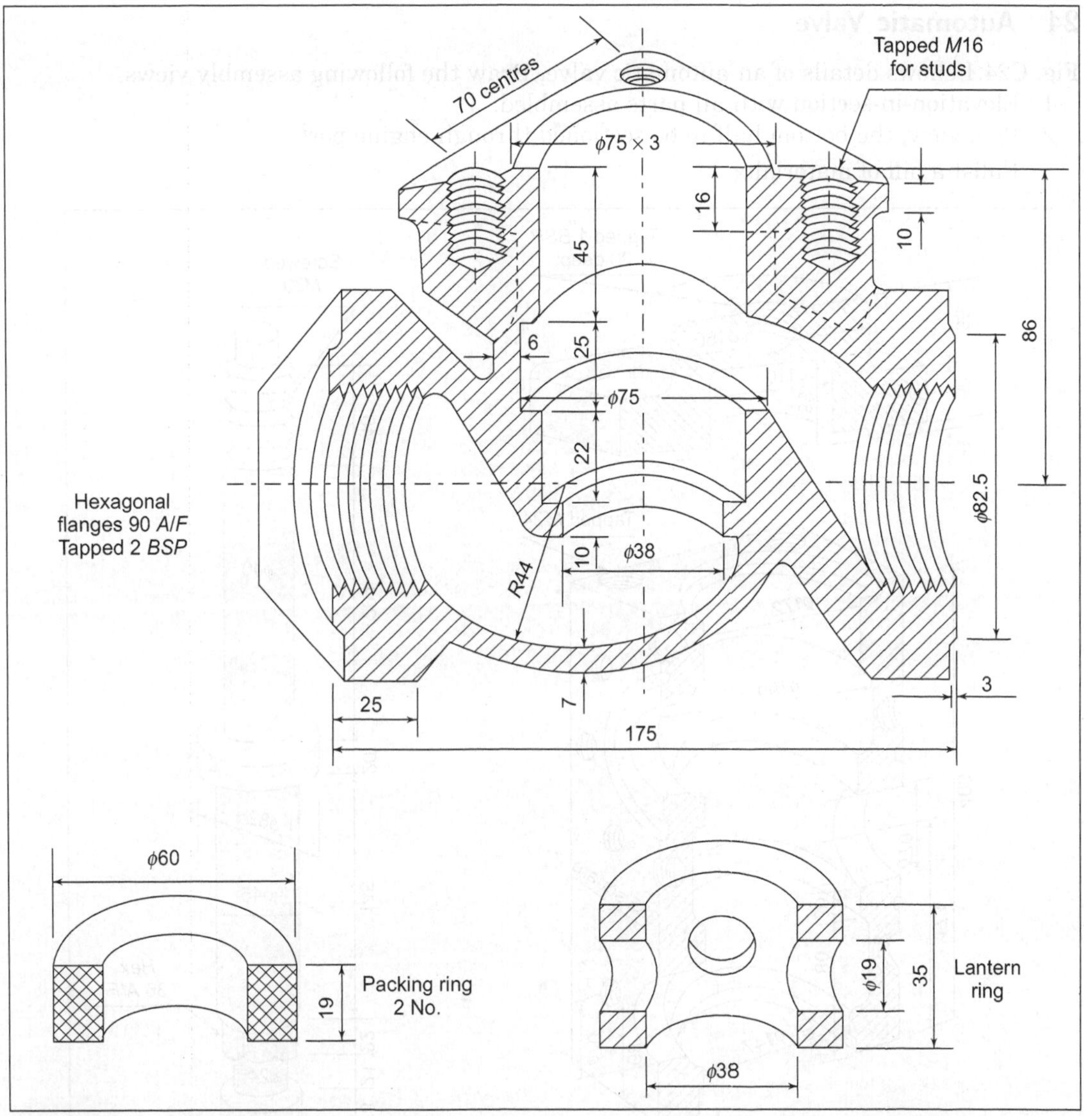

Fig. C23.1 Plunger type stop valve

Sl. No.	*Part*	*Quantity*	*Material*
1	Body	1	Cast steel
2	Cover	1	Cast steel
3	Piston	1	Monel metal
4	Split nut	1	Bronze
5	Lantern bush	1	Bronze
6	Studs with nuts	4 Sets	MS
7	Hand wheel	1	Malleable iron
8	Packing ring	2	Bronze

24 Automatic Valve

Fig. C24.1 shows details of an automatic valve. Draw the following assembly views.

1. Elevation-in-section with all parts assembled.
2. Plan view, the bottom half to be sectioned through engine port.

   Enlist a bill of material.

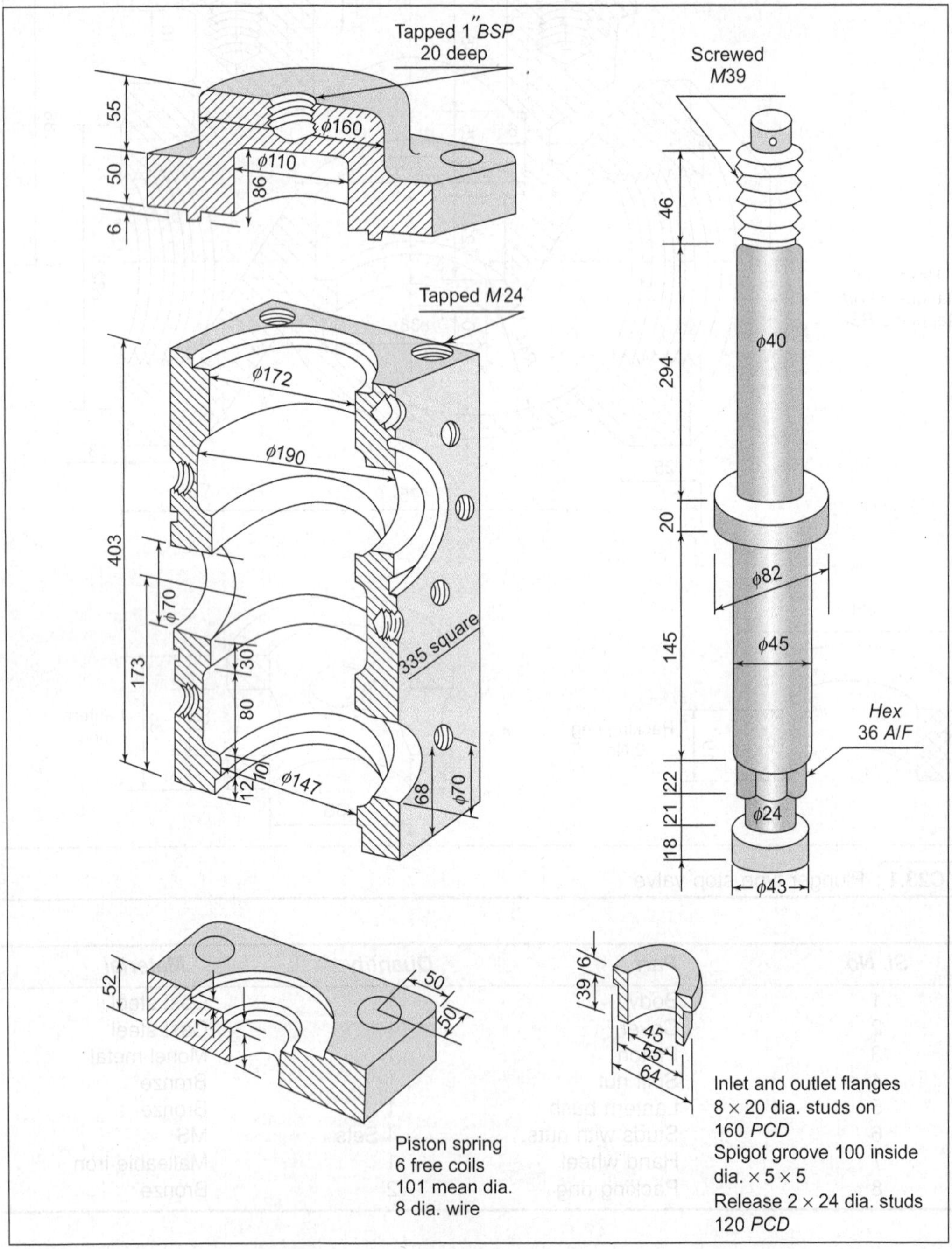

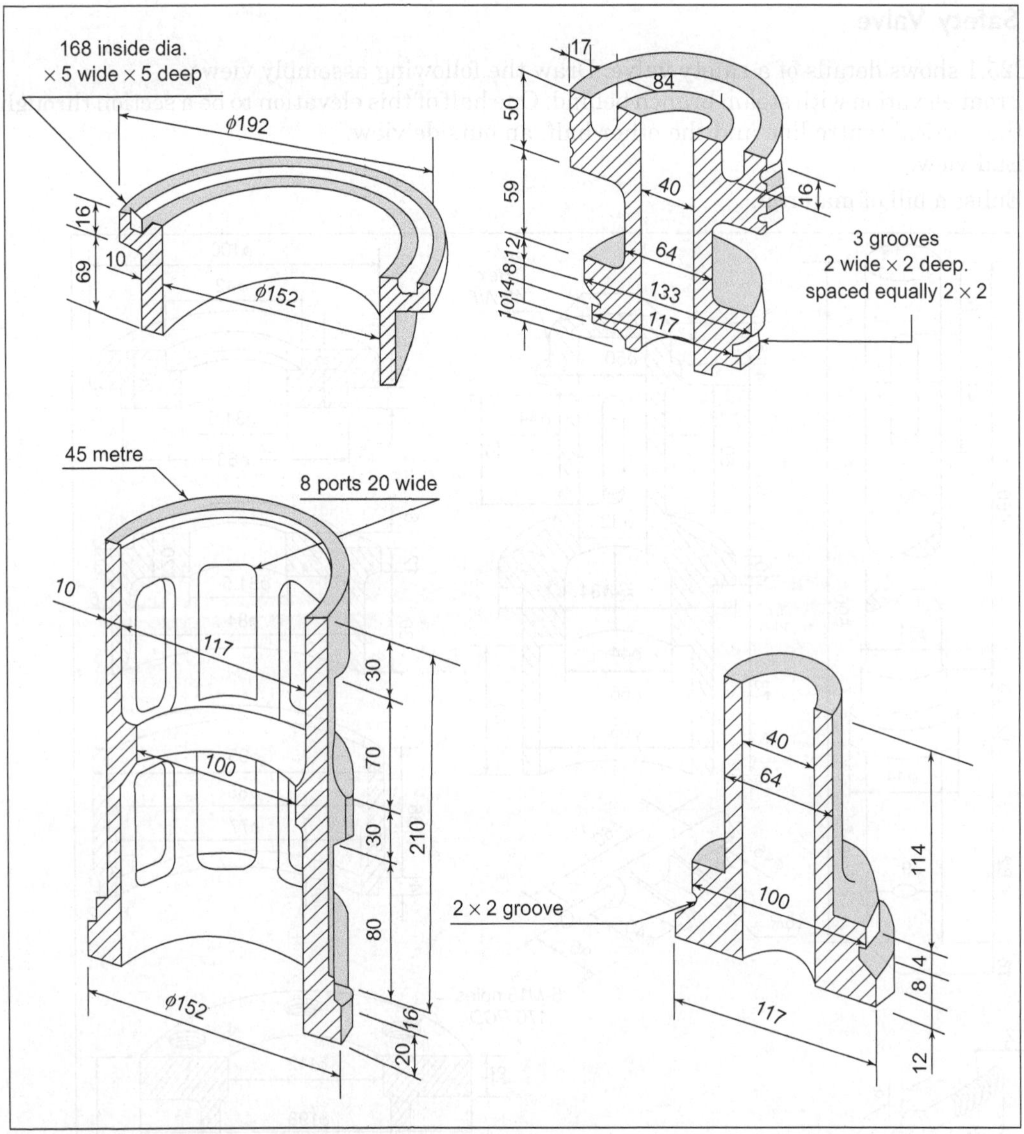

Fig. C24.1 Automatic valve

Sl. No.	Part	Quantity	Material
1	Body	1	MS
2	Cover	1	MS
3	Bushes valve and seats	2 sets	GM
4	Spindle	1	MS
5	Spring	1	Spring steel
6	Studs	4	MS

25 Safety Valve

Fig. C25.1 shows details of a safety valve. Draw the following assembly views.

1. Front elevation with steam branch behind. One half of this elevation to be a section through the vertical centre line and the other half, an outside view.
2. End view.

   Enlist a bill of material.

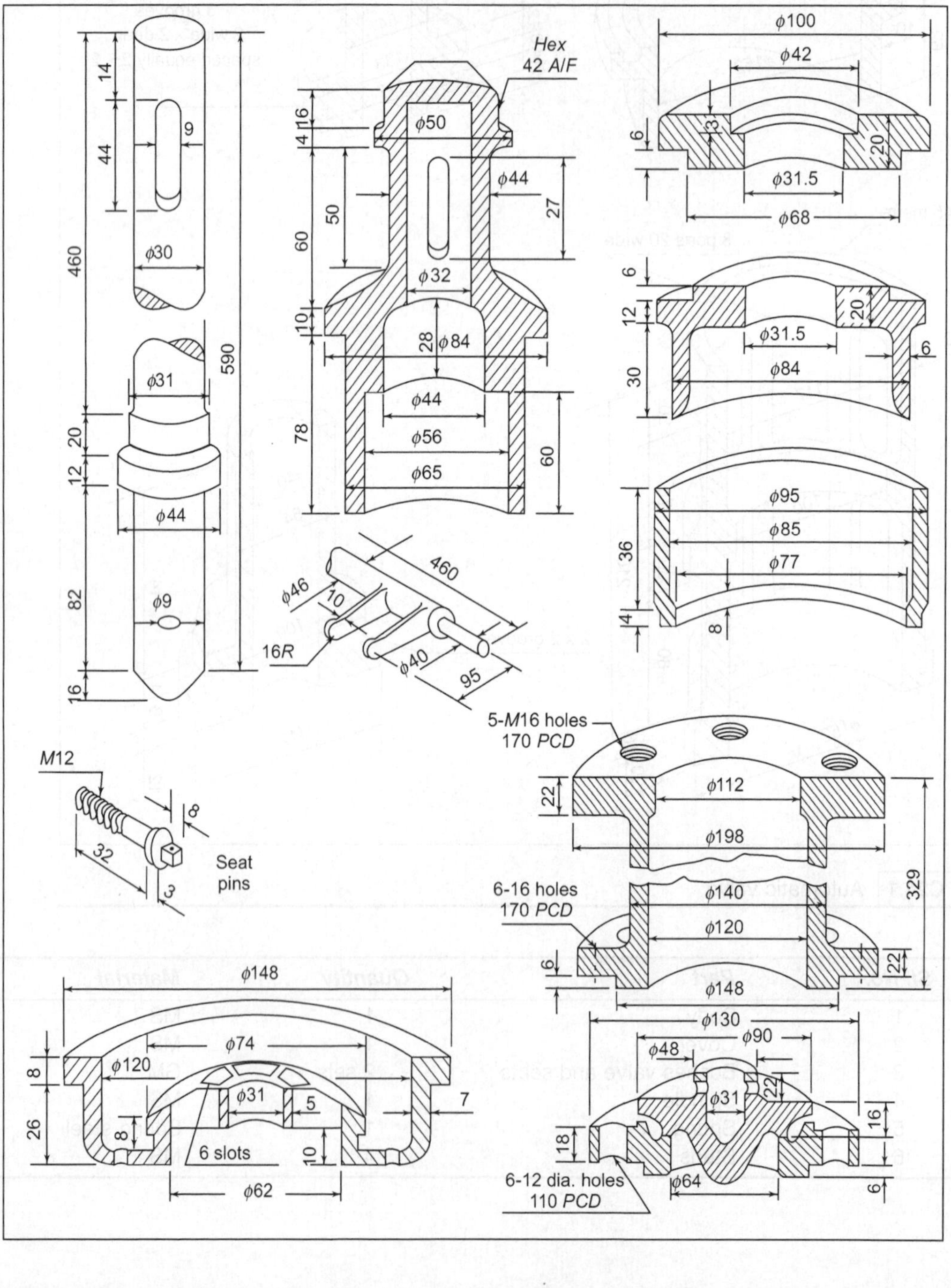

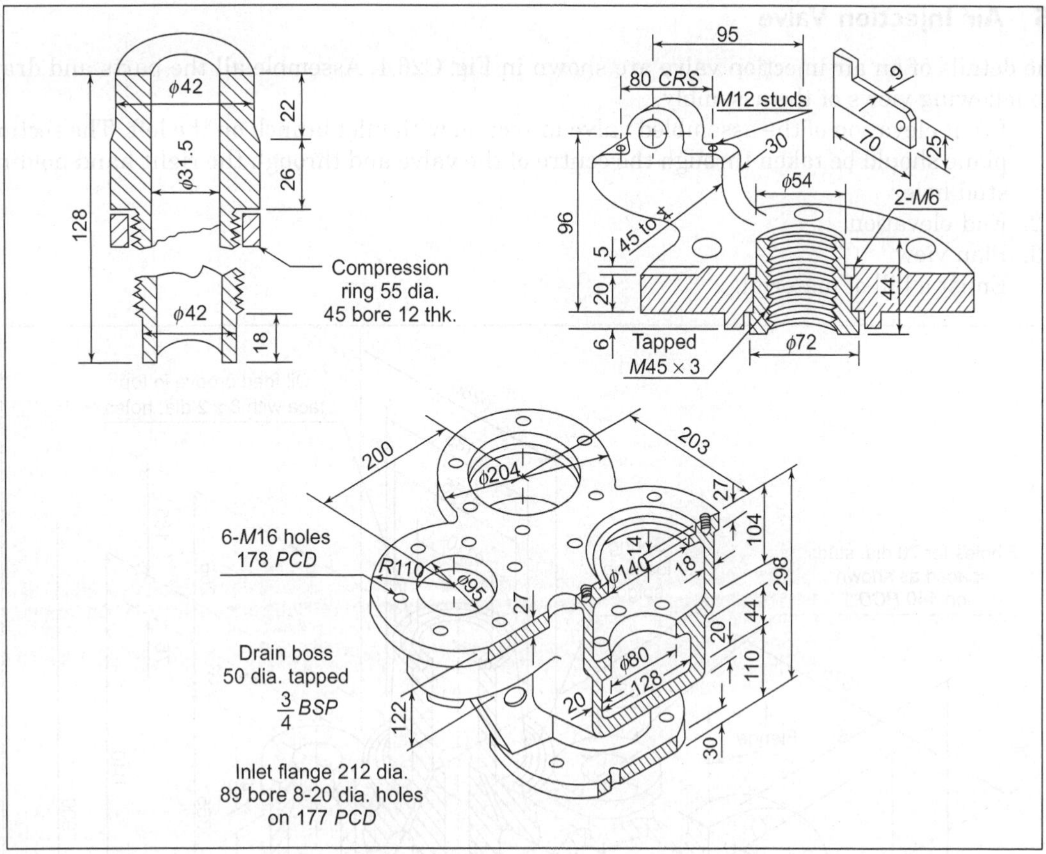

Fig. C25.1 Safety valve

Sl. No.	Part	Quantity	Material
1	Chest	1	Cast steel
2	Casing	1	Cast iron
3	Cover	1	Cast iron
4	Valve and seat	1 set	Aluminium bronze
5	Diaphragm	1	Gun metal
6	Loose ring	1	Gun metal
7	Piston	1	Gun metal
8	Spindle	1	En 8 (alloy steel)
9	Spring	1	Spring steel
10	Cover bush	1	Brass
11	Compression ring	1	Brass
12	Easing shaft	1	Mild steel

26 Air Injection Valve

The details of an air injection valve are shown in Fig. C26.1. Assemble all the parts and draw the following views of the assembly.

1. Front elevation of the assembled valve in section with inlet branch on the left. The section plane should be taken through the centre of the valve and through the right hand holding stud hole.
2. End elevation
3. Plan view

Enlist a bill of material.

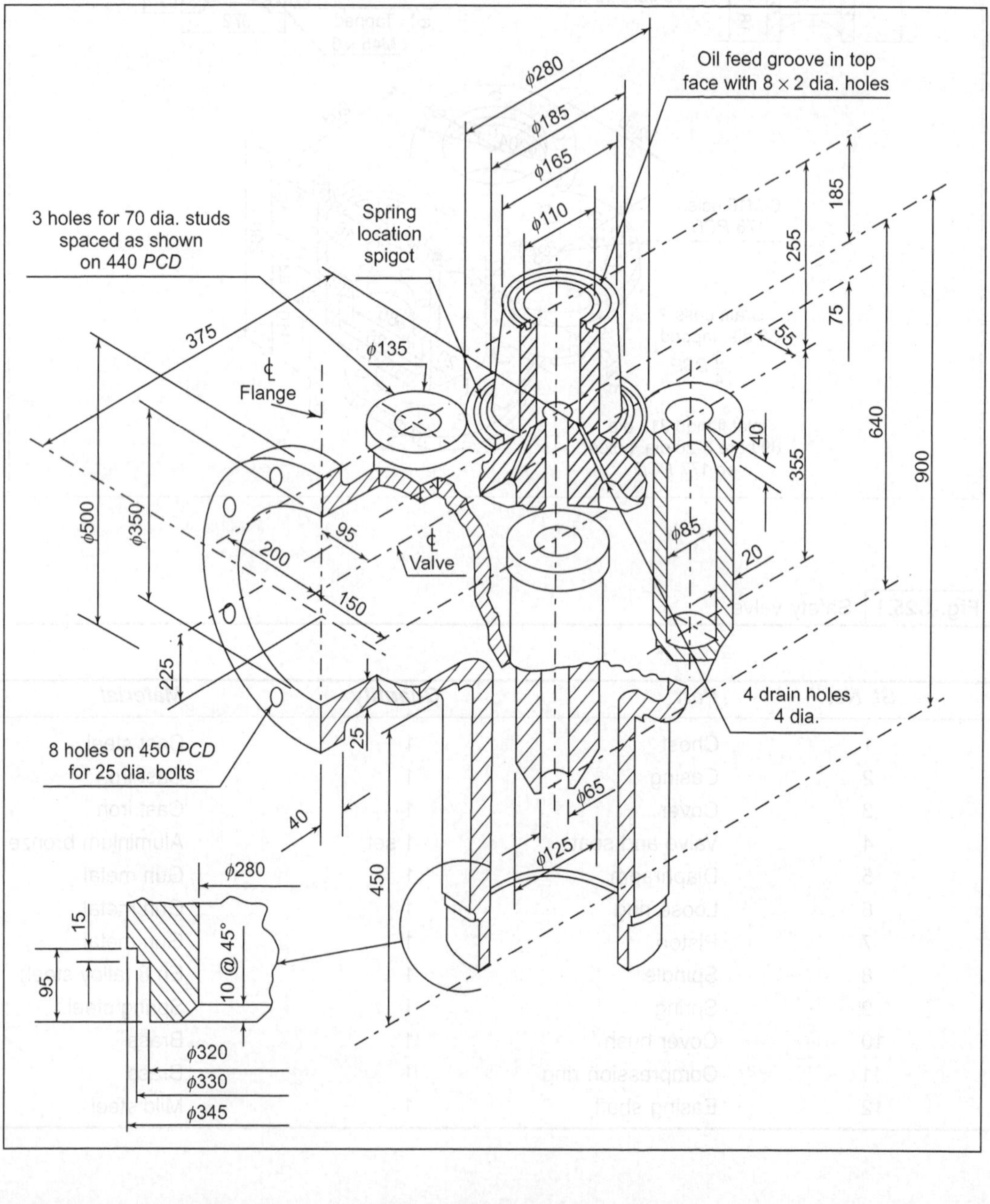

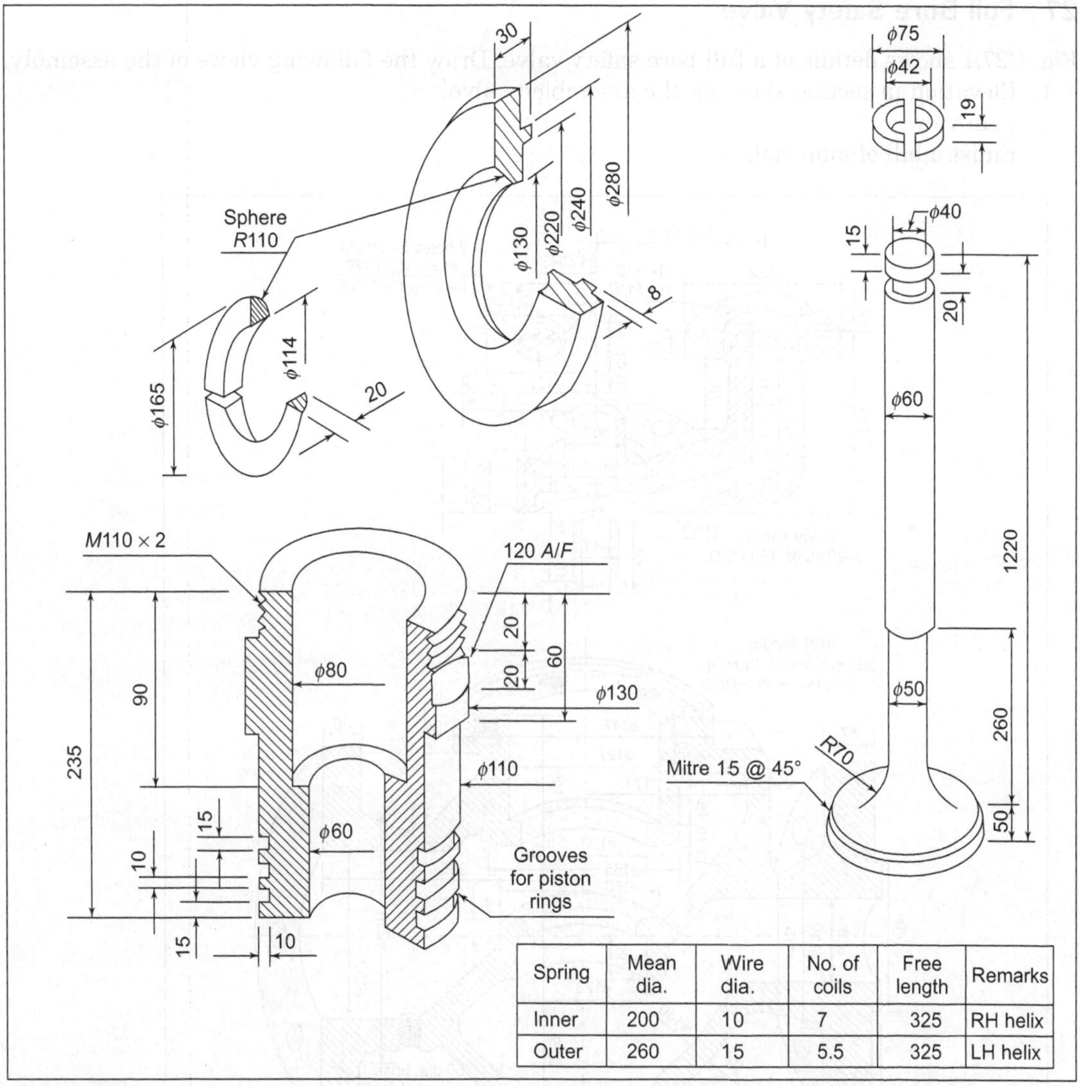

Spring	Mean dia.	Wire dia.	No. of coils	Free length	Remarks
Inner	200	10	7	325	RH helix
Outer	260	15	5.5	325	LH helix

Fig. C26.1 Air injection valve

Sl. No.	*Part*	*Quantity*	*Material*
1	Valve body	1	Cast iron
2	Valve spindle	1	Heat resisting steel EN 55
3	Piston	1	MS
4	Spring seat	1	MS
5	Split collar	1	MS
6	Split ring	1	MS
7	Piston rings	3	Cast iron
8	Spring	2	Spring steel

27 Full Bore Safety Valve

Fig. C27.1 shows details of a full bore safety valve. Draw the following views of the assembly.

1. Elevation in section showing the assembled valve.
2. Plan view
   Enlist a bill of material.

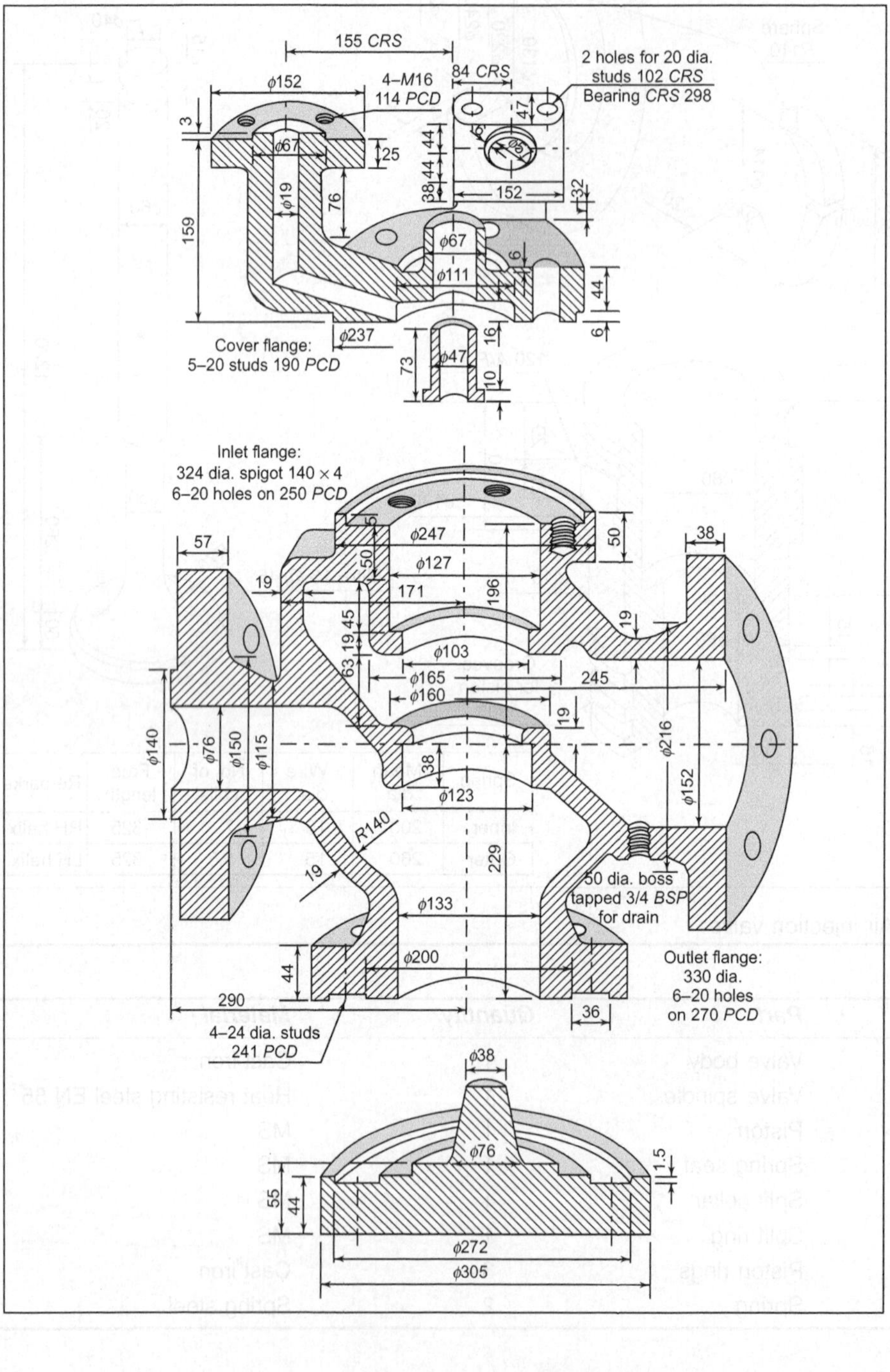

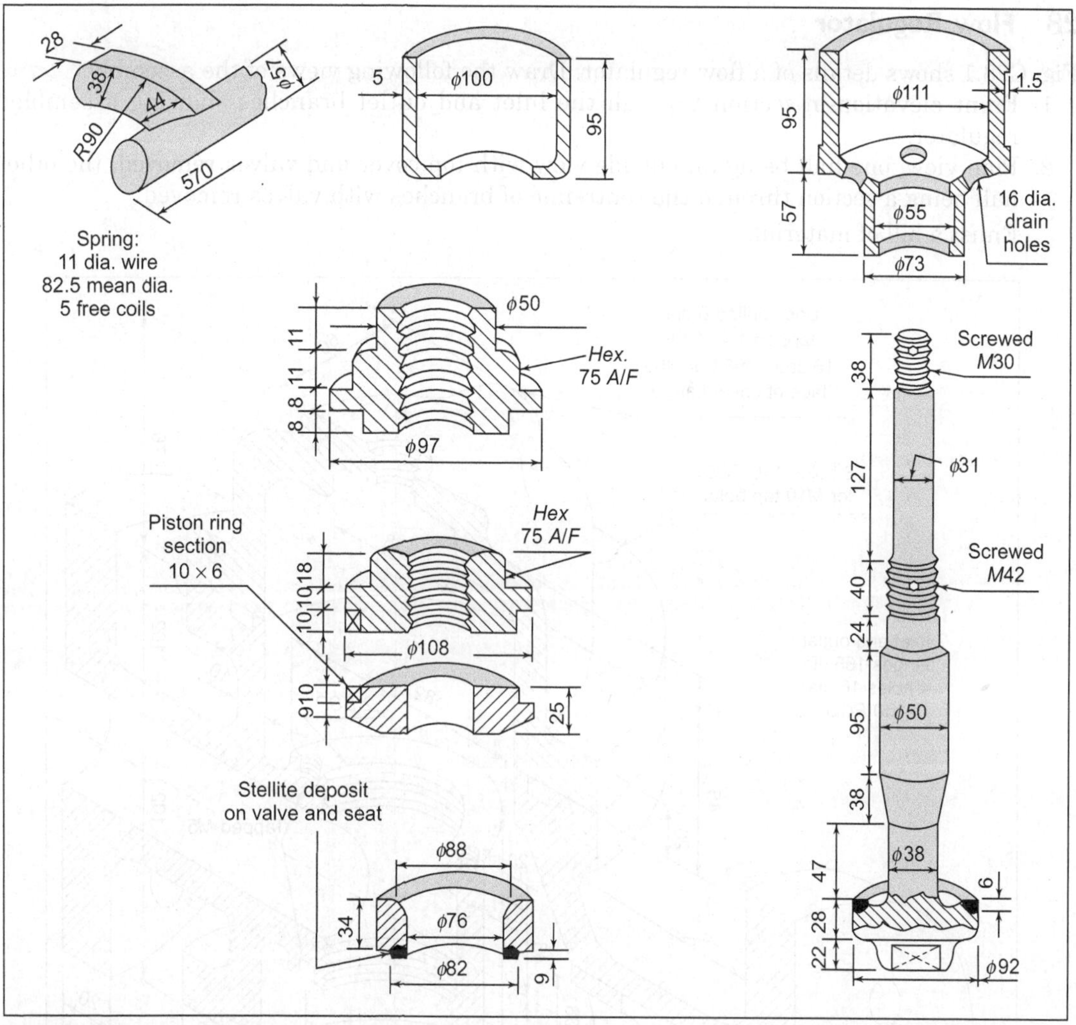

Fig. C27.1 Full bore safety valve

Sl. No.	*Part*	*Quantity*	*Material*
1	Valve body	1	Cast steel
2	Top cover	1	Cast steel
3	Bottom cover	1	Cast iron
4	Valve spindle	1	Stainless steel
5	Valve seat	1	Nickel steel
6	Pistons	1	Monel metal
7	Piston rings	1	Phos bronze
8	Neck bush	1	Brass
9	Easing gear bearings	1	Cast iron
10	Easing gear shaft	1	Mild steel

28 Flow Regulator

Fig. C28.1 shows details of a flow regulator. Draw the following views of the assembly.

1. Front elevation-in-section through the inlet and outlet branches showing assembled regulator.
2. Plan view, one half being an outside view with top cover and valves removed, the other half being a section through the centreline of branches with valves removed.

   Enlist a bill of material.

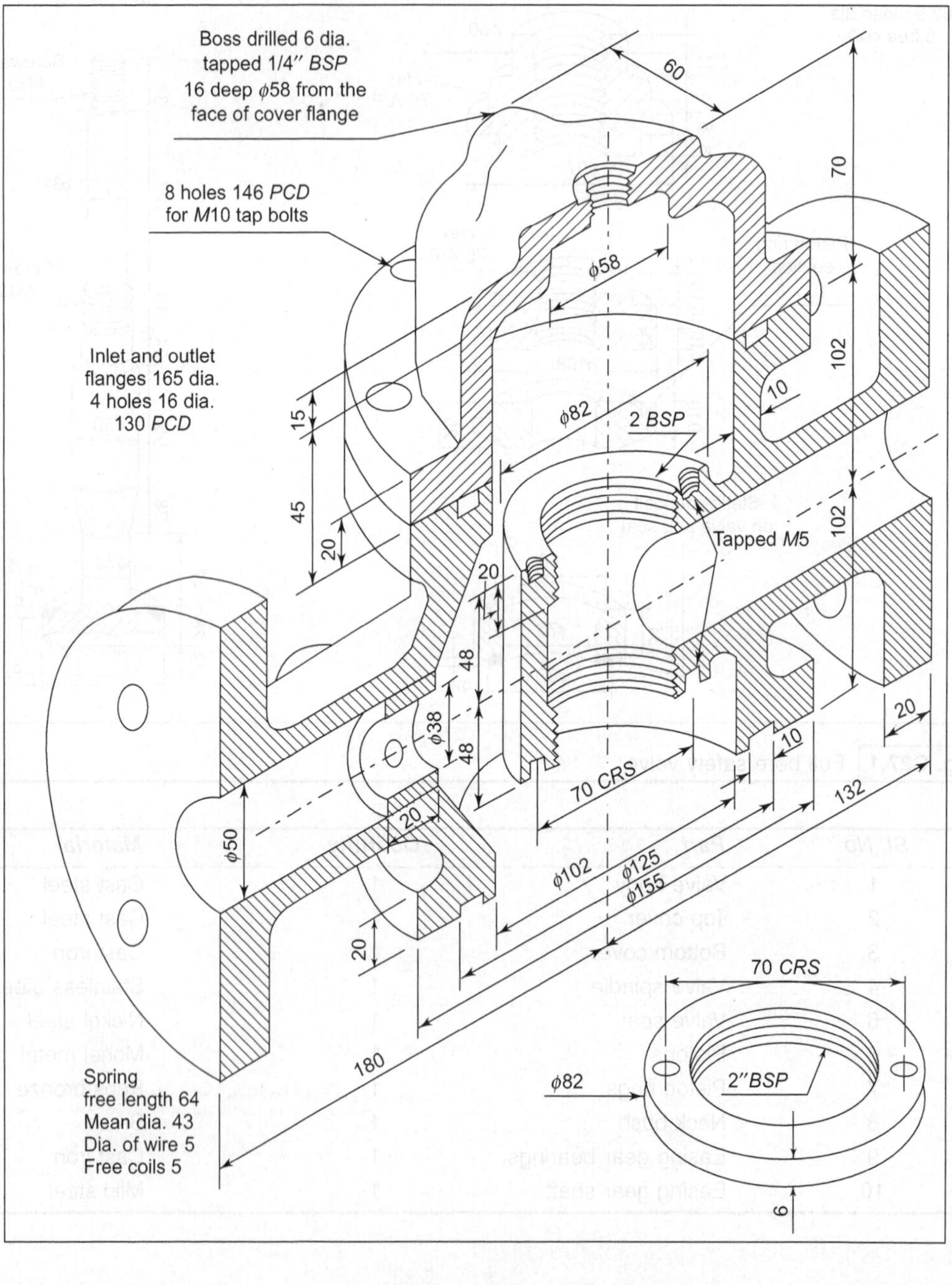

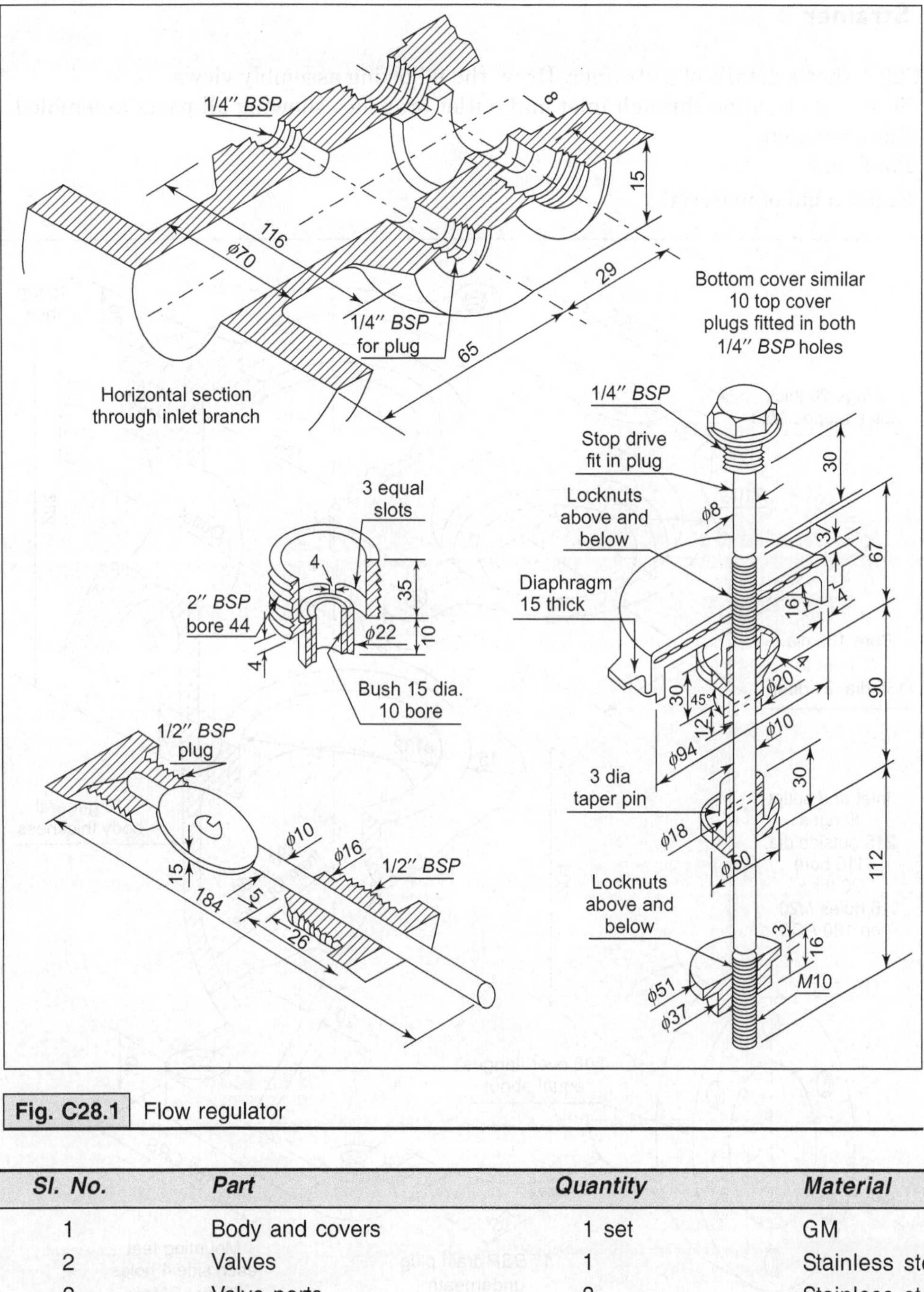

Fig. C28.1 Flow regulator

Sl. No.	Part	Quantity	Material
1	Body and covers	1 set	GM
2	Valves	1	Stainless steel
3	Valve ports	3	Stainless steel
4	Spindles	1	Stainless steel
5	Butterfly	1	Stainless steel
6	Piston and retaining plate	1 set	Stainless steel
7	Diaphragm	1	Nitrile

29 Strainer

Fig. C29.1 shows details of a strainer. Draw the following assembly views.

1. Sectional elevation through inlet and outlet branches showing all parts assembled.
2. End elevation
3. Plan view

   Enlist a bill of material.

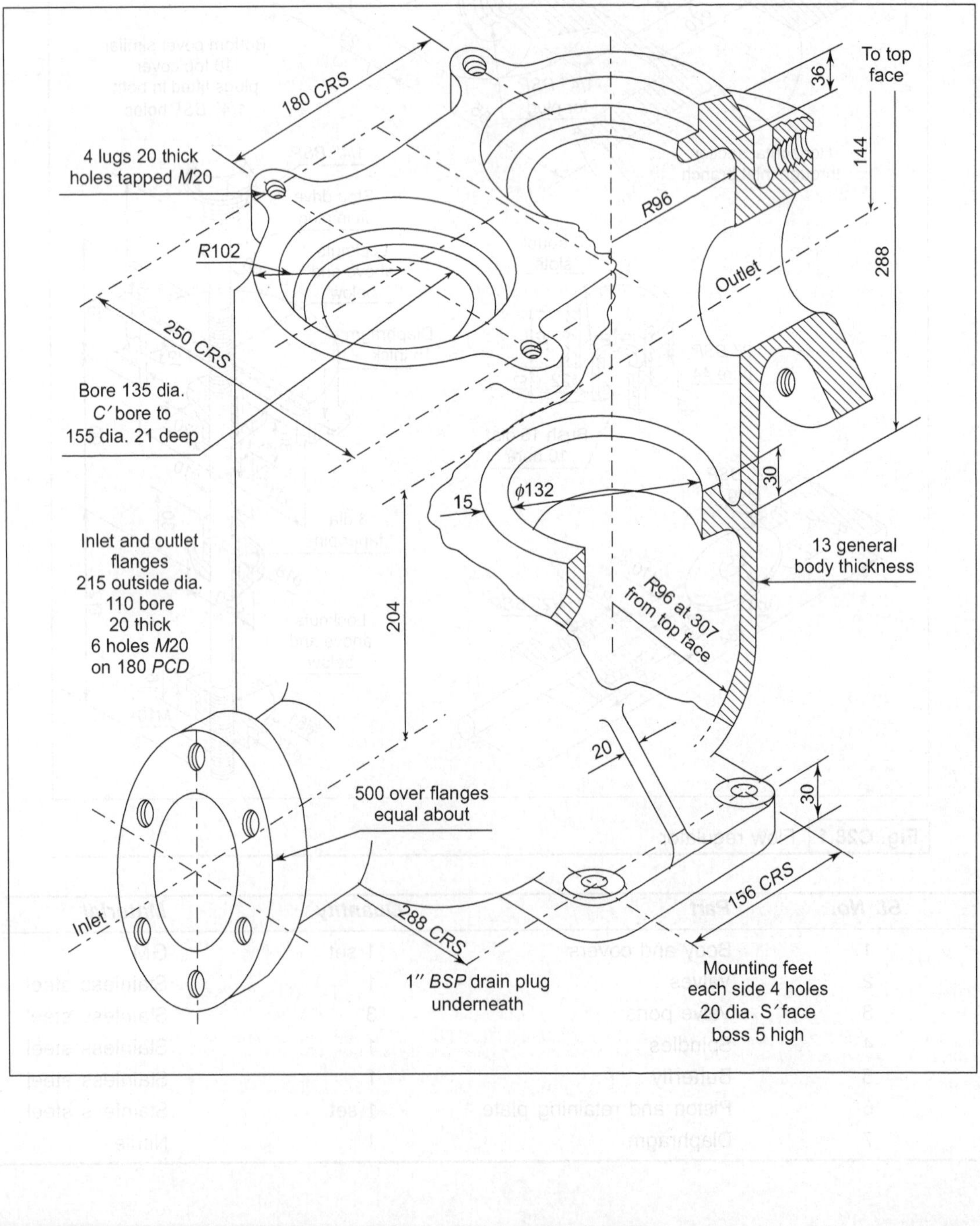

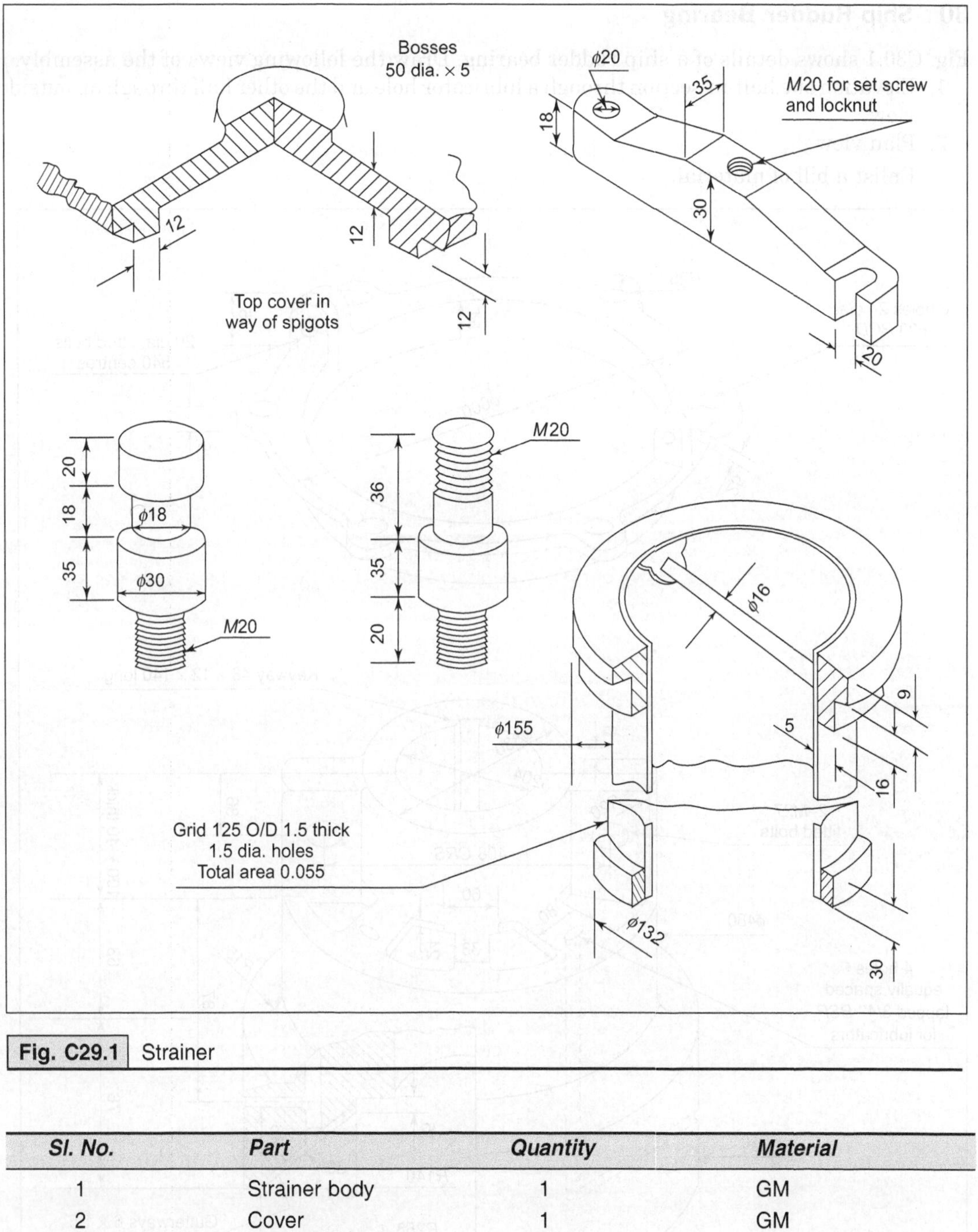

Fig. C29.1 Strainer

Sl. No.	*Part*	*Quantity*	*Material*
1	Strainer body	1	GM
2	Cover	1	GM
3	Filter elements	2	MS (Fabricated)
4	Strong backs	1	Forged steel
5	Pins and studs	4 sets	MS
6	Drain plug	1	GM

30 Ship Rudder Bearing

Fig. C30.1 shows details of a ship rudder bearing. Draw the following views of the assembly.

1. Elevation one half-in-section through a lubricator hole and the other half through an outside view.
2. Plan view

   Enlist a bill of material.

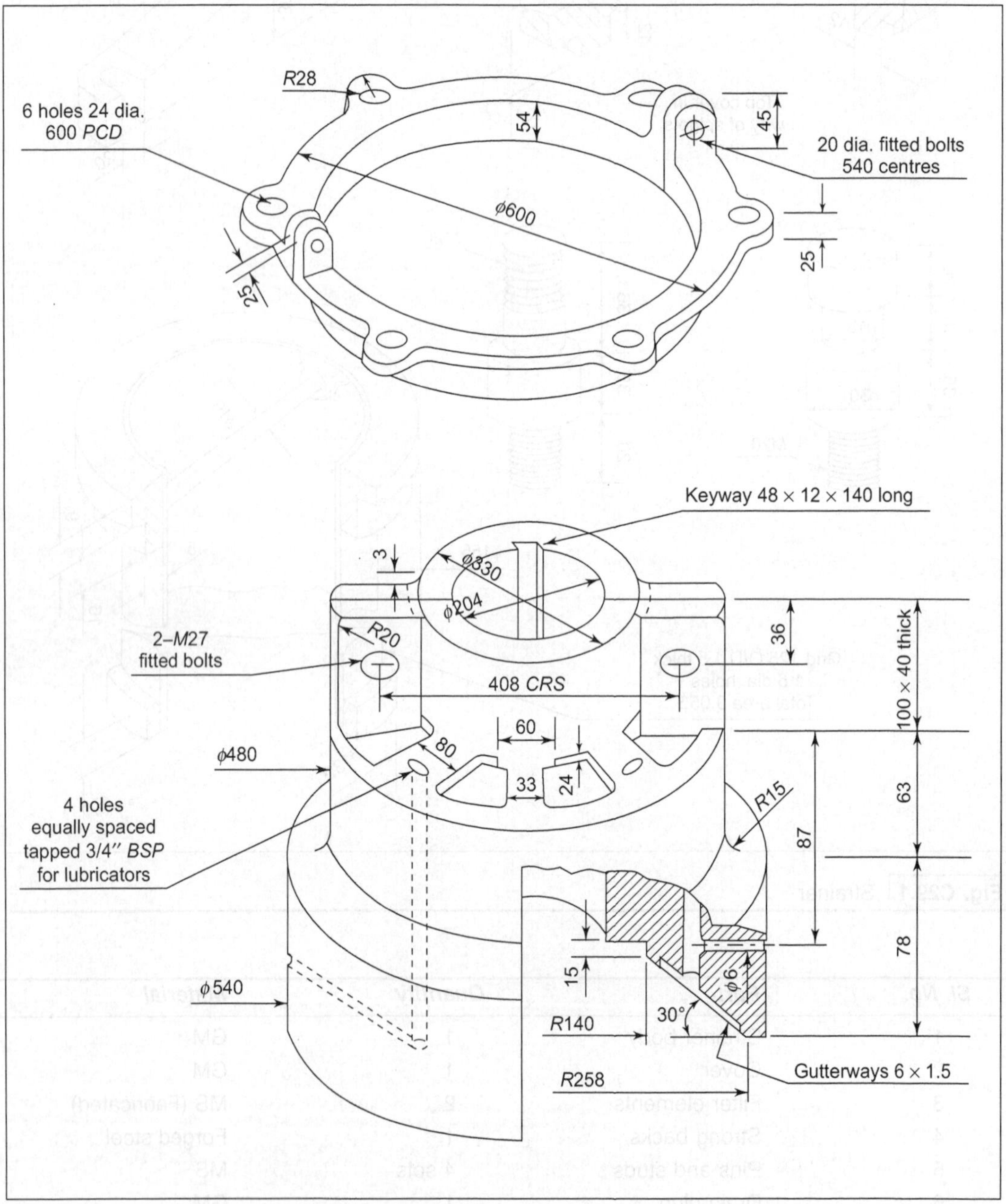

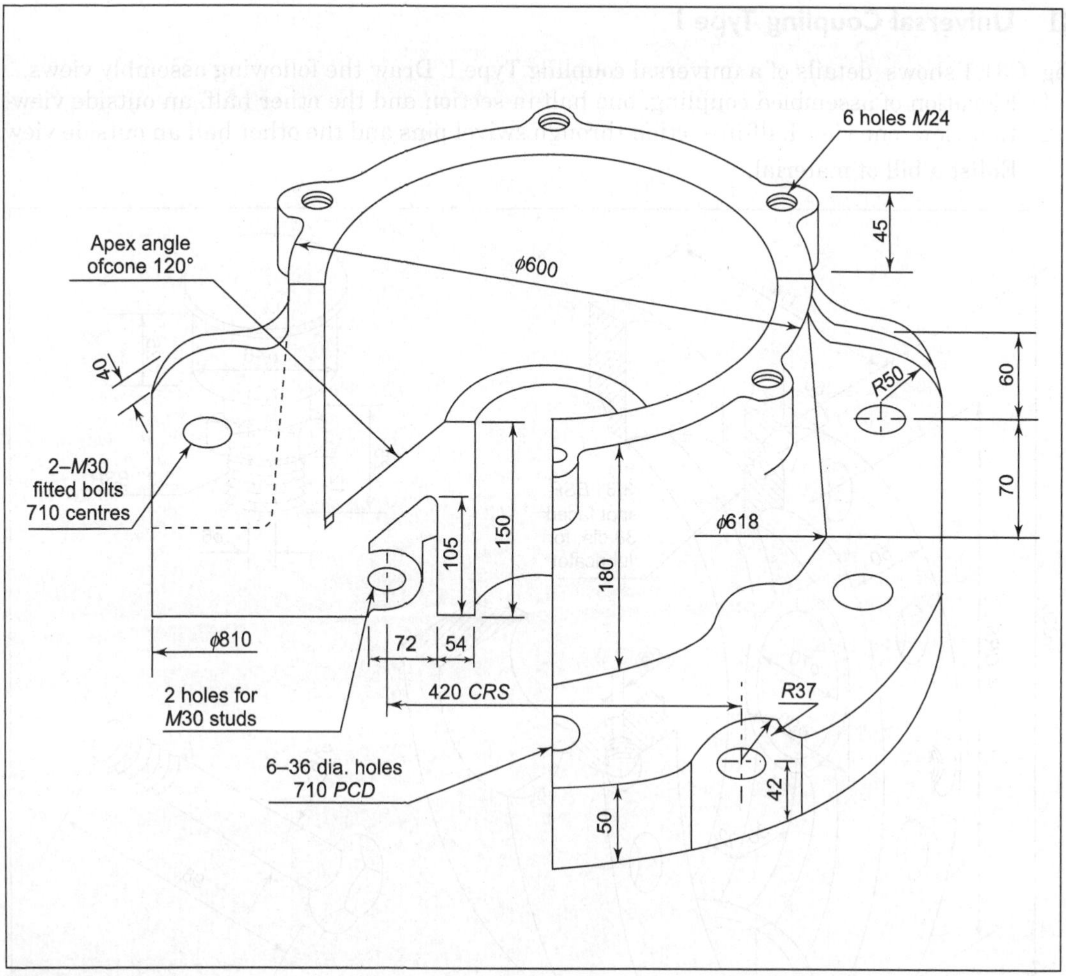

Fig. C30.1 Ship rudder bearing

Sl. No.	*Part*	*Quantity*	*Material*
1	Moving cone	1	Cast iron
2	Fixed cone	1	Cast iron
3	Gland	1	GM
4	Key	1	Key steel
5	Bolts	12 sets	MS

31 Universal Coupling Type I

Fig. C31.1 shows details of a universal coupling Type I. Draw the following assembly views.

1. Elevation of assembled coupling, one half-in-section and the other half, an outside view.
2. End view, one view half-in-section through swivel pins and the other half an outside view.
   Enlist a bill of material.

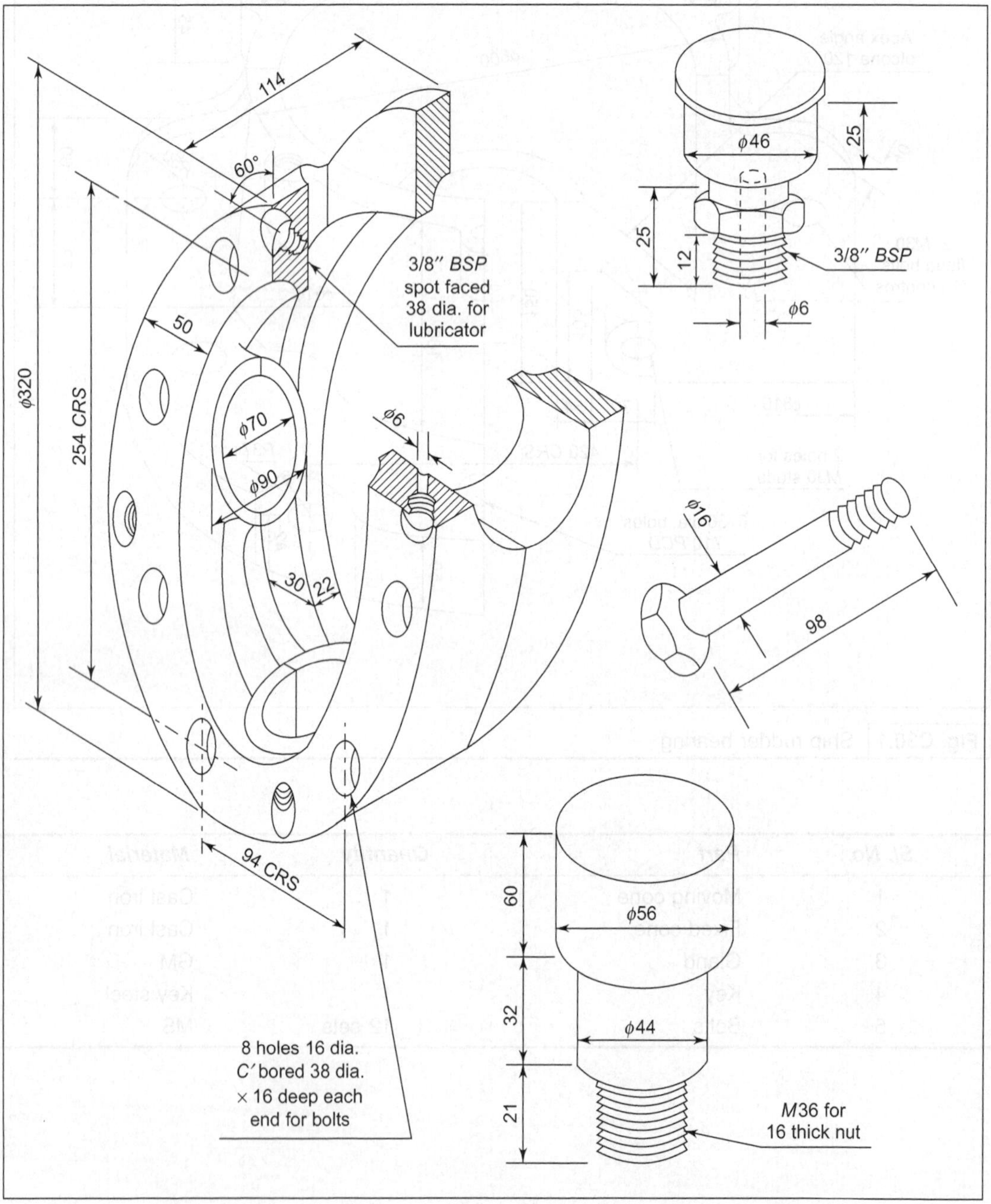

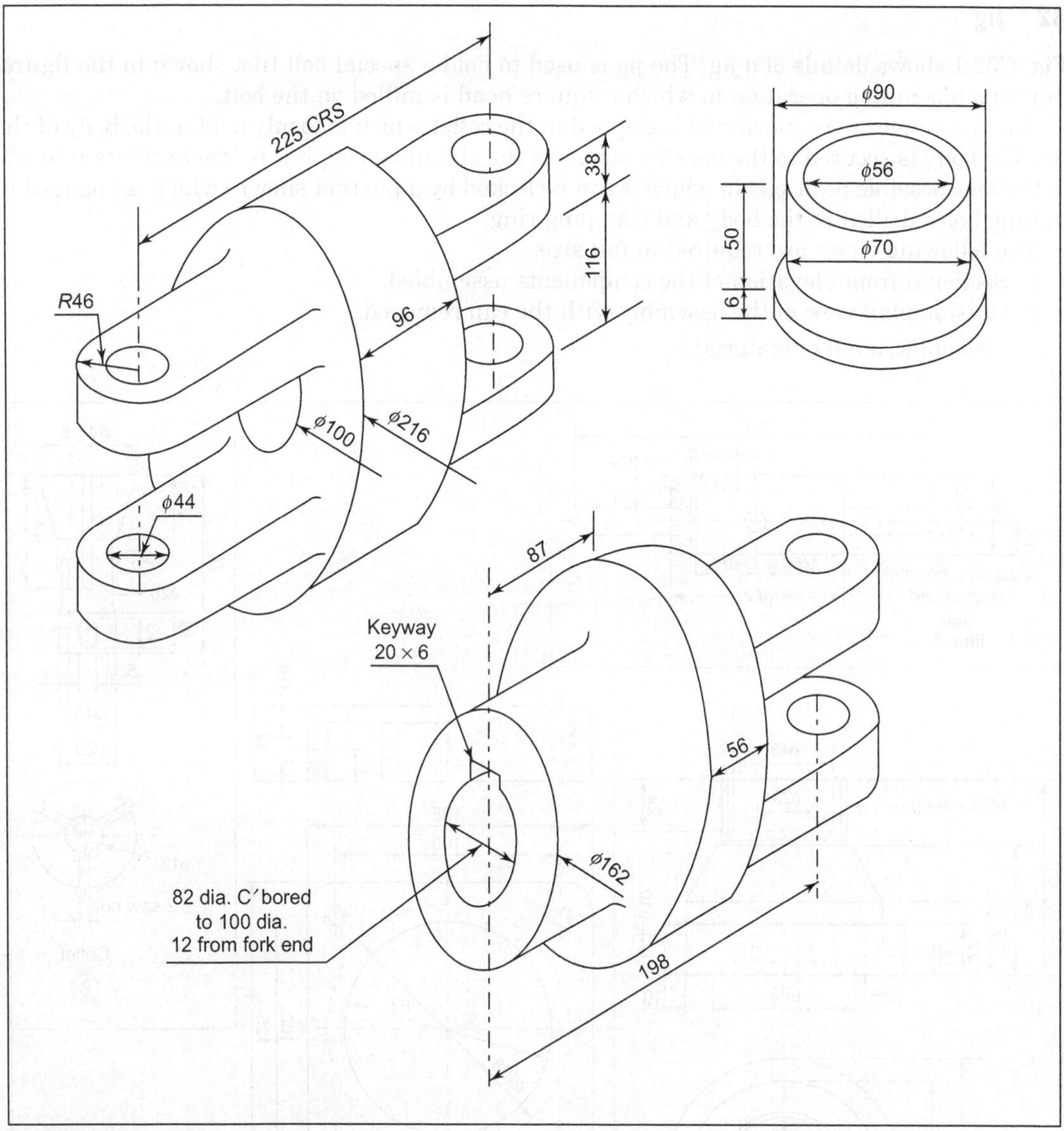

Fig. C31.1 Universal coupling type I

Sl. No.	*Part*	*Quantity*	*Material*
1	Coupling body	1	Cast steel
2	Forked ends	1	Forged steel
3	Forked link	1	Forged steel
4	Pins	1	MS
5	Bushes	4	Brass
6	Bolts	8 sets	MS

32 Jig

Fig. C32.1 shows details of a jig. The jig is used to hold a special bolt (not shown in the figure) during a machining operation in which a square head is milled on the bolt.

The component to be machined is gripped in the collet which is firmly held in the body of the jig. The body is secured to the base by means of the clamping ring but is free to rotate into one of the four possible positions in which it can be locked by a pin (not shown) which is engaged in mating holes drilled in the body and clamping ring.

The following views are required in full size.

1. Sectional front elevation of the components assembled.
2. Outside plan view of the assembly with the cap removed.

Also enlist a bill of material.

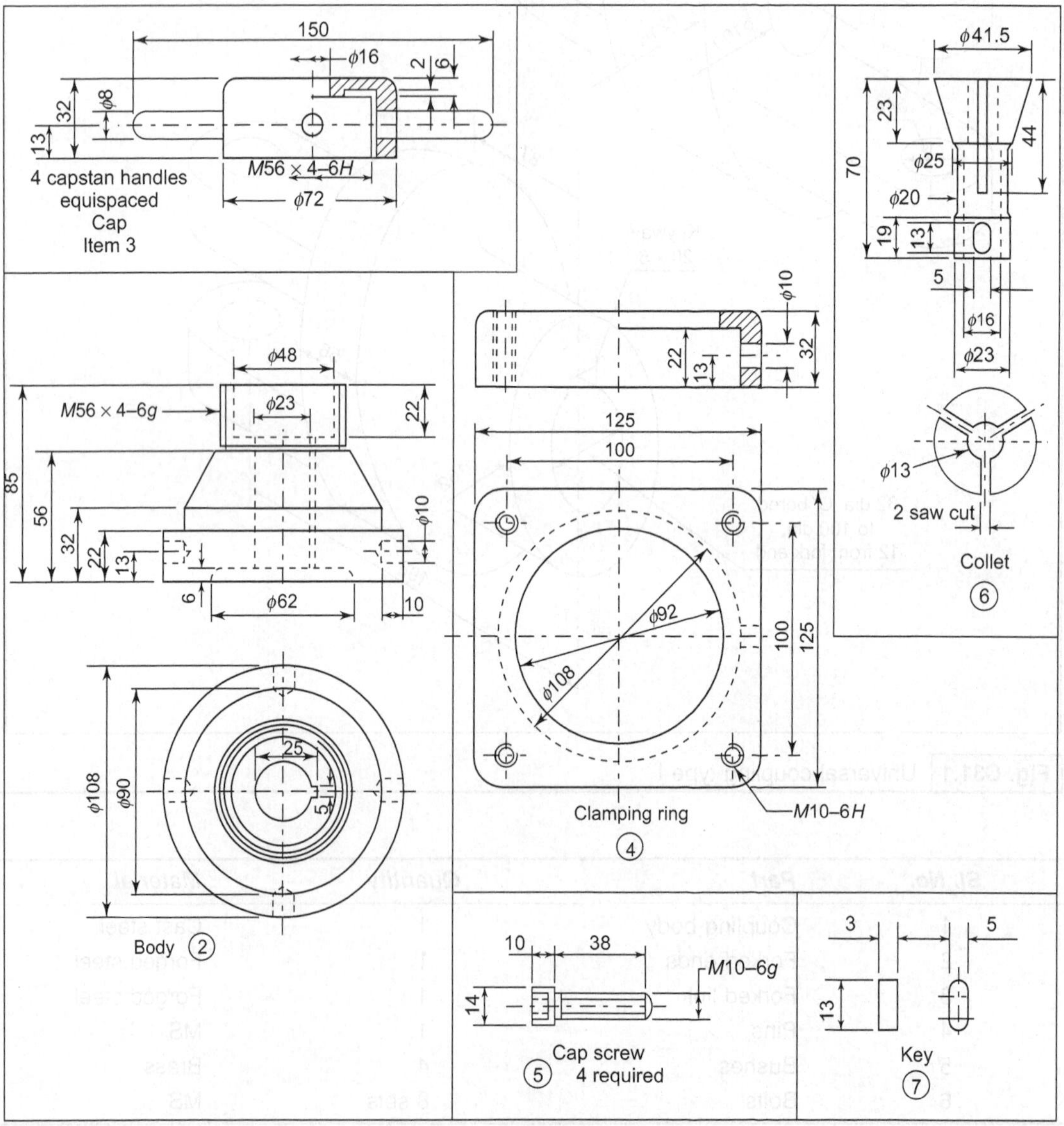

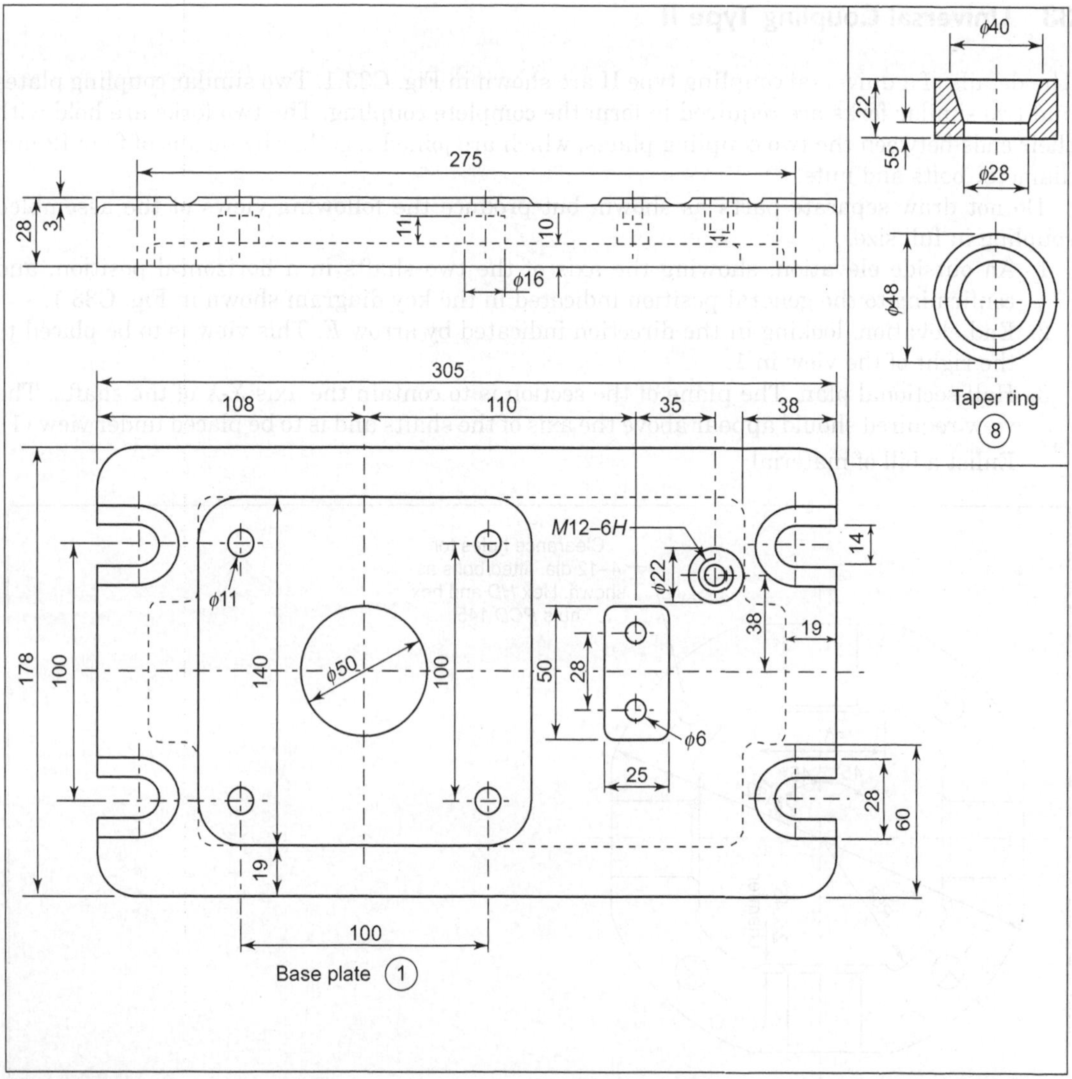

Fig. C32.1 Jig

Sl. No.	*Part*	*Quantity*	*Material*
1	Base plate	1	MS
2	Body	1	MS
3	Cap	1	MS
4	Clamping ring	1	MS
5	Cap screw	1	MS
6	Collet	4	Spring steel (Alloy steel)
7	Key	1	Key steel
8	Taper ring	1	Alloy steel

33 Universal Coupling Type II

The details of a universal coupling type II are shown in Fig. C33.1. Two similar coupling plates and two similar forks are required to form the complete coupling. The two forks are held with their ends between the two coupling plates, which are joined together by means of four 12 mm diameter bolts and nuts.

Do not draw separate parts as shown, but produce the following views of the assembled coupling in full size.

1. An outside elevation, showing the axis of the two shafts in a horizontal position, and confirming to the general position indicated in the key diagram shown in Fig. C33.1.
2. End elevation, looking in the direction indicated by arrow *E*. This view is to be placed to the right of the view in 1.
3. Half-sectional plan. The plane of the section is to contain the axis *X-X* of the shafts. The view required should appear above the axis of the shafts and is to be placed under view (1). Enlist a bill of material.

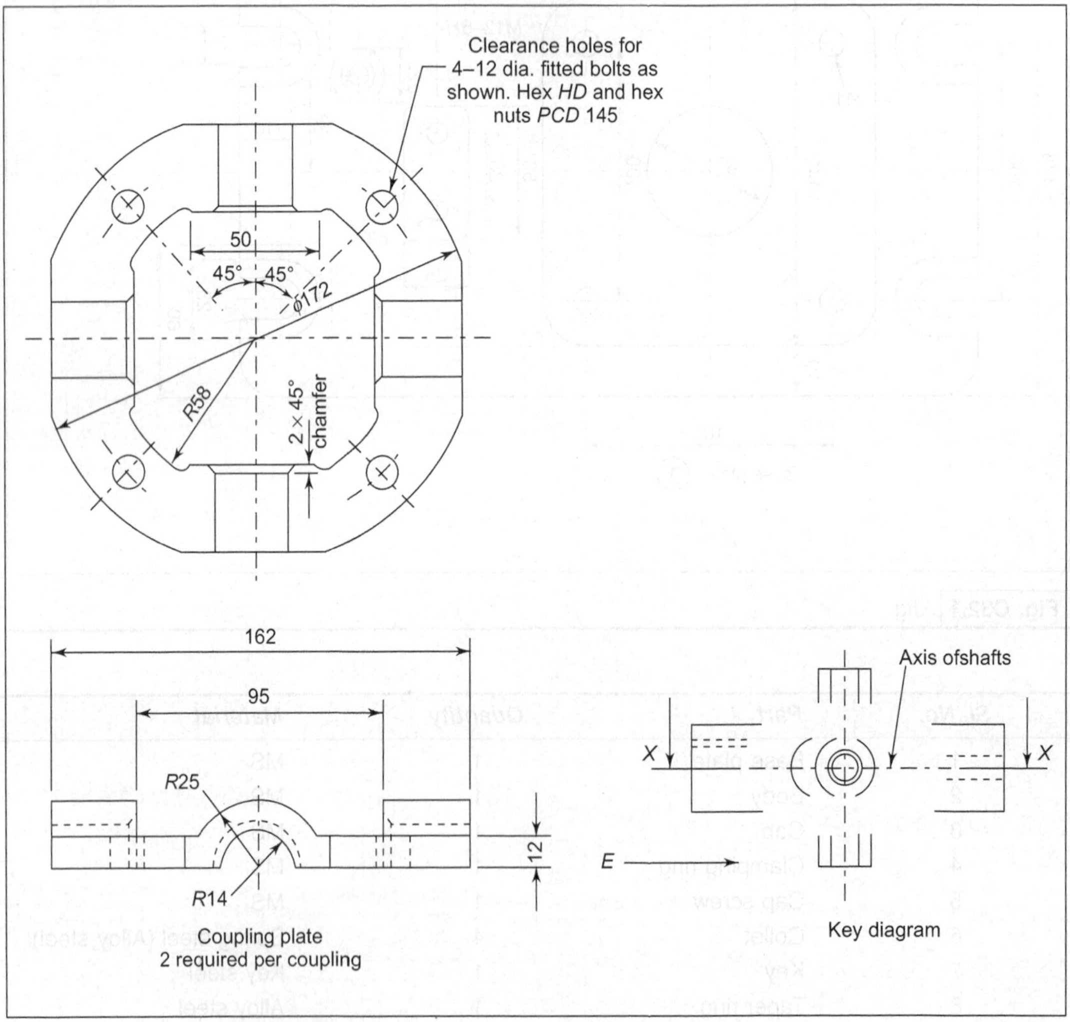

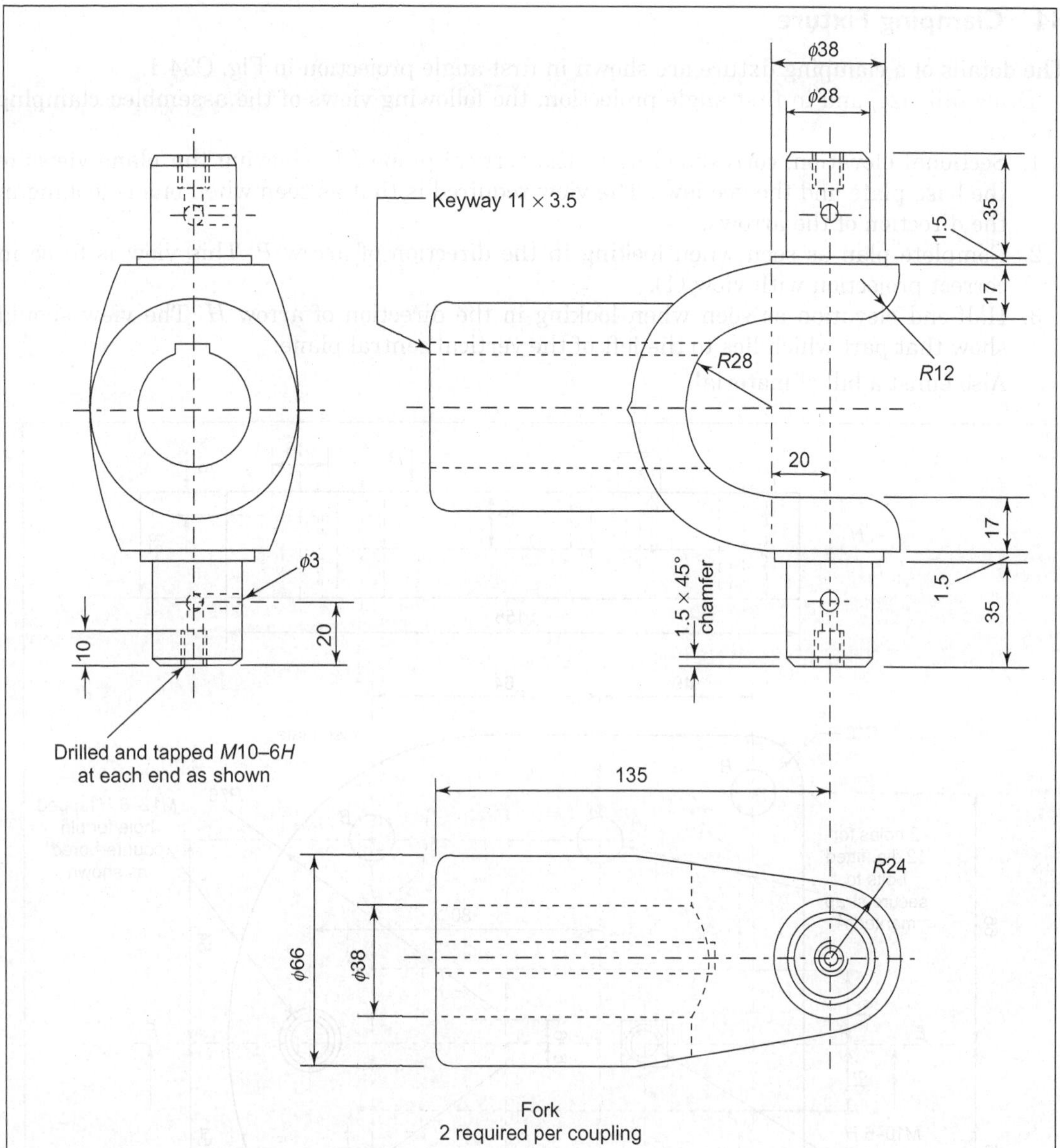

Fig. C33.1 Universal coupling type II

Sl. No.	*Part*	*Quantity*	*Material*
1	Fork	2	Forge steel
2	Coupling plate	2	MS
3	Key	1	Key steel
4	Nuts and bolts *M*12	4 sets	MS

34 Clamping Fixture

The details of a clamping fixture are shown in first angle projection in Fig. C34.1.

Draw full size, and in first angle projection, the following views of the assembled clamping device:

1. Sectional elevation, corresponding to the vertical plane *EF*, showing the plane views of the base plate and the vee jaw. The view required is that as seen when one is looking in the direction of the arrows.
2. Complete plan as seen when looking in the direction of arrow *P*. This view is to be in correct projection with view (1).
3. Half-end elevation as seen when looking in the direction of arrow *H*. The view should show that part which lies to the left of the vertical central plane.

Also enlist a bill of material.

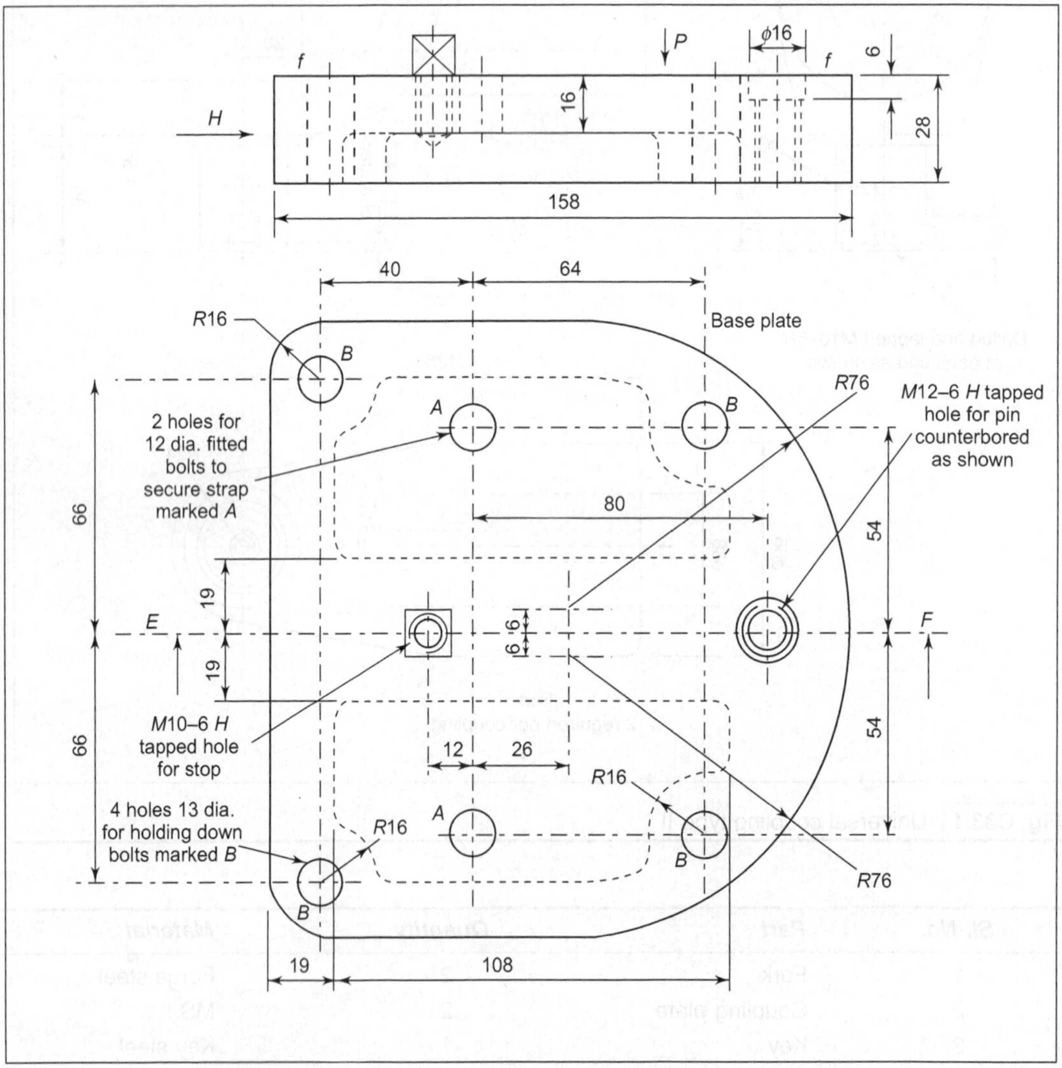

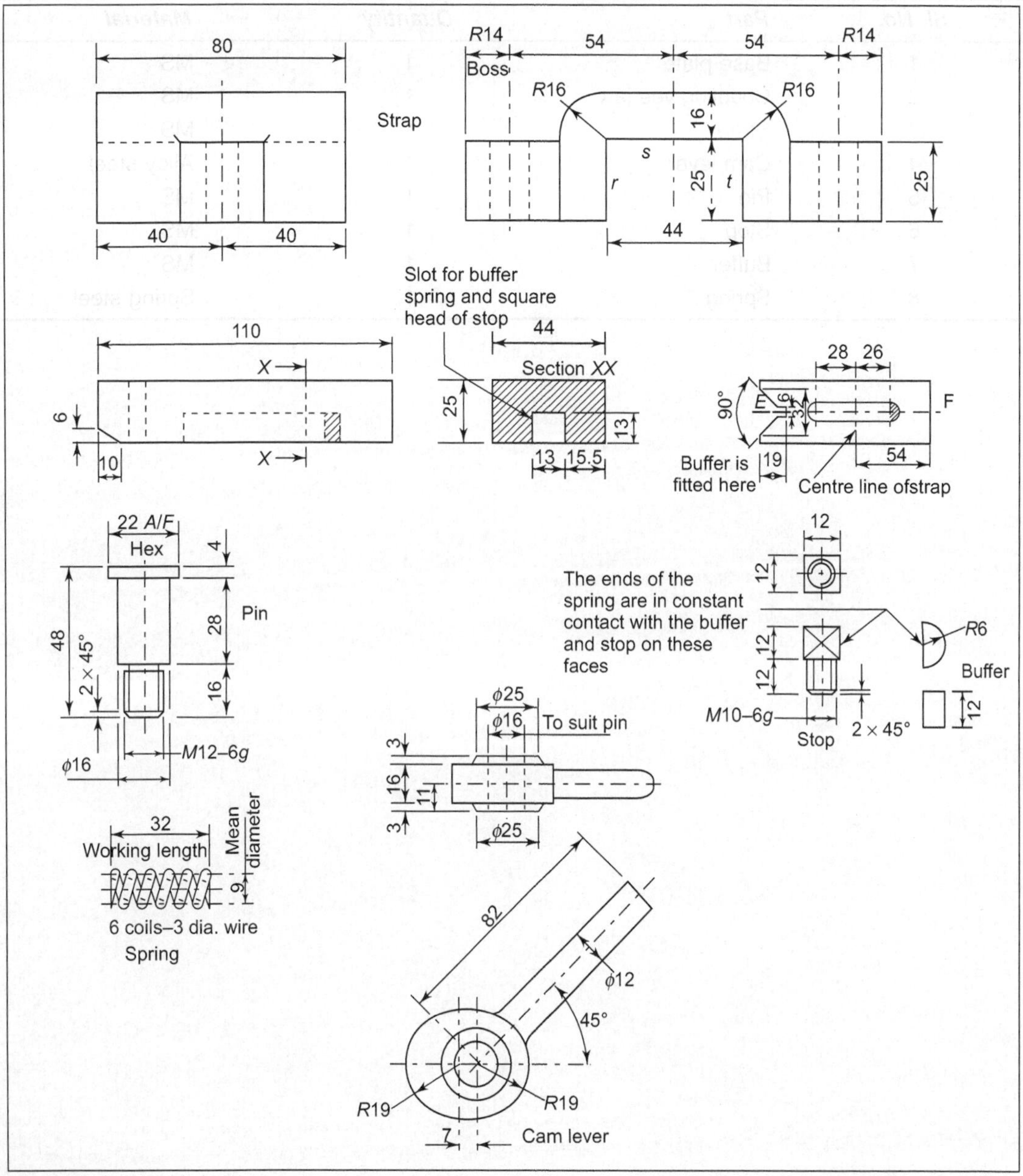

Fig. C34.1 Clamping fixture

Sl. No.	Part	Quantity	Material
1	Base plate	1	MS
2	Coupling vee jaw	1	MS
3	Strap	1	MS
4	Cam lever	1	Alloy steel
5	Pin	1	MS
6	Stop	1	MS
7	Buffer	1	MS
8	Spring	1	Spring steel

35 Non-Return Valve

Fig. C35.1 shows details of a non-return valve.

Draw full size, using first angle projection, the following views of the assembled valve in closed position.

1. Sectional elevation, the plane of the section to be along and in the direction of *BB*.
2. Sectional plan projected from view *A*, the plane of the section to be along *CC*.

Also enlist a bill of material.

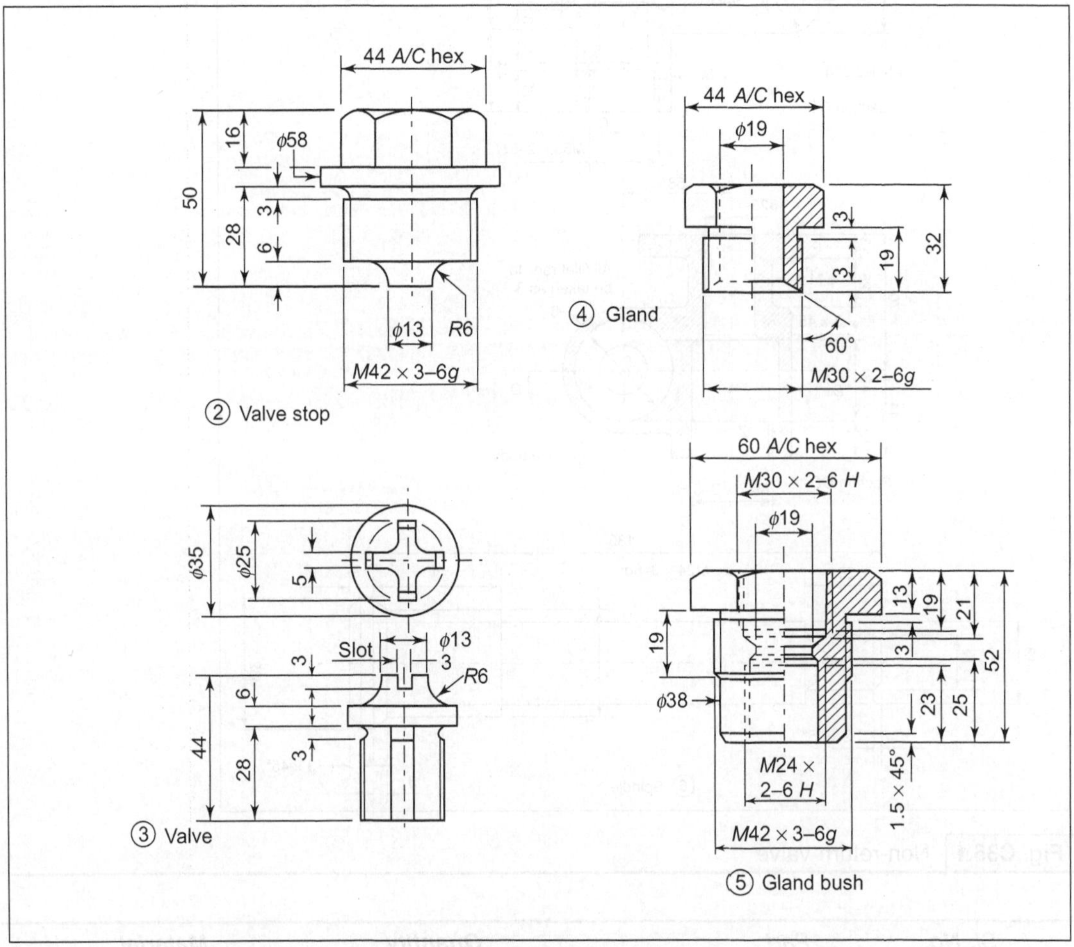

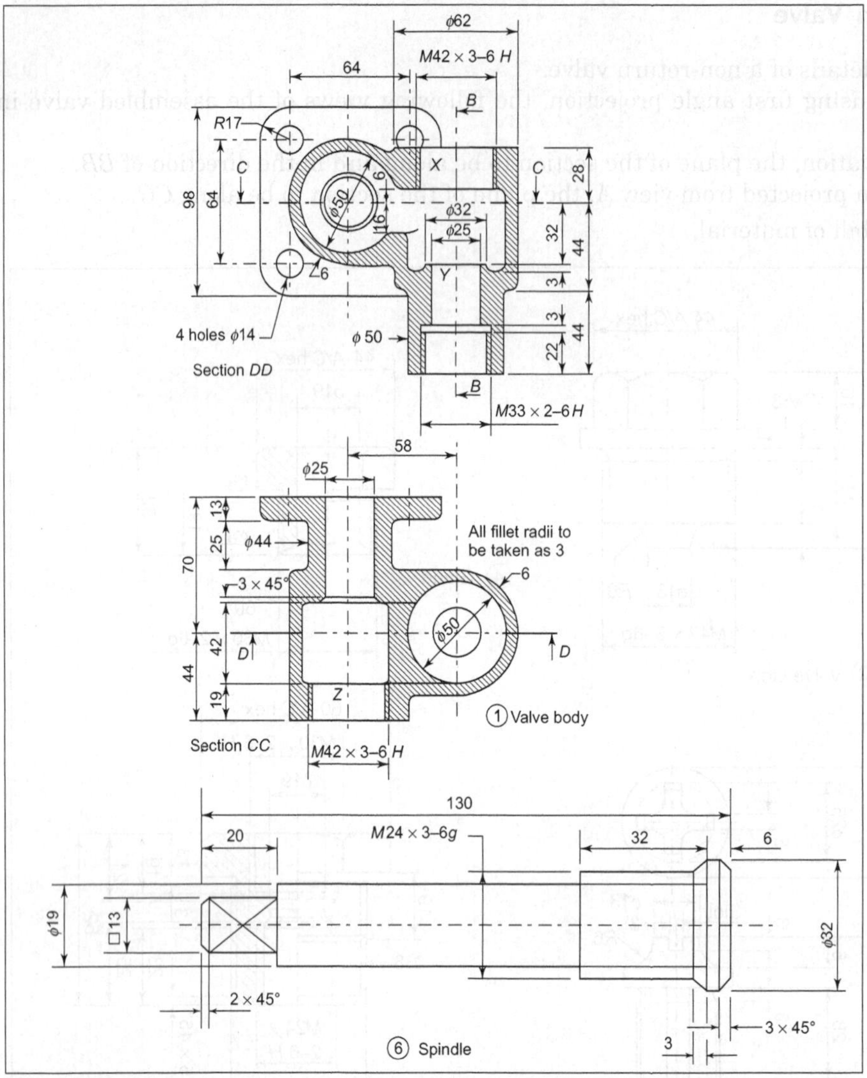

Fig. C35.1 Non-return valve

Sl. No.	Part	Quantity	Material
1	Valve body	1	Forging
2	Valve stop	1	MS
3	Valve	1	Brass
4	Gland	1	Brass
5	Gland bush	1	Brass
6	Spindle	1	SS

36 Diaphragm Regulator

Fig. C36.1 shows details of a part of a diaphragm regulator.

Draw full size, using third angle projection, the following views of a diaphragm regulator with the parts assembled in their correct relative positions.

1. Sectional elevation, the plane of the section and the direction of the required view being shown at *EE*.
2. Elevation looking in the direction of arrow *F*.

Also enlist a bill of material.

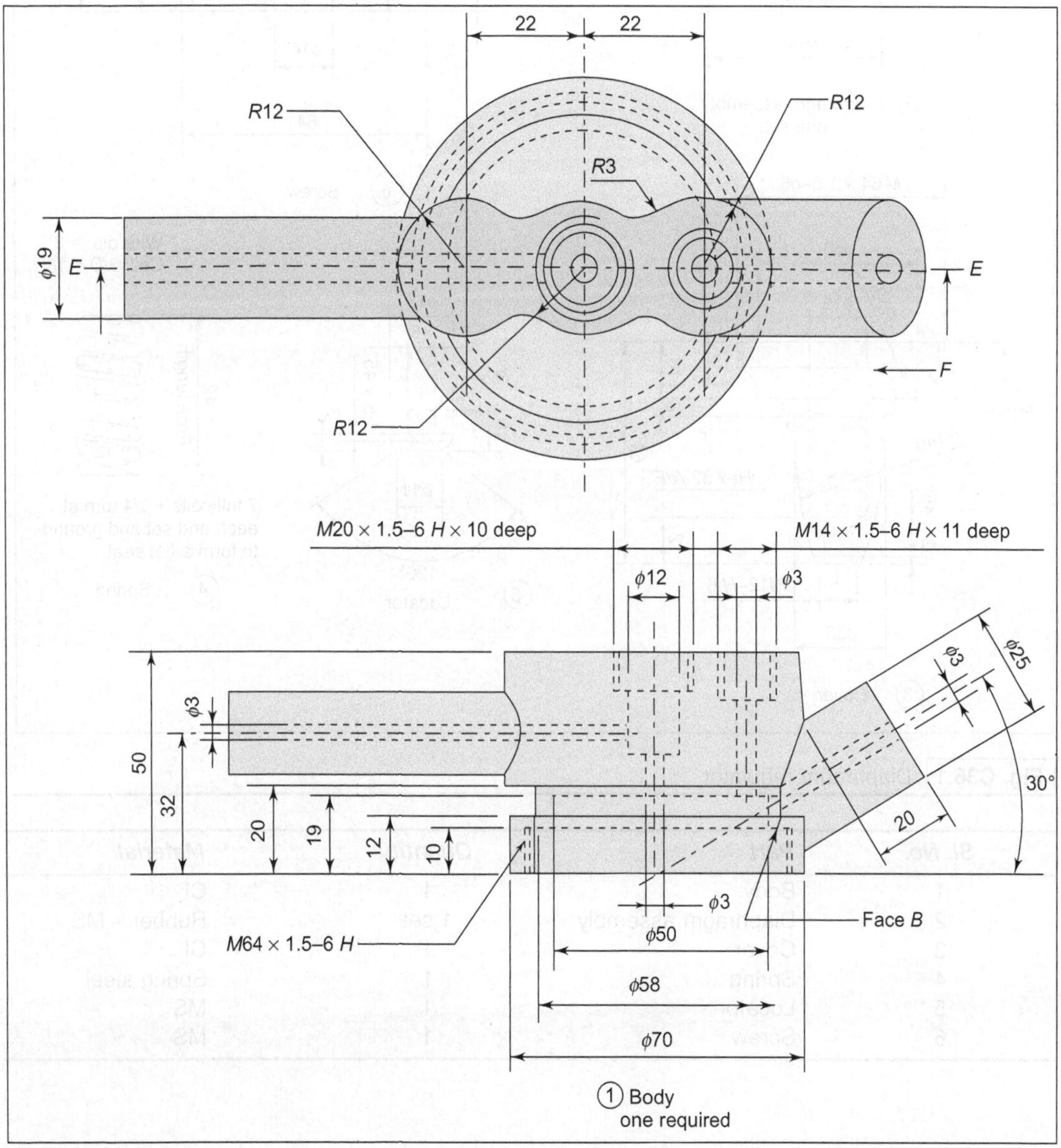

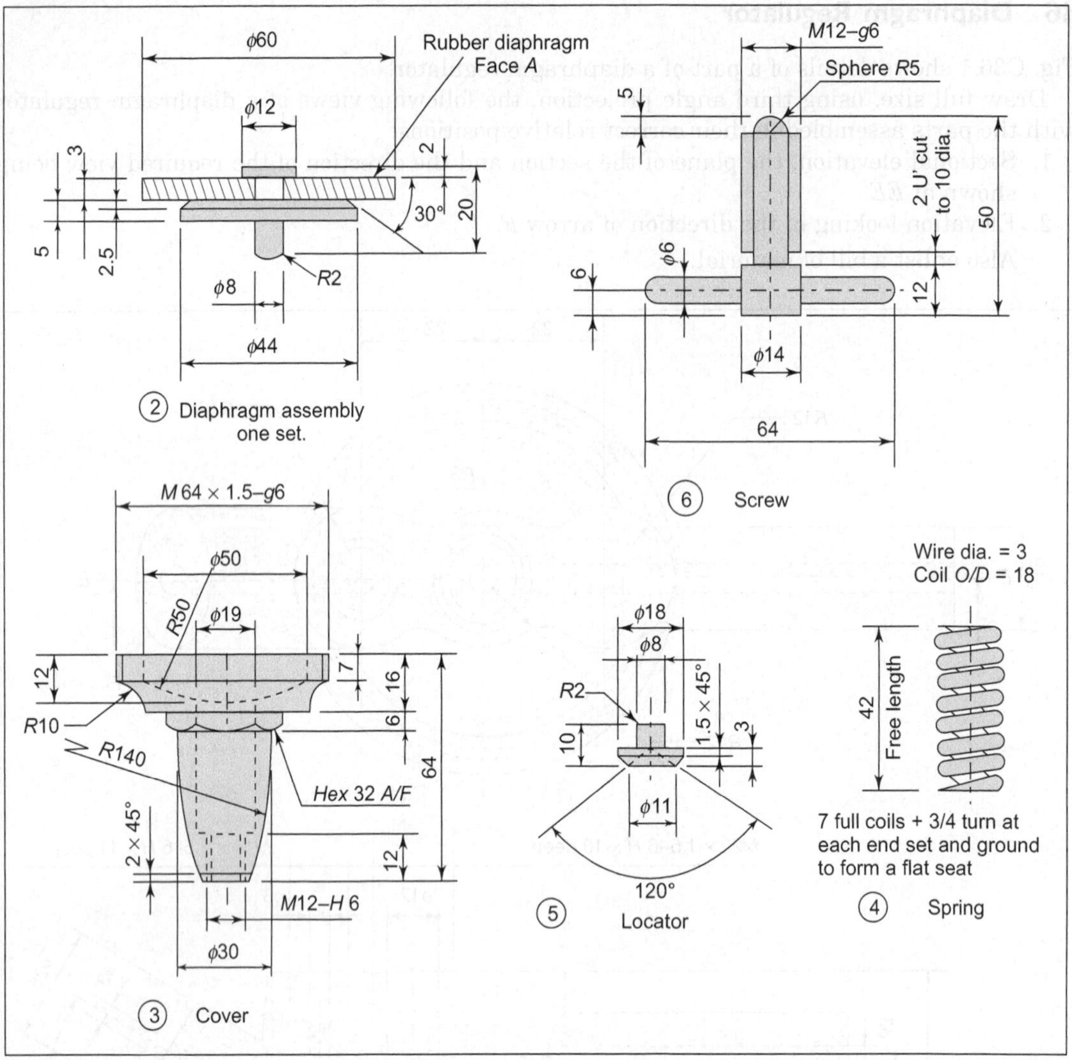

Fig. C36.1 Diaphragm regulator

Sl. No.	*Part*	*Quantity*	*Material*
1	Body	1	CI
2	Diaphragm assembly	1 set	Rubber + MS
3	Cover	1	CI
4	Spring	1	Spring steel
5	Locator	1	MS
6	Screw	1	MS

37 Gate Valve

The details of the component parts of a gate valve are shown in Fig. C37.1.
Draw full size, using first angle projection, the following views of the assembled valve.

1. Sectional elevation corresponding to the section plane *XX* as shown on the given views of the valve chest.
2. Outside elevation projected to the right hand side of view (1).

Enlist a bill of material.

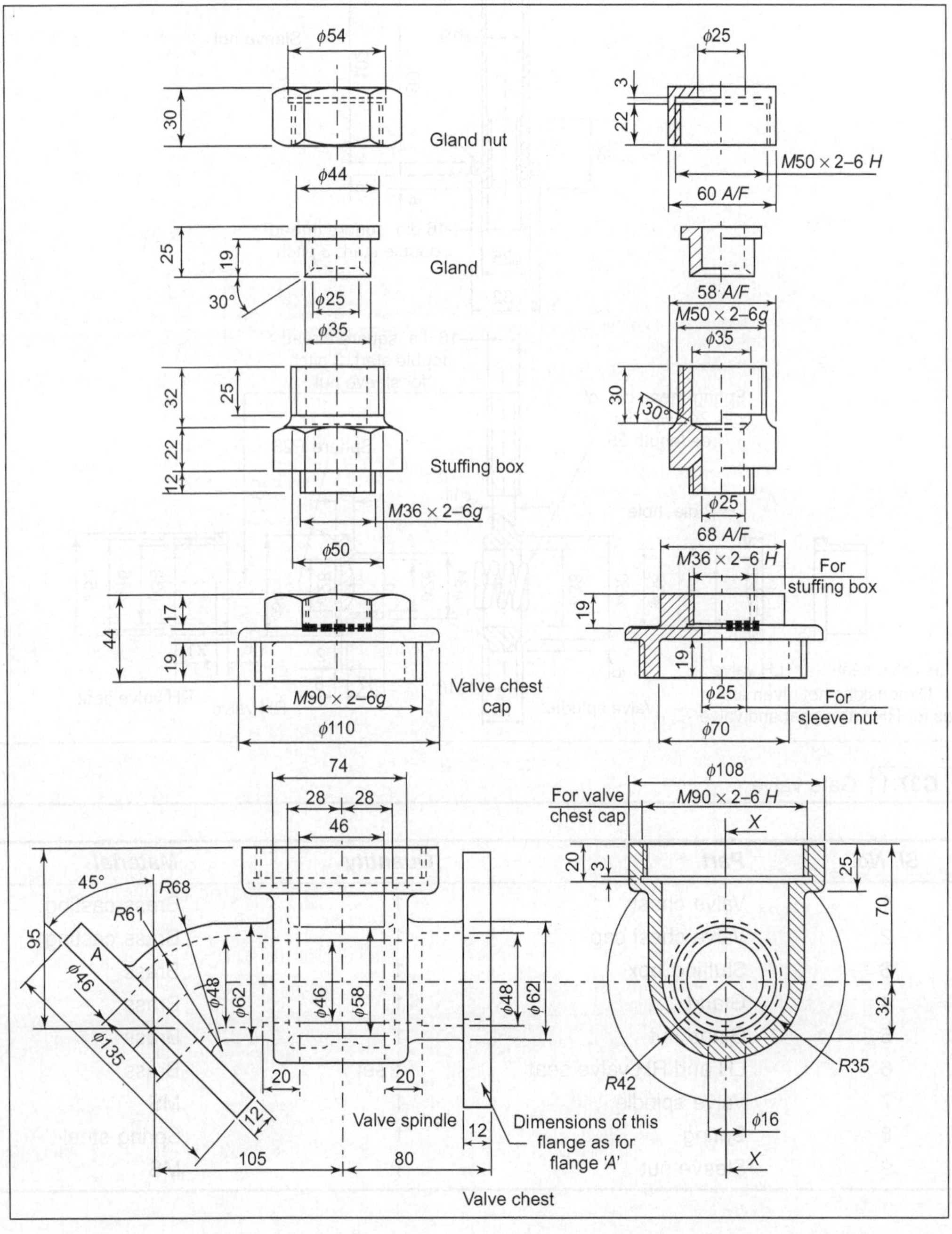

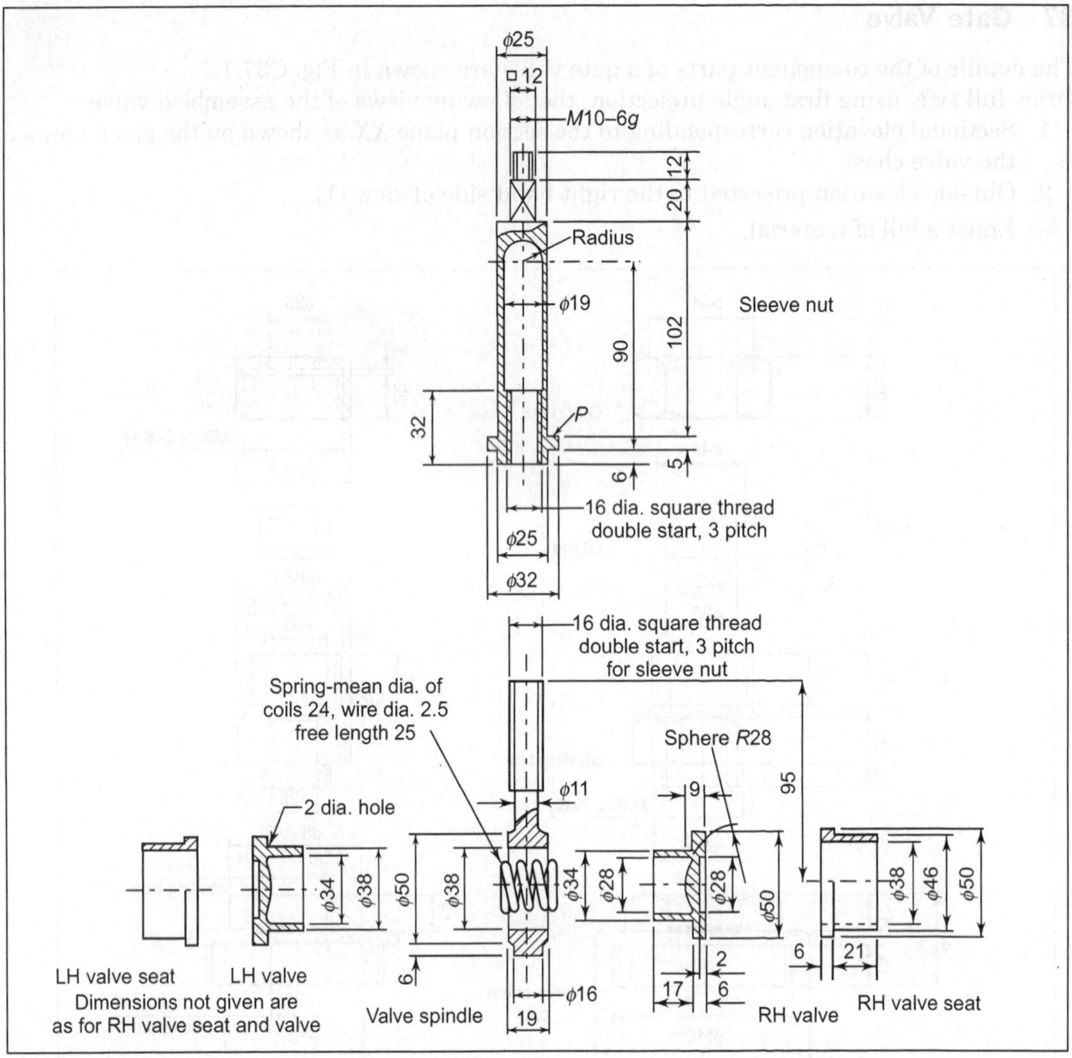

Fig. C37.1 Gate valve

Sl. No.	*Part*	*Quantity*	*Material*
1	Valve chest	1	Brass casting
2	Valve chest cap	1	Brass casting
3	Stuffing box	1	Brass
4	Gland	1	Brass
5	Gland nut	1	Brass
6	LH and RH valve seat	1 set	Brass
7	Valve spindle	1	MS
8	Spring	1	Spring steel
9	Sleeve nut	1	MS

38 3-Jaw Chuck

Figs C38.1 and C38.2 shows details of a 3-jaw chuck for a lathe.

1. Sketch freehand and proportionate sketch for a 3-jaw chuck.
2. Showing the parts in their respective working positions, draw two views.
   (a) The side view of the assembled chuck and flange, that is, the view corresponding with the one seen in the direction of arrow *P*.
   (b) The view seen when looking from the right hand side of (a).

Also enlist a bill of material for the assembly components.

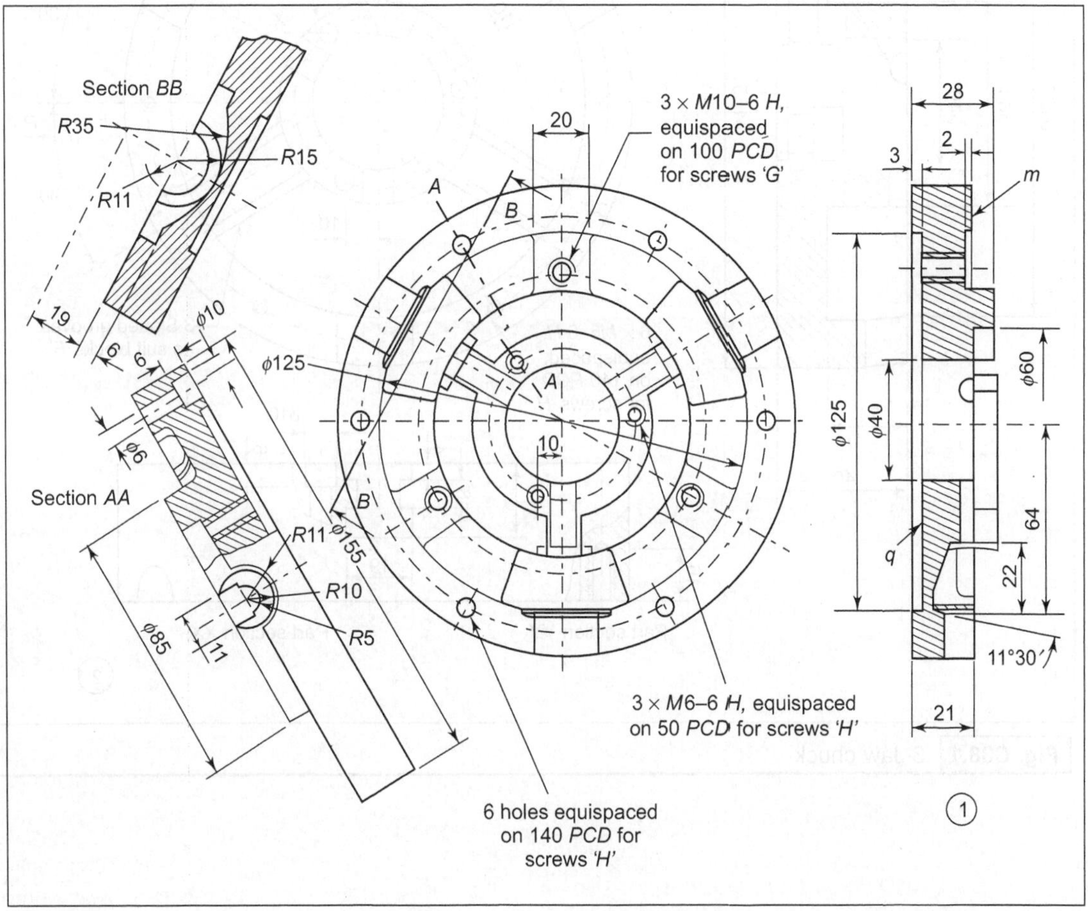

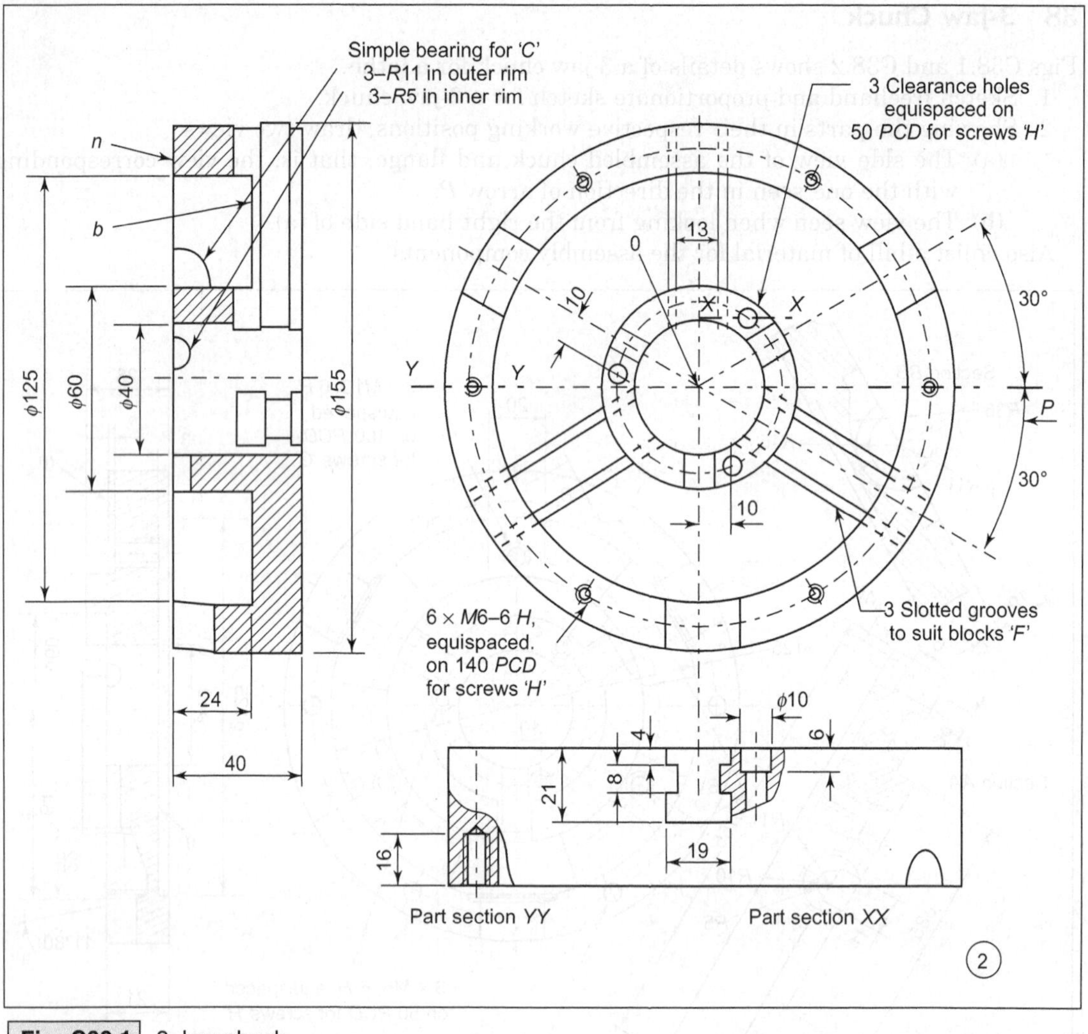

Fig. C38.1 3-Jaw chuck

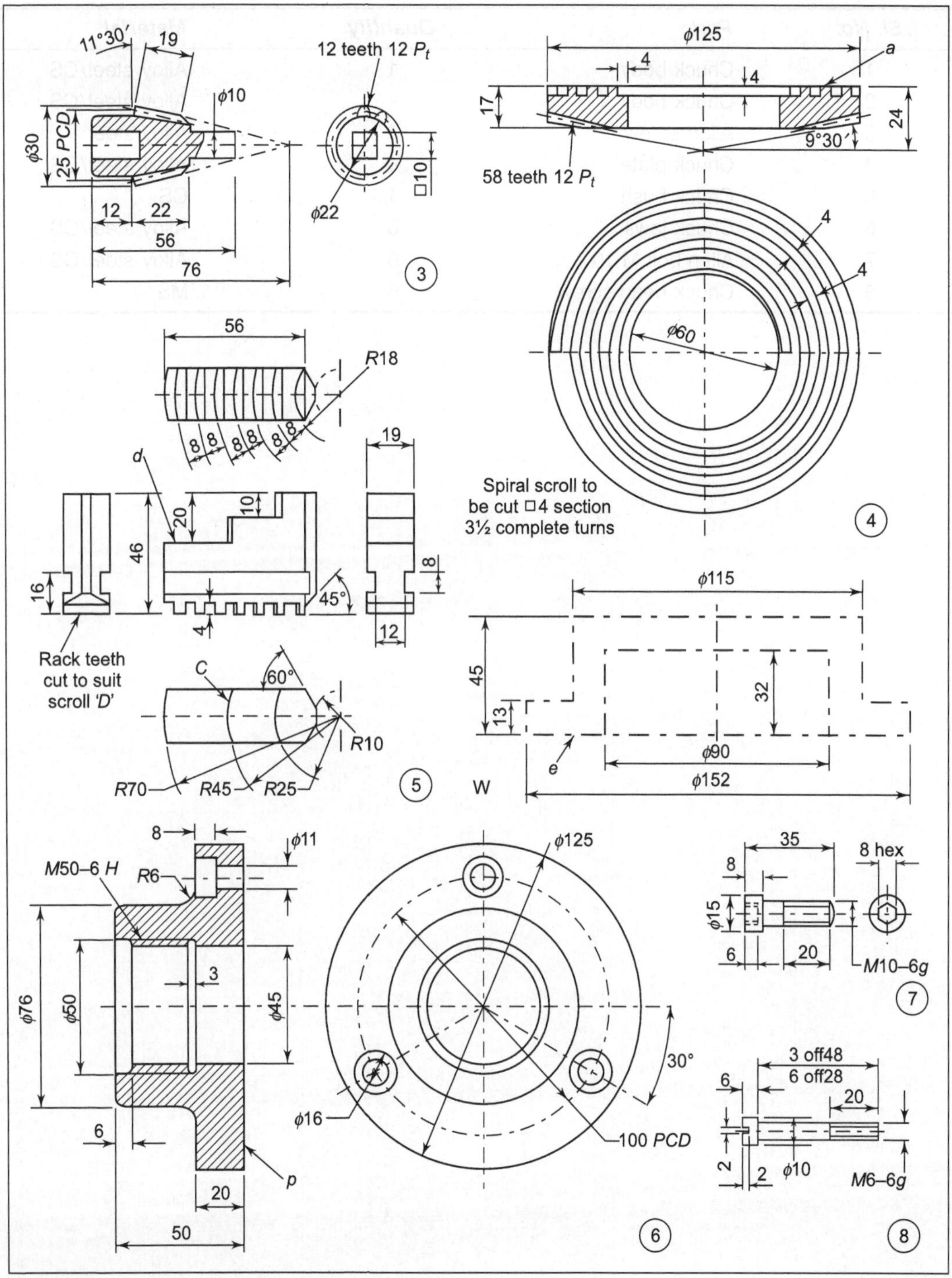

Fig. C38.2 Details of 3-jaw chuck

Sl. No.	Part	Quantity	Material
1	Chuck body	1	Alloy steel/CS
2	Chuck body	–	Alloy steel/CS
3	Allen screw	3	Alloy steel
4	Chuck plate	1	Alloy steel/CS
5	Chuck bush	1	CS
6	Chuck jaws	3	Alloy steel/CS
7	Allen bolt M 10	6	Alloy steel/CS
8	Chuck head screw	6	MS

39 Water Circulating Pump

The details components of a small water circulating pump are shown in Figs. C39.1 and C39.2.

Draw to the scale of full size, the following views of the assembled pump.

1. Sectional end elevation on *CC* viewed in the direction of the arrows.
2. Sectional elevation on *AA* as seen in the dirction indicated.
3. Outside plan view projected below.

Also enlist a bill of material for the assembly components.

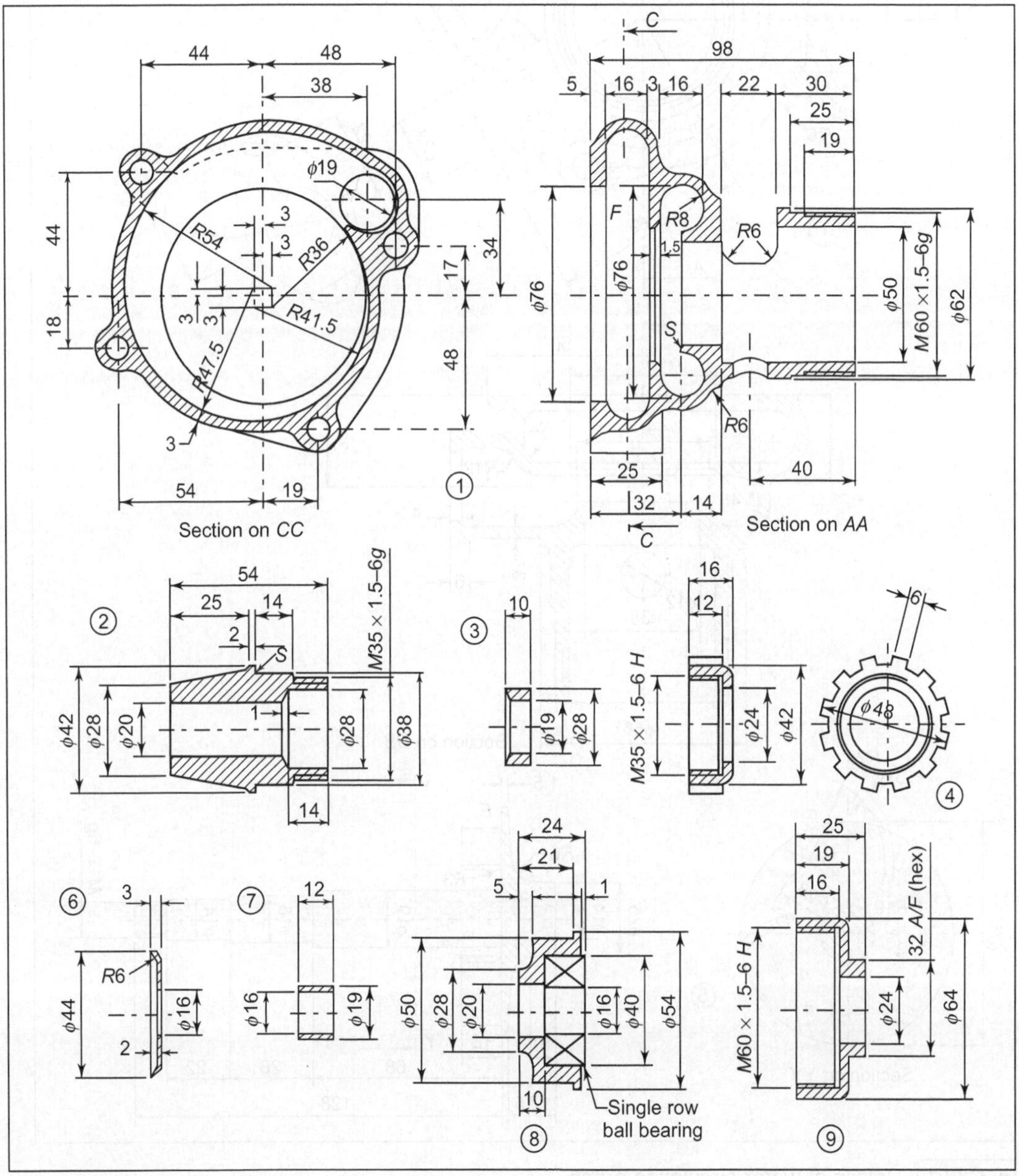

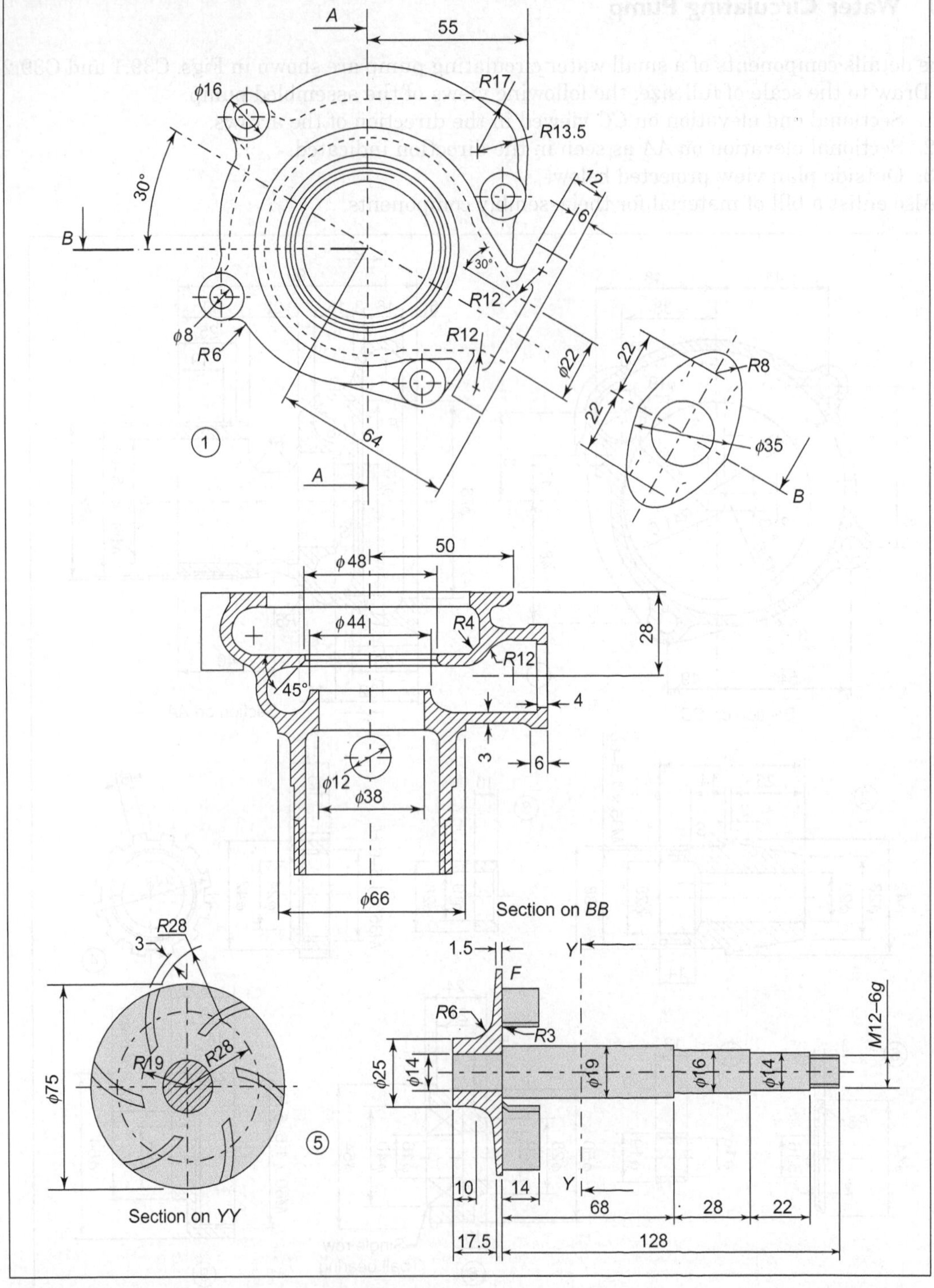

Fig. C39.1 Details of water circulating pump

Sl. No.	*Part*	*Quantity*	*Material*
1	Pamp casing	1	CS
2	Reducer	1	MS
3	Bush	1	GM
4	Splined bush	1	MS
5	Impeller shaft	1	Forged steel
6	Disk	1	GM
7	Bush	1	GM
8	Ball bearing with housing	1 set	Brg. steel
9	Casing nut	1	CS

40 Manual Drilling Machine

The details of the component parts of a hand drill are given in Fig. C40.1.

Draw the assembly views, in full size.

1. External elevation of the assembled parts.
2. From this view project orthographically a sectional elevation, taking the section along the longitudinal centre line. Consider the views carefully to ensure that those chosen will illustrate the object to the best advantage.

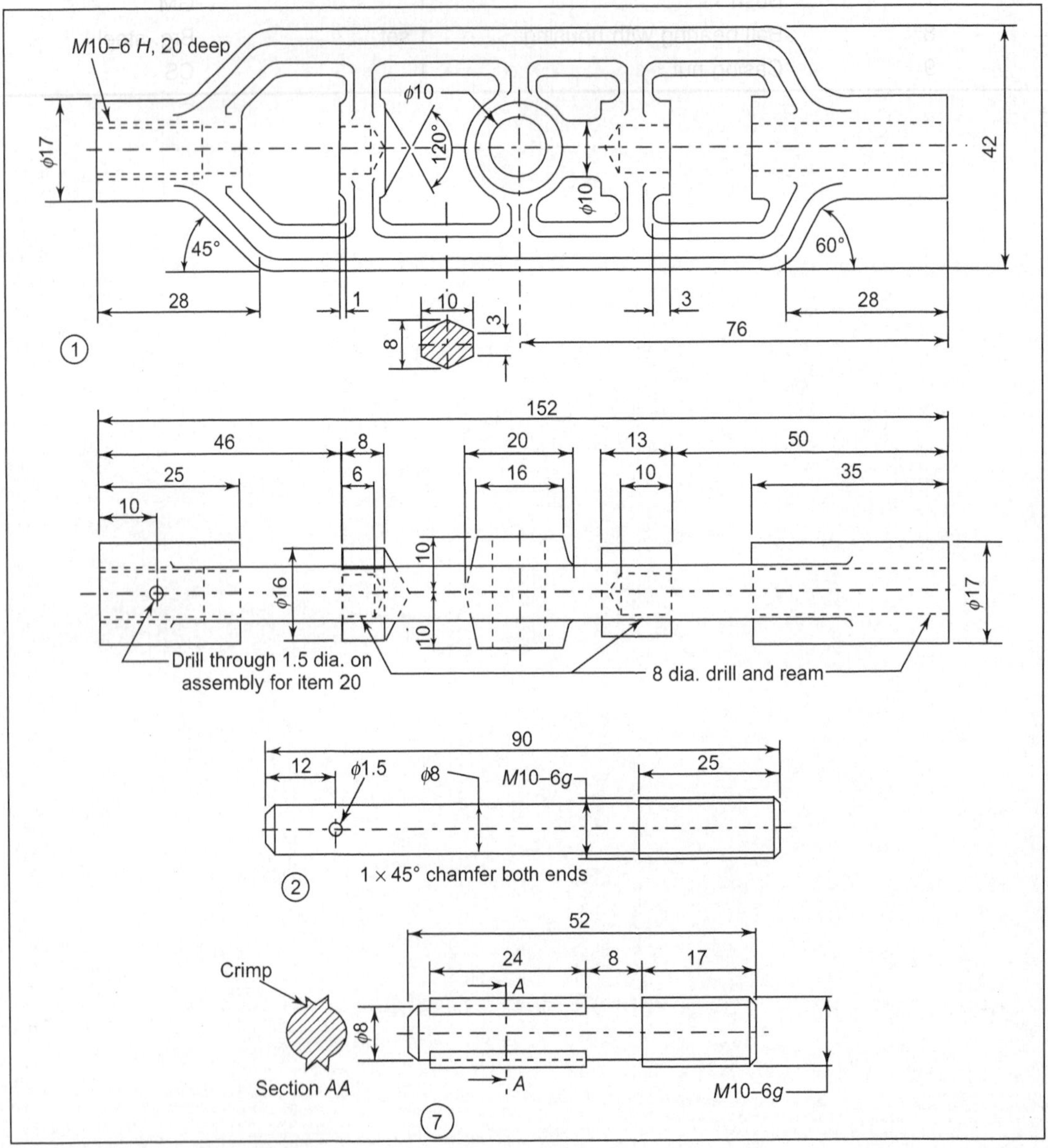

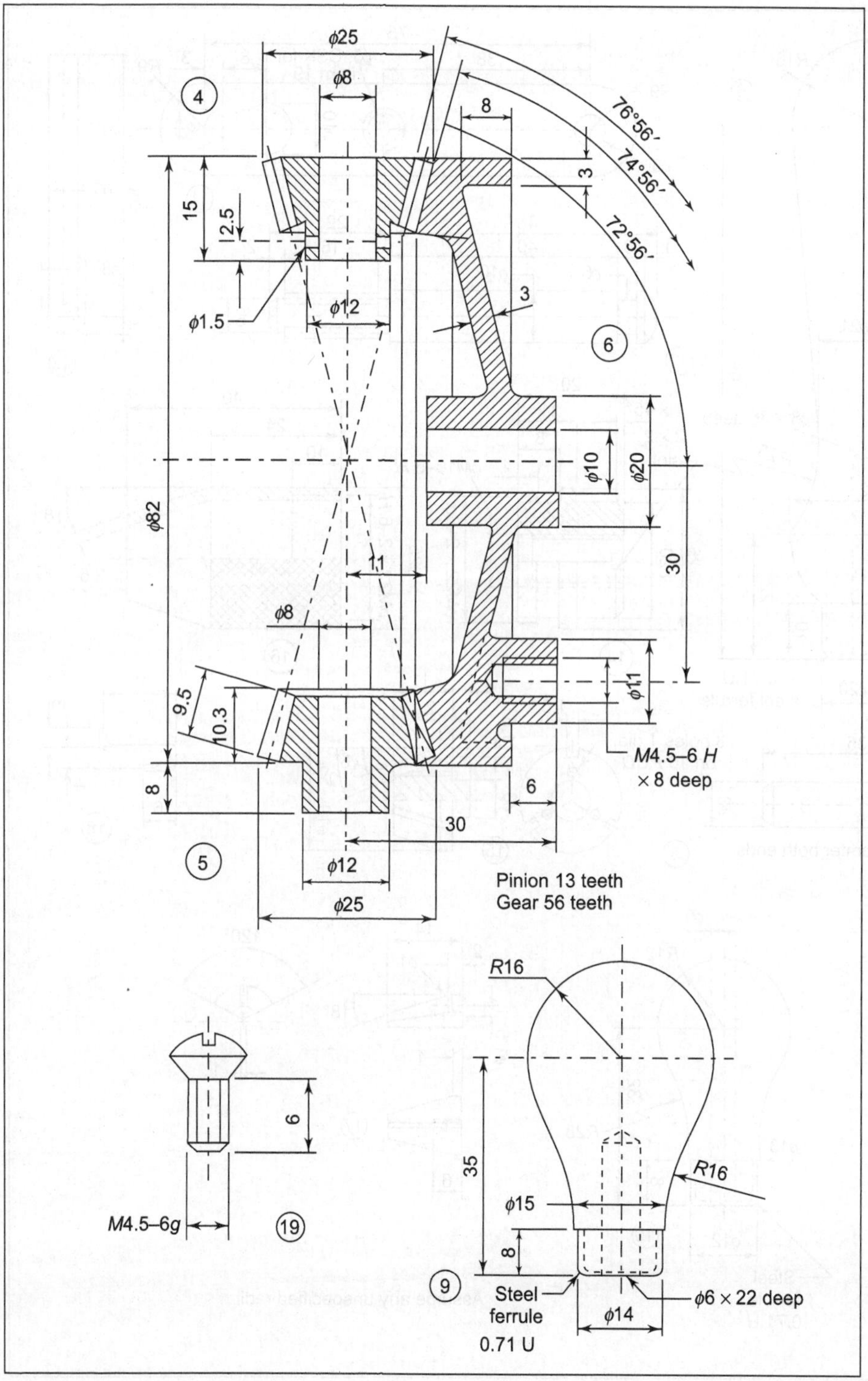

ϕ25
ϕ8
4
8
76°56′
74°56′
72°56′
3
15
2.5
ϕ1.5
ϕ12
3
6
ϕ10
ϕ20
ϕ82
11
30
ϕ8
9.5
10.3
ϕ11
M4.5–6 H
× 8 deep
8
6
30
5
ϕ12
ϕ25
Pinion 13 teeth
Gear 56 teeth
R16
6
35
R16
ϕ15
M4.5–6g
19
8
9
Steel
ferrule
ϕ6 × 22 deep
ϕ14
0.71 U

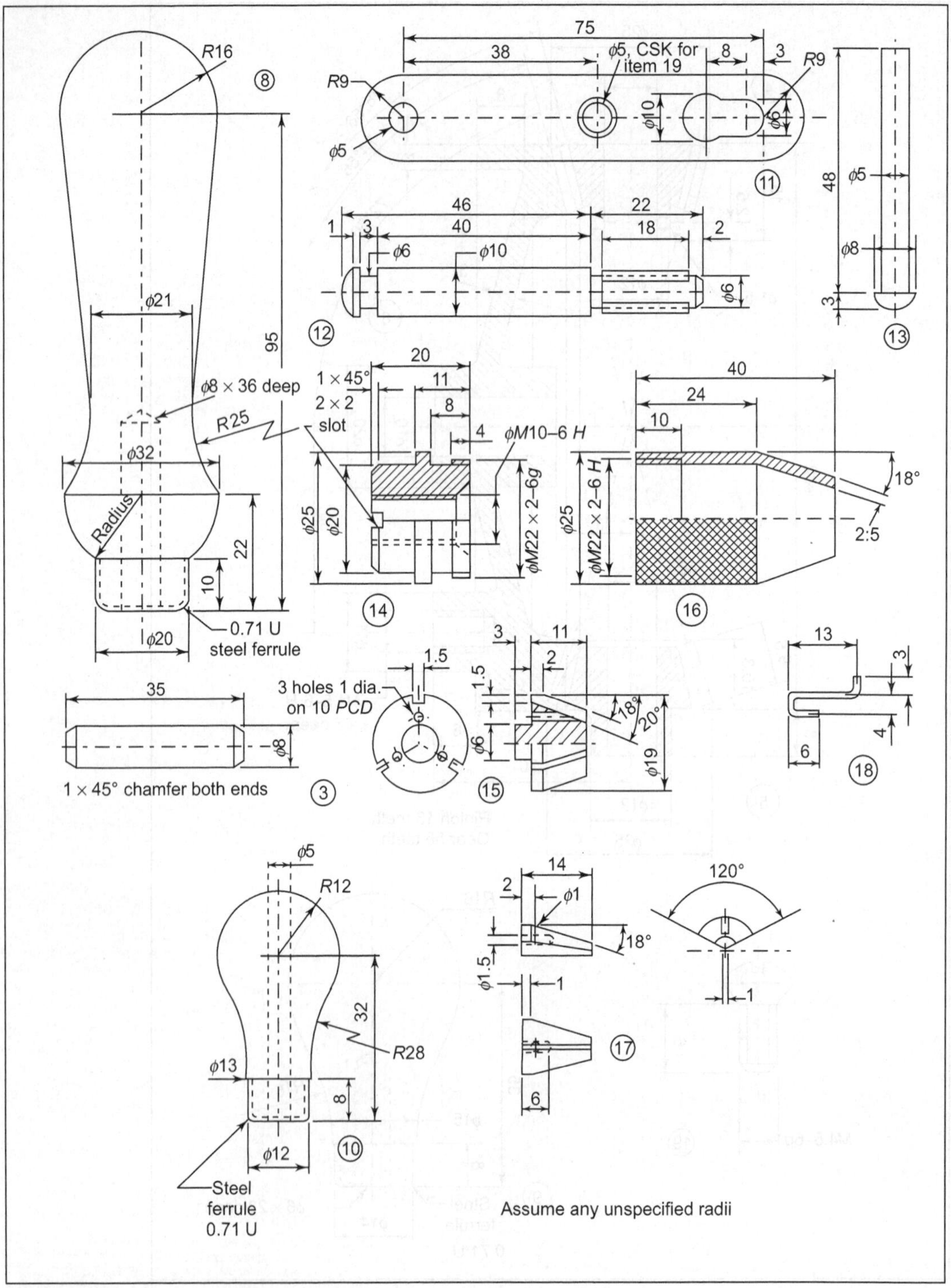

Fig. C40.1 Manual drilling machine

Sl. No.	Part	Quantity	Material
1	Housing	1	CI
2	Shaft driving pinion	1	MS
3	Shaft idler pinion	1	MS
4	Pinion driving	1	CI
5	Pinion idler	1	CI
6	Gear wheel	1	CI
7	Shaft handle	1	MS
8	Handle	1	Wood
9	Knob gear	1	Wood
10	Knob crank	1	Wood
11	Crank	1	MS
12	Shaft gear	1	MS
13	Shaft crank knob	1	MS
14	Back plate chuck	1	MS
15	Compressor chuck	1	MS
16	Cap chuck	1	MS
17	Jaw chuck	3	CS
18	Spring chuck	3	Wire
19	Screw c/sunk RSD head	1	MS
20	Pins taper	2	MS

HINTS TO QUESTIONS ON ASSEMBLY DRAWING

1. Gear Detacher It is used for the application of pulling out a gear assembled on the shaft with the help of keys.

Hints for Answer

1. Study the number of components given in a set of components for the question.
2. Study the name of the assembly.
3. Prepare a bill of material for the assembly.
4. Draw horizontal and vertical axis for elevation.
5. Start assembled elevation with the part–screw, *MS* $M12 \times 2$. Assemble rest of the components with screw *MS* and complete the assembly.
6. Take projection of the assembled elevation to draw the remaining views which are asked in the question.

2. Drill Machine Vice It is used for clamping the job on the bench of a drilling machine for drilling, tapping and such similar operations.

Hints for Answer

1. Study the number of components given in a set of components for the question.
2. Study the name of the assembly.
3. Prepare a bill of material for the assembly.
4. Draw the horizontal and vertical axis for component no. 1. (Base)
5. Start with part no. 1 as base, for elevation.
6. Take the projection of elevation for other views.

3. Crane Truck It is used for the movement of the crane hook attached to the trolley and is assembled on the arm of jib crane. With this arrangement, shifting of load from one location to another is possible.

Hints for Answer

1. Study the number of components given in a set of components for the question.
2. Study the name of the assembly.
3. Prepare a bill of material for the assembly.
4. Draw the elevation with part no. 1, that is, trolley frame.
5. Complete the answer by drawing other views.

4. Pipe Holding Device It is the assembly used for pipe and pipe fitting work in a workshop, especially in plumbing section.

Hints for Answer

1. Study the number of components given in a set of components for the question.
2. Study the name of the assembly.
3. Prepare a bill of material for the assembly.
4. Start the drawing for the elevation of assembly with part no. 1, that is, frame of pipe vice.
5. Take the projection of elevation for other views.

5. Screw Operated Lifting Machine It is hand operated equipment used in workshop for hoisting heavy parts. In order to get enough space, it is normally mounted on wall mounting structures either mobile or stationary.

Hints for Answer

1. Study the number of components given in a set of components for the question.
2. Study the name of the assembly.
3. Prepare a bill of material for the assembly.
4. Begin the assembly elevation with part no. 1, that is, base frame for screw hoist.
5. Complete the other views by taking projection of the elevation drawn.

6. Lifting Machine Hook It is used as a supporting device and for lifting and shifting heavy loads from one place to another. These are of different types and their shape and size depends on the capacity of the work to be carried out.

Hints for Answer

1. Study the number of components given in a set of components for the question.
2. Study the name of the assembly.
3. Prepare a bill of material for the assembly.
4. Start assembly elevation with part no. 5, that is, main hook of crane.
5. Assemble all the remaining parts with respect to hook and proceed with projection of other views.

7. Tool Carrier It is used for holding single point cutting tool on a machine. It could be any general purpose machine like lathe, milling, shaping, slotting m/c.

Hints for Answer

1. Study the number of components given in a set of components for the question.
2. Study the name of the assembly.
3. Prepare a bill of material for the assembly.
4. Body of tool holder, that is, part no. 1 helps in assembling rest of the components for the given question.
5. Assemble all the parts with part no. 1 and draw rest of the views by taking projection of elevation drawn.

8. Shaping Machine Tool Holder It is a general purpose machine used in work shops. Shaping machine tool-head is an assembly which holds tool-post or tool holder for holding single point cutting tool. It is generally used for sizing of the raw-material.

Hints for Answer

1. Study the number of components given in a set of components for the question.
2. Study the name of the assembly.
3. Prepare a bill of material for the assembly.
4. For drawing the elevation of the assembled tool-head, part no. 1 and 2 are required. Assemble all the parts in reference to the important dimensions with part no. 1 and 2.
5. Complete the elevation of the assembly and take the projection for rest of the views asked in the question.

9. Elastic Film Regulator It is used as a non-return valve and it regulates the supply of fluid for the given system.

Hints for Answer

1. Study the number of components given in a set of components for the question.
2. Study the name of the assembly.
3. Prepare a bill of material for the assembly.
4. Body part no. 1 of the diaphregn regulator is the main component of the assembly and with reference to the dimensions of this part assemble rest of the parts. Assembled elevation is obtained.
5. From the elevation, take the projection for obtaining the answer for rest of the views asked in the question.

10. Connecting Rod for Small Engine It is one of the important component in operating mechanism of internal cumbusion engine. Reciprocating motion of the piston is converted into rotary motion of crank-shaft by joining in between, the link named, connecting rod.

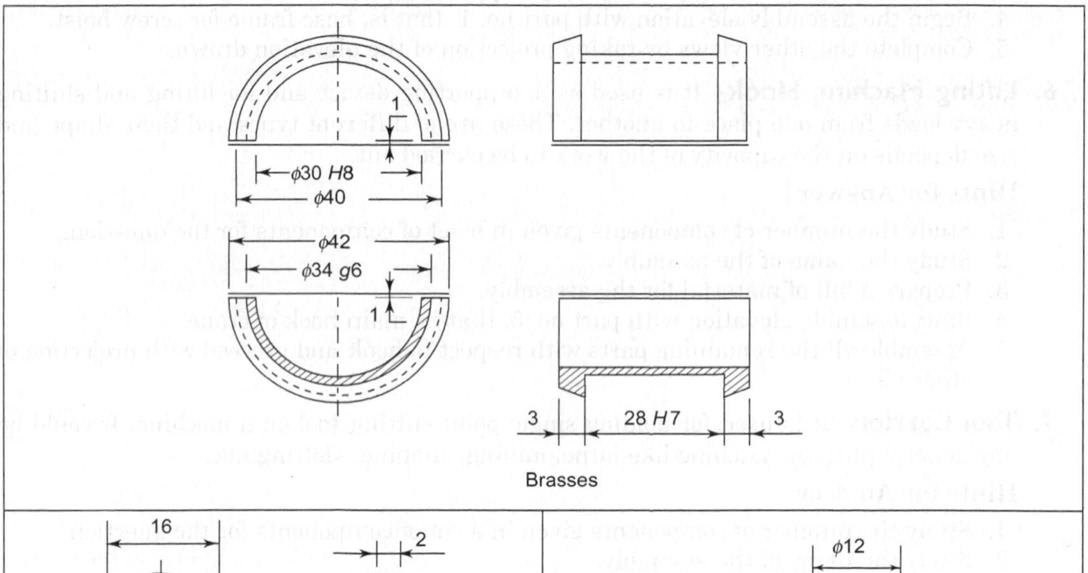

Hints for Answer

1. Study the number of components given in set a of components for the question.
2. Study the name of the assembly.
3. Prepare a bill of material for the assembly.
4. Draw the elevation of the assembly by taking care of part marked A, that is, rod.
5. Take the projection of above view to get rest of the views asked in question.

11. Connecting Rod for Big Engine (As per hint under 10 except where part no. 4 replaces A by 1).

12. Spark Plug It is used for introducing spark to ignite the compressed fuel air in petrol engine. Spark is produced by means of electrodes through coil and battery system. Two electrode are insulated and separated to avoid the short circuiting. A gap is maintained proportionately in between the electrodes.

Hints for Answers

1. Study the number of components given in a set of components for the question.
2. Study the name of the assembly.
3. Prepare a bill of material for the assembly.
4. Start the assembly elevation with the elevation of the part plug body. Assemble rest of the parts with plug body as reference and complete the assembled elevation.
5. Take the projection of elevation to get rest of the views asked in given question.
6. Complete the answer.

13. Variable Supply Fuel Pump It is also known as plunger pump. It is driven by the crankshaft of the engine and is used to compress the diesel oil in the engine cylinder to enable its injection through the injector in variable quantity, in order to suit the changing load requirement of the engine.

Hints for Answers

1. Study the number of components given in a set of components for the question.
2. Study the name of the assembly.
3. Prepare a bill of material for the assembly.
4. Start assembled elevation with the part no. 1, that is, pump body. Assemble rest of the parts on the pump body with reference to the dimensions of each part and complete the elevation of the assembly.
5. Project this elevation to get the other views asked in the given question.

14. Fuel Injector It is used alongwith fuel pump in I.C. engine to atomise the fuel particles into fine droplets in order to obtain a homogeneous mixture of fuel with air. This completes combustion.

Hints for Answer

1. Study the number of components given in a set of components for the question.
2. Study the name of the assembly.
3. Prepare a bill of material for the assembly.
4. Start assembled elevation with part no. 1, that is, fuel injector body. Assemble rest of the parts with reference to the dimensions.
5. Project this elevation to set other views.

15. Job Holding Device It is used for clarifying, holding and locating the part which is to be machined for any machining process on the respective machine in the workshop, like lathe, milling, slotting and so on. By introducing machine vice and damping it on the machine table, safety provisions are adhered to.

Hints for Answer

1. Study the number of components given in a set of components for the question.
2. Study the name of the assembly.
3. Prepare a bill of material for the assembly.
4. Draw the assembled elevation by taking reference of the part named base plate. Remaining parts can be assembled by taking the reference of its dimensions.
5. Assemble all the parts, draw the elevation, and project, for obtaining other views asked in given question.

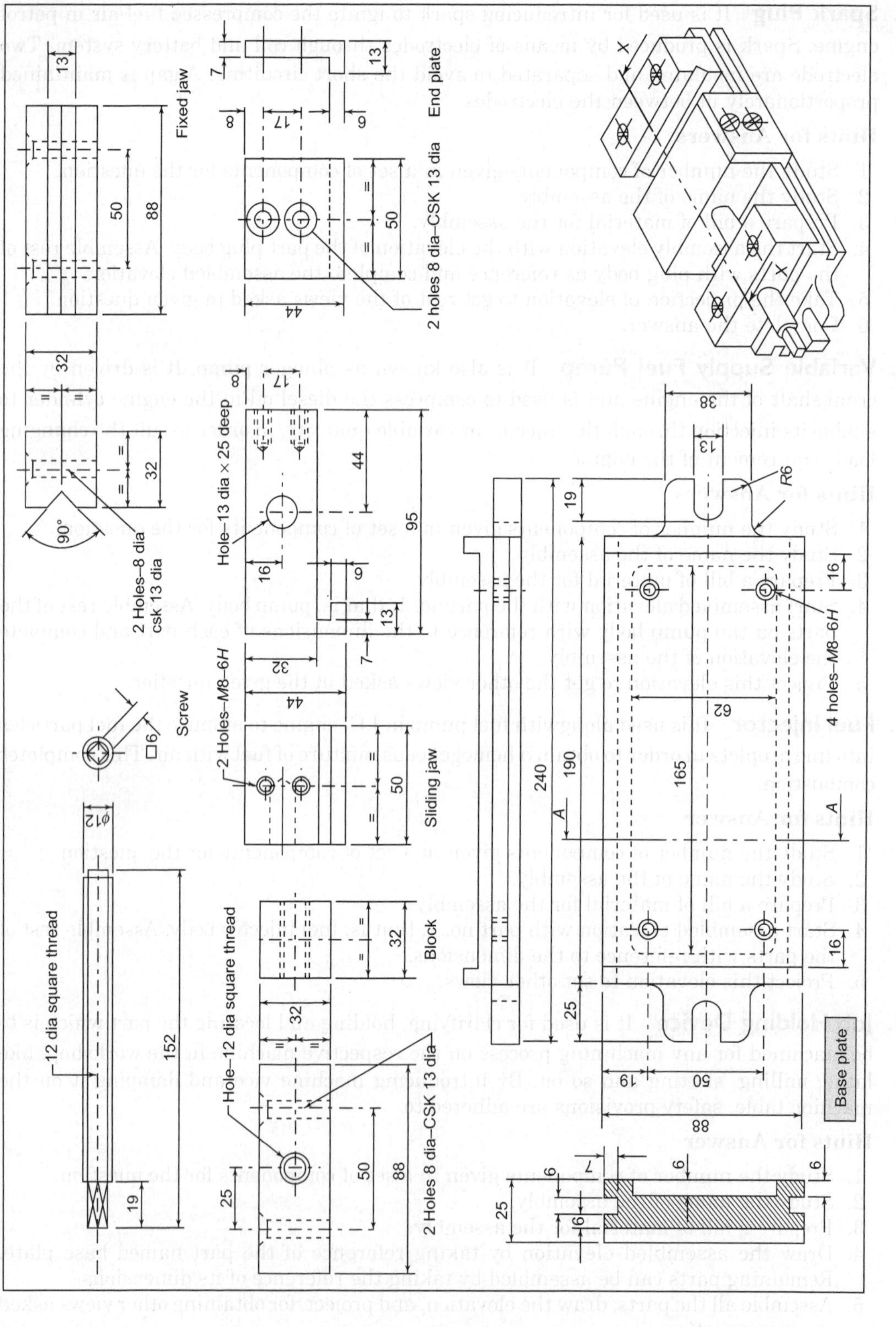

Fixed jaw
End plate
2 holes 8 dia–CSK 13 dia
2 Holes–8 dia csk 13 dia
90°
Hole-13 dia × 25 deep
2 Holes–M8–6H
Screw
□6
ϕ12
Sliding jaw
12 dia square thread
Hole–12 dia square thread
Block
2 Holes 8 dia–CSK 13 dia
Base plate
4 holes–M8–6H
R6
A
A

16. Force Pump It is a type of reciprocating pump. It is suitable for water applications. Due to life of the plunger, water rushes into the valve chest and water is delivered. When plunger moves down, suction gets closed and delivery of water take place through discharge valve.

Hint for Answer

1. Study the number of components given in a set of components for the question.
2. Study the name of the assembly.
3. Prepare a bill of material for the assembly.
4. Draw assembled elevation by drawing elevation of components 1, 3, 13, and 23. Study the functional dimensions of these parts and assemble rest of the parts.
5. Project this view to get other views as asked in the given question.

17. Feed Check Valve It is used as boiler mounting. It is used to supply feed water to boiler, under control. It is a non-return type valve.

Hints for Answer

1. Study the number of components given in a set of components for the question.
2. Study the name of the assembly.
3. Prepare a bill of material for the assembly.
4. Draw assembled elevation by considering valve body as the initial part, that is, part 1. By studying the dimensions of the remaining parts, start assembling those parts with reference to the valve body and complete assembled elevation.
5. Project elevation to get other views as asked in the question.

18. Flow Check Valve It is use as boiler mounting. It is used to supply feed water to boiler, under control. It is a non-return type valve.

Hints for Answer

1. Study the number of component given in set of components for the question.
2. Study the name of the assembly.
3. Prepare a bill of material for the assembly.
4. Draw assembled elevation by considering valve body as the initial part, that is, part 1. By studying the dimensions of the remaining parts, start assembling those parts with reference to the valve body and complete assembled elevation.
5. Project elevation to get other views as asked in the question.

19. Blow Off Cock It is used as boiler mounting. Blow-off refers to discharge water from Boiling water drum that takes place through blow-off cock. Water discharge is important for internal inspection and cleansing of the parts of the boiler. Provision of a plug is made at the bottom of the blow-off cock, in order to discharge sediments collected at the bottom of it.

Hints for Answer

1. Study the number of components given in a set of components for the question.
2. Study the name of the assembly.
3. Prepare a bill of material for the assembly.
4. Draw assembled elevation by considering the body of the valve. Take care of all important functioning dimensions of the valve body and assemble rest of the parts with it. Complete the drawing of assembled elevation.
5. Take projection of elevation to complete the other views.

20. Stop Valve It is also used as boiler mounting. It can also be called as control-valve. It controls the flow of steam from boiler to the engine. Stop valve is normally bolted to the steam pipes.

Hints for Answer

1. Study the number of components given in a set of components for the question.
2. Study the name of the assembly.
3. Prepare a bill of material for the assembly.
4. Start assembled elevation with the body of the blow-off cock. Identify functional dimensions of the body and assemble rest of the parts with the body of the valve. Complete the assembled elevation.
5. Project assembled elevation for obtaining other views asked in the given question.

21. 125 mm Stop Valve It is also used as boiler mounting. It is also called as control valve. It controls the flow of steam from boiler to the engine. Stop valve is normally bolted to the steam pipes.

Hints for Answer

1. Study the number of component given in set of components for the question.
2. Study the name of the assembly.
3. Prepare a bill of material for the assembly.
4. Start assembled elevation with the body of the blow-off cock. Identify the functional dimensions of body and assemble rest of the parts with the body of valve. Complete assembled elevation.
5. Project assembled elevation for obtaining other views asked in the given question.

22. Screw Down Stop Valve Screw down stop valve is used as a safety valve for high pressure fluid and steam lines. It allows passage for the volume of fluid to the capacity of full bore size of the pipe lines to which it is connected. Maintenance of this valve is a little difficult part during working condition.

Hints for Answer

1. Study the number of components given in a set of component for the question.
2. Study the name of the assembly.
3. Prepare a bill of material for the assembly.
4. Draw assembled elevation of the parallel slide stop valve by drawing elevation of the part named as valve body. Study the dimensions of valve body and rest of the components given in the question. Assemble all the parts with the valve body.
5. Project assembled elevation to project rest of the views.

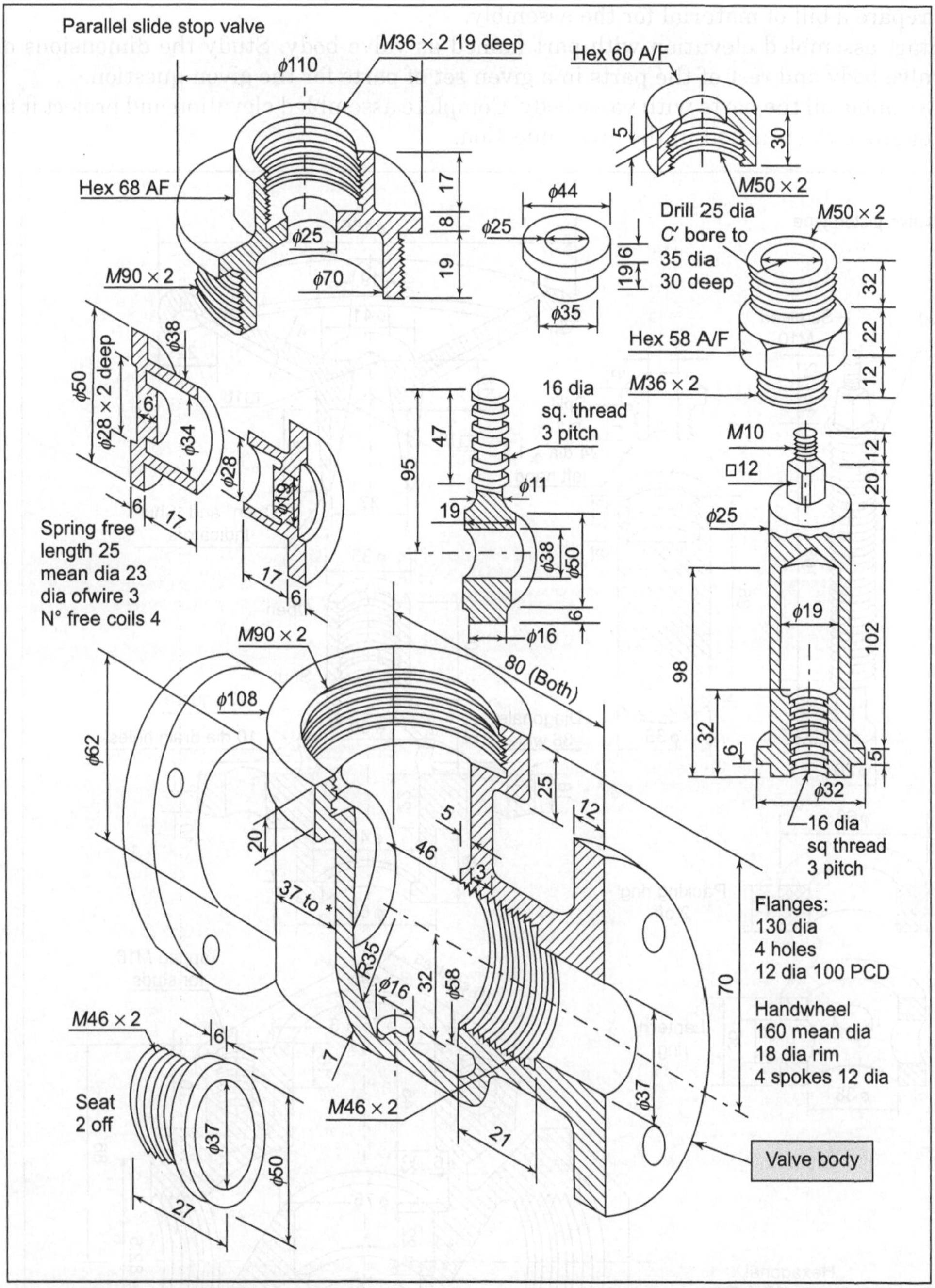

23. Plunger Type Stop Valve It is a screw-down non-return valve. It can be used for both, high and low pressure fluid lines. Easy maintenance is one of the added advantage.

Hints for Answer

1. Study the number of components given in a set of component for the question.
2. Study the name of the assembly.

3. Prepare a bill of material for the assembly.
4. Start assembled elevation with part named as valve body. Study the dimensions of valve body and rest of the parts in a given set of parts for the given question.
5. Assemble all the parts with valve body. Complete assembled elevation and project it to get other views asked in the given question.

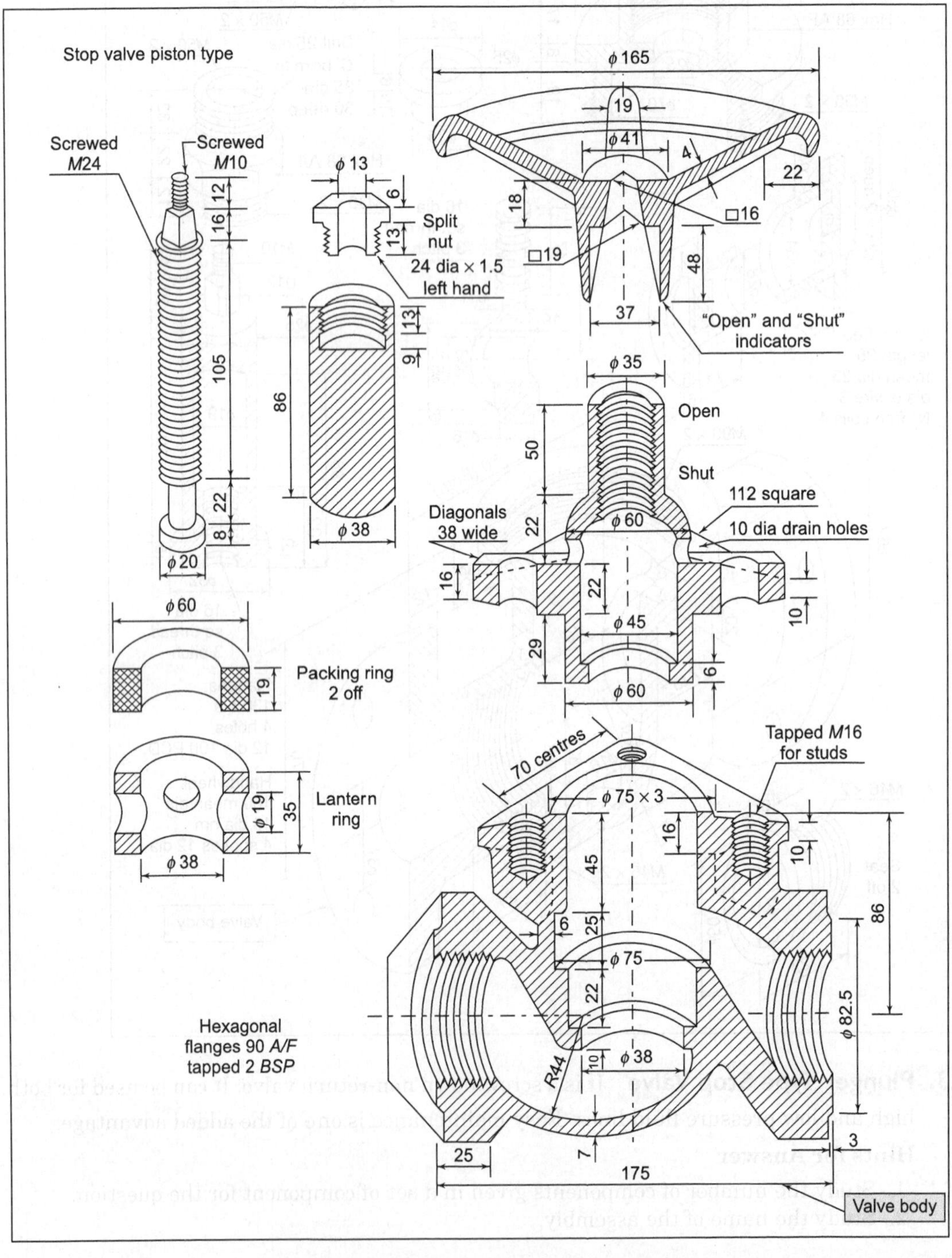

24. Automatic Valve It is used in a diesel air starting system alongwith pilot valve. These valves are used with diesel engines. Starting air is admitted through the pilot valve on the top of the piston. At the same time, air is admitted from the main air receiver to the valve.

Hints for Answer

1. Study the number of components given in a set of component for the question.
2. Study the name of the assembly.
3. Prepare a bill of material for the assembly.
4. Draw assembled elevation starting with part named as body of the automatic valve. Study the dimension of this part and rest of the parts which are given in a set of parts in the given question.
5. Assemble all the parts with body of the automatic valve. Complete assembled elevation and project it to get rest of the views as given in the question.

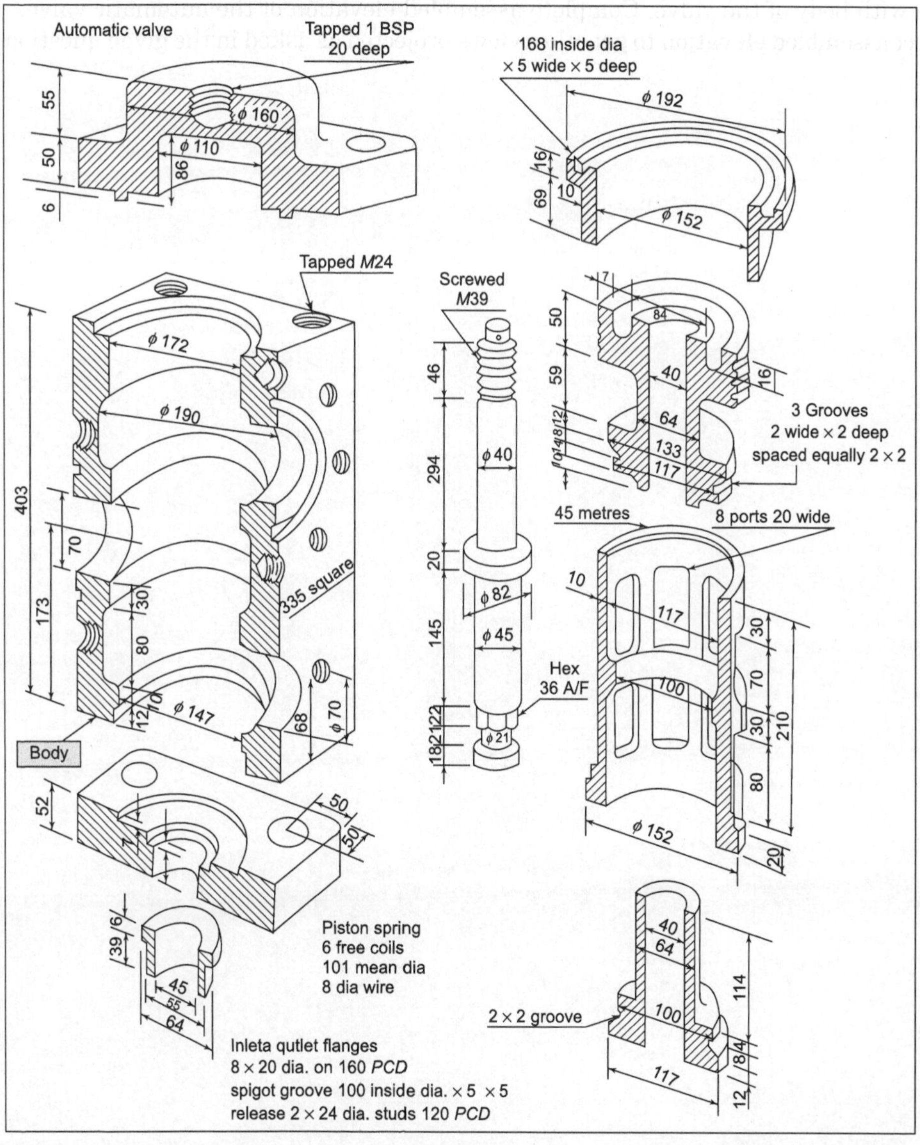

25. Safety Valve It is also used as boiler mounting and is statutory requirement of all boiler installations. Steam with pressure from the boiler exerts force on the valve from down below. The force is overcome by the compression of the spring on the valve spindle. The compression of the spring is adjustable through the nut and ring of the top of spindle. If the spring pressure is more than the steam pressure then the valve will remain close and if it is less than the value, it will open and flow of the steam will take place to release the pressure.

Hints for Answer

1. Study the number of components given in a set of components for the question.
2. Study the name of the assembly.
3. Prepare a bill of material for the assembly.
4. Assembled elevation can be drawn by taking care of part named as body of valve. For arranging other parts in sequence, study the dimensions of other parts and assemble them with body of the valve. Complete assembled elevation of the automatic valve.
5. Project assembled elevation to get other views projected as asked in the given question.

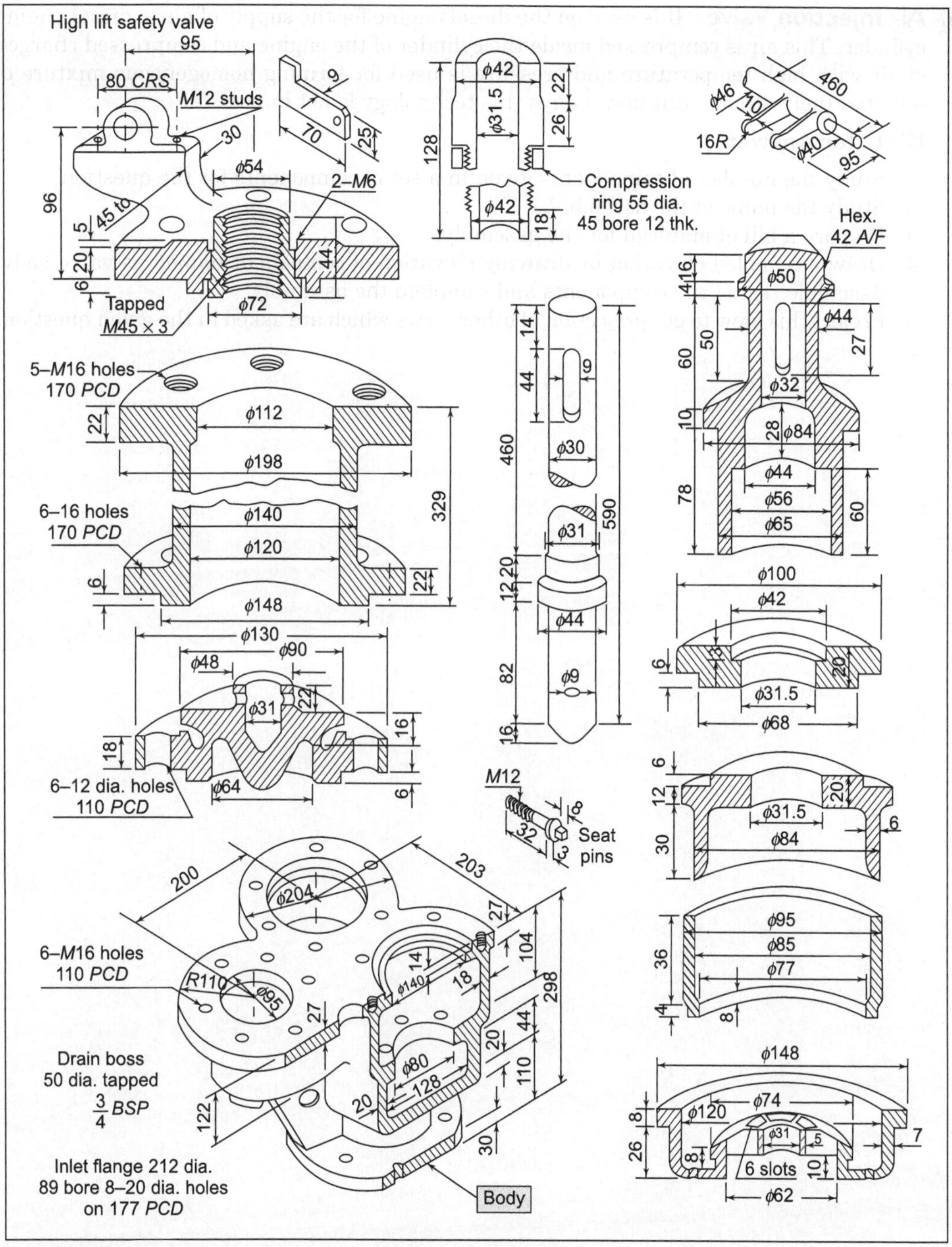
High lift safety valve
95
80 CRS
M12 studs
30
9
70
25
ϕ42
22
ϕ31.5
26
128
ϕ54
2–M6
Compression
ring 55 dia.
45 bore 12 thk.
ϕ42
18
460
ϕ46
10
16R
ϕ40
95
96
45 to
5
20
6
44
Tapped
M45 × 3
ϕ72
Hex.
42 A/F
416
ϕ50
5–M16 holes
170 PCD
22
ϕ112
ϕ198
329
6–16 holes
170 PCD
ϕ140
ϕ120
6
22
ϕ148
14
44
9
460
ϕ30
590
ϕ31
12 20
ϕ44
82
ϕ9
16
60
50
ϕ44
27
ϕ32
10
28
ϕ84
78
ϕ44
ϕ56
ϕ65
60
ϕ130
ϕ90
ϕ48
ϕ31
22
16
18
6
6–12 dia. holes
110 PCD
ϕ64
ϕ100
ϕ42
6
13
20
ϕ31.5
ϕ68
M12
8
32
3
Seat
pins
6
12
20
ϕ31.5
6
30
ϕ84
200
203
ϕ204
27
104
6–M16 holes
110 PCD
R110
ϕ95
14
ϕ140
18
298
27
44
20
110
ϕ80
128
20
122
30
Drain boss
50 dia. tapped
3/4 BSP
Inlet flange 212 dia.
89 bore 8–20 dia. holes
on 177 PCD
Body
36
ϕ95
ϕ85
ϕ77
4
8
ϕ148
ϕ74
8
ϕ120
ϕ31
5
7
26
8
10
6 slots
ϕ62

26. Air Injection Valve It is used on the diesel engine for the supply of air to diesel engine cylinder. This air is compressed inside the cylinder of the engine and compressed charged of air with high temperature and pressure is used for forming homogeneous mixture of inverted fuel with air. But now a days this technology is not in use.

Hints for Answer

1. Study the number of components given in a set of components for the question.
2. Study the name of the assembly.
3. Prepare a bill of material for the assembly.
4. Draw assembled elevation by drawing elevation of component named as valve body. Assemble rest of the components and complete the assembly.
5. Project this view to get projection of other views which are asked in the given question.

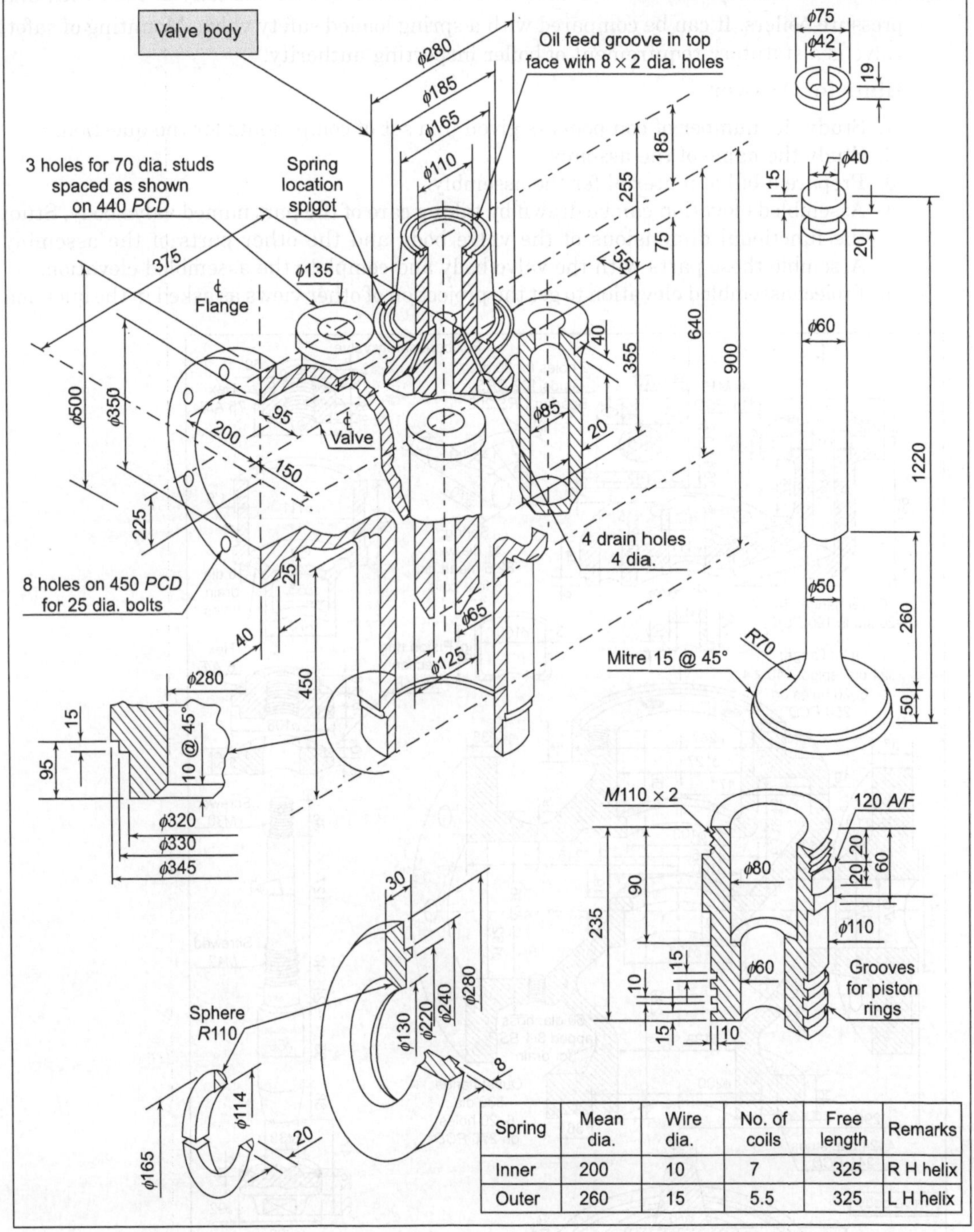

Spring	Mean dia.	Wire dia.	No. of coils	Free length	Remarks
Inner	200	10	7	325	R H helix
Outer	260	15	5.5	325	L H helix

27. Full Bore Safety Valve It is used as boiler mounting and basically is used with high pressure boilers. It can be compared with a spring loaded safety valve. Mounting of safety valve is a statutory requirement of boiler inspecting authority.

Hints for Answer

1. Study the number of components given in a set of components for the question.
2. Study the name of the assembly.
3. Prepare a bill of material for the assembly.
4. Assembled elevation can be drawn by taking care of the part named valve body. Study the functional dimensions of the valve body and the other parts of the assembly. Assemble these parts with the valve body and complete the assembled elevation.
5. Project assembled elevation to get the projection of other views as asked in the question.

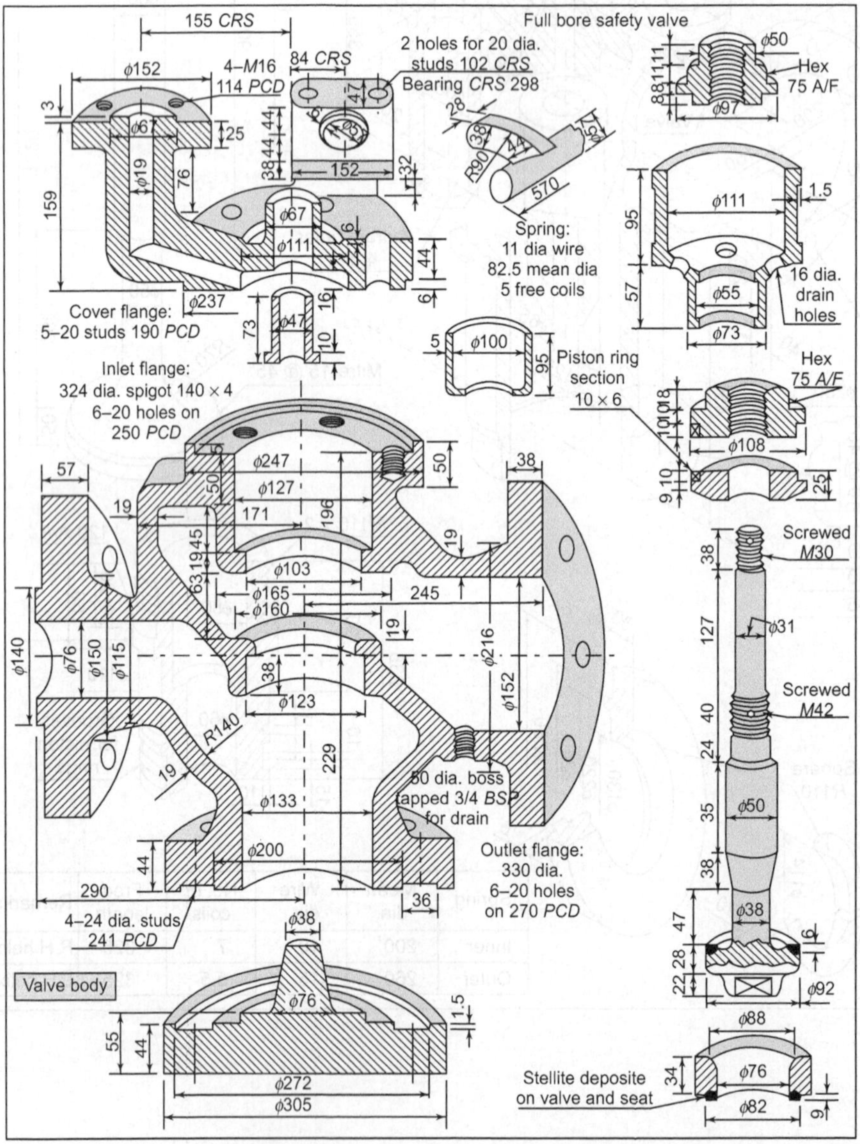

28. Flow Regulator It is used to regulate the flow of the fluid and can be compared to an accelerator used in regulating the supply of fuel to internal combustion engine of an automobile.

Hint for Answer

1. Study the number of components given in a set of components for the question.
2. Study the name of the assembly.
3. Prepare a bill of material for the assembly.
4. Concentrate on the component named as body of flow regulator as marked in the given question statement. Study the dimensions of this part and rest of the parts and assemble the parts with body. Complete assembly of parts in elevation.
5. Project the elevation to get other views asked in the given question.

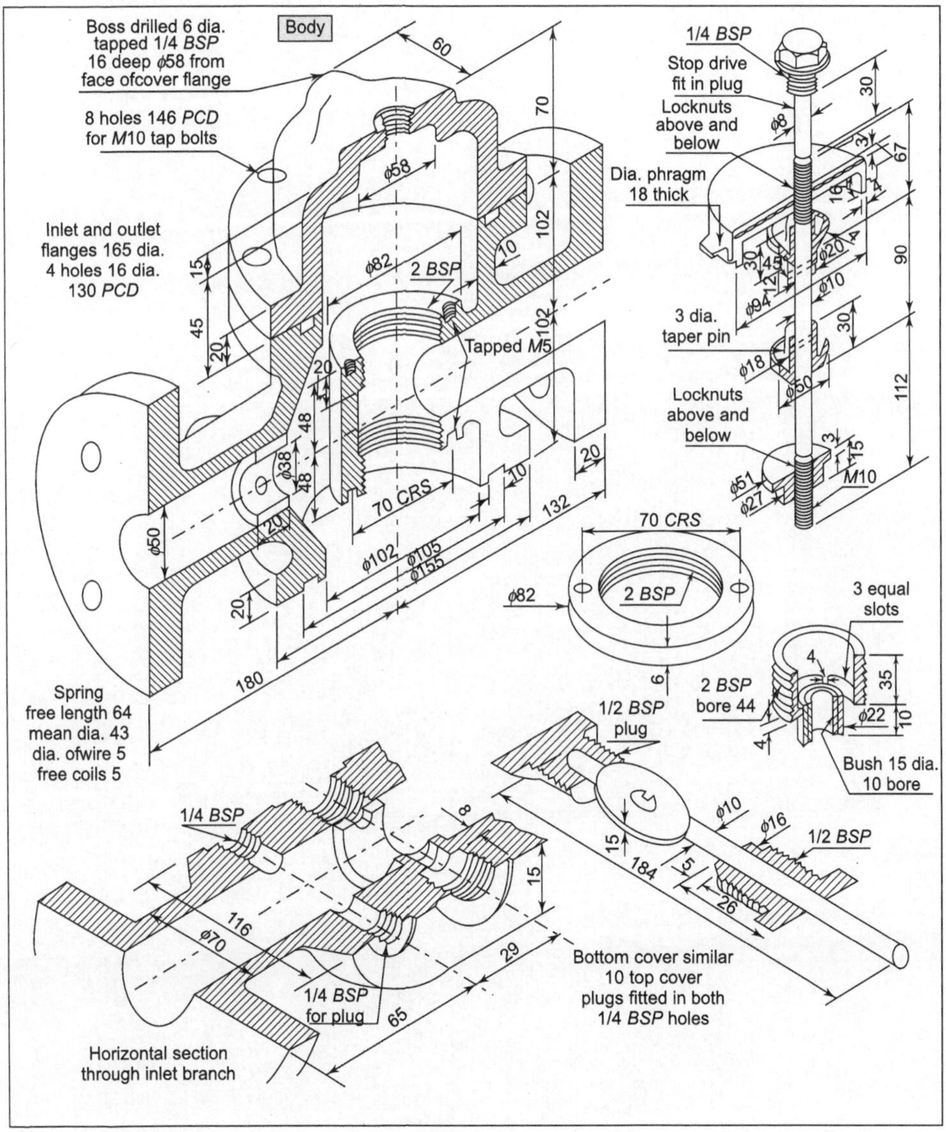

29. Strainer It is used in the supply line of fuel oil to separate unwanted fine particles of waste materials from the fuel. Impurities may cause damage to the system.

Hints for Answer

1. Study the number of components given in a set of components for the question.
2. Study the name of the assembly.
3. Prepare a bill of material for the assembly.
4. Draw assembled elevation by drawing elevation of strainer body. Study the dimensions of strainer body and try to assemble rest of the components given in the question. Complete assembled elevation.
5. Project assembled elevation to projected rest of the views as asked in the given question.

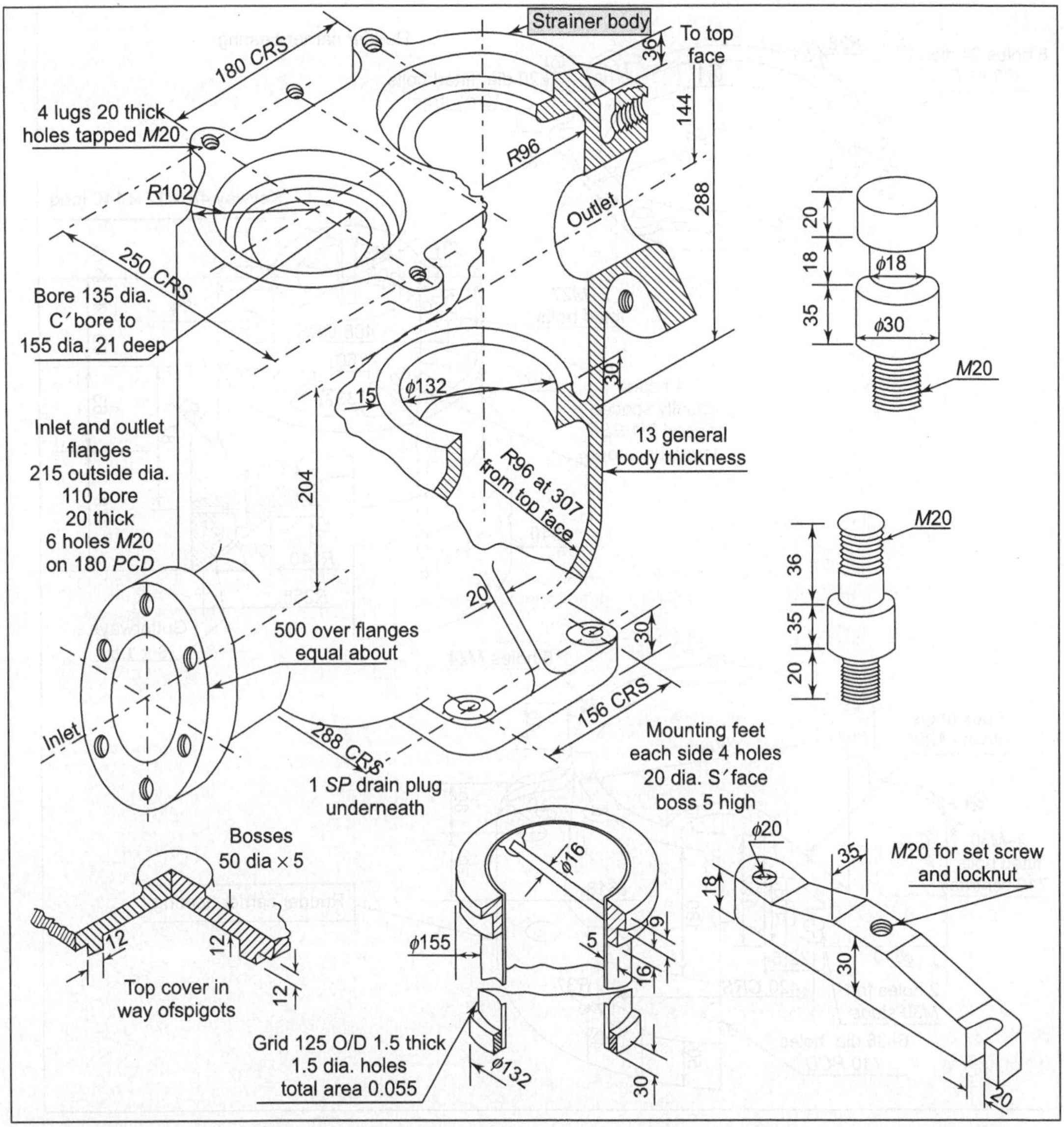

30. Ship Rudder Bearing It is the assembly used for guiding rudder shaft of the ship. It is also used for providing bearing surface to the rudder shaft. Rudder and Rudder shaft are used to give direction to the ship.

Hint for Answer

1. Study the number of components given in a set of components for the question.
2. Study the name of the assembly.
3. Prepare a bill of material for the assembly.
4. Start assembly drawing with part no. 1 named as rudder carrier housing. Assemble remaining parts with it and complete assembled elevation.
5. Project elevation drawn, to get the projection of other views.

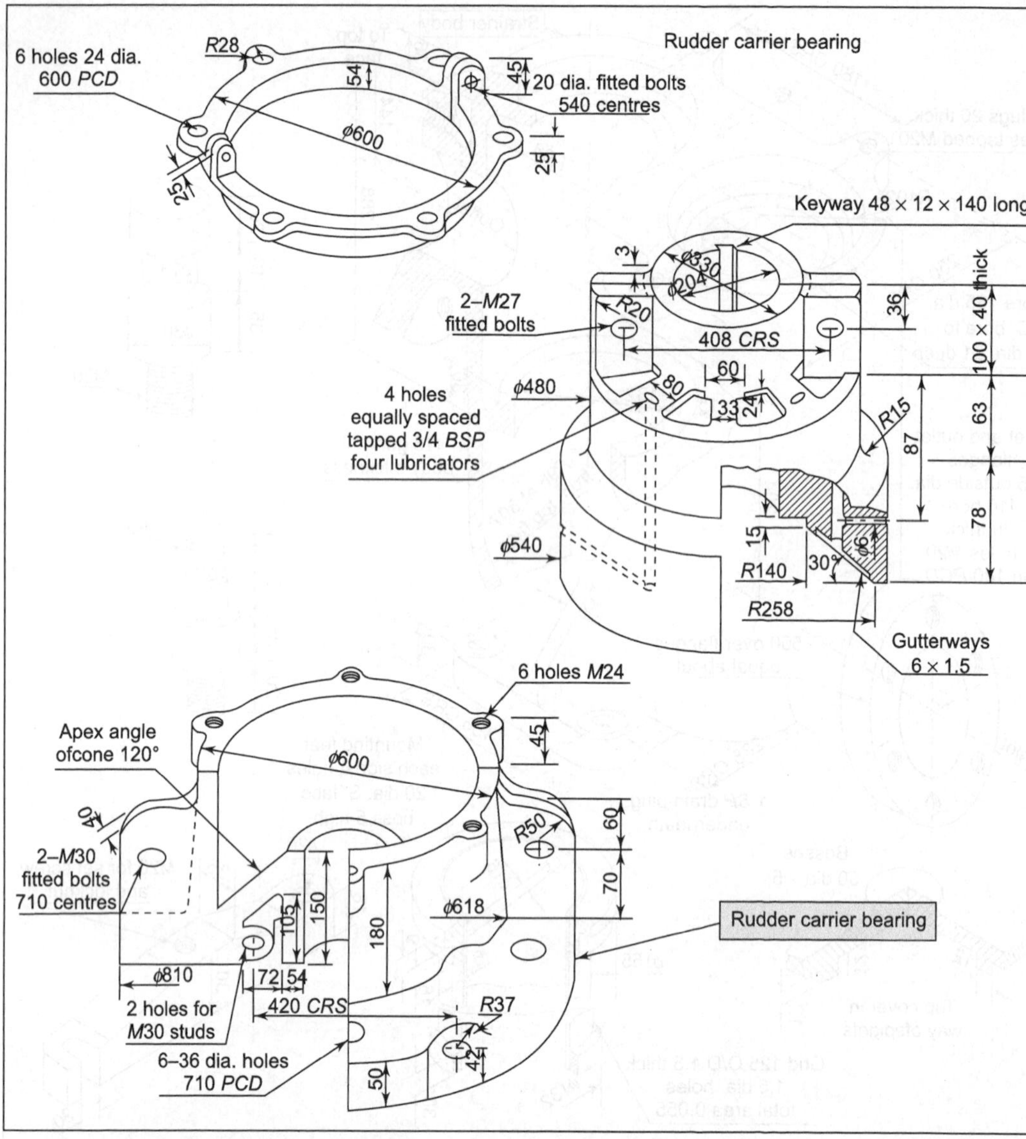

31. Universal Coupling Type I It is similar to knuckle joint. It is used for coupling two shafts where the joint is flexible and motion transmission also takes place. Coupling is used to couple two shafts whose axes are not in line with each other, but intersect at a point. With this coupling, angle between two shafts may be varied even when the shafts are in motion.

Hint for Answer

1. Study the number of components given in a set of components for the question.
2. Study the name of the assembly.
3. Prepare a bill of material for the assembly.
4. Start elevation of the assembly with forks. Refer the dimensions of fork and the other parts. Assemble other parts with forks and complete assembled elevation of the universal coupling.

5. Project this view to get the projection of other views as asked in the question.

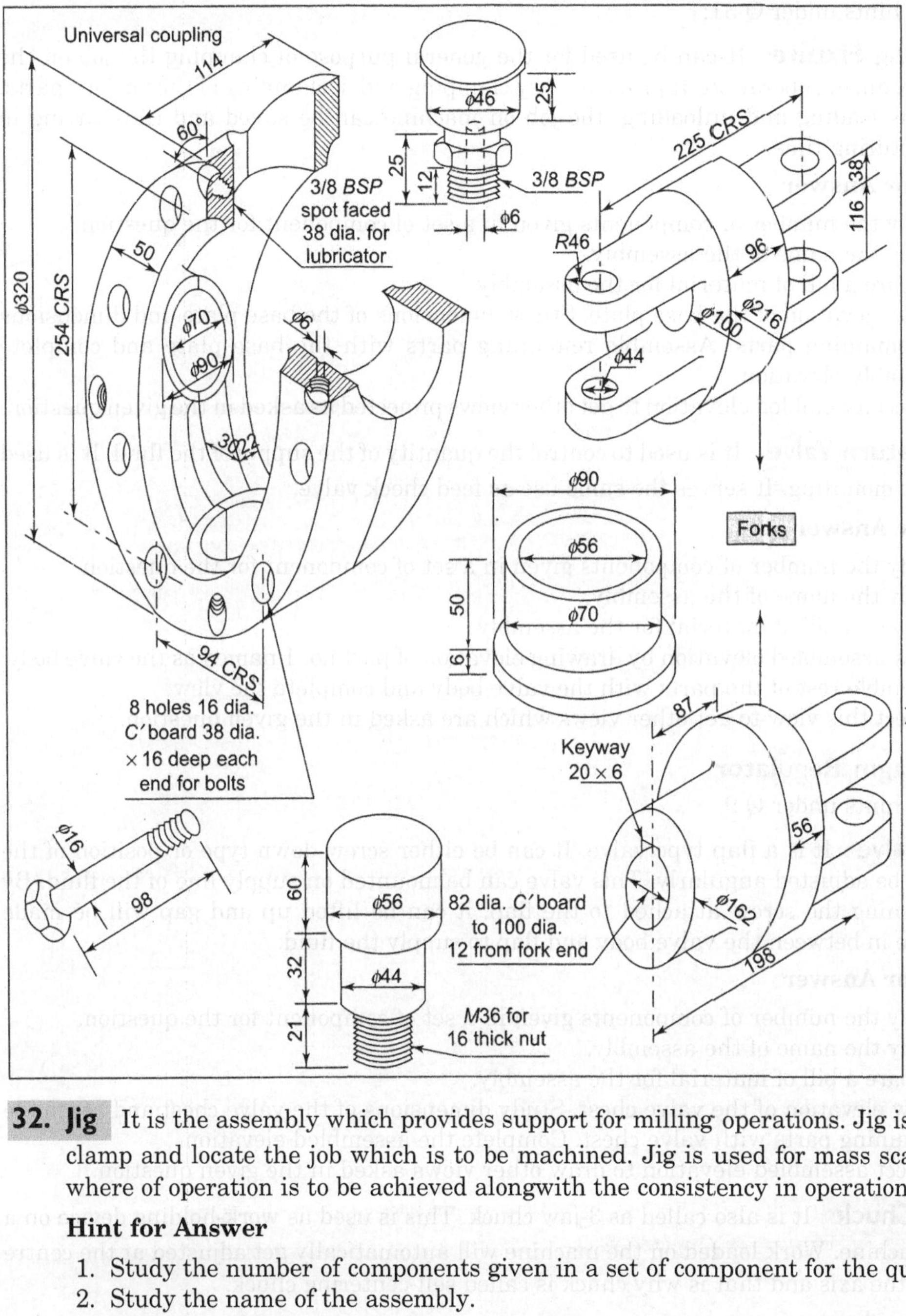

32. Jig It is the assembly which provides support for milling operations. Jig is used to hold, clamp and locate the job which is to be machined. Jig is used for mass scale production where of operation is to be achieved alongwith the consistency in operation.

Hint for Answer

1. Study the number of components given in a set of component for the question.
2. Study the name of the assembly.
3. Prepare a bill of material for the assembly.
4. Draw assembled elevation by starting with the drawing of the base plate. Study dimensions of the base plate and assemble rest of the parts with base plate.
5. Project assembled elevation to get other views projected as asked in the question.

33. Universal Coupling Type II

Refer to hints under Q 31.

34. Clamping Fixture

It can be used for the general purpose of clamping the job on the machine, during operation. It provides easy clamping and declamping of the various parts. With this loading and unloading, the job on machine can be saved and thus saving in manufacturing time.

Hints for Answer

1. Study the number of components given in a set of component for the question.
2. Study the name of the assembly.
3. Prepare a bill of material for the assembly.
4. Draw elevation of the base plate. Study dimensions of the base plate and dimensions of remaining parts. Assemble remaining parts with the base plate and complete assembly elevation.
5. Project assembled elevation to get other views projected as asked in the given question.

35. Non-Return Valve

It is used to control the quantity of the supply of the fluid. It is used as boiler mounting. It serves the same use as feed check valve.

Hint for Answer

1. Study the number of components given in a set of component for the question.
2. Study the name of the assembly.
3. Prepare a bill of material for the assembly.
4. Draw assembled elevation by drawing elevation of part no. 1 named as the valve body. Assemble rest of the parts with the valve body and complete the view.
5. Project this view to get other views which are asked in the given question.

36. Diaphragm Regulator

Refer to hints under Q 9.

37. Gate Valve

It is a flap type valve. It can be either screw-down type or position of the flap can be adjusted angularly. This valve can be mounted on supply line of the fluid. By untightening the screw attached to the flap, it can be lifted up and gap will be made available in between the valve body and flap to supply the fluid.

Hints for Answer

1. Study the number of components given in a set of component for the question.
2. Study the name of the assembly.
3. Prepare a bill of material for the assembly.
4. Draw elevation of the valve chest. Study dimensions of the valve chest and assemble remaining parts with valve chest. Complete the assembled elevation.
5. Project assembled elevation to draw other views asked in the given question.

38. 3-Jaw Chuck

It is also called as 3-jaw chuck. This is used as work-holding device on a lathe machine. Work loaded on the machine will automatically get adjusted at the centre of the lathe axis and that is why chuck is called self-centering chuck.

Hints for Answer

1. Study the number of components given in a set of component for the question.
2. Study the name of the assembly.
3. Prepare a bill of material for the assembly.

4. Draw assembled elevation of chuck assembly by starting with part A named as chuck body. All components will be there in the final assembly. By keeping in mind the function of chuck, try to assemble rest of the parts. Complete assembled elevation of the chuck.
5. Project this view to get projection of other views which are asked in the given question.

39. Water Circulating Pump It is also known as centrifugal pump. It is used for lifting water and pumping it to a height of designated water head. This pump can also be used for transfering water from one place to another through pipelines.

Hint for Answer

1. Study the number of components given in a set of component for the question.
2. Study the name of the assembly.
3. Prepare a bill of material for the assembly.
4. Draw assembled elevation by drawing elevation of part no. 1, that is, pump body, Study dimensions of the pump body and assemble rest of the parts to get an assembled elevation.
5. Project assembled elevation to get rest of the views as asked in the given question.

40. Manual Drilling Machine It is used for hand drilling or manual drilling operation especially in carpentary work. With this, machine drill of different sizes, mainly straight shank drills upto the size of $\phi \frac{1}{2}''$ can be easily mounted and held with the help of 3 jaws chuck. The chuck can be locked with geared key. Handle is given to the machine for applying manual effort. By locating drill centre, drilling operation can be performed manually by this machine.

Hints for Answer

1. Study the number of components given in a set of component for the question.
2. Study the name of the assembly.
3. Prepare a bill of material for the assembly.
4. Draw assembled elevation by drawing the elevation of part no. 1 named as housing of the machine. In final assembly there are 20 parts. Try to concentrate on main functioning dimensions of each part and complete the assembled elevation by joining all parts together.
5. Project assembled elevation to get other views of the assembly as asked in the question.

Bibliography

1. Bhatt, N. D. and V. M. Panchal (2000), *Machine Drawing,* Anand: Charotar Publishing House.
2. Burrows, Nevillen (1983), *Manufacturing Technology Level 2,* Holt, Rinehart and Winston Ltd.
3. Central Instructional Media Institute, Chennai, *Fitter First Year Trade Theory* (1999), Hyderabad: Indo-German Project, Universities Press Ltd.
4. Engineering Drawing for Marine Engineers, Reed's Volume 11.
5. Gerling, Heinrich and Karl H. Heller (1960), *All about M/C Tools,* Wiley Eastern Ltd.
6. Gupta R. B., (2001), *Engineering Drawing,* New Delhi: Satya Prakashan, Tech India Publications.
7. Haideri, Farazdak (1998), *Machine Drawing and Computer Graphics,* Pune: Nirau Prakashan.
8. Hiran E. Grant (1971), *Jigs and Fixtures Non-standard Clamping Devices,* 10th Ed., New Delhi: Tata McGraw-Hill Publishing Company Ltd.
9. Johnes G. Holburn and John J. Seacton (1978), *Drawing Book for Marine Engineers,* 8th Ed., Jaines Munro and Company Limited, Glasgow.
10. K. R. Hart, *Engineering Drawing,* 2nd Ed., Plessey Company Limited.
11. McGeorge, H. D. (1995), *Marine Auxiliary Machinery,* Reed Educational and Professional Publishing Ltd.
12. Sidheswar, N. P. Kannaiah, V. V. S. Sastry (1980), *Machine Drawing,* Tata McGraw-Hill Publishing Company Limited.
13. Spotts, M. F. (1991), *Design of Machine Elements,* 5th Ed., Pearson Education.
14. Sharma, P. C. and Dr. D. K. Aggarwal (1999), *Machine Design,* 9th Ed., Delhi: S. K. Kataria and Sons.

Index